Electromagnetics and Antenna Technology

For a complete listing of titles in the
Artech Antenna and Electromagnetics Analysis Library,
turn to the back of this book.

Electromagnetics and Antenna Technology

Alan J. Fenn

ARTECH
HOUSE

BOSTON | LONDON
artechhouse.com

Library of Congress Cataloging-in-Publication Data
A catalog record for this book is available from the U.S. Library of Congress.

British Library Cataloguing in Publication Data
A catalogue record for this book is available from the British Library.

Cover design by John Gomes

ISBN 13: 978-1-63081-374-1

10 9 8 7 6 5 4 3 2 1

To my family

Contents

Preface

Electromagnetics and antennas are enabling technologies for ground-based, airborne, and spaceborne communications systems and sensing systems. This monograph is written primarily for practicing antenna engineers, radar engineers, communications engineers, researchers, faculty, staff, undergraduate and graduate students in electrical engineering, and describes some of the research on electromagnetics and antennas that I have performed with my colleagues at MIT Lincoln Laboratory. No prior knowledge of electromagnetics or antennas is needed to understand the content of this book, although some basic background in electromagnetic field theory, antennas, radar, and communications will be helpful to the reader.

Chapter 1 begins with the fundamentals of transmission lines, electromagnetic theory and antennas, and detailed derivations of important equations are provided. Chapter 2 provides the fundamentals of phased array antennas including a formulation of the method of moments that is the principal technique used in simulating the performance of example antennas throughout the book. The rest of the chapters in the book concentrate on details of electromagnetics and antenna technology that can have broad applications in sensing and communications systems. In Chapter 3, an impedance matching example for a wideband dual-polarized dipole antenna over ground plane is explored in detail, including a derivation of the series line transformer formulation. In Chapter 4, a wideband monocone antenna is investigated with simulations and measurements. Wire discone, monopole array, loop, slotted cylinder, and turnstile antennas are also discussed. A vector sensor antenna consisting of crossed dipoles, loops, and a monopole with an application for small satellites for radio astronomy and other applications is described in Chapter 5. In Chapter 6, the theory of lumped element impedance matching for antennas is reviewed, and an example impedance matched handset antenna is considered. Chapter 7 provides an overview of the high-frequency (HF) band Chain Home dipole array antennas including theory simulations that take account of multipath effects due to reflection from the earth's surface. In Chapter 8, the theory and application of an open-end circular waveguide antenna for a portable radar system is described. In Chapter 9, the theory for circular patch antennas is described and a nonplanar 3D-printed circular patch array antenna is investigated with simulations and measurements. In Chapter 10, simulations of a Gregorian dual reflector array-fed antenna, with application to deployment from a small satellite, are presented. Chapter 11 describes ultrawideband dipole array technology. Chapter 12 describes antenna and electromagnetic field measurement techniques with an emphasis on indoor anechoic chamber testing. Problems are provided in several of the chapters to reinforce the materials presented.

The author is grateful to numerous individuals at MIT Lincoln Laboratory for management support, technical discussions, mechanical design, fabrication, and measurements support. Special thanks are extended to the Artech editorial staff and to an anonymous reviewer provided by Artech for in-depth technical review of draft versions of the manuscript.

1

Electromagnetics and Antenna Theory

1.1 Introduction

Electromagnetics [1–14] and antenna technology [15–29] are well developed and they have numerous practical applications. Radiofrequency (RF) electromagnetic waves can be produced by natural and manufactured processes. There are numerous antenna designs that can be considered for radar and communications applications depending on frequency band, type of platform, and many other considerations to be discussed in this book. This book provides a detailed review of fundamental electromagnetics and antenna theory and development of practical implementations of antenna technology including phased arrays [30–47]. As part of the development of antenna theory, detailed discussions of transmission lines and antennas as a load impedance are provided. For some of the antennas described in this book, details of impedance matching between the antenna and transmission line are given based on established methods.

An antenna is a transducer that can convert both an incident polarized RF electromagnetic wave to a time-varying signal voltage (or waveform) on a transmission line and a time-varying signal voltage on a transmission line to a transmitted polarized electromagnetic wave [1–28] as depicted in Figure 1.1. An antenna can be characterized by a number of parameters including bandwidth, impedance, scattering parameters, radiation pattern gain, directivity, beamwidth, polarization, and sidelobes. A phased array antenna system [30–47] consists of two or more antenna elements that typically transmit or receive signals with a proper timing (phase relation) and amplitude relationship between the elements. Arrays operate with analog or

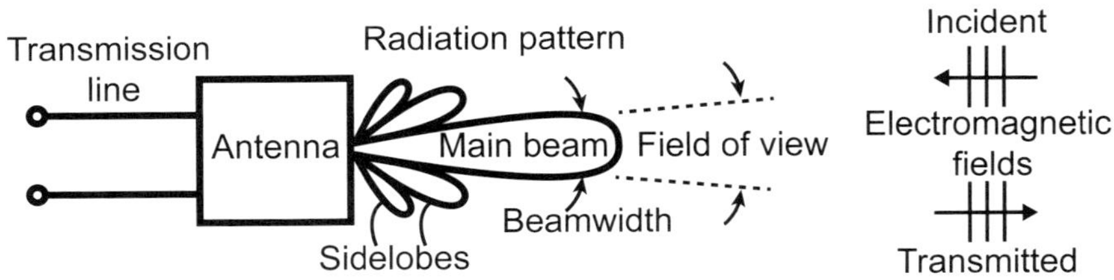

Figure 1.1 Simplified diagram showing a transmission line connected to an antenna that can transmit and receive electromagnetic fields.

digital beamforming either nonadaptively or adaptively [48–61] such that the combination of the array element received signals provides desired radiation pattern characteristics in particular directions over a band of frequencies.

Antennas have applications over a broad range of frequency bands, which have various designations. For example, frequency bands are defined by the International Telecommunications Union (ITU), which is a United Nations agency for information and communications technology. The telecommuncations bands as defined by the ITU are:

- 300 Hz to 3 kHz (ultra-low-frequency (ULF) band)
- 3 kHz to 30 kHz (very-low-frequency (VLF) band)
- 30 kHz to 300 kHz (low-frequency (LF) band)
- 300 kHz to 3 MHz (medium-frequency (MF) band)
- 3 MHz to 30 MHz (high-frequency (HF) band)
- 30 MHz to 300 MHz (very-high-frequency (VHF) band)
- 300 MHz to 3 GHz (ultra-high-frequency (UHF) band)
- 3 GHz to 30 GHz (super-high-frequency (SHF) band)
- 30 GHz to 300 GHz (extremely-high-frequency (EHF) band)

Frequency bands are also defined by the Institute for Electronic and Electrical Engineers (IEEE). The IEEE radar frequency bands are as follows:

- 3 MHz to 30 MHz (high-frequency (HF) band)
- 30 MHz to 300 MHz (very-high-frequency (VHF) band)
- 300 MHz to 1 GHz (ultra-high-frequency (UHF))
- 1 GHz to 2 GHz (L band)
- 2 GHz to 4 GHz (S band)
- 4 GHz to 8 GHz (C band)
- 8 GHz to 12 GHz (X band)
- 12 GHz to 18 GHz (Ku band)
- 18 GHz to 27 GHz (K band)

- 27 GHz to 40 GHz (Ka band)
- 40 GHz to 75 GHz (V band)
- 75 GHz to 110 GHz (W band)
- 110 GHz to 300 GHz (mm band)

The broad term *radiofrequency* (RF or Rf) is typically used for frequencies in the range of 3 kHz to 300 GHz. The term microwave is typically used for frequencies in the range of 300 MHz to 30 GHz. The term millimeter wave is often used for frequencies in the range of 30 GHz to 300 GHz.

An antenna and the surrounding structures can consist of electrically conducting materials, dielectric materials, magnetic materials, and absorbing materials, arranged in such a way as to provide desired transmission and reception of electromagnetic fields over a specified bandwidth. An antenna typically has the same transmit and receive characteristics except when active electronics such as power amplifiers and low-noise amplifiers are included in the antenna design. In developing antenna technology, design parameters as well as testing methods need to be considered as listed below:

- RF radiation parameters
  - Field of view, scan sector, bandwidth (in-band, out-of-band), polarization, gain, half-power beamwidth (HPBW), beamwidth between first nulls (BWFN), sidelobe levels, front-to-back ratio, null depth, reflection coefficient, noise figure (NF), effective isotropic radiated power (EIRP)
- RF components
  - Antenna radiating elements
  - Feed cables, transmission lines, connectors, combiners, hybrids, couplers, amplifiers, filters, circulators, switches, attenuators, phase shifters, time-delay units
- Mechanical parameters
  - Volume and mass budget, durability/survivability, fabrication technique(s) and materials, ground plane and support structure
- Ground-based testing
  - Anechoic chamber measurements (far-field (rectangular chamber, tapered chamber, compact range), near-field)
  - Thermal, vacuum, shake and vibration measurements
  - Outdoor field testing
- Flight testing (airborne, spaceborne)
  - In-flight radiation patterns and calibration

In the general development of an antenna, numerous topics often are important considerations as listed below in Table 1.1 — many of these topics are addressed in this book.

Table 1.1
Some Important Topics in Electromagnetics and Antenna Technology

Antenna types
Transmission lines
Reflection coefficient, input impedance, impedance matching
Maxwell's equations
Electromagnetic simulations
Materials (conductivity, permittivity, permeability)
Bandwidth
Polarization
Radiation patterns, directivity, gain
Near field, far field
Propagation, reflection, scattering, diffraction
Resonant antennas
Electrically small antennas
Finite, infinite arrays
Periodic, random arrays, sparse arrays
Linear, planar, conformal arrays
Array mutual coupling
Beamforming
Grating lobes, blind spots, surface waves
Interference, low sidelobes, adaptive nulling
Direction finding
Subarrays, phase centers
Electronic scanning
Passive electronically scanned array (PESA)
Active electronically scanned array (AESA)
Transmit/Receive modules
Phase shifters, time delays
High-power amplifiers, low noise amplifiers
Array design, fabrication, measurements
Array calibration
Radomes
Radar cross-section control
Field testing

The rest of this chapter is organized as follows. Section 1.2 reviews transmission line theory as it relates to antennas. In Section 1.3, the generation of electromagnetic radiation and Maxwell's equations are described. Since the goal of electromagnetic analysis often is to compute the radiation pattern of antennas, a derivation of the fields radiated by current sources is given in Section 1.4. Boundary conditions for the relations between the field quantities at the interface between two arbitrary media are derived in Section 1.5. Section 1.6 describes the wave equation and the complex propagation constant of electromagnetic waves in conducting media. In Section 1.7, a derivation of the electromagnetic wave energy flow and Poynting vector is given. Sections 1.8 and 1.9 consider the derivation of near-field and far-field expressions for Hertzian (short) dipole and small loop antennas. In Section 1.10, discussion of polarization of electromagnetic waves and receive antennas is given. The topics of bandwidth and antenna quality factor are discussed in Section 1.11. Antenna radiation intensity is discussed in Section 1.12. Section 1.13 discusses the topic of antenna radiation patterns and directivity with examples. Section 1.14 discusses the topics of antenna gain, realized gain, transmit power, and receive power. A brief discussion of different types of antennas is given in Section 1.15. Section 1.16 provides a summary and Section 1.17 provides a problem set.

1.2 Some Basics: Transmission Lines and Antennas as a Load

1.2.1 Introduction

As depicted earlier in Figure 1.1, at some point an antenna will be connected to a transmission line. As far as transmission line analysis [62] is concerned, the antenna behaves effectively as a frequency-dependent complex load impedance. In order to provide efficient transfer of microwave signal between the transmission line and an arbitrary load, an impedance matching network is often used [62–75] as depicted in Figure 1.2. Similarly, an impedance matching network can be inserted at a position along the transmission line to improve the antenna performance in terms of the efficiency in transmitting and receiving microwave signals as depicted in Figure 1.3. The impedance matching network can take on a number of forms including ladder networks of low-loss inductors and capacitors, stub tuners, line transformers, amplifiers, and other circuits.

1.2.2 Transmission Line Impedance

A radiofrequency or microwave transmission line typically can be coaxial, microstrip, twin-lead, or a general wave-guiding structure. A fundamental frequency-dependent quantity is referred to as the antenna input impedance $Z_A(f)$, which is a complex quantity with units of ohms defined at the terminals of the antenna at frequency f with units of cycles per second. Let $V(f)$ be the complex voltage drop in volts across the terminals of the antenna and let $I(f)$ be the complex electric current in amperes flowing through the terminals of the antenna, and then the antenna complex input impedance is equal to the ratio of the terminal voltage to current as

$$Z_A(f) = \frac{V(f)}{I(f)} = R_A(f) + jX_A(f) \tag{1.1}$$

where R_A in the input resistance (real part), j is the imaginary number, and X_A is the input reactance (imaginary part) of the antenna input impedance, respectively. Note that if the antenna is located in a phased array of antennas, then the antenna input impedance is also scan angle dependent as will be considered in Chapter 2. The antenna complex input admittance Y_A with units of siemens is simply the inverse of the input impedance as

$$Y_A(f) = \frac{1}{Z_A} = G_A(f) + jB_A(f) \tag{1.2}$$

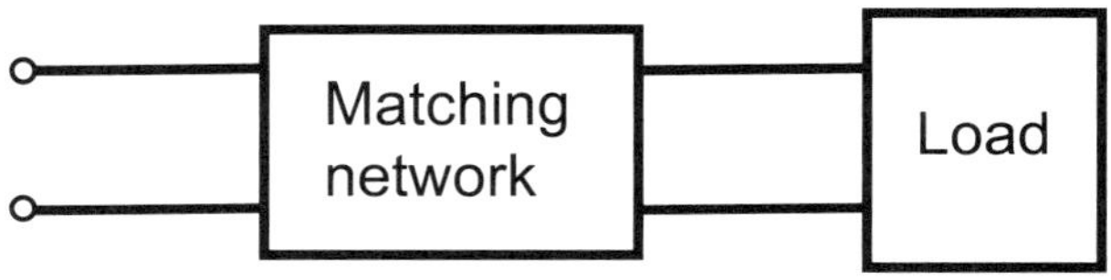

Figure 1.2 Simplified diagram showing a matching circuit connected to a load that can be an antenna.

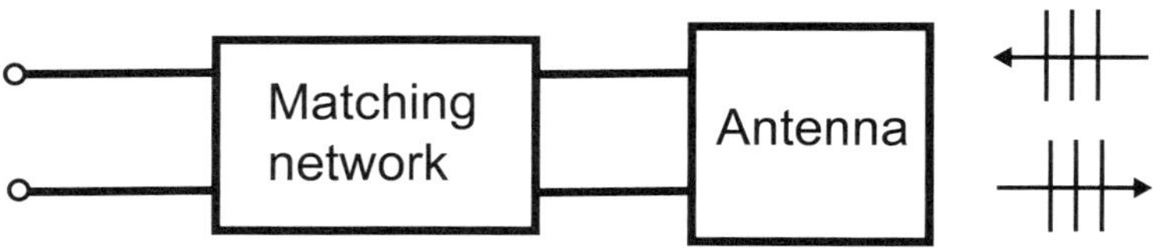

Figure 1.3 Simplified diagram showing a matching network connected to an antenna that receives and transmits electromagnetic fields.

where G_A in the input conductance (real part) and B_A is the input susceptance (imaginary part) of the antenna input admittance, respectively. Note that it is common to also define antenna parameters in terms of the radian frequency (denoted ω) as

$$\omega = 2\pi f \tag{1.3}$$

with units of radians per second.

Consider now the general case of a transmission line connected to an antenna represented by the complex load impedance Z_L as shown in Figure 1.4. The transmission line can, in general, be lossy and will be described by the characteristic impedance Z_o, attenuation constant α, and phase constant β. The complex propagation constant (denoted γ) is expressed in terms of the attenuation constant and phase constant as

$$\gamma = \alpha + j\beta \tag{1.4}$$

Assume that an RF source has generated voltage $V(z)$ and current $I(z)$ waves along the transmission line. It is desired to quantify the voltage and current at any point z along the transmission line. The voltage and current waves on a lossy transmission line can be expressed in terms of waves traveling in the plus z (denoted V_o^+, I_o^+) and minus z (denoted V_o^-, I_o^-) directions as

$$V(z) = V_o^+ e^{-\gamma z} + V_o^- e^{\gamma z} \tag{1.5}$$

$$I(z) = I_o^+ e^{-\gamma z} + I_o^- e^{\gamma z} \tag{1.6}$$

Assume now that the transmission line is lossless (zero attenuation versus distance); that is, $\alpha = 0$ so that $\gamma = j\beta$, where $\beta = 2\pi/\lambda$, where λ is the wavelength for the transmission line. For the lossless transmission line $V(z)$ and $I(z)$ become

$$V(z) = V_o^+ e^{-j\beta z} + V_o^- e^{j\beta z} \tag{1.7}$$

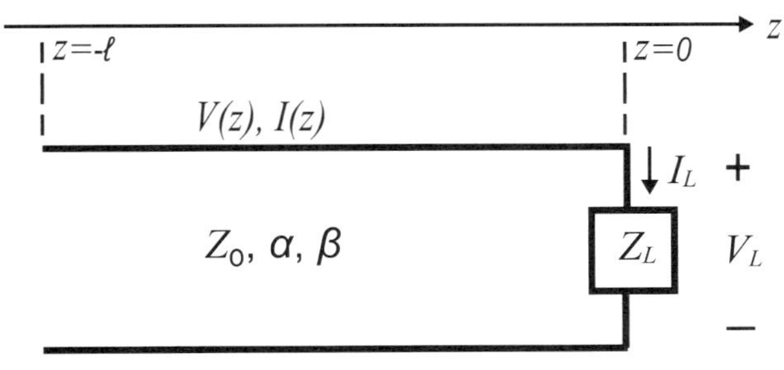

Figure 1.4 Simplified diagram showing a transmission line connected to a frequency-dependent complex load impedance.

$$I(z) = I_o^+ e^{-j\beta z} + I_o^- e^{j\beta z} \tag{1.8}$$

The characteristic impedance of the transmission line is computed from the ratio of the voltage and current waves in the positive and negative traveling wave directions as

$$Z_o = \frac{V_o^+}{I_o^+} = \frac{-V_o^-}{I_o^-} \tag{1.9}$$

which yields

$$I_o^+ = \frac{V_o^+}{Z_o} \tag{1.10}$$

$$I_o^- = \frac{-V_o^-}{Z_o} \tag{1.11}$$

Note that the minus sign for the negative traveling current wave is required due to the opposing vector orientation of the negative traveling current with respect to the vector orientation of the positive traveling current. Using Equations (1.10) and (1.11), the current wave $I(z)$ in Equation (1.8) can be expressed now as

$$I(z) = \frac{V_o^+}{Z_o} e^{-j\beta z} + \frac{-V_o^-}{Z_o} e^{j\beta z} \tag{1.12}$$

Taking the ratio of the voltage wave given by Equation (1.5) and the current wave given by Equation (1.12) at the load position ($z = 0$) yields

$$Z_L = \frac{V_L(0)}{I_L(0)} = \frac{V_o^+ + V_o^-}{V_o^+ - V_o^-} Z_o \tag{1.13}$$

Solving for V_o^- yields

$$V_o^- = \frac{Z_L - Z_o}{Z_L + Z_o} V_o^+ \tag{1.14}$$

The complex voltage reflection coefficient Γ is defined as the ratio of the voltage wave reflected from the load to the voltage wave incident at the load; that is,

$$\Gamma = \frac{V_o^-}{V_o^+} = \frac{Z_L - Z_o}{Z_L + Z_o} \tag{1.15}$$

The voltage transmission coefficient T_v is defined as the voltage across the load divided by the incident voltage as

$$T_v = \frac{V_o^+ + V_o^-}{V_o^+} = 1 + \frac{V_o^-}{V_o^+} = 1 + \Gamma \tag{1.16}$$

Now the average incident and reflected powers are given by

$$P^+ = \frac{1}{2} |V_o^+|^2 / Z_o \tag{1.17}$$

$$P^- = \frac{1}{2}|V_o^-|^2/Z_o \tag{1.18}$$

The average power actually delivered to the load (same as the average power accepted by the load or the average power actually transmitted by a lossless antenna) is equal to the difference between the average incident power and the average reflected power; that is,

$$P_L = P^+ - P^- = \frac{1}{2}|V_o^+|^2/Z_o - \frac{1}{2}|V_o^-|^2/Z_o \tag{1.19}$$

Now dividing the average power delivered to the load by the average incident power, which is referred to here as the transmission mismatch loss (ML) efficiency η_{ML}, yields

$$P_L/P^+ = \eta_{\text{ML}} = \frac{|V_o^+|^2/Z_o - |V_o^-|^2/Z_o}{|V_o^+|^2/Z_o} \tag{1.20}$$

which reduces to

$$\eta_{\text{ML}} = 1 - \frac{|V_o^-|^2}{|V_o^+|^2} = 1 - |\Gamma|^2 \tag{1.21}$$

The parameter η_{ML} can be described as the mismatch loss efficiency (with values ranging from 0 to 1) in transferring power from the transmission line to the load or antenna, and it represents either power loss or gain loss, depending on its use. The antenna mismatch loss or realized gain loss due to mismatch between the antenna load and the transmission line in decibels (dB) as a positive quantity is quantified as

$$\text{ML}_{\text{dB}} = -10\log_{10}\eta_{\text{ML}} = -10\log_{10}\left[1 - |\Gamma|^2\right] \tag{1.22}$$

Using Equation (1.15) in Equations (1.7) and (1.8) it follows that

$$V(z) = V_o^+(e^{-j\beta z} + \Gamma e^{j\beta z}) \tag{1.23}$$

$$I(z) = \frac{V_o^+}{Z_o}(e^{-j\beta z} - \Gamma e^{j\beta z}) \tag{1.24}$$

From Equation (1.23) it can be shown that the maximum and minimum voltages on the transmission line are expressed as

$$V_{\max} = |V_o^+|(1 + |\Gamma|) \tag{1.25}$$

$$V_{\min} = |V_o^+|(1 - |\Gamma|) \tag{1.26}$$

The voltage standing wave ratio (VSWR) is defined as

$$\text{VSWR} = \frac{V_{\text{max}}}{V_{\text{min}}} = \frac{1 + |\Gamma|}{1 - |\Gamma|} \tag{1.27}$$

Solving Equation (1.27) for the magnitude of the reflection coefficient yields

$$|\Gamma| = \frac{\text{VSWR} - 1}{\text{VSWR} + 1} \tag{1.28}$$

Next, using Equations (1.23) and (1.24), the input impedance Z_{in} at a point $z = -l$ along the transmission line is now expressed as

$$Z_{\text{in}}(-l) = \frac{V(-l)}{I(-l)} = \frac{V_o^+(e^{j\beta l} + \Gamma e^{-j\beta l})}{V_o^+(e^{j\beta l} - \Gamma e^{-j\beta l})} Z_o = \frac{1 + \Gamma e^{-j2\beta l}}{1 - \Gamma e^{-j2\beta l}} Z_o \tag{1.29}$$

or as a normalized impedance $z_{\text{in}}(-l)$ as

$$z_{\text{in}}(-l) = Z_{\text{in}}(-l)/Z_o = \frac{1 + \Gamma e^{-j2\beta l}}{1 - \Gamma e^{-j2\beta l}} \tag{1.30}$$

Subsituting Equation (1.15) in Equation (1.29) yields after simplification

$$Z_{\text{in}}(-l) = Z_o \frac{Z_L + jZ_o \tan \beta l}{Z_o + jZ_L \tan \beta l} \tag{1.31}$$

In the general case with frequency dependence, if an antenna with complex input impedance Z_A is connected to a transmission line with characteristic impedance Z_o, then the antenna's frequency-dependent complex reflection coefficient Γ_A is given by

$$\Gamma_A(f) = \frac{Z_A(f) - Z_o(f)}{Z_A(f) + Z_o(f)} \tag{1.32}$$

It should be noted that the characteristic impedance of a coaxial transmission line or microstrip line is frequency independent, whereas in a rectangular or circular waveguide the characteristic impedance is frequency dependent.

Impedance matching is a critical technique that is used to improve the performance of antennas. Figure 1.2 shows the situation in which a two-port matching network is connected at a position along the transmission that is connected to the antenna. To reduce impedance mismatch effects between the transmission line and the antenna, the parameters of the matching device can be adjusted so that the reflection coefficient magnitude is minimized. A significant amount of impedance matching theory has been developed by prior researchers. In this book, some general theory of impedance matching theory is given as well as specific applied examples for narrowband and wideband antennas. The terms wideband and broadband are considered equivalent here.

1.2.3 Smith Chart Theory

In this book, impedance data are presented in a rectangular plot either versus frequency or with the horizontal axis as the real part of the impedance and the vertical axis as the imaginary part of the impedance, or in a Smith chart [63–66] as is described in this section. In a Smith chart, the impedance data are typically normalized by the charateristic impedance of a transmission line connected to a load. A derivation of the Smith chart equations is now given.

The derivation starts by normalizing the right-hand side (top and bottom) of the voltage reflection coefficient Equation (1.15) by Z_o with the result

$$\Gamma = \frac{z_L - 1}{z_L + 1} \tag{1.33}$$

where $z_L = Z_L/Z_o$ is the normalized impedance. Now solving Equation (1.33) for z_L, and using

$$\Gamma = |\Gamma| e^{j\Phi} \tag{1.34}$$

where Φ is the phase angle of the reflection coefficient yields

$$z_L = \frac{1+\Gamma}{1-\Gamma} = \frac{1+|\Gamma|e^{j\Phi}}{1-|\Gamma|e^{j\Phi}} \tag{1.35}$$

Expressing the normalized load impedance and the reflection cofficient in terms of real and imaginary components as $z_L = r_L + jx_L$ and $\Gamma = \Gamma_r + j\Gamma_i$ Equation (1.35) becomes

$$z_L = r_L + jx_L = \frac{1+\Gamma}{1-\Gamma} = \frac{(1+\Gamma_r) + j\Gamma_i}{(1-\Gamma_r) - j\Gamma_i} \tag{1.36}$$

Now, multiplying the top and bottom of the right-hand side of Equation (1.36) by the complex conjugate of the denominator and then simplifying and equating real and imaginary components with r_L and x_L yields

$$r_L = \frac{(1-\Gamma_r^2) - \Gamma_i^2}{(1-\Gamma_r)^2 + \Gamma_i^2} \tag{1.37}$$

$$x_L = \frac{2\Gamma_i}{(1-\Gamma_r)^2 + \Gamma_i^2} \tag{1.38}$$

By rearranging terms and factoring, Equations (1.37) and (1.38) can be put into the standard form for constant resistance circles as

$$\left(\Gamma_r - \frac{r_L}{r_L+1}\right)^2 + \Gamma_i^2 = \left(\frac{1}{r_L+1}\right)^2 \tag{1.39}$$

which is centered at

$$\Gamma_r = \frac{r_L}{r_L + 1} \qquad \Gamma_i = 0 \tag{1.40}$$

and has radius

$$\frac{1}{r_L + 1} \tag{1.41}$$

and for constant reactance circles as

$$(\Gamma_r - 1)^2 + \left(\Gamma_i - \frac{1}{x_L}\right)^2 = \left(\frac{1}{x_L}\right)^2 \tag{1.42}$$

which is centered at

$$\Gamma_r = 1 \qquad \Gamma_i = \frac{1}{x_L} \tag{1.43}$$

and has radius

$$\frac{1}{x_L} \tag{1.44}$$

Similarly, in a Smith chart, the admittance data are typically normalized by the charateristic admittance of a transmission line. The derivation starts with Equation (1.33) and then using $z_L = 1/y_L$ with the result

$$\Gamma = \frac{z_L - 1}{z_L + 1} = \frac{1/y_L - 1}{1/y_L + 1} = \frac{1 - y_L}{1 + y_L} \tag{1.45}$$

where y_L is the normalized impedance. Now solving Equation (1.45) for y_L yields

$$y_L = \frac{1 - \Gamma}{1 + \Gamma} \tag{1.46}$$

Expressing the normalized load impedance and the reflection coefficient in terms of real and imaginary components as $y_L = g_L + jb_L$ and $\Gamma = \Gamma_r + j\Gamma_i$ Equation (1.46) becomes

$$y_L = g_L + jb_L = \frac{1 - \Gamma}{1 + \Gamma} = \frac{(1 - \Gamma_r) - j\Gamma_i}{(1 + \Gamma_r) + j\Gamma_i} \tag{1.47}$$

Now, multiplying the top and bottom of the right-hand side of Equation (1.47) by the complex conjugate of the denominator, and then simplifying and equating real and imaginary components with g_L and b_L yields

$$g_L = \frac{(1 - \Gamma_r^2) - \Gamma_i^2}{(1 + \Gamma_r)^2 + \Gamma_i^2} \tag{1.48}$$

$$b_L = -\frac{2\Gamma_i}{(1 - \Gamma_r)^2 + \Gamma_i^2} \tag{1.49}$$

By rearranging terms and factoring, Equations (1.49) and (1.37) can be put into the standard form for constant conductance circles as

$$\left(\Gamma_r + \frac{g_L}{g_L+1}\right)^2 + \Gamma_i^2 = \left(\frac{1}{g_L+1}\right)^2 \tag{1.50}$$

which is centered at

$$\Gamma_r = -\frac{g_L}{g_L+1} \qquad \Gamma_i = 0 \tag{1.51}$$

and has radius

$$\frac{1}{g_L+1} \tag{1.52}$$

and for constant susceptance circles as

$$(\Gamma_r + 1)^2 + \left(\Gamma_i + \frac{1}{b_L}\right)^2 = \left(\frac{1}{b_L}\right)^2 \tag{1.53}$$

which is centered at

$$\Gamma_r = -1 \qquad \Gamma_i = -\frac{1}{b_L} \tag{1.54}$$

and has radius

$$\frac{1}{b_L} \tag{1.55}$$

A standard Smith chart normalized by the characteristic impedance or characteristic admitttance of the transmission line is shown in Figure 1.5. This type of Smith chart is referred to here as a Z-type Smith chart. The same Z-type of Smith chart is repeated in Figure 1.6, but now with constant resistance ($r = 0.3, 1, 2$) and positive constant reactance ($x = 1, 1.8, 3$) circles highlighted. Figure 1.7 shows a normalized Smith chart (Z-type) with example circles showing a few values of VSWR, reflection coefficient magnitude $|\Gamma|$, return loss (RL=20 $\log_{10}|\Gamma|$), and mismatch loss (ML).

As an example of the use of the Smith chart, in Figure 1.7 a solid line is drawn from the center of the Smith chart to a point on the rim where the magnitude of the reflection coefficient is unity $|\Gamma| = 1$ and the phase angle is $\theta = -40°$. On the outer rim of the Smith chart, the corresponding wavelength distance is approximately 0.306λ. Now, assume that a load impedance is connected to a 50-ohm lossless transmission line such that the reflection coefficient has magnitude 0.33 and phase angle $-40°$. The solid line intersects the $|\Gamma| = 0.33$ circle at a point where the normalized impedance

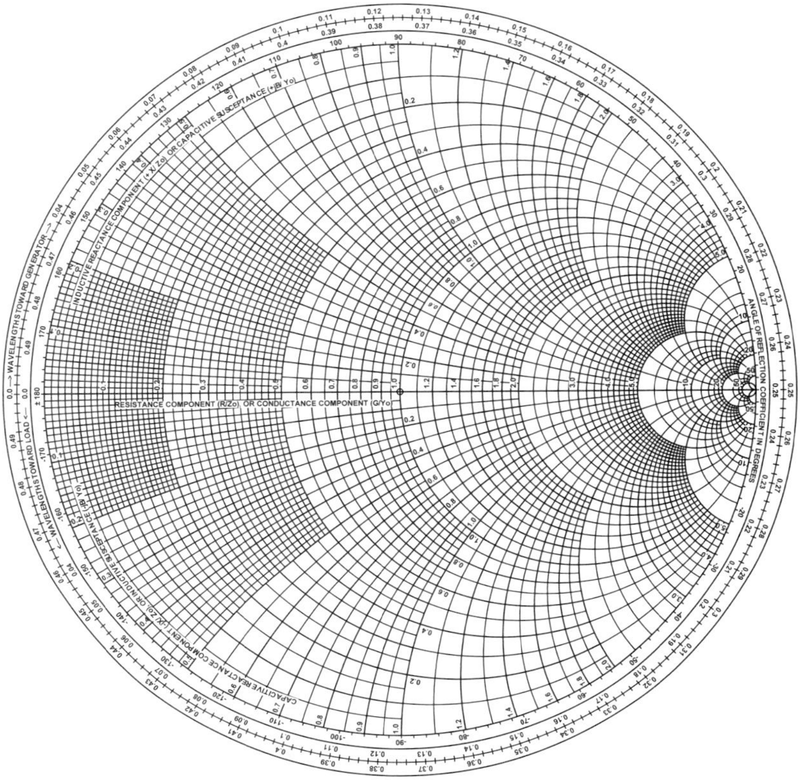

Figure 1.5 Smith chart (Z-type) normalized by the characteristic impedance or characteristic admitttance of the transmission line. ©IEEE [66].

z_L can be observed directly on the Smith chart or it can be computed from Equation (1.35) as

$$z_L = \frac{1 + |\Gamma|e^{j\Phi}}{1 - |\Gamma|e^{j\Phi}} = \frac{1 + 0.33e^{-j40^\circ}}{1 - 0.33e^{-j40^\circ}} = 1.477 - j0.703 \tag{1.56}$$

For a 50-ohm reference impedance the absolute impedance is then

$$Z_L = 50z_L = 73.8 - j35.15 \text{ ohms} \tag{1.57}$$

To transform the load impedance to an input impedance at a distance $l = 0.139\lambda$ toward the generator (source), a dashed line is drawn from the center of the Smith chart to the outer rim intersecting the wavelengths toward generator value of 0.445λ. Since the transmission line is assumed lossless,

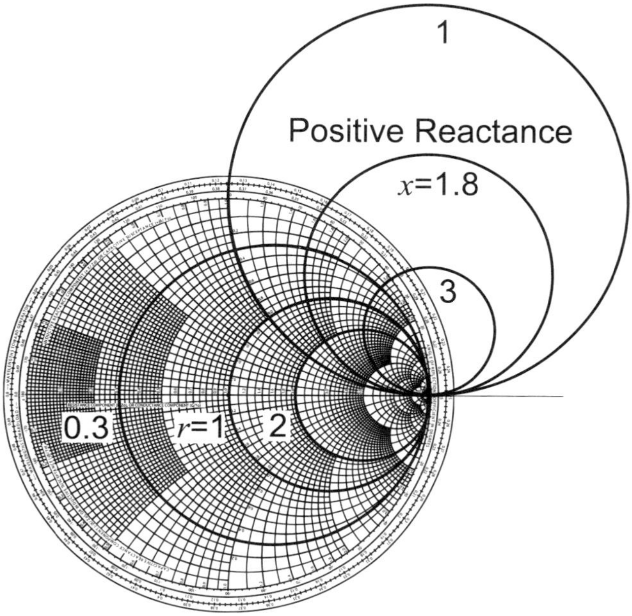

Figure 1.6 Smith chart (Z-type) with a few positive constant resistance ($r = 0.3, 1, 2$) and positive constant reactance ($x = 1, 1.8, 3$) circles highlighted.

the load impedance value on the Smith chart then rotates clockwise toward the generator on the constant VSWR=2 circle. Then either the normalized input impedance at $l = 0.445\lambda$ is observed directly at the intersection point of the dashed line and VSWR=2 circle on the Smith chart, or it is computed from Equation (1.30), as $z_{\text{in}} = 0.552 - j0.262$ and as the absolute input impedance for a 50-ohm reference as $Z_{\text{in}} = 50z_{\text{in}} = 27.59 - j13.11$ ohms.

As mentioned, the standard Z-type Smith chart can be used to analyze impedance and admittance data. When impedance values are plotted in a Z-type Smith chart, the corresponding admittance values are positioned 180° away from the impedance values on a constant VSWR circle. An alternate method for working with admittance values, is to physically rotate the Smith chart by 180°. A 180° rotated Smith chart normalized by the characteristic impedance or characteristic admitttance of the transmission line is shown in Figure 1.8. This type of Smith chart is referred to here as a Y-type Smith chart (some authors [72] refer to this type of Smith chart presentation as an overlay), because the chart is used directly with the admittance circles. The same Y-type Smith chart is repeated in Figure 1.9, but now with constant conductance ($g = 0.3, 1, 2$) and constant susceptance ($b = 1, 1.8, 3$) circles

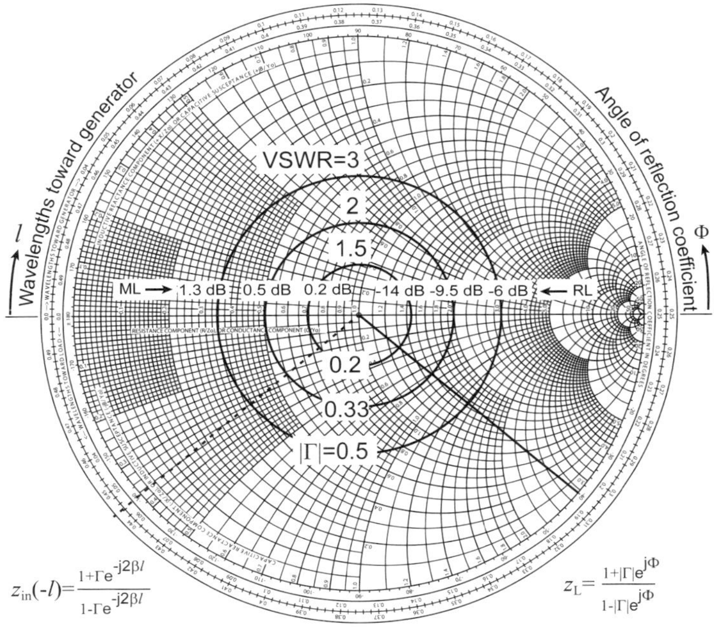

Figure 1.7 Smith chart (Z-type) with example circles showing values of reflection coefficient magnitude $|\Gamma|$ =0.2, 0.33, 0.5, and the corresponding voltage standing wave ratio (VSWR), return loss (RL), and mismatch loss (ML).

highlighted. Note that the MATLAB™ RF Toolbox (www.mathworks.com) provides Smith charts of the following types: Z-type, Y-type, ZY-type (Z is the primary chart with Y the overlay), and YZ-type (Y is the primary chart with Z the overlay).

1.2.4 Impedance Matching Circuits

Impedance matching [62–75] is an important part of antenna technology, as matching can be used to improve the transmit and receive characteristics of antennas over a bandwidth of interest. To improve the impedance match of an antenna to a transmission line of characteristic impedance Z_o, it is common to use one or more low loss inductors and/or capacitors in a series or parallel circuit with the antenna. These circuit elements are placed as close to the antenna terminals as possible to reduce dispersive effects and allow improved impedance matching over a given bandwidth.

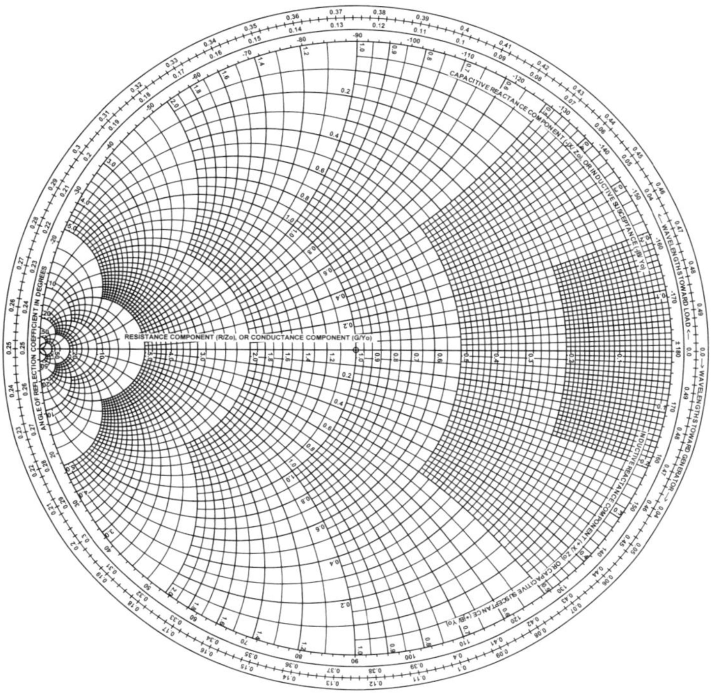

Figure 1.8 180° rotated Smith chart (Y-type) normalized by the characteristic impedance or characteristic admitttance of the transmission line.

1.2.4.1 Single-Stage Matching

In this section, only a single element (inductor or capacitor) is used to match the antenna load to the transmission line at a single frequency. Let L be the inductance (typically in nH) of an inductor and let C be the capacitance (typically in pF) of a capacitor. Let Z_A be the antenna input impedance. It follows then that

- A series inductor provides an impedance equal to $j\omega L$ and a series capacitor provides an impedance equal to $1/(j\omega C)$.
- A parallel inductor provides an admittance equal to $1/(j\omega L)$ and a parallel capacitor provides an admittance equal to $j\omega C$.

Figure 1.10 summarizes the input impedance Z_{in} for these four circuit types with either a single inductor or single capacitor in series or parallel to the antenna (load) impedance Z_A. The corresponding Smith charts in Figure 1.11 show the effect of tuning to a perfect match ($|\Gamma| = 0$) at a single frequency with these four circuit types.

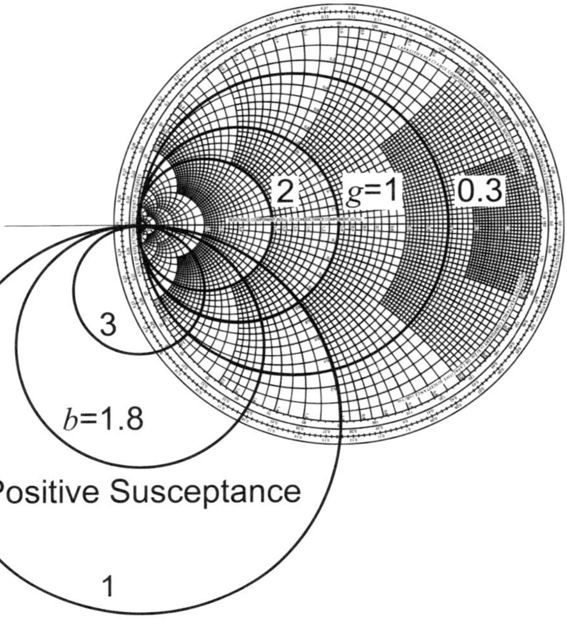

Figure 1.9 180° rotated normalized Smith chart (Y-type) with a few positive constant conductance ($g = 0.3, 1, 2$) and constant susceptance ($b = 1, 1.8, 3$) circles highlighted.

- With a series inductor, the impedance rotates clockwise along circles of constant resistance on the impedance (Z-type) Smith chart (Figure 1.11(a)).
- With a series capacitor, the impedance rotates counter-clockwise along circles of constant resistance on the impedance (Z-type) Smith chart (Figure 1.11(b)).
- With a parallel inductor, the admittance rotates counter-clockwise along circles of constant conductance on the admittance (Y-type) Smith chart (Figure 1.11(c)).
- With a parallel capacitor, the admittance rotates clockwise along circles of constant conductance on the admittance (Y-type) Smith chart (Figure 1.11(d)).

In cases where single stage matching cannot achieve the desired match, two or more stages can be considered.

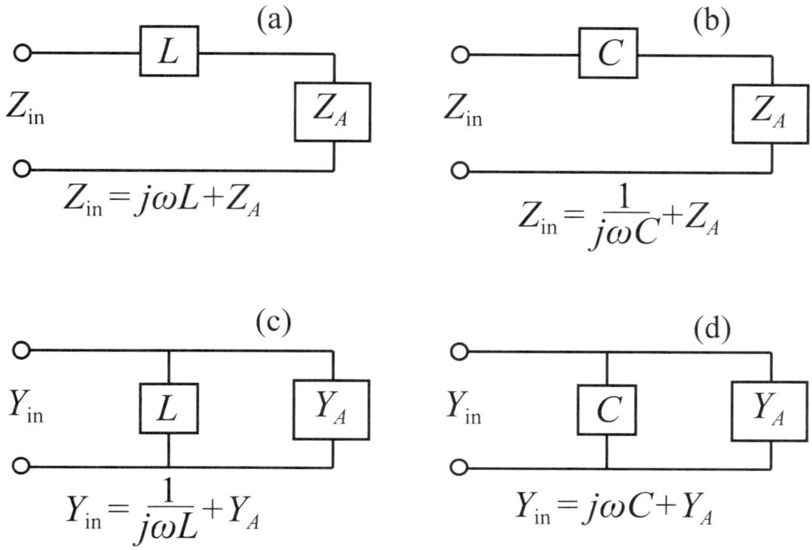

Figure 1.10 Four single element circuit types for impedance matching an antenna with input impedance Z_A or input admittance Y_A. (a) Series inductor, (b) series capacitor, (c) parallel inductor, and (d) parallel capacitor.

1.2.4.2 Two-Stage Matching

In Figure 1.12, the effect of two-stage tuning to a perfect match ($|\Gamma| = 0$) at a single frequency with four series/parallel circuit types is shown. In these four examples, since the first element is in series and the second element is in parallel with the antenna load, first a Z-type Smith chart rotation on a constant resistance circle is implemented which is then followed by a Y-type Smith chart rotation on the $g = 1$ constant conductance circle. The input admittance equation is indicated in each of the diagrams in Figure 1.12.

- With an initial series inductor L, the impedance rotates clockwise along a circle of constant resistance on the impedance (Z-type) Smith chart. The rotation stops when the impedance reaches the point indicated by a prime $'$ on the $g = 1$ constant conductance circle. Then with a parallel capacitor C and switching to the Y-type Smith chart overlay the admittance rotates clockwise on the $g = 1$ circle to reach the center of the Smith chart for a perfect match (Figure 1.12(a)).
- With an initial series capacitor C, the impedance rotates counter-clockwise along a circle of constant resistance on the impedance (Z-type) Smith chart. The rotation stops when the impedance reaches

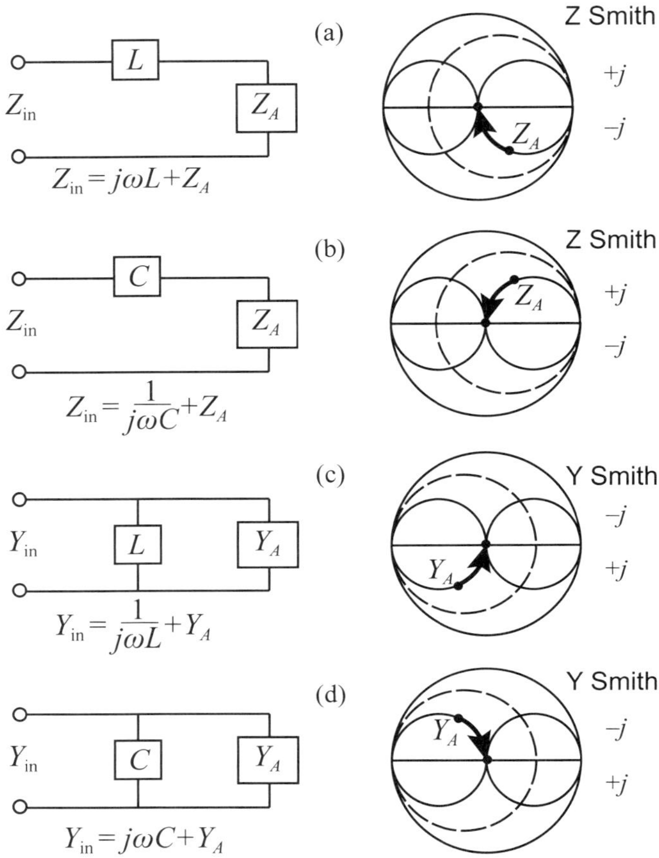

Figure 1.11 Series L or C circuits with corresponding Z and Y Smith chart diagrams. (a) Series inductor, (b) series capacitor, (c) parallel inductor, and (d) parallel capacitor.

the point indicated by a prime ′ on the $g = 1$ constant conductance circle. Then with a parallel inductor L and switching to the Y-type Smith chart overlay the admittance rotates counterclockwise on the $g = 1$ circle to reach the center of the Smith chart for a perfect match (Figure 1.12(b)).

- With an initial series inductor L_2, the impedance rotates clockwise along a circle of constant resistance on the impedance (Z-type) Smith chart. The rotation stops when the impedance reaches the point

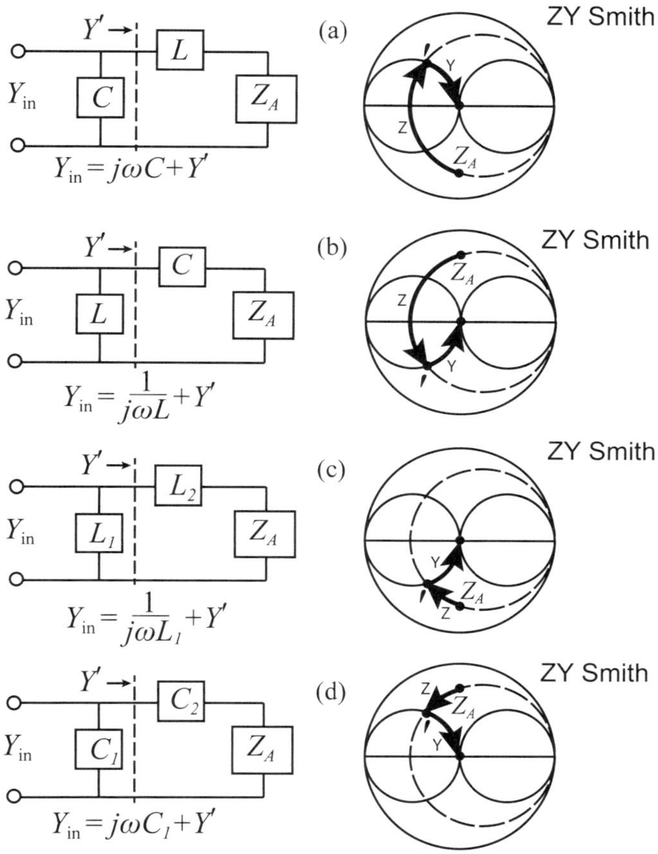

Figure 1.12 Series/parallel L-section matching circuits with corresponding Smith chart diagrams in Z then Y overlay format. The initial matching element is in series with the antenna load impedance Z_A. (a) Series inductor followed by parallel capacitor, (b) series capacitor followed by parallel inductor, (c) series inductor followed by parallel inductor, and (d) series capacitor followed by parallel capacitor.

indicated by a prime $'$ on the $g = 1$ constant conductance circle. Then with a parallel inductor L_1 and switching to the Y-type Smith chart overlay the admittance rotates counterclockwise on the $g = 1$ circle to reach the center of the Smith chart for a perfect match (Figure (1.12(c)).

- With an initial series capacitor C_2, the impedance rotates counterclockwise along a circle of constant resistance on the impedance (Z-type) Smith chart. The rotation stops when the impedance reaches the point indicated by a prime $'$ on the $g = 1$ constant conductance circle. Then with a parallel capacitor C_1 and switching to the Y-type Smith chart overlay the admittance rotates clockwise on the $g = 1$ circle to reach the center of the Smith chart for a perfect match (Figure (1.12d)).

In Figure 1.13, the effect of two-stage tuning (with so-called L-sections) to a perfect match ($|\Gamma| = 0$) at a single frequency with four parallel/series circuit types is shown. In these four examples, since the first element is in parallel and the second element is in series with the antenna load, first a Y-type Smith chart rotation on a constant conductance circle is implemented which is then followed by a Z-type Smith chart rotation on the $r = 1$ constant resistance circle. The input impedance equation is indicated in each of the diagrams in Figure 1.13.

- With an initial parallel inductor L, the impedance rotates counterclockwise along a circle of constant conductance on the admittance (Y-type) Smith chart. The rotation stops when the admittance reaches the point indicated by a prime $'$ on the $r = 1$ constant resistance circle. Then with a series capacitor C and switching to the Z-type Smith chart overlay the impedance rotates counterclockwise on the $r = 1$ circle to reach the center of the Smith chart for a perfect match (Figure (1.13a)).
- With an initial parallel capacitor C, the admittance rotates clockwise along a circle of constant conductance on the admittance (Y-type) Smith chart. The rotation stops when the admittance reaches the point indicated by a prime $'$ on the $r = 1$ constant resistance circle. Then with a series inductor L and switching to the Z-type Smith chart overlay the impedance rotates clockwise on the $r = 1$ circle to reach the center of the Smith chart for a perfect match (Figure 1.13(b)).
- With an initial parallel inductor L_2, the admittance rotates counterclockwise along a circle of constant admittance on the admittance (Y-type) Smith chart. The rotation stops when the admittance reaches the point indicated by a prime $'$ on the $r = 1$ constant resistance circle. Then with a series inductor L_1 and switching to the Z-type Smith chart overlay the impedance rotates clockwise on the $r = 1$ circle to reach the center of the Smith chart for a perfect match (Figure 1.13(c)).
- With an initial parallel capacitor C_2, the admittance rotates clockwise along a circle of constant conductance on the admittance (Y-type)

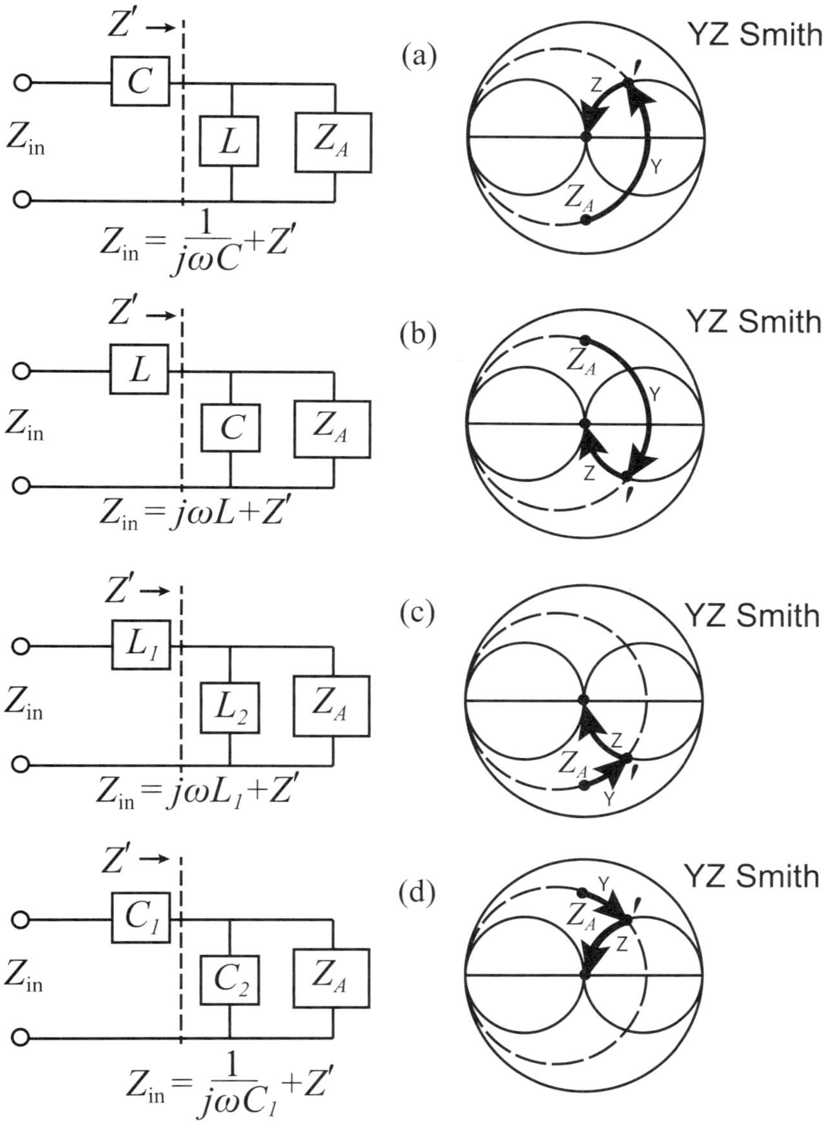

Figure 1.13 Parallel/series L-section matching circuits with corresponding Smith chart diagrams in Y then Z overlay format. The initial matching element is in parallel with the antenna load impedance Z_A. (a) Parallel inductor followed by series capacitor, (b) parallel capacitor followed by series inductor, (c) parallel inductor followed by series inductor, and (d) parallel capacitor followed by series capacitor.

Smith chart. The rotation stops when the admittance reaches the point indicated by a prime $'$ on the $r = 1$ constant resistance circle. Then with a series capacitor C_1 and switching to the Z-type Smith chart overlay the impedance rotates counterclockwise on the $r = 1$ circle to reach the center of the Smith chart for a perfect match (Figure 1.13(d)).

1.2.4.3 Practical Wideband Impedance Matching

The examples shown in Figures 1.11 to 1.13 are ideal cases where a perfect impedance match ($|\Gamma| = 0$) is achieved at a single frequency. For practical antennas (loads) operating over a band of frequencies, a perfect match at all frequencies is not feasible. Instead, the antenna impedance match will provide a maximum allowed reflection coefficient magnitude or equivalently a maximum allowed voltage standing wave ratio (VSWR) over the desired frequency band. Impedance matching over a band of frequencies can be achieved for example with L-section networks, pi networks, T networks, and ladder networks consisting of multiple L, pi, and/or T networks, short-circuit and open-circuit stub tuners, and line transformers. Impedance matching can be achieved using graphical methods manually on a Smith chart, or by commercial software that performs circuit synthesis and optimization such as the MATLAB RF Toolbox (www.mathworks.com), or Optenni Lab (www.optenni.com). For simple L-section networks, the values of the inductor(s) and/or capacitor(s) can be computed in closed form [62] by first deriving the input impedance equation for the particular L-section network. Next, the real part of the input impedance is set equal to the characteristic impedance of the transmission line, and the imaginary part is set to zero. The resulting equation is then rearranged and separated into real and imaginary terms which gives two simultaneous equations. These two simultaneous equations are then solved for the values of the two reactive circuit components. In practice, for implementation of impedance matching with measurements and analysis of RF circuits, it is necessary to consider S-parameters as discussed in the next section.

1.2.5 S-Parameters

S-parameters, or scattering parameters [62, 76–79] are commonly used to quantify the reflection and transmission of voltage and current waves in matching networks and antennas. In general, the S-parameters are quantified for an N-port system as an $N \times N$ complex matrix. A network analyzer is typically used in measuring S-parameters versus frequency. Consider Figure 1.14 which shows a transmission line with a two-port network. Voltage waves a_1 and a_2 are incident at ports 1 and 2, respectively. Similarly, voltage waves b_1 and b_2 are reflected at ports 1 and 2, respectively. In this case, the S-matrix or scattering matrix is a 2 x 2 complex matrix that is expressed as

$$\boldsymbol{S} = \begin{bmatrix} S_{11} & S_{12} \\ S_{21} & S_{22} \end{bmatrix} \tag{1.58}$$

where the S-parameters are related to the voltage waves as

$$b_1 = S_{11}a_1 + S_{12}a_2 \tag{1.59}$$

$$b_2 = S_{21}a_1 + S_{22}a_2 \tag{1.60}$$

Now when $a_2 = 0$ it follows that $S_{11} = b_1/a_1$, and $S_{21} = b_2/a_1$, and when $a_1 = 0$ it follows that $S_{12} = b_1/a_2$, and $S_{22} = b_2/a_2$. In the case of a two-port network, the scattering matrix $\boldsymbol{S}$ provides a relation between the reflected wave parameters at the two ports and the incident wave paramters at the two port. Figure 1.15 shows a signal flow diagram [62] for the scattering matrix $\boldsymbol{S}$.

Refer now to the previous Figure 1.3, where the matching network is quantified by the scattering matrix S and the antenna (load) is quantified by the reflection coefficient Γ_L. Figure 1.16 shows a signal flow diagram for the scattering matrix matching network where the load impedance characterized by reflection coefficient Γ_L has been added. We want to determine the input reflection coefficient $\Gamma_{\text{in}} = b_1/a_1$. It is observed that two signal paths

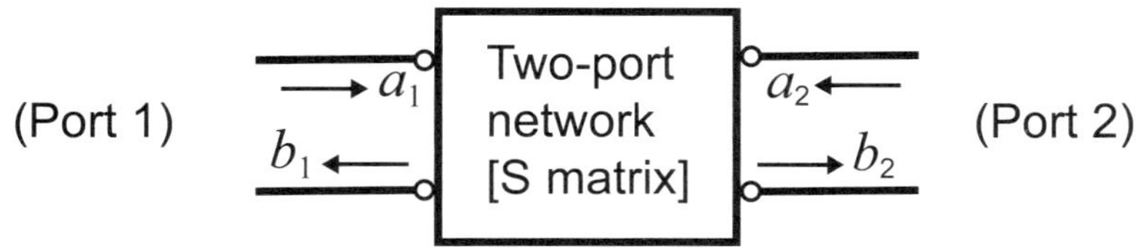

Figure 1.14 Two-port network quantified by the scattering matrix $\boldsymbol{S}$.

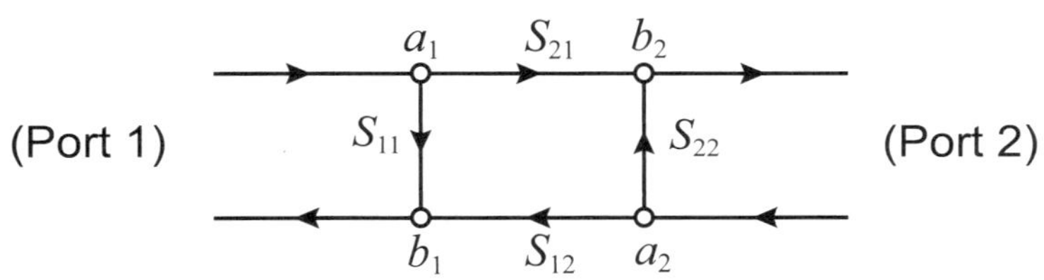

Figure 1.15 Signal flow diagram for the scattering matrix $\boldsymbol{S}$.

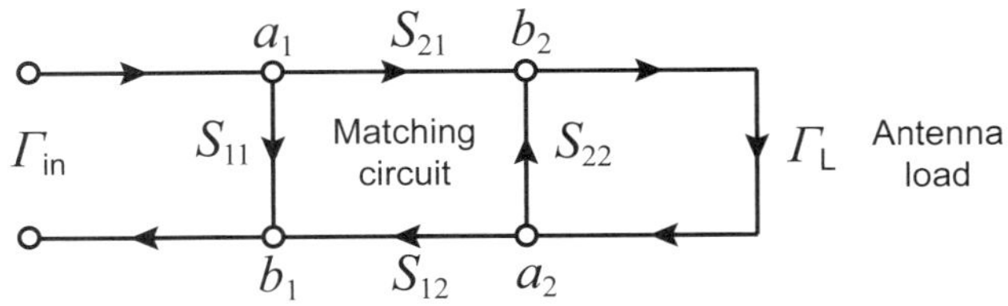

Figure 1.16 Signal flow diagram for the scattering matrix $\boldsymbol{S}$ with load characterized by reflection coefficient Γ_L.

contribute to the voltage wave at node b_2. At node b_2 the load reflection coefficient acts as a feedback signal equal to $S_{22}\Gamma_L$ Thus, the flow diagram in Figure 1.16 can be reduced to the equivalent flow diagram shown in Figure 1.17, and it follows that

$$b_2 = a_1 S_{21} + b_2 S_{22}\Gamma_L \tag{1.61}$$

and solving for b_2 yields

$$b_2 = a_1 \frac{S_{21}}{1 - S_{22}\Gamma_L} \tag{1.62}$$

It follows then that the flow diagram in Figure 1.17 can be reduced finally to Figure 1.18. It immediately follows that the input reflection coefficient can be expressed as

$$\Gamma_{\text{in}} = \frac{b_1}{a_1} = S_{11} + \frac{S_{12}S_{21}\Gamma_L}{1 - S_{22}\Gamma_L} \tag{1.63}$$

S-parameters are used directly when only a single network is utilized in

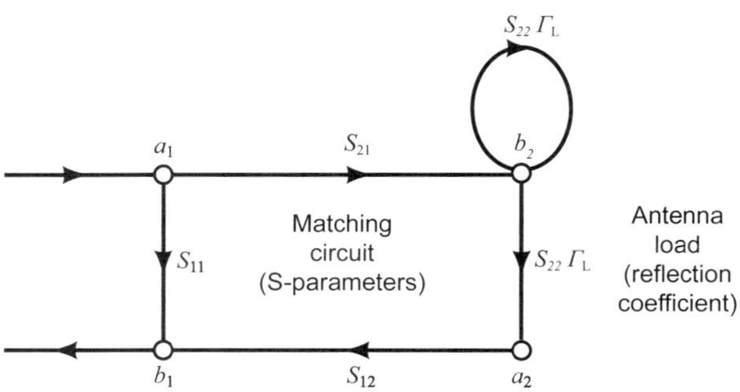

Figure 1.17 Signal flow diagram for the scattering matrix S with the antenna load reflection coefficient included as a feedback loop at node b_2.

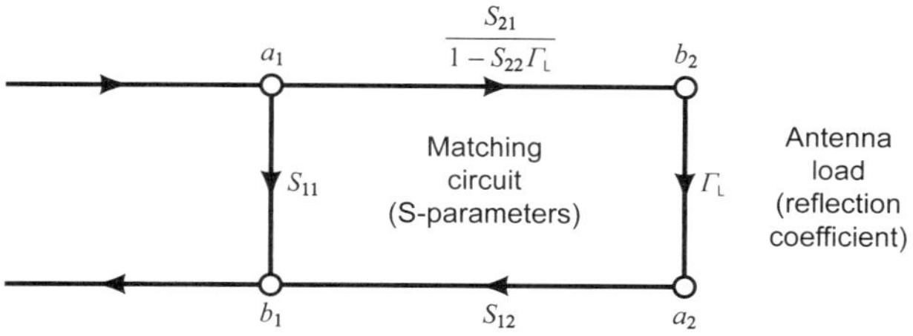

Figure 1.18 Final signal flow diagram for the scattering matrix $\boldsymbol{S}$ combined with the antenna load impedance characterized by the reflection coefficient Γ.

matching a transmission line to a load. In the case where cascaded networks are involved, it is necessary to convert the S-parameters to T-chain scattering parameters (T-parameters) as discussed in the next section.

1.2.6 T-Chain Scattering Matrices for Cascaded Networks

T-chain (transfer) scattering matrices [77–79] can be used when multiple matching circuit networks are cascaded. A block diagram of a single T-chain scattering matrix is depicted in Figure 1.19. The T-chain scattering matrix $\boldsymbol{T}$ provides a relation between the incident and reflected waves (a_1 and b_1, respectively) of the input port to the incident and reflected waves (a_2 and b_2, respectively) of the output port. Figure 1.20 shows a block diagram where two T matrices ($\boldsymbol{T}^A$ and $\boldsymbol{T}^B$) are cascaded.

The T-chain scattering matrix is related to the voltage wave parameters as

$$a_1 = T_{11}b_2 + T_{12}a_2 \tag{1.64}$$

$$b_1 = T_{21}b_2 + T_{22}a_2 \tag{1.65}$$

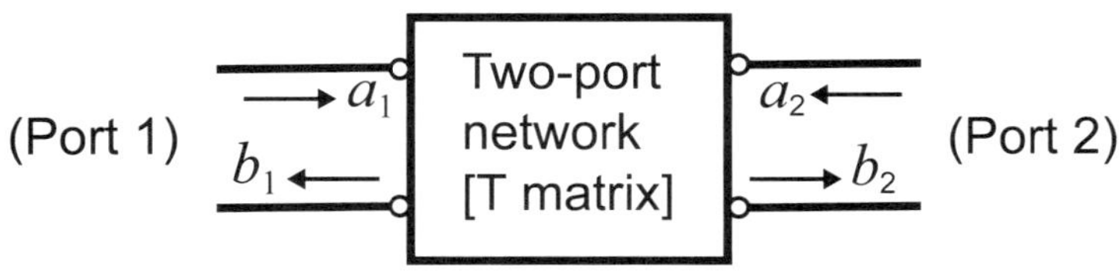

Figure 1.19 Block diagram for a T-chain scattering matrix.

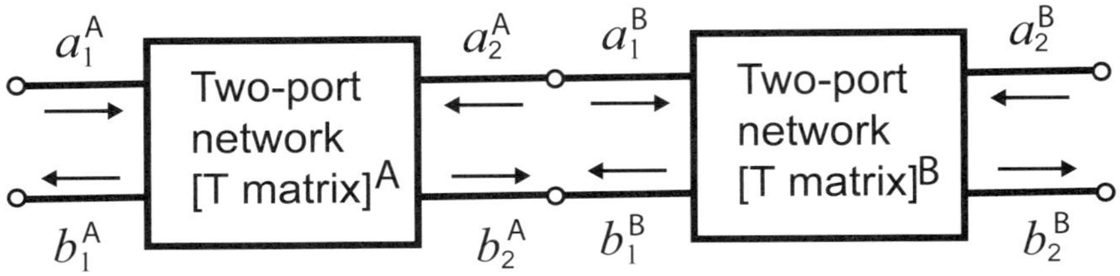

Figure 1.20 Block diagram for cascaded T-chain scattering matrices.

The T-chain scattering matrix parameters can be determined from the S-parameters as follows. In the case where $a_2 = 0$ it follows from Equations (1.59), (1.60), and (1.64) that

$$T_{11} = \frac{1}{S_{21}} \tag{1.66}$$

$$T_{21} = \frac{S_{11}}{S_{21}} \tag{1.67}$$

When $a_1 = 0$ it follows from Equation (1.60) and (1.65) that

$$T_{12} = -\frac{S_{22}}{S_{21}} \tag{1.68}$$

$$T_{22} = \frac{S_{12}S_{21} - S_{11}S_{22}}{S_{21}} \tag{1.69}$$

Similarly, the S-parameters can be determined from the T-chain scattering parameters as

$$S_{11} = \frac{T_{21}}{T_{11}} \tag{1.70}$$

$$S_{21} = \frac{1}{T_{11}} \tag{1.71}$$

$$S_{12} = \frac{T_{11}T_{22} - T_{12}T_{21}}{T_{11}} \tag{1.72}$$

$$S_{22} = -\frac{T_{12}}{T_{11}} \tag{1.73}$$

Referring now to Figure 1.20 the T-matrices $\boldsymbol{T}^A$ and $\boldsymbol{T}^B$ are expressed in terms of voltage waves as

$$\begin{bmatrix} a_1^A \\ b_1^A \end{bmatrix} = \begin{bmatrix} T_{11}^A & T_{12}^A \\ T_{21}^A & T_{22}^A \end{bmatrix} \begin{bmatrix} b_2^A \\ a_2^A \end{bmatrix} \tag{1.74}$$

$$\begin{bmatrix} a_1^B \\ b_1^B \end{bmatrix} = \begin{bmatrix} T_{11}^B & T_{12}^B \\ T_{21}^B & T_{22}^B \end{bmatrix} \begin{bmatrix} b_2^B \\ a_2^B \end{bmatrix} \tag{1.75}$$

but since

$$\begin{bmatrix} b_2^A \\ a_2^A \end{bmatrix} = \begin{bmatrix} a_1^B \\ b_1^B \end{bmatrix} \tag{1.76}$$

it follows that the cascaded T-chain matrices can be expressed as

$$\begin{bmatrix} a_1^A \\ b_1^A \end{bmatrix} = \begin{bmatrix} T_{11}^A & T_{12}^A \\ T_{21}^A & T_{22}^A \end{bmatrix} \begin{bmatrix} T_{11}^B & T_{12}^B \\ T_{21}^B & T_{22}^B \end{bmatrix} \begin{bmatrix} b_2^B \\ a_2^B \end{bmatrix} \tag{1.77}$$

Thus, the cascaded T-matrix product is given by

$$\boldsymbol{T}^C = \boldsymbol{T}^A \boldsymbol{T}^B \tag{1.78}$$

Once the cascaded T-matrix product $\boldsymbol{T}^C$ has been computed, the S-parameters for the cascaded network can then be determined by Equations (1.70) to (1.73). The MATLAB RF Toobox includes T-parameter computations.

1.2.6.1 Discussion

In Section 1.2 an antenna has been treated in a simplified manner as a load impedance connected to a transmission line, and in this case the antenna was not considered as radiating or receiving microwave radiation in the volume of space surrounding the antenna. To analyze electromagnetic radiation in the context of antennas, propagation, or scattering, it is necessary to utilize Maxwell's equations as discussed in the next section.

1.3 Electromagnetic Radiation: Maxwell's Equations

This section begins with a brief discussion of accelerated charges, time-varying currents, and photons, with the goal of providing a fundamental understanding of the mechanisms of electromagnetic radiation as described mathematically by Maxwell's equations.

Electromagnetic waves are known to be generated by accelerated electron charges or, equivalently, by time-varying currents [1-28]. For example, the time-varying vector current density $\boldsymbol{J}(t)$ with units of amperes/m^2 at a particular cross section of an electrically conducting antenna can be expressed in terms of the volume electric charge density ρ (coulombs/m^3) and the velocity $\boldsymbol{v}(t)$ (m/s) of the charge as

$$\boldsymbol{J}(t) = \rho \boldsymbol{v}(t). \tag{1.79}$$

Thus, the electric current density has units of coulombs per m^2 per second. In Equation (1.79), the local charge density is related to the electron charge (denoted as q_e) times the number of electrons (N_e) accelerated through the local cross section of the antenna, that is,

$$\rho = q_e N_e. \tag{1.80}$$

Taking the derivative of $\boldsymbol{J}(t)$ with respect to time yields

$$\frac{d\boldsymbol{J}(t)}{dt} = \rho \frac{d\boldsymbol{v}(t)}{dt} = \rho \boldsymbol{a}(t) \tag{1.81}$$

where $d\boldsymbol{v}(t)/dt = \boldsymbol{a}(t)$ is the acceleration of the charge with units of m/s^2. Thus, the units of the time derivative of the current density are coulombs per meters squared per second per second, and, as will be shown, radiation occurs when the time derivative of the current density $\boldsymbol{J}$ is nonzero and equivalently when the charge acceleration $\boldsymbol{a}$ is nonzero. The time derivative of the current density and acceleration of charge are of interest in explaining the source of the electromagnetic radiation for antenna systems.

A complete characterization of electromagnetic fields can be determined by means of Maxwell's equations. However, in the process of generating electromagnetic waves, an intermediate process, referred to in the literature as photon emission, occurs and ultimately produces the electromagnetic wave. A photon can be described as a discrete packet of electromagnetic radiation that includes X-rays, visible light, and microwaves. The term ionizing radiation refers to radiation that has enough energy to knock electrons out of atoms and includes X-rays, protons, heavy ions, and neutrons that are generated by accelerators. Nonionizing radiation includes visible light, ultraviolet (UV) radiation, infrared (IR) radiation, and microwaves. In the case of an antenna receiving an electromagnetic wave, photon absorption and scattering occur with some probability that produces accelerated charges or, equivalently, time-varying currents on the antenna. Here, a brief discussion of the generation and reception of photons and electromagnetic waves is given. A complete mathematical treatment of photon emission and photon absorption is beyond the scope of this book, and the reader is referred to other sources mentioned in the paragraphs below. One situation in which photon/electron interaction is important is in determining the effect of electromagnetic waves on human tissues [80, 81] as a potential safety issue. When the electromagnetic wave frequency is sufficiently high, as is the case for X-ray radiation, the energy in the electromagnetic photons is large enough to break the atom-electron bond and knock out electrons causing damage to the DNA of tissue. Low-level (nonthermal) microwave energy does not contain sufficient photon energy to knock out electrons from tissue, but if the power density is sufficiently high and is applied for a sufficient time interval, it is capable of heating and damaging tissues [82, 83]. The interaction of microwave energy and a wide range of dielectric materials has been investigated in the literature [84].

For RF transmission, when a time-varying signal voltage (e.g., a sinusoidal or pulsed waveform) on a transmission line is applied across the terminals of an antenna, electrons (negatively charged particles) distributed across the metallic lattice portion of the antenna experience a time-varying vector force, and they accelerate and generate a time-varying electric current. An electron has a negative electric charge equal to $q_e = -1.602 \times 10^{-19}$ coulombs and a rest mass of $m_{oe} = 9.109 \times 10^{-31}$ kg. The RF-induced acceleration of the electrons alters their vector momentum and energy state, which with some probability gives rise to the emission of RF photons, which can be described in a dual nature as particles or waves with zero rest mass ($m_{op} = 0$) and zero electric charge ($q_p = 0$), as described by Cohen-Tannoudji [85]. The shape of a photon has been described in terms of its quantum wavefunction expressed as a hologram by Chrapkiewicz [86]. Each photon has an energy level E (equal to the change in energy level between the excited state and the ground state of an electron) as computed by

$$E = hf \tag{1.82}$$

where $h = 6.626 \times 10^{-34}$J-s is Planck's constant and f is the frequency in cycles per second or Hertz (Hz). For example, the energy of a single microwave photon packet at 300 MHz is computed and rounded as

$$E = (6.626 \times 10^{-34}\text{J-s})(300 \times 10^{6}\text{s}^{-1}) = 2 \times 10^{-25}\text{J}. \tag{1.83}$$

Suppose an antenna is radiating 1 watt or 1 joule per second of microwave power, then the number of 300 MHz microwave photons generated per second is computed as

$$N_{\text{photons per second}} = \frac{1\text{J/s}}{2 \times 10^{-25}\text{J/photon}} = 5 \times 10^{24}. \tag{1.84}$$

Microwave photons are polarized according to the design of the antenna, which can be linear, circular, or elliptically polarized. In free space, RF photons travel at the speed of light, and they have an energy-momentum vector $\boldsymbol{p}$ with amplitude equal to

$$p = \frac{h}{\lambda} = \frac{hf}{c} \tag{1.85}$$

where $\lambda = c/f$ is the wavelength of the photon in free space. It should be noted that the energy-momentum vector $\boldsymbol{p}$ has units of J-s/m.

As microwave power is continuously pumped into the antenna terminals, RF photons are continuously generated and radiated outward from the transmitting antenna. When a sufficient amount and spatial distribution of photons are generated from the radiating antenna, a propagating wavefront, described by an electromagnetic field, is created. Electromagnetic fields (and photons) are characterized by Maxwell's equations, and these fields travel at the speed of light in vacuum and slower than the speed of light in a dielectric medium.

The characteristics of the electromagnetic-field wavefront, such as the spatial distribution of amplitude and phase, signal bandwidth, power density, and polarization, depend on the design of the antenna and the characteristics of the transmitting source. The radiating field will produce a force on any electrical charges that it may encounter. The total time-varying electromagnetic force vector $\boldsymbol{F}(\boldsymbol{r}, t)$ at vector position $\boldsymbol{r}$ and time t applied to an electric charge q is given by the Lorentz force equation as described by Jackson [87, p. 3]:

$$\boldsymbol{F}(\boldsymbol{r}, t) = q\boldsymbol{E}(\boldsymbol{r}, t) + q\boldsymbol{v}(\boldsymbol{r}, t) \times \boldsymbol{B}(\boldsymbol{r}, t) \tag{1.86}$$

where q is the charge of the particle (in this case an electron), $\boldsymbol{E}$ is the electric-field vector, $\boldsymbol{v}$ is the vector velocity of the charged particle, $\boldsymbol{B}$ is the magnetic flux density, and the symbol $\times$ means vector cross product. Such charges will exist, for example, as electrons in a distant metallic receive antenna for communications or within a metallic or other scattering target in a radar application.

For RF signal reception, when a time-varying electromagnetic field is incident on a metallic antenna, the electrons in the metal experience a time-varying force, as computed by the Lorentz force equation, which accelerates them according to the classical equation

$$\boldsymbol{F}(t) = m\boldsymbol{a}(t) \tag{1.87}$$

or

$$\boldsymbol{a}(t) = \frac{\boldsymbol{F}(t)}{m} \tag{1.88}$$

where $\boldsymbol{a}(t)$ is the acceleration as a function of time and m is the mass of the particle. Equivalently, the photon particles in the incident electromagnetic field are absorbed with some probability by the electrons in the metal, altering the electron momentum and causing acceleration of the electrons. (The interaction of the moving, drifting electrons in the metal is a complex process, but here we only care about the net effect.) For low velocities ($v << c$), using Equation (1.86), the force on the electrons in the metal can be approximated by $\boldsymbol{F}(t) = q\boldsymbol{E}(t)$, and in the case of a sinusoidally varying electric field from Equation (1.88), the force and charge acceleration are time varying in a sinusoidal fashion as well.

The acceleration of the negatively charged electrons induces a time-varying electric current in the metallic antenna, which produces a signal voltage across the antenna terminals connected to the transmission line that is connected to a receiver. The time variation of the electric current in the receive antenna tends to be matched to the time variation of the incident field, depending on the design of the antenna.

The remainder of this section describes the mathematical foundation of Maxwell's equations for computation of electromagnetic fields of antennas. In computing electromagnetic fields, the fundamental photon/electron interaction is not used, but rather the current distribution induced on the antenna or scatterer is the required source of the fields.

To quantify the propagation of an electromagnetic wave from or to an antenna, it is useful to review certain fundamental equations that govern the field characteristics. Assume an isotropic medium that is characterized by a permittivity, denoted ϵ, and a permeability μ. As described in many textbooks on electromagnetic theory, electromagnetic fields can be analyzed by Maxwell's equations [1–14, 87, 88], which provide the relationships between electric and magnetic fields, electric charge, and electric current. Depending on the application, Maxwell's equations can take on many different forms, including differential, integral, source driven, and source free, and take account of many different types of materials both isotropic and anisotropic. An overview of the development of Maxwell's equations is given by Paris and Hurd [4, pp. 45–55]. In time-dependent differential form in the general case for antennas in isotropic media, Maxwell's equations follow from Faraday's law, Ampere's circuital law, Gauss's law for electric fields, and Gauss's law for magnetic fields, respectively, as given by the following curl (denoted $\boldsymbol{\nabla}\times$) and divergence (denoted $\boldsymbol{\nabla}\cdot$) equations:

$$\boldsymbol{\nabla}\times\boldsymbol{E} = -\boldsymbol{M} - \frac{\partial \boldsymbol{B}}{\partial t} \tag{1.89}$$

$$\boldsymbol{\nabla}\times\boldsymbol{H} = \boldsymbol{J} + \frac{\partial \boldsymbol{D}}{\partial t} \tag{1.90}$$

$$\boldsymbol{\nabla}\cdot\boldsymbol{D} = \rho \tag{1.91}$$

$$\nabla \cdot \boldsymbol{B} = \rho_m \tag{1.92}$$

where $\boldsymbol{E} = \boldsymbol{E}(\boldsymbol{r}, t)$ is the vector electric field with units of volts/meter, $\boldsymbol{H} = \boldsymbol{H}(\boldsymbol{r}, t)$ is the vector magnetic field with units of amperes/meter, $\boldsymbol{J} = \boldsymbol{J}(\boldsymbol{r}, t)$ is the electric current density with units of amperes/m^2, $\boldsymbol{M} = \boldsymbol{M}(\boldsymbol{r}, t)$ is the equivalent magnetic current density with units of volts/m^2, ρ is the volume electric charge density with units of coulombs/m^3, and ρ_m is the equivalent volume magnetic charge density with units of webers/m^3, which is usually taken to be zero since magnetic monopoles (charges) have not been observed to date. Bold lettering or bold symbols are used here to represent vectors or vector operations. Bold lettering with an overbar is used to refer to a dyadic (tensor rank two) quantity. Vector and dyadic analyses presented here make use of the theoretical work developed by prior researchers such as summarized by Tai [90 - 92]. In Equation (1.90), the term $\partial \boldsymbol{D}/\partial t$ was introduced as a displacement current density by Maxwell. The previous Maxwell equations take into account the general properties of the linear isotropic material, radiating sources, and fields, all of which can have both spatial and time dependence. The inclusion (for mathematical convenience) of the equivalent magnetic current density $\boldsymbol{M}$ and equivalent volume magnetic charge density ρ_m allows for analysis of aperture fields and other general electromagnetics problems requiring these equivalent quantities. In the prior form of the Maxwell's equations, the time dependence of the electromagnetic fields is arbitrary. For arbitrary antenna geometries, this time-dependent form of the Maxwell equations can be solved via numerical methods such as the finite difference time domain technique. An overview of computational methods for solving Maxwell's equations for the electromagnetic fields is given for example in [2]. For the case in which the material properties are linear, isotropic, and nondispersive, the following constitutive relations hold:

$$\boldsymbol{D}(\boldsymbol{r}, t) = \epsilon \boldsymbol{E}(\boldsymbol{r}, t) \tag{1.93}$$

where $\boldsymbol{D}$ is the vector electric flux density with units of coulombs/m^2, and ϵ is the permittivity of the material

$$\boldsymbol{B}(\boldsymbol{r}, t) = \mu \boldsymbol{H}(\boldsymbol{r}, t) \tag{1.94}$$

where $\boldsymbol{B}$ is the vector magnetic flux density with units of webers/m^2 or Teslas, and μ is the permeability of the material. Equations (1.93) and (1.94) are referred to as the constitutive relations, and in this isotropic case, $\boldsymbol{D}$ is parallel to $\boldsymbol{E}$ and $\boldsymbol{B}$ is parallel to $\boldsymbol{H}$. In general, the material properties ϵ and μ are frequency dependent.

In the case in which the material is anisotropic, the constitutive relations are expressed in terms of the material dyadic (tensor) quantities $\overline{\boldsymbol{\epsilon}}$ and $\overline{\boldsymbol{\mu}}$ to convert the vector fields to vector flux densities as

$$\boldsymbol{D}(\boldsymbol{r},t) = \overline{\boldsymbol{\epsilon}}\cdot\boldsymbol{E}(\boldsymbol{r},t) \tag{1.95}$$

$$\boldsymbol{B}(\boldsymbol{r},t) = \overline{\boldsymbol{\mu}}\cdot\boldsymbol{H}(\boldsymbol{r},t) \tag{1.96}$$

where the overbar is used to indicate a dyadic quantity (refer to Section 1.6 for a detailed discussion of dyadics), and the symbol $\cdot$ means dot product. For the anisotropic case, the vector directions of the vector flux densities $\boldsymbol{D}$ and $\boldsymbol{B}$ can be different from the vector directions of the vector fields $\boldsymbol{E}$ and $\boldsymbol{H}$, respectively.

In spherical coordinates, with $\boldsymbol{A} = A_r\hat{\boldsymbol{r}} + A_\theta\hat{\boldsymbol{\theta}} + A_\phi\hat{\boldsymbol{\phi}}$, the curl operation $\boldsymbol{\nabla}\times\boldsymbol{A}$ is given in terms of the unit vectors $\hat{\boldsymbol{r}}, \hat{\boldsymbol{\theta}}, \hat{\boldsymbol{\phi}}$ in convenient determinant form as

$$\boldsymbol{\nabla}\times\boldsymbol{A} = \frac{1}{r^2\sin\theta}\begin{vmatrix} \hat{\boldsymbol{r}} & r\hat{\boldsymbol{\theta}} & r\sin\theta\hat{\boldsymbol{\phi}} \\ \frac{\partial}{\partial r} & \frac{\partial}{\partial\theta} & \frac{\partial}{\partial\phi} \\ A_r & rA_\theta & r\sin\theta A_\phi \end{vmatrix}. \tag{1.97}$$

Performing the determinant operation in Equation (1.97), the curl operation in expanded form is then given by

$$\begin{aligned}\boldsymbol{\nabla}\times\boldsymbol{A} \;=\;& \frac{\hat{\boldsymbol{r}}}{r\sin\theta}\left[\frac{\partial}{\partial\theta}(A_\phi\sin\theta) - \frac{\partial A_\theta}{\partial\phi}\right] + \frac{\hat{\boldsymbol{\theta}}}{r}\left[\frac{1}{\sin\theta}\frac{\partial A_r}{\partial\phi} - \frac{\partial}{\partial r}(rA_\phi)\right] \\ & + \frac{\hat{\boldsymbol{\phi}}}{r}\left[\frac{\partial}{\partial r}(rA_\theta) - \frac{\partial A_r}{\partial\theta}\right].\end{aligned} \tag{1.98}$$

The divergence operation $\boldsymbol{\nabla}\cdot\boldsymbol{A}$ is given by

$$\boldsymbol{\nabla}\cdot\boldsymbol{A} = \frac{1}{r^2}\frac{\partial}{\partial r}(r^2A_r) + \frac{1}{r\sin\theta}\frac{\partial}{\partial\theta}(A_\theta\sin\theta) + \frac{1}{r\sin\theta}\frac{\partial}{\partial\phi}(A_\phi). \tag{1.99}$$

Additionally, note that the vector cross product in spherical coordinates is given by

$$\boldsymbol{A}\times\boldsymbol{B} = \begin{bmatrix} \hat{\boldsymbol{r}} & \hat{\boldsymbol{\theta}} & \hat{\boldsymbol{\phi}} \\ A_r & A_\theta & A_\phi \\ B_r & B_\theta & B_\phi \end{bmatrix}. \tag{1.100}$$

Substituting Equations (1.93) and (1.94) into Equations (1.89) and (1.90), respectively, yields an alternate representation of Maxwell's equations:

$$\boldsymbol{\nabla}\times\boldsymbol{E} = -\boldsymbol{M} - \mu\frac{\partial\boldsymbol{H}}{\partial t} \tag{1.101}$$

$$\boldsymbol{\nabla}\times\boldsymbol{H} = \boldsymbol{J} + \epsilon\frac{\partial\boldsymbol{E}}{\partial t} \tag{1.102}$$

$$\nabla \cdot \boldsymbol{D} = \rho \tag{1.103}$$

$$\nabla \cdot \boldsymbol{B} = \rho_m. \tag{1.104}$$

An equation for expressing the continuity of electrical charge can be derived by taking the divergence of Maxwell's curl equation for the magnetic field, $(\nabla \cdot \nabla \times \boldsymbol{H})$, and applying Gauss's law for electric fields (Equation (1.103)), and using the property that the divergence of the curl of any vector is zero, with the result

$$\nabla \cdot \boldsymbol{J}(\boldsymbol{r}, t) = -\frac{\partial \rho(\boldsymbol{r}, t)}{\partial t}. \tag{1.105}$$

The continuity equation given by Equation (1.105) indicates that the divergence of the electric current density $\boldsymbol{J}$ depends on the time rate of change of the charge density. Similarly, an equation for expressing the continuity of equivalent magnetic charge can be derived by taking the divergence of Maxwell's curl equation for the electric field, applying Gauss's law for magnetic fields, and using the property that the divergence of the curl of any vector is zero, with the result

$$\nabla \cdot \boldsymbol{M}(\boldsymbol{r}, t) = -\frac{\partial \rho_m(\boldsymbol{r}, t)}{\partial t}. \tag{1.106}$$

The continuity equation given by Equation (1.106) indicates that the divergence of the equivalent magnetic current density $\boldsymbol{M}$ depends on the time rate of change of the equivalent magnetic charge density ρ_m.

The electric current density $\boldsymbol{J}$ in Equation (1.102) can be composed of a source current $\boldsymbol{J}_{\text{source}}$ and a conduction current $\boldsymbol{J}_c$ as

$$\boldsymbol{J} = \boldsymbol{J}_{\text{source}} + \boldsymbol{J}_c. \tag{1.107}$$

For a conducting medium with electrical conductivity σ having units of Siemens/meter (amperes per volt-meter), the conducting current density $\boldsymbol{J}_c$ and electric field $\boldsymbol{E}$ are related as

$$\boldsymbol{J}_c = \sigma \boldsymbol{E}, \tag{1.108}$$

which is Ohm's law for conducting current density. Note that electrical conductivity is equal to the reciprocal of electrical resistivity. In free space, it should be noted that $\sigma = 0$.

Similarly, the equivalent magnetic current density $\boldsymbol{M}$ in Equation (1.101) can be composed of an equivalent magnetic source current $\boldsymbol{M}_{\text{source}}$ and an equivalent magnetic conduction current expressed as

$$\boldsymbol{M}_c = \sigma_m \boldsymbol{H} \tag{1.109}$$

where σ_m is the magnetic conductivity that has units of ohms per meter.

Thus, in the case of a lossy dielectric material, the electric current density can be expressed as

$$\boldsymbol{J} = \boldsymbol{J}_{\text{source}} + \sigma \boldsymbol{E}. \tag{1.110}$$

Similarly, in the case of a lossy magnetic material, the magnetic current density can be written as

$$\boldsymbol{M} = \boldsymbol{M}_{\text{source}} + \sigma_m \boldsymbol{H}. \tag{1.111}$$

In summary, the source currents $\boldsymbol{J}_{\text{source}}$ and $\boldsymbol{M}_{\text{source}}$ produce the electromagnetic fields, which then induce a corresponding conduction current density.

Substituting Equations (1.110) and (1.111) into Maxwell's curl equations given by Equations (1.101) and (1.102) yields an alternate form of Maxwell's equations as

$$\nabla \times \boldsymbol{E} = -\boldsymbol{M}_{\text{source}} - \sigma_m \boldsymbol{H} - \mu \frac{\partial \boldsymbol{H}}{\partial t} \tag{1.112}$$

$$\nabla \times \boldsymbol{H} = \boldsymbol{J}_{\text{source}} + \sigma \boldsymbol{E} + \epsilon \frac{\partial \boldsymbol{E}}{\partial t} \tag{1.113}$$

$$\nabla \cdot \boldsymbol{D} = \rho \tag{1.114}$$

$$\nabla \cdot \boldsymbol{B} = \rho_m. \tag{1.115}$$

The instantaneous electric and magnetic fields, as a function of spatial position (x, y, z) or position vector $(\boldsymbol{r})$ and time t, are real functions, and for the time-harmonic case, they are sinusoidal (or cosinusoidal) and can be expressed as

$$\boldsymbol{E}(\boldsymbol{r}, t) = |\boldsymbol{E}(\boldsymbol{r})| \cos(\omega t + \psi(r)) = |\boldsymbol{E}(\boldsymbol{r})| \cos(\omega t - kr) \tag{1.116}$$

$$\boldsymbol{H}(\boldsymbol{r}, t) = |\boldsymbol{H}(\boldsymbol{r})| \cos(\omega t + \psi(r)) = |\boldsymbol{H}(\boldsymbol{r})| \cos(\omega t - kr) \tag{1.117}$$

where $\psi(r) = -kr$ is the spatial phase variation of the electric and magnetic fields, with $k = 2\pi/\lambda = \omega\sqrt{\mu\epsilon}$ the wavenumber, and $\omega = 2\pi f$ is the radian frequency, where f is the frequency. The time-harmonic fields can be extended to most waveforms by utilizing standard Fourier transform techniques. Equivalently, the instantaneous electric and magnetic fields can be written in a more convenient exponential form in terms of the non-time-varying complex electric and magnetic fields $\boldsymbol{E}(\boldsymbol{r})$ and $\boldsymbol{H}(\boldsymbol{r})$ as

$$\boldsymbol{E}(\boldsymbol{r}, t) = \text{Re}[\boldsymbol{E}(\boldsymbol{r}) e^{j\omega t}] \tag{1.118}$$

$$\boldsymbol{H}(\boldsymbol{r},t) = \text{Re}[\boldsymbol{H}(\boldsymbol{r})e^{j\omega t}] \tag{1.119}$$

where Re[·] indicates the real part. Note that the $e^{j\omega t}$ convention used here throughout is common to the electrical engineering convention; however, in cases for which the $e^{-i\omega t}$ convention would be used, the equations presented here can be converted by simply interchanging $j\omega t$ with $-i\omega t$. Similarly, the instantaneous electric current density $\boldsymbol{J}$ and equivalent magnetic current density $\boldsymbol{M}$ can be written as

$$\boldsymbol{J}(\boldsymbol{r},t) = \text{Re}[\boldsymbol{J}(\boldsymbol{r})e^{j\omega t}] \tag{1.120}$$

$$\boldsymbol{M}(\boldsymbol{r},t) = \text{Re}[\boldsymbol{M}(\boldsymbol{r})e^{j\omega t}]. \tag{1.121}$$

The time-dependent fields can also be expressed in terms of the magnitude and phase of the fields as

$$\boldsymbol{E}(\boldsymbol{r},t) = \text{Re}[|\boldsymbol{E}(\boldsymbol{r})|e^{j(\omega t+\psi(r))}] \tag{1.122}$$

$$\boldsymbol{H}(\boldsymbol{r},t) = \text{Re}[|\boldsymbol{H}(\boldsymbol{r})|e^{j(\omega t+\psi(r))}]. \tag{1.123}$$

If the Re[·] operation in Equations (1.122) and (1.123) is dropped and the $e^{j\omega t}$ factor is suppressed, then the following phasor designations for the electric and magnetic fields can also be used:

$$\boldsymbol{E}(\boldsymbol{r}) = |\boldsymbol{E}(\boldsymbol{r})|e^{j\psi(r)} \tag{1.124}$$

$$\boldsymbol{H}(\boldsymbol{r}) = |\boldsymbol{H}(\boldsymbol{r})|e^{j\psi(r)}. \tag{1.125}$$

In Equations (1.124) and (1.125), the magnitudes $|\boldsymbol{E}(\boldsymbol{r})|$ and $|\boldsymbol{H}(\boldsymbol{r})|$ represent the envelope of the electric and magnetic fields, respectively.

In an electrically conducting medium, the electric flux density $\boldsymbol{D}$ is given by

$$\boldsymbol{D}(\boldsymbol{r},t) = \epsilon'\boldsymbol{E}(\boldsymbol{r},t) \tag{1.126}$$

where ϵ' is the real part of the complex permittivity of the medium. Similarly, in a magnetically conducting medium, the magnetic flux density $\boldsymbol{B}$ is given by

$$\boldsymbol{B}(\boldsymbol{r},t) = \mu'\boldsymbol{H}(\boldsymbol{r},t) \tag{1.127}$$

where μ' is the real part of the complex permeability of the medium.

With the prior relations, it is now possible to convert Maxwell's equations to a time-harmonic form as follows. This form of the Maxwell equations is particularly amenable to solution via numerical methods such as the method of moments (discussed in Chapter 2). Substituting Equations (1.118), (1.119), (1.126), (1.127), and (1.120) into Equations (1.101) and (1.102) yields

$$\nabla\times\text{Re}[\boldsymbol{E}(\boldsymbol{r})e^{j\omega t}] = -\text{Re}[\boldsymbol{M}(\boldsymbol{r})e^{j\omega t}] - \frac{\partial}{\partial t}\text{Re}[\mu'\boldsymbol{H}(\boldsymbol{r})e^{j\omega t}] \tag{1.128}$$

$$\nabla\times\text{Re}[\boldsymbol{H}(\boldsymbol{r})e^{j\omega t}] = \text{Re}[\boldsymbol{J}(\boldsymbol{r})e^{j\omega t}] + \frac{\partial}{\partial t}\text{Re}[\epsilon'\boldsymbol{E}(\boldsymbol{r})e^{j\omega t}] \tag{1.129}$$

and computing the partial derivatives, and noting that the real operation commutes with the partial derivative operation, yields

$$\text{Re}\nabla\times[\boldsymbol{E}(\boldsymbol{r})e^{j\omega t}] = -\text{Re}[\boldsymbol{M}(\boldsymbol{r})e^{j\omega t}] - \text{Re}[j\omega\mu'\boldsymbol{H}(\boldsymbol{r})e^{j\omega t}] \tag{1.130}$$

$$\text{Re}\nabla\times[\boldsymbol{H}(\boldsymbol{r})e^{j\omega t}] = \text{Re}[\boldsymbol{J}(\boldsymbol{r})e^{j\omega t}] + \text{Re}[j\omega\epsilon'\boldsymbol{E}(\boldsymbol{r})e^{j\omega t}]. \tag{1.131}$$

The previous two equations can be simplified by dropping the real (Re) operation and suppressing the exponential function $e^{j\omega t}$ that are common to all terms, with the result for Maxwell's equations in time-harmonic (frequency domain) form

$$\nabla\times\boldsymbol{E} = -\boldsymbol{M} - j\omega\mu'\boldsymbol{H} \tag{1.132}$$

$$\nabla\times\boldsymbol{H} = \boldsymbol{J} + j\omega\epsilon'\boldsymbol{E} \tag{1.133}$$

$$\nabla\cdot\boldsymbol{D} = \rho \tag{1.134}$$

$$\nabla\cdot\boldsymbol{B} = \rho_m. \tag{1.135}$$

In the prior Maxwell's equations, the real part of the complex permeability can be expressed as

$$\mu' = \mu_r'\mu_o \tag{1.136}$$

where μ_r' is the relative permeability and μ_o is the permeability of free space. The real part of the complex permittivity is expressed as

$$\epsilon' = \epsilon_r'\epsilon_o \tag{1.137}$$

where ϵ_r' is the dielectric constant (relative permittivity) and ϵ_o is the permittivity of free space. In free space, the speed of light c (also the speed of electromagnetic wave propagation) is related to ϵ_o and μ_o as

$$c = \frac{1}{\sqrt{\mu_o\epsilon_o}} \tag{1.138}$$

and has a measured value of approximately 2.99792458×10^8 m/s $\approx 3.0 \times 10^8$ m/s. The value of ϵ_o can be measured by an experiment involving Coulomb's force law for two charges separated by a distance [4, p. 61]. The established value for the permittivity of free space is $\epsilon_o = \frac{1}{36\pi} \times 10^{-9} = 8.8542 \times 10^{-12}$ farads/meter (farads have units of amperes-seconds/volts). The established value for the permeability of free space is $\mu_o = 4\pi \times 10^{-7} = 1.257 \times 10^{-6}$ henries/meter (henries have units of volts-seconds/ampere).

To determine the speed of electromagnetic wave propagation, consider now Equations (1.116) and (1.117) and follow a constant phase point along the electromagnetic wavefront. Then taking the derivative with respect to time of the argument of the cosine function set equal to a constant yields

$$\frac{d}{dt}(\omega t - kr) = \omega - k\frac{dr}{dt} = 0 \tag{1.139}$$

or in free space

$$\frac{dr}{dt} = \frac{\omega}{k} = \frac{\omega}{\omega\sqrt{\mu_o \epsilon_o}} = c, \tag{1.140}$$

which shows that the time-harmonic electromagnetic fields travel at the speed of light.

For an electrically conducting medium, the electric conducting current density $\boldsymbol{J}_c$ and electric field $\boldsymbol{E}$ are related by Equation (1.108). When a source $\boldsymbol{J}_{\text{source}}$ is present, substituting Equations (1.107) and (1.108) into Equation (1.133) yields

$$\nabla \times \boldsymbol{H} = \boldsymbol{J}_{\text{source}} + (\sigma + j\omega\epsilon')\boldsymbol{E}. \tag{1.141}$$

In Equation (1.141), the quantity $(\sigma + j\omega\epsilon')$ can be expressed as a complex electrical conductivity σ_c as

$$\sigma + j\omega\epsilon' = \sigma' + j\sigma'' = \sigma_c \tag{1.142}$$

where it is observed that the real part of σ_c is $\sigma' = \sigma$ and the imaginary part is $\sigma'' = \omega\epsilon'$.

The electrical conductivity σ of the dielectric medium can be expressed in terms of the imaginary component (denoted ϵ'') of the complex permittivity as

$$\sigma = \omega\epsilon'' = \omega\epsilon_o\epsilon_r'' \tag{1.143}$$

where ϵ_r'' is the relative imaginary component of the complex permittivity. Thus, using Equations (1.137) and (1.143), it follows that Equation (1.141) can be expressed as

$$\nabla \times \boldsymbol{H} = \boldsymbol{J}_{\text{source}} + (j\omega\epsilon_o\epsilon_r' + \omega\epsilon_o\epsilon_r'')\boldsymbol{E}, \tag{1.144}$$

which can be written in an alternate form (by factoring the quantity $j\omega\epsilon_o$) as

$$\nabla \times \boldsymbol{H} = \boldsymbol{J}_{\text{source}} + j\omega\epsilon_o(\epsilon_r' - j\epsilon_r'')\boldsymbol{E}. \tag{1.145}$$

In Equation (1.145), the complex permittivity of the conducting medium can be expressed in terms of real and imaginary components as

$$\epsilon_c = \epsilon' - j\epsilon'' = \epsilon_o(\epsilon_r' - j\epsilon_r''). \tag{1.146}$$

Similarly, the complex permeability can be defined in terms of real and imaginary components as

$$\mu_c = \mu' - j\mu'' = \mu_o(\mu_r' - j\mu_r''). \tag{1.147}$$

Thus, Equation (1.145) can be written in a simplified form in terms of the complex permittivity given by Equation (1.146) as

$$\boldsymbol{\nabla}\times\boldsymbol{H} = \boldsymbol{J}_{\text{source}} + j\omega\epsilon_c\boldsymbol{E}. \tag{1.148}$$

By using the above mathematical development and a similar development for equivalent magnetic current, an explicit representation of Maxwell's equations in time-harmonic (frequency domain) form can be written in terms of source currents as

$$\boldsymbol{\nabla}\times\boldsymbol{E} = -\boldsymbol{M}_{\text{source}} - j\omega\mu_c\boldsymbol{H} \tag{1.149}$$

$$\boldsymbol{\nabla}\times\boldsymbol{H} = \boldsymbol{J}_{\text{source}} + j\omega\epsilon_c\boldsymbol{E} \tag{1.150}$$

$$\boldsymbol{\nabla}\cdot\boldsymbol{D} = \rho \tag{1.151}$$

$$\boldsymbol{\nabla}\cdot\boldsymbol{B} = \rho_m. \tag{1.152}$$

In the above form of Maxwell's equations, the electric conduction current has effectively been folded into the complex permittivity quantity.

By comparing the Maxwell curl of $\boldsymbol{H}$ equations given by Equations (1.141) and (1.150), the following relation can be determined:

$$j\omega\epsilon_c = \sigma + j\omega\epsilon' \tag{1.153}$$

and solving for ϵ_c yields

$$\epsilon_c = \epsilon' - j\frac{\sigma}{\omega} = \epsilon' - j\epsilon''. \tag{1.154}$$

Thus,

$$\epsilon'' = \frac{\sigma}{\omega} \tag{1.155}$$

and it follows from Equation (1.146) that

$$\epsilon_c = \epsilon_o(\epsilon_r' - j\frac{\sigma}{\omega\epsilon_o}), \tag{1.156}$$

which can be rewritten as

$$\epsilon_c = \epsilon_o \epsilon_r'(1 - j\frac{\sigma}{\omega \epsilon_o \epsilon_r'}) \tag{1.157}$$

or as

$$\epsilon_c = \epsilon_o \epsilon_r'(1 - j\frac{\sigma}{\omega \epsilon'}). \tag{1.158}$$

Next, defining the loss tangent as

$$\tan \delta = \frac{\sigma}{\omega \epsilon'}, \tag{1.159}$$

the complex permittivity can be expressed as

$$\epsilon_c = \epsilon_o \epsilon_r'(1 - j \tan \delta). \tag{1.160}$$

1.4 Fields from Time-Varying Electric and Magnetic Current Sources

In this section, Maxwell's equations (described in the previous section) are used to formulate the radiated electromagnetic fields from time-varying electric and magnetic current sources. A detailed derivation is given for the case of electric current sources, and the result is simply stated for magnetic current sources.

Maxwell's equations as given by Equations (1.101–1.104) can be expressed in the time domain in terms of the current densities $\boldsymbol{J}$ and $\boldsymbol{M}$ as

$$\nabla \times \boldsymbol{E}(\boldsymbol{r}, t) = -\boldsymbol{M}(\boldsymbol{r}, t) - \mu \frac{\partial \boldsymbol{H}(\boldsymbol{r}, t)}{\partial t} \tag{1.161}$$

$$\nabla \times \boldsymbol{H}(\boldsymbol{r}, t) = \boldsymbol{J}(\boldsymbol{r}, t) + \epsilon \frac{\partial \boldsymbol{E}(\boldsymbol{r}, t)}{\partial t} \tag{1.162}$$

$$\nabla \cdot \boldsymbol{D}(\boldsymbol{r}, t) = \rho \tag{1.163}$$

$$\nabla \cdot \boldsymbol{B}(\boldsymbol{r}, t) = \rho_m. \tag{1.164}$$

It was mentioned at the start of the previous section that the source of electromagnetic radiation is the time-varying current, which is equal to the acceleration of charges. To demonstrate this fact, the following derivation can be made by using Maxwell's curl equations and a vector identity. Starting with the curl of Maxwell's curl of $\boldsymbol{E}$ equation with the equivalent magnetic current source equal to zero, that is, $\boldsymbol{M}_{\text{source}} = 0$ and the equivalent magnetic charge equal to zero ($\rho_m = 0$), and assuming free space, that is, $\mu = \mu_o$, $\epsilon = \epsilon_o$, $\sigma = 0$, and $\boldsymbol{J}_c = 0$ so $\boldsymbol{J} = \boldsymbol{J}_{\text{source}}$ and then Equation (1.161) becomes

$$\boldsymbol{\nabla}\times\boldsymbol{\nabla}\times\boldsymbol{E} = -\mu_o\frac{\partial}{\partial t}\boldsymbol{\nabla}\times\boldsymbol{H} \tag{1.165}$$

and substituting Maxwell's curl of $\boldsymbol{H}$ equation expressed as

$$\boldsymbol{\nabla}\times\boldsymbol{H} = \boldsymbol{J} + \epsilon_o\frac{\partial\boldsymbol{E}}{\partial t} \tag{1.166}$$

yields

$$\boldsymbol{\nabla}\times\boldsymbol{\nabla}\times\boldsymbol{E} - \mu_o\epsilon_o\frac{\partial^2\boldsymbol{E}}{\partial t^2} = -\mu_o\frac{\partial}{\partial t}\boldsymbol{J}. \tag{1.167}$$

Now use the vector identity

$$\boldsymbol{\nabla}\times\boldsymbol{\nabla}\times\boldsymbol{E} = \boldsymbol{\nabla}(\boldsymbol{\nabla}\cdot\boldsymbol{E}) - \boldsymbol{\nabla}^2\boldsymbol{E} \tag{1.168}$$

where

$$\boldsymbol{\nabla}^2 = \boldsymbol{\nabla}\cdot\boldsymbol{\nabla} \tag{1.169}$$

is the Laplacian operator, such that Equation (1.167) becomes

$$\boldsymbol{\nabla}(\boldsymbol{\nabla}\cdot\boldsymbol{E}) - \boldsymbol{\nabla}^2\boldsymbol{E} - \mu_o\epsilon_o\frac{\partial^2\boldsymbol{E}}{\partial t^2} = -\mu_o\frac{\partial}{\partial t}\boldsymbol{J}. \tag{1.170}$$

It should be noted that the quantity $\boldsymbol{\nabla}^2\boldsymbol{E} = \boldsymbol{\nabla}\cdot\boldsymbol{\nabla}\boldsymbol{E}$ yields a vector. The quantity $\boldsymbol{\nabla}\boldsymbol{E}$ yields a dyadic (tensor of order two), and then taking the divergence of the dyadic yields a vector. The left-hand side of Equation (1.170) contains expressions involving the electric field $\boldsymbol{E}$ and the right-hand side is a function of the derivative of the time-varying current density $\boldsymbol{J}$. Thus, Equation (1.170) is inhomogeneous and it explicitly shows that the time derivative of the electric current density $\boldsymbol{J}(t)$ is a source of electromagnetic radiation. Substituting Equation (1.81) in Equation (1.170) yields

$$\boldsymbol{\nabla}(\boldsymbol{\nabla}\cdot\boldsymbol{E}) - \boldsymbol{\nabla}^2\boldsymbol{E} - \mu_o\epsilon_o\frac{\partial^2\boldsymbol{E}}{\partial t^2} = -\mu_o\rho\boldsymbol{a}, \tag{1.171}$$

which explicitly shows that the charge ρ with acceleration $\boldsymbol{a}$ is a source of electromagnetic radiation, as desired. Furthermore, in free space, using Equation (1.138) in Equation (1.171) yields the following inhomogeneous equation in terms of the speed of light with the accelerated charge as the driving function for the operations on the electric field:

$$\boldsymbol{\nabla}(\boldsymbol{\nabla}\cdot\boldsymbol{E}) - \boldsymbol{\nabla}^2\boldsymbol{E} - \frac{1}{c^2}\frac{\partial^2\boldsymbol{E}}{\partial t^2} = -\mu_o\rho\boldsymbol{a}. \tag{1.172}$$

When both $\rho = 0$ and $\boldsymbol{J} = 0$, that is, there are no local sources, Equations (1.170) to (1.172) reduce to the homogeneous wave equation form as

$$\nabla^2 \boldsymbol{E} - \frac{1}{c^2}\frac{\partial^2 \boldsymbol{E}}{\partial t^2} = 0, \tag{1.173}$$

and similarly it can be shown that

$$\nabla^2 \boldsymbol{H} - \frac{1}{c^2}\frac{\partial^2 \boldsymbol{H}}{\partial t^2} = 0. \tag{1.174}$$

Equations (1.173) and (1.174) indicate that in free space the electromagnetic field exists and is self-sustaining and propagating at the speed of light in the absence of local driving sources.

While Equations (1.170) to (1.172) indicate a source of the electromagnetic radiation, to compute the electromagnetic fields from the electric current density $\boldsymbol{J}$ for a practical antenna, it is necessary to determine an expression for $\boldsymbol{E}$ in terms of $\boldsymbol{J}$.

Consider now, in the frequency domain, the derivation of the electromagnetic fields caused by an electric current source $\boldsymbol{J}$ radiating in an infinite space described by the complex parameters $\mu = \mu_c$ and $\epsilon = \epsilon_c$ (note that although represented here as scalar quantities, μ and ϵ can also be dyadic (tensor) quantities). There are two ways to arrive at the desired expression for $\boldsymbol{E}$ in terms of a volume integral involving $\boldsymbol{J}$. One way starts with the traditional magnetic vector potential approach, as derived in the time-harmonic form [2], and with this approach, the scalar Green's function for isotropic media is used. The other approach is to use the dyadic Green's function approach as described by Tai [89] and Kong [3]. The Green's function characterizes the response to a point source and essentially allows the current density distribution to be integrated in such a way as to determine the electromagnetic field distribution in the surrounding space. Both approaches using the scalar Green's function and the dyadic Green's function are described, respectively, in the next two sections.

1.4.1 Magnetic Vector Potential and the Scalar Green's Function

Starting with the Maxwell divergence equation for the time-harmonic magnetic field (Equation (1.135)) with the equivalent magnetic charge density equal to zero yields

$$\nabla \cdot \boldsymbol{B} = \nabla \cdot \mu \boldsymbol{H} = 0, \tag{1.175}$$

that is,

$$\nabla \cdot \boldsymbol{H} = 0. \tag{1.176}$$

Since the divergence of the vector magnetic field is zero, the vector magnetic field has only circulation and is called a solenoidal field, which can be expressed in terms of the curl of another function by a vector identity. Now, define the magnetic flux density $\boldsymbol{B}$ in terms of the curl of a magnetic vector potential function $\boldsymbol{A}$ as

$$\boldsymbol{B} = \mu \boldsymbol{H} = \boldsymbol{\nabla} \times \boldsymbol{A}. \tag{1.177}$$

Thus, the magnetic field $\boldsymbol{H}$ in terms of $\boldsymbol{A}$ is just

$$\boldsymbol{H} = \frac{1}{\mu} \boldsymbol{\nabla} \times \boldsymbol{A}. \tag{1.178}$$

Substituting this result in Maxwell's curl of $\boldsymbol{E}$ Equation (1.132) yields

$$\boldsymbol{\nabla} \times \boldsymbol{E} = -j\omega\mu \boldsymbol{H} = -j\omega \boldsymbol{\nabla} \times \boldsymbol{A}, \tag{1.179}$$

which can be rewritten as

$$\boldsymbol{\nabla} \times [\boldsymbol{E} + j\omega \boldsymbol{A}] = 0. \tag{1.180}$$

Making use of the general vector identity

$$\boldsymbol{\nabla} \times \boldsymbol{\nabla} \Phi = 0 \tag{1.181}$$

where Φ is defined as the electric scalar potential, allows Equation (1.180) to be written as

$$\boldsymbol{\nabla} \times [\boldsymbol{E} + j\omega \boldsymbol{A}] = -\boldsymbol{\nabla} \times \boldsymbol{\nabla} \Phi \tag{1.182}$$

or as

$$\boldsymbol{E} + j\omega \boldsymbol{A} = -\boldsymbol{\nabla} \Phi, \tag{1.183}$$

and solving for $\boldsymbol{E}$ as

$$\boldsymbol{E} = -j\omega \boldsymbol{A} - \boldsymbol{\nabla} \Phi. \tag{1.184}$$

Now use the vector identity

$$\boldsymbol{\nabla} \times \boldsymbol{\nabla} \times \boldsymbol{A} = \boldsymbol{\nabla}(\boldsymbol{\nabla} \cdot \boldsymbol{A}) - \boldsymbol{\nabla}^2 \boldsymbol{A} \tag{1.185}$$

such that from Equation (1.177)

$$\boldsymbol{\nabla} \times (\mu \boldsymbol{H}) = \boldsymbol{\nabla}(\boldsymbol{\nabla} \cdot \boldsymbol{A}) - \boldsymbol{\nabla}^2 \boldsymbol{A} \tag{1.186}$$

and for a homogeneous medium

$$\mu \boldsymbol{\nabla} \times \boldsymbol{H} = \boldsymbol{\nabla}(\boldsymbol{\nabla} \cdot \boldsymbol{A}) - \boldsymbol{\nabla}^2 \boldsymbol{A}. \tag{1.187}$$

Now using the Maxwell curl equation for the magnetic field, that is,

$$\boldsymbol{\nabla}\times\boldsymbol{H} = \boldsymbol{J} + j\omega\epsilon\boldsymbol{E}, \tag{1.188}$$

Equation (1.187) can now be written as

$$\mu\boldsymbol{J} + j\omega\mu\epsilon\boldsymbol{E} = \boldsymbol{\nabla}(\boldsymbol{\nabla}\cdot\boldsymbol{A}) - \boldsymbol{\nabla}^2\boldsymbol{A}. \tag{1.189}$$

Substituting Equation (1.184) in Equation (1.189) yields

$$\boldsymbol{\nabla}^2\boldsymbol{A} + \omega^2\mu\epsilon\boldsymbol{A} = -\mu\boldsymbol{J} + \boldsymbol{\nabla}(\boldsymbol{\nabla}\cdot\boldsymbol{A} + j\omega\mu\epsilon\Phi). \tag{1.190}$$

Next, applying the divergence operation on the electric field vector in Equation (1.184) as $\boldsymbol{\nabla}\cdot\boldsymbol{E}$ and using Maxwell's equation $\boldsymbol{\nabla}\cdot\boldsymbol{E} = \rho/\epsilon$ yields

$$\boldsymbol{\nabla}^2\Phi + j\omega\boldsymbol{\nabla}\cdot\boldsymbol{A} = -\rho/\epsilon. \tag{1.191}$$

In order to uncouple, simplify, and produce a wave-equation form of Equations (1.190) and (1.191), the Lorenz condition [87, p. 240] can be invoked, that is,

$$\boldsymbol{\nabla}\cdot\boldsymbol{A} = -j\omega\mu\epsilon\Phi. \tag{1.192}$$

It should be noted that the Lorenz condition is consistent with the conservation of electric current and charge density as in the continuity equation from Equation (1.105) expressed in time-harmonic form as (see the discussion following Equation (1.205))

$$\boldsymbol{\nabla}\cdot\boldsymbol{J} = -j\omega\mu\rho. \tag{1.193}$$

Substituting the Lorenz condition (Equation (1.192)) in Equation (1.190) yields

$$\boldsymbol{\nabla}^2\boldsymbol{A}(\boldsymbol{r}) + \omega^2\mu\epsilon\boldsymbol{A}(\boldsymbol{r}) = -\mu\boldsymbol{J}(\boldsymbol{r}), \tag{1.194}$$

which is an inhomogeneous vector wave equation. Similarly, it follows that substituting Equation (1.192) in Equation (1.191) yields

$$\boldsymbol{\nabla}^2\Phi + \omega^2\mu\epsilon\Phi = -\rho/\epsilon, \tag{1.195}$$

which is an inhomogeneous scalar wave equation. Again by using the Lorenz condition given by Equation (1.192), Equation (1.184) becomes

$$\boldsymbol{E} = -j\omega\boldsymbol{A} - \frac{j}{\omega\mu\epsilon}\boldsymbol{\nabla}(\boldsymbol{\nabla}\cdot\boldsymbol{A}). \tag{1.196}$$

Equation (1.196) is an exact relation between the vector electric field $\boldsymbol{E}$ and the magnetic vector potential $\boldsymbol{A}$ and applies in the near field and far field of a radiating source. Thus, to determine the vector electric field $\boldsymbol{E}$, it is necessary to determine the magnetic vector potential $\boldsymbol{A}$.

Consider now Figure 1.21, which shows the coordinate system for a radiating source and field observation point. To solve the inhomogeneous equation given by Equation (1.194), the Green's function method is used [93]. Let L_{op} be a linear differential operator such that

$$L_{\text{op}}y(x) = f(x). \tag{1.197}$$

It is desired to determine the Green's function $G(x, x')$ that is given by the solution to

$$L_{\text{op}}G(x, x') = \delta(x - x') \tag{1.198}$$

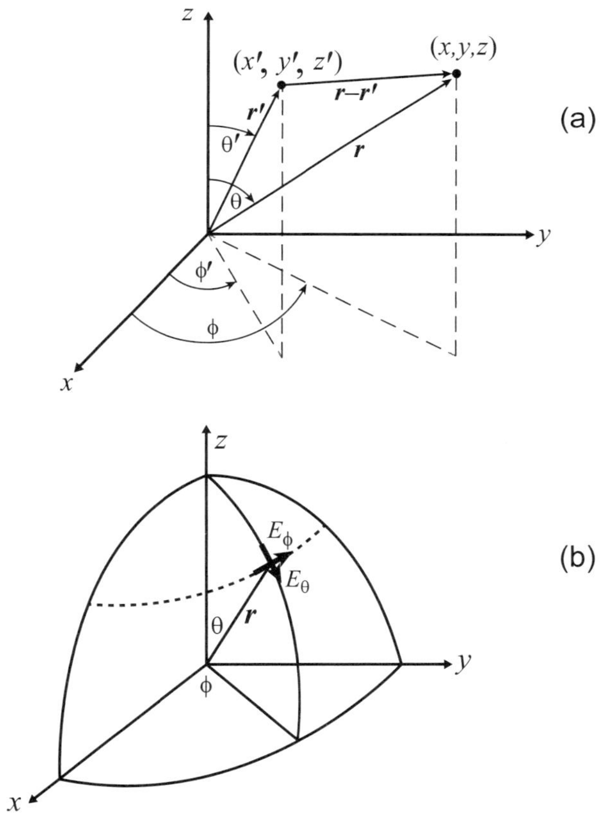

Figure 1.21 (a) Coordinate system for radiating source at $\boldsymbol{r}'$ and field observation point at $\boldsymbol{r}$, and (b) electric field components E_θ and E_ϕ.

where δ is the delta function with $\delta(0) = 1$ and δ is equal to zero otherwise, and boundary conditions must be satisfied. A solution for $y(x)$ is then

$$y(x) = \int G(x, x')f(x')\, dx', \tag{1.199}$$

which follows since

$$L_{\text{op}}y(x) = \int L_{\text{op}}G(x, x')f(x')\, dx' = \int \delta(x - x')f(x')\, dx' = f(x) \tag{1.200}$$

where it is noted that $\delta(x - x')$ acts as a sampling function and L_{op} was moved inside the integral since this operator depends only on unprimed coordinates. Assume that the vector electric current source $\boldsymbol{J}$ is a function of the primed coordinates (x', y', z') and the position vector $\boldsymbol{R} = \boldsymbol{r} - \boldsymbol{r}'$ is a function of the observation position (x, y, z) and the source coordinates. Then it follows from Equation (1.199), which applies to each vector component in homogeneous space, that the solution of Equation (1.194) is

$$\boldsymbol{A}(\boldsymbol{r}) = \mu \int_V \boldsymbol{J}(\boldsymbol{r}')\frac{e^{-jk|\boldsymbol{r}-\boldsymbol{r}'|}}{4\pi|\boldsymbol{r} - \boldsymbol{r}'|}\, dv' \tag{1.201}$$

where

$$k = \omega\sqrt{\mu\epsilon} \tag{1.202}$$

is the wavenumber. Thus, in homogeneous space, the magnetic vector potential $\boldsymbol{A}$ is collinear with the source current $\boldsymbol{J}$, which allows the use of the scalar Green's function method to solve Equation (1.194). The magnetic vector potential is also commonly expressed as

$$\boldsymbol{A}(\boldsymbol{r}) = \mu \int_V \boldsymbol{J}(\boldsymbol{r}')g(\boldsymbol{r}, \boldsymbol{r}')\, dv'. \tag{1.203}$$

In Equation (1.203), the quantity

$$g(\boldsymbol{r}, \boldsymbol{r}') = \frac{e^{-jk|\boldsymbol{r}-\boldsymbol{r}'|}}{4\pi|\boldsymbol{r} - \boldsymbol{r}'|} \tag{1.204}$$

is referred to as the scalar Green's function solution to the vector wave equation (Equation (1.194)), which is derived later in Section 1.4.2 (starting with Equation (1.315)). By solving Equation (1.195) using the Green's function method, it can be shown that the electric scalar potential Φ is given by

$$\Phi(\boldsymbol{r}) = \frac{1}{\epsilon} \int_V \rho(\boldsymbol{r}')g(\boldsymbol{r}, \boldsymbol{r}')\, dv'. \tag{1.205}$$

Note that by substituting Equations (1.203) and (1.205) into the Lorenz condition given by Equation (1.192), it is clear that the continuity equation (Equation (1.193)) is satisfied when multiplied by the scalar Green's function and integrated over a volume.

Using Equations (1.201) and (1.203), Equation (1.196) can be written as

$$\boldsymbol{E} = -j\omega\mu \int_V \boldsymbol{J}(\boldsymbol{r}')g(\boldsymbol{r},\boldsymbol{r}')\,dv' - \frac{j}{\omega\epsilon}\boldsymbol{\nabla}[\boldsymbol{\nabla}\cdot \int_V \boldsymbol{J}(\boldsymbol{r}')g(\boldsymbol{r},\boldsymbol{r}')\,dv'] \quad (1.206)$$

or as

$$\boldsymbol{E} = -j\omega\mu \int_V \boldsymbol{J}(\boldsymbol{r}')\frac{e^{-jk|\boldsymbol{r}-\boldsymbol{r}'|}}{4\pi|\boldsymbol{r}-\boldsymbol{r}'|}\,dv' - \frac{j}{\omega\epsilon}\boldsymbol{\nabla}[\boldsymbol{\nabla}\cdot \int_V \boldsymbol{J}(\boldsymbol{r}')\frac{e^{-jk|\boldsymbol{r}-\boldsymbol{r}'|}}{4\pi|\boldsymbol{r}-\boldsymbol{r}'|}\,dv'], \quad (1.207)$$

which can be reduced to the following integral form

$$\boldsymbol{E} = -j\omega\mu \left[\int_V \boldsymbol{J}(\boldsymbol{r}')\frac{e^{-jk|\boldsymbol{r}-\boldsymbol{r}'|}}{4\pi|\boldsymbol{r}-\boldsymbol{r}'|}\,dv' + \frac{1}{k^2}\boldsymbol{\nabla}[\boldsymbol{\nabla}\cdot \int_V \boldsymbol{J}(\boldsymbol{r}')\frac{e^{-jk|\boldsymbol{r}-\boldsymbol{r}'|}}{4\pi|\boldsymbol{r}-\boldsymbol{r}'|}\,dv']\right] \quad (1.208)$$

where $k^2 = \omega^2\mu\epsilon$. Equation (1.208) can be expressed in a simplified form by using a dyadic notation (discussed in Section 1.4.2) with $\overline{\boldsymbol{I}}$ the identity dyadic (refer to Equation (1.246)):

$$\boldsymbol{E}(\boldsymbol{r}) = -j\omega\mu[\overline{\boldsymbol{I}} + \frac{1}{k^2}\boldsymbol{\nabla}\boldsymbol{\nabla}]\cdot \int_V \boldsymbol{J}(\boldsymbol{r}')\frac{e^{-jk|\boldsymbol{r}-\boldsymbol{r}'|}}{4\pi|\boldsymbol{r}-\boldsymbol{r}'|}\,dv' \quad (1.209)$$

which is a general form of the electric field caused by an electric current source $\boldsymbol{J}$.

A similar wave equation can be derived in terms of the electric vector potential $\boldsymbol{F}$ and the magnetic current density $\boldsymbol{M}$ as

$$\boldsymbol{\nabla}^2\boldsymbol{F} + \omega^2\mu\epsilon\boldsymbol{F} = -\epsilon\boldsymbol{M} \quad (1.210)$$

and for the magnetic field $\boldsymbol{H}$ in terms of $\boldsymbol{F}$ as

$$\boldsymbol{H} = -j\omega\boldsymbol{F} - \frac{j}{\omega\mu\epsilon}\boldsymbol{\nabla}(\boldsymbol{\nabla}\cdot\boldsymbol{F}). \quad (1.211)$$

The solution for $\boldsymbol{F}$ in terms of $\boldsymbol{M}$ is given by

$$\boldsymbol{F}(\boldsymbol{r}) = \epsilon \int_V \boldsymbol{M}(\boldsymbol{r}')\frac{e^{-jk|\boldsymbol{r}-\boldsymbol{r}'|}}{4\pi|\boldsymbol{r}-\boldsymbol{r}'|}\,dv'. \quad (1.212)$$

The fields radiated by an antenna are in general spherical waves. In the far field, the spherical waves more closely resemble planar (plane) waves. The far-field range distance R from an aperture antenna is typically computed as

$$R = 2D^2/\lambda \tag{1.213}$$

where D is the antenna diameter and λ is the wavelength. In the far field, using spherical coordinates, it can be shown (see Equation (1.283)) that only the first term in Equation (1.196) involving the $\hat{\boldsymbol{\theta}}$ and $\hat{\boldsymbol{\phi}}$ components is significant; thus,

$$\boldsymbol{E}_{\text{FF}} = -j\omega\left(A_\theta\hat{\boldsymbol{\theta}} + A_\phi\hat{\boldsymbol{\phi}}\right). \tag{1.214}$$

In the far zone in spherical coordinates, the radial component of the electric and magnetic fields is zero, and the far-zone magnetic field can be computed from the electric field in free space as (see Equation (1.285))

$$\boldsymbol{H}_{\text{FF}} = \frac{\hat{\boldsymbol{r}}\times\boldsymbol{E}_{\text{FF}}}{\eta_o} \tag{1.215}$$

where η_o is the wave impedance of free space as given by

$$\eta_o = \sqrt{\frac{\mu_o}{\epsilon_o}}. \tag{1.216}$$

Similarly, it can be shown that the far-zone magnetic field $\boldsymbol{H}_{\text{FF}}$ can be computed from the electric vector potential $\boldsymbol{F}$, for example, in spherical coordinates, as

$$\boldsymbol{H}_{\text{FF}} = -j\omega\left(F_\theta\hat{\boldsymbol{\theta}} + F_\phi\hat{\boldsymbol{\phi}}\right) \tag{1.217}$$

from which the far-zone electric field can be computed as

$$\boldsymbol{E}_{\text{FF}} = -\eta_o\hat{\boldsymbol{r}}\times\boldsymbol{H}_{\text{FF}}. \tag{1.218}$$

Once the current density and field quantities have been determined, to convert between frequency and time domains, the Fourier transform pairs between time domain and frequency domain can be applied, for example, for $\boldsymbol{J}$ and $\boldsymbol{E}$, as

$$\boldsymbol{J}(r,\omega) = \int_{-\infty}^{\infty}\boldsymbol{J}(r,t)e^{-j\omega t}dt \tag{1.219}$$

$$\boldsymbol{J}(r,t) = \frac{1}{2\pi}\int_{-\infty}^{\infty}\boldsymbol{J}(r,\omega)e^{j\omega t}d\omega \tag{1.220}$$

$$\boldsymbol{E}(r,\omega) = \int_{-\infty}^{\infty}\boldsymbol{E}(r,t)e^{-j\omega t}dt \tag{1.221}$$

$$\boldsymbol{E}(r,t) = \frac{1}{2\pi}\int_{-\infty}^{\infty}\boldsymbol{E}(r,\omega)e^{j\omega t}d\omega. \tag{1.222}$$

1.4.2 Dyadic Green's Function

Dyadic Green's functions have been investigated by Tai [89] for analysis of a number of geometries involving electromagnetic fields including free space, rectangular and circular waveguides, circular and elliptical cylinders, flat earth, perfectly conducting wedges, spheres and cones, and inhomogeneous media. In this section, the free-space dyadic Green's function is derived following along the lines of a derivation given by Kong [3].

It was mentioned earlier in Section 1.4.1 that a direct dyadic relationship between the source current and electric field can be determined. To see this fact, Equations (1.196) and (1.201) can be combined and re-expressed by using a dyadic notation as was given in Equation (1.209). Some background on vector and dyadic relations is necessary to derive the desired expression. Given two vectors, $\boldsymbol{U}$ and $\boldsymbol{V}$, their dot product is expressed as $\boldsymbol{U} \cdot \boldsymbol{V}$, which is also referred to as an inner product. The inner product is equal to the product of a row vector $\boldsymbol{U}^T$ ($\boldsymbol{U}$ transpose) and a column vector $\boldsymbol{V}$, which yields a scalar. Thus, $\boldsymbol{U} \cdot \boldsymbol{V} = \boldsymbol{U}^T \boldsymbol{V}$ is a scalar quantity. The dyadic $\overline{\boldsymbol{D}} = \boldsymbol{U}\boldsymbol{V}$ is an outer product of two vectors and is the product of a column vector and a row vector, that is, $\boldsymbol{U}\boldsymbol{V}$ is interpreted as $\boldsymbol{U}\boldsymbol{V}^T$. Thus, $\overline{\boldsymbol{D}} \cdot \boldsymbol{C} = \boldsymbol{U}\boldsymbol{V} \cdot \boldsymbol{C} = \boldsymbol{U}(\boldsymbol{V}^T \boldsymbol{C})$, that is, a vector weighted by a scalar. Let the unit vectors in a given coordinate system, such as Cartesian, cylindrical, or spherical, be expressed as $\hat{\boldsymbol{e}_1}$, $\hat{\boldsymbol{e}_2}$, $\hat{\boldsymbol{e}_3}$. Now, let the two vectors $\boldsymbol{U}$ and $\boldsymbol{V}$ be represented in terms of the unit vectors as

$$\boldsymbol{U} = U_1 \hat{\boldsymbol{e}_1} + U_2 \hat{\boldsymbol{e}_2} + U_3 \hat{\boldsymbol{e}_3} \tag{1.223}$$

$$\boldsymbol{V} = V_1 \hat{\boldsymbol{e}_1} + V_2 \hat{\boldsymbol{e}_2} + V_3 \hat{\boldsymbol{e}_3}, \tag{1.224}$$

which can also be written as the column vectors

$$\boldsymbol{U} = \begin{bmatrix} U_1 \\ U_2 \\ U_3 \end{bmatrix} \tag{1.225}$$

$$\boldsymbol{V} = \begin{bmatrix} V_1 \\ V_2 \\ V_3 \end{bmatrix}. \tag{1.226}$$

The dyadic matrix $\overline{\boldsymbol{D}} = \boldsymbol{U}\boldsymbol{V}$ is formed as

$$\overline{\boldsymbol{D}} = \boldsymbol{U}\boldsymbol{V}^T = \begin{bmatrix} U_1 \\ U_2 \\ U_3 \end{bmatrix} \begin{bmatrix} V_1 & V_2 & V_3 \end{bmatrix} = \begin{bmatrix} U_1V_1 & U_1V_2 & U_1V_3 \\ U_2V_1 & U_2V_2 & U_2V_3 \\ U_3V_1 & U_3V_2 & U_3V_3 \end{bmatrix}. \tag{1.227}$$

The dyadic $\overline{\boldsymbol{D}} = \boldsymbol{U}\boldsymbol{V}^T$ can be expressed in terms of the unit vectors as

$$\overline{\boldsymbol{D}} = \sum_i \sum_j U_i V_j \hat{\boldsymbol{e}}_i \hat{\boldsymbol{e}}_j \tag{1.228}$$

and the summation can be expanded as

$$\begin{aligned} \overline{\boldsymbol{D}} = \boldsymbol{U}\boldsymbol{V}^T \quad = \quad & U_1 V_1 \hat{\boldsymbol{e}}_1 \hat{\boldsymbol{e}}_1 + U_1 V_2 \hat{\boldsymbol{e}}_1 \hat{\boldsymbol{e}}_2 + U_1 V_3 \hat{\boldsymbol{e}}_1 \hat{\boldsymbol{e}}_3 \\ & + U_2 V_1 \hat{\boldsymbol{e}}_2 \hat{\boldsymbol{e}}_1 + U_2 V_2 \hat{\boldsymbol{e}}_2 \hat{\boldsymbol{e}}_2 + U_2 V_3 \hat{\boldsymbol{e}}_2 \hat{\boldsymbol{e}}_3 \\ & + U_3 V_1 \hat{\boldsymbol{e}}_3 \hat{\boldsymbol{e}}_1 + U_3 V_2 \hat{\boldsymbol{e}}_3 \hat{\boldsymbol{e}}_2 + U_3 V_3 \hat{\boldsymbol{e}}_3 \hat{\boldsymbol{e}}_3. \end{aligned} \tag{1.229}$$

The product of a dyadic $\overline{\boldsymbol{D}}$ and a vector $\boldsymbol{A}$ yields another vector $\boldsymbol{X}$, that is,

$$\overline{\boldsymbol{D}} \cdot \boldsymbol{A} = \boldsymbol{X}. \tag{1.230}$$

Now the dyadic $\overline{\boldsymbol{D}}$ can be written in matrix form as

$$\overline{\boldsymbol{D}} = \begin{bmatrix} D_{11} & D_{12} & D_{13} \\ D_{21} & D_{22} & D_{23} \\ D_{31} & D_{32} & D_{33} \end{bmatrix}. \tag{1.231}$$

For the vectors $\boldsymbol{U}$ and $\boldsymbol{V}$ discussed earlier, in Equation (1.231), the dyadic matrix elements (dyads) D_{ij} are equal to $U_i V_j$. In general, the dyadic $\overline{\boldsymbol{D}}$ can be expressed in terms of the unit vectors as

$$\overline{\boldsymbol{D}} = \sum_i \sum_j D_{ij} \hat{\boldsymbol{e}}_i \hat{\boldsymbol{e}}_j. \tag{1.232}$$

The vector $\boldsymbol{A}$ above can be written as a column matrix as

$$\boldsymbol{A} = \begin{bmatrix} A_1 \\ A_2 \\ A_3 \end{bmatrix} \tag{1.233}$$

and it can also be written in terms of the unit vectors as

$$\boldsymbol{A} = \sum_k A_k \hat{\boldsymbol{e}}_k. \tag{1.234}$$

It follows then that

$$\overline{\boldsymbol{D}} \cdot \boldsymbol{A} = \left[\sum_i \sum_j D_{ij} \hat{\boldsymbol{e}}_i \hat{\boldsymbol{e}}_j \right] \cdot \left[\sum_k A_k \hat{\boldsymbol{e}}_k \right] \tag{1.235}$$

or

$$\overline{\boldsymbol{D}} \cdot \boldsymbol{A} = \sum_i \sum_j \sum_k D_{ij} A_k \left[\hat{\boldsymbol{e}}_i \hat{\boldsymbol{e}}_j \cdot \hat{\boldsymbol{e}}_k \right] \tag{1.236}$$

$$\overline{\boldsymbol{D}}\cdot\boldsymbol{A} = \sum_i \sum_j \sum_k D_{ij} A_k \left[\hat{\boldsymbol{e}}_i \delta_{jk}\right] \tag{1.237}$$

where δ_{jk} is the Kronecker delta function, that is, $\delta_{jk} = 0$, if $j \neq k$ and $\delta_{jk} = 1$, if $j = k$. It follows then that

$$\overline{\boldsymbol{D}}\cdot\boldsymbol{A} = \sum_i \hat{\boldsymbol{e}}_i \left[\sum_j D_{ij} A_j\right], \tag{1.238}$$

which can be expressed as

$$\begin{aligned} \overline{\boldsymbol{D}}\cdot\boldsymbol{A} \quad = \quad & \hat{\boldsymbol{e}}_1(D_{11}A_1 + D_{12}A_2 + D_{13}A_3) \\ & + \hat{\boldsymbol{e}}_2(D_{21}A_1 + D_{22}A_2 + D_{23}A_3) \\ & + \hat{\boldsymbol{e}}_3(D_{31}A_1 + D_{32}A_2 + D_{33}A_3). \end{aligned} \tag{1.239}$$

For the spherical coordinate geometry under consideration, with coordinate variables r, θ, ϕ, the unit vectors are $\hat{\boldsymbol{e}}_1 = \hat{\boldsymbol{r}}$, $\hat{\boldsymbol{e}}_2 = \hat{\boldsymbol{\theta}}$, $\hat{\boldsymbol{e}}_3 = \hat{\boldsymbol{\phi}}$. The metric coefficients in the spherical coordinate system are $1, r, r\sin\theta$. The mixed dyadic/vector calculus needed to solve electromagnetic problems is clearly more involved than vector calculus. In some cases, it is necessary to differentiate the unit vectors, for example, in spherical coordinates, where the nonzero partial derivatives are given by $\partial\hat{\boldsymbol{r}}/\partial\theta = \hat{\boldsymbol{\theta}}$, $\partial\hat{\boldsymbol{r}}/\partial\phi = \sin\theta\hat{\boldsymbol{\phi}}$, $\partial\hat{\boldsymbol{\theta}}/\partial\theta = -\hat{\boldsymbol{r}}$, $\partial\hat{\boldsymbol{\theta}}/\partial\phi = \cos\theta\hat{\boldsymbol{\phi}}$, and $\partial\hat{\boldsymbol{\phi}}/\partial\phi = -\sin\theta\hat{\boldsymbol{r}} - \cos\theta\hat{\boldsymbol{\theta}}$, and the partial derivatives equal to zero are $\partial\hat{\boldsymbol{r}}/\partial r = \partial\hat{\boldsymbol{\theta}}/\partial r = \partial\hat{\boldsymbol{\phi}}/\partial r = \partial\hat{\boldsymbol{\phi}}/\partial\theta = 0$. These unit vector derivatives are readily derived by differentiating the spherical unit vectors in terms of the rectangular unit vector components as expressed as

$$\hat{\boldsymbol{r}} = \hat{\boldsymbol{x}}\sin\theta\cos\phi + \hat{\boldsymbol{y}}\sin\theta\sin\phi + \hat{\boldsymbol{z}}\cos\theta \tag{1.240}$$

$$\hat{\boldsymbol{\theta}} = \hat{\boldsymbol{x}}\cos\theta\cos\phi + \hat{\boldsymbol{y}}\cos\theta\sin\phi - \hat{\boldsymbol{z}}\sin\theta \tag{1.241}$$

$$\hat{\boldsymbol{\phi}} = -\hat{\boldsymbol{x}}\sin\phi + \hat{\boldsymbol{y}}\cos\phi. \tag{1.242}$$

Furthermore, the rectangular unit vectors can be expressed in terms of the spherical unit vectors as

$$\hat{\boldsymbol{x}} = \hat{\boldsymbol{r}}\sin\theta\cos\phi + \hat{\boldsymbol{\theta}}\cos\theta\cos\phi - \hat{\boldsymbol{\phi}}\sin\phi \tag{1.243}$$

$$\hat{\boldsymbol{y}} = \hat{\boldsymbol{r}}\sin\theta\sin\phi + \hat{\boldsymbol{\theta}}\cos\theta\sin\phi + \hat{\boldsymbol{\phi}}\cos\phi \tag{1.244}$$

$$\hat{\boldsymbol{z}} = \hat{\boldsymbol{r}}\cos\theta - \hat{\boldsymbol{\theta}}\sin\theta. \tag{1.245}$$

The identity dyadic $\overline{\boldsymbol{I}}$ in spherical coordinates can be expressed in tensor summation form as

$$\overline{\boldsymbol{I}} = \hat{\boldsymbol{r}}\hat{\boldsymbol{r}} + \hat{\boldsymbol{\theta}}\hat{\boldsymbol{\theta}} + \hat{\boldsymbol{\phi}}\hat{\boldsymbol{\phi}} \tag{1.246}$$

or in matrix form as

$$\overline{\boldsymbol{I}} = \begin{bmatrix} 1 & 0 & 0 \\ 0 & 1 & 0 \\ 0 & 0 & 1 \end{bmatrix}. \quad (1.247)$$

The product of the identity dyadic $\overline{\boldsymbol{I}}$ and any vector, say $\boldsymbol{A}$, is just the vector $\boldsymbol{A}$, that is,

$$\overline{\boldsymbol{I}}\cdot\boldsymbol{A} = \boldsymbol{A}\cdot\overline{\boldsymbol{I}} = \boldsymbol{A}. \quad (1.248)$$

Next, it is observed that the quantity $\boldsymbol{\nabla}(\boldsymbol{\nabla}\cdot\boldsymbol{A})$ in Equation (1.196) is the gradient of the divergence of $\boldsymbol{A}$, which is a vector quantity. In spherical coordinates, the vector $\boldsymbol{A}$ is expressed as

$$\boldsymbol{A} = A_r\hat{\boldsymbol{r}} + A_\theta\hat{\boldsymbol{\theta}} + A_\phi\hat{\boldsymbol{\phi}}. \quad (1.249)$$

Furthermore, the del operator in spherical coordinates is expressed as

$$\boldsymbol{\nabla} = \hat{\boldsymbol{r}}\frac{\partial}{\partial r} + \hat{\boldsymbol{\theta}}\frac{1}{r}\frac{\partial}{\partial\theta} + \hat{\boldsymbol{\phi}}\frac{1}{r\sin\theta}\frac{\partial}{\partial\phi}. \quad (1.250)$$

Special care must be used in evaluating mixed vector/dyadic expressions involving the del differential operator $\boldsymbol{\nabla}$ [91]. From Tai [90], given an orthogonal coordinate system with unit vectors $\hat{\boldsymbol{e_1}}, \hat{\boldsymbol{e_2}}, \hat{\boldsymbol{e_3}}$, coordinate variables v_1, v_2, v_3, and metric coefficients h_1, h_2, h_3, the gradient, divergence, and curl differential operations for the vector function $\boldsymbol{A}$ are given by, respectively,

$$\boldsymbol{\nabla}\boldsymbol{A} = \sum_i \frac{\hat{\boldsymbol{e}}_i}{h_i}\frac{\partial\boldsymbol{A}}{\partial v_i} \quad (1.251)$$

$$\boldsymbol{\nabla}\cdot\boldsymbol{A} = \sum_i \frac{\hat{\boldsymbol{e}}_i}{h_i}\cdot\frac{\partial\boldsymbol{A}}{\partial v_i} \quad (1.252)$$

$$\boldsymbol{\nabla}\times\boldsymbol{A} = \sum_i \frac{\hat{\boldsymbol{e}}_i}{h_i}\times\frac{\partial\boldsymbol{A}}{\partial v_i}. \quad (1.253)$$

In spherical coordinates, the coordinate variables are $v_1 = r, v_2 = \theta, v_3 = \phi$, the unit vectors are $\boldsymbol{e}_1 = \hat{\boldsymbol{r}}, \boldsymbol{e}_2 = \hat{\boldsymbol{\theta}}, \boldsymbol{e}_3 = \hat{\boldsymbol{\phi}}$, and the metric coefficients are $h_1 = 1, h_2 = r, h_3 = r\sin\theta$.

It is worthwhile utilizing the del operator $\boldsymbol{\nabla}$ as follows. A dyadic such as $\boldsymbol{\nabla}\boldsymbol{A}$ can be formed either as a summation by Equation (1.251) or by matrix operation as

$$\boldsymbol{\nabla}\boldsymbol{A} = \boldsymbol{\nabla}\boldsymbol{A}^T = \begin{bmatrix} (\hat{\boldsymbol{r}})\frac{\partial}{\partial r} \\ (\hat{\boldsymbol{\theta}})\frac{1}{r}\frac{\partial}{\partial\theta} \\ (\hat{\boldsymbol{\phi}})\frac{1}{r\sin\theta}\frac{\partial}{\partial\phi} \end{bmatrix} \begin{bmatrix} (\hat{\boldsymbol{r}})A_r & (\hat{\boldsymbol{\theta}})A_\theta & (\hat{\boldsymbol{\phi}})A_\phi \end{bmatrix}. \quad (1.254)$$

The nine terms of the tensor summation of Equation (1.254) are computed using the chain rule for the unit vectors. For example, the $\hat{\boldsymbol{r}}\hat{\boldsymbol{r}}$ dyad term is given by

$$(\boldsymbol{\nabla A})_{rr} = \hat{\boldsymbol{r}}[\hat{\boldsymbol{r}}\frac{\partial}{\partial r}A_r + A_r\frac{\partial \hat{\boldsymbol{r}}}{\partial r}] \tag{1.255}$$

and noting $\frac{\partial \hat{\boldsymbol{r}}}{\partial r} = 0$. The remaining eight terms of Equation (1.254) are computed similarly, with the result expressed as the tensor summation

$$\begin{aligned}\boldsymbol{\nabla A} \quad = \quad & \tfrac{\partial A_r}{\partial r}\hat{\boldsymbol{r}}\hat{\boldsymbol{r}} + \tfrac{\partial A_\theta}{\partial r}\hat{\boldsymbol{r}}\hat{\boldsymbol{\theta}} + \tfrac{\partial A_\phi}{\partial r}\hat{\boldsymbol{r}}\hat{\boldsymbol{\phi}} \\ & +(\tfrac{1}{r}\tfrac{\partial A_r}{\partial \theta} - \tfrac{A_\theta}{r})\hat{\boldsymbol{\theta}}\hat{\boldsymbol{r}} + (\tfrac{1}{r}\tfrac{\partial A_\theta}{\partial \theta} + \tfrac{A_r}{r})\hat{\boldsymbol{\theta}}\hat{\boldsymbol{\theta}} + \tfrac{1}{r}\tfrac{\partial A_\phi}{\partial \theta}\hat{\boldsymbol{\theta}}\hat{\boldsymbol{\phi}} \\ & +(\tfrac{1}{r\sin\theta}\tfrac{\partial A_r}{\partial \phi} - \tfrac{A_\phi}{r})\hat{\boldsymbol{\phi}}\hat{\boldsymbol{r}} + (\tfrac{1}{r\sin\theta}\tfrac{\partial A_\theta}{\partial \phi} - \cot\theta\tfrac{A_\phi}{r})\hat{\boldsymbol{\phi}}\hat{\boldsymbol{\theta}} \\ & +(\tfrac{1}{r\sin\theta}\tfrac{\partial A_\phi}{\partial \phi} + \cot\theta\tfrac{A_\theta}{r} + \tfrac{A_r}{r})\hat{\boldsymbol{\phi}}\hat{\boldsymbol{\phi}}.\end{aligned} \tag{1.256}$$

Earlier it was stated (see Equation (1.209)) that the electric field caused by the source current density $\boldsymbol{J}$ is given by

$$\boldsymbol{E}(\boldsymbol{r}) = -j\omega\mu[\overline{\boldsymbol{I}} + \frac{1}{k^2}\boldsymbol{\nabla\nabla}]\cdot\int_V \boldsymbol{J}(\boldsymbol{r}')\frac{e^{-jk|\boldsymbol{r}-\boldsymbol{r}'|}}{4\pi|\boldsymbol{r}-\boldsymbol{r}'|}dv' \tag{1.257}$$

and now it is desired to solve this equation under far-field conditions. Using the vector identity

$$(\boldsymbol{\nabla\nabla})\cdot\boldsymbol{A} = \boldsymbol{\nabla}(\boldsymbol{\nabla}\cdot\boldsymbol{A}), \tag{1.258}$$

it follows that the gradient of the divergence of a vector function $\boldsymbol{A}$ is [90, p. 126]

$$\boldsymbol{\nabla}(\boldsymbol{\nabla}\cdot\boldsymbol{A}) = \sum_i \frac{\hat{\boldsymbol{e}}_i}{h_i}\frac{\partial}{\partial v_i}\sum_j \frac{\hat{\boldsymbol{e}}_j}{h_j}\cdot\frac{\partial \boldsymbol{A}}{\partial v_j}, \tag{1.259}$$

and then performing the partial derivative operation using the chain rule yields

$$\boldsymbol{\nabla}(\boldsymbol{\nabla}\cdot\boldsymbol{A}) = \sum_i\sum_j \frac{\hat{\boldsymbol{e}}_i}{h_i}\left[\frac{\partial}{\partial v_i}\left(\frac{\hat{\boldsymbol{e}}_j}{h_j}\right)\cdot\frac{\partial \boldsymbol{A}}{\partial v_j} + \frac{\hat{\boldsymbol{e}}_j}{h_j}\cdot\frac{\partial^2 \boldsymbol{A}}{\partial v_i \partial v_j}\right]. \tag{1.260}$$

Using the spherical coordinate parameters in Equation (1.260) yields for the $i = 1, j = 1$ term

$$\boldsymbol{\nabla}(\boldsymbol{\nabla}\cdot\boldsymbol{A})_{i=1,j=1} = \hat{\boldsymbol{r}}\left[\frac{\partial}{\partial r}(\hat{\boldsymbol{r}})\cdot\frac{\partial \boldsymbol{A}}{\partial r} + \hat{\boldsymbol{r}}\cdot\frac{\partial^2 \boldsymbol{A}}{\partial r^2}\right], \tag{1.261}$$

and since $\partial\hat{\boldsymbol{r}}/\partial r = 0$, it follows that

$$\boldsymbol{\nabla}(\boldsymbol{\nabla}\cdot\boldsymbol{A})_{i=1,j=1} = \hat{\boldsymbol{r}}\hat{\boldsymbol{r}}\cdot\frac{\partial^2 \boldsymbol{A}}{\partial r^2}. \tag{1.262}$$

Similarly, Equation (1.260) yields for the $i = 1, j = 2$ term

$$\nabla(\nabla \cdot \boldsymbol{A})_{i=1,j=2} = \hat{\boldsymbol{r}} \left[\frac{\partial}{\partial r} \left(\frac{\hat{\boldsymbol{\theta}}}{r} \right) \cdot \frac{\partial \boldsymbol{A}}{\partial \theta} + \frac{\hat{\boldsymbol{\theta}}}{r} \cdot \frac{\partial^2 \boldsymbol{A}}{\partial r \partial \theta} \right], \tag{1.263}$$

and since $\frac{\partial}{\partial r}(\hat{\boldsymbol{\theta}}/r) = -\hat{\boldsymbol{\theta}}/r^2$, it follows that

$$\nabla(\nabla \cdot \boldsymbol{A})_{i=1,j=2} = \hat{\boldsymbol{r}}\hat{\boldsymbol{\theta}} \left[-\frac{1}{r^2} \cdot \frac{\partial \boldsymbol{A}}{\partial \theta} + \frac{1}{r} \cdot \frac{\partial^2 \boldsymbol{A}}{\partial r \partial \theta} \right]. \tag{1.264}$$

Next, for the $i = 1, j = 3$ term,

$$\nabla(\nabla \cdot \boldsymbol{A})_{i=1,j=3} = \hat{\boldsymbol{r}}\hat{\boldsymbol{\phi}} \left[-\frac{1}{r^2 \sin\theta} \cdot \frac{\partial \boldsymbol{A}}{\partial \phi} + \frac{1}{r \sin\theta} \cdot \frac{\partial^2 \boldsymbol{A}}{\partial r \partial \phi} \right]. \tag{1.265}$$

In solving for the far-zone radiation pattern function of an antenna, only a $1/r$ variation is retained with all other terms involving $1/r^2$, $1/r^3$ ignored. All of the remaining terms of $\nabla(\nabla \cdot \boldsymbol{A})_{i,j}$ for $i = 2, 3$ have higher-order dependence on inverse r and can be neglected. Thus, in the far field, the gradient of the divergence of the magnetic vector potential can be expressed as

$$\nabla(\nabla \cdot \boldsymbol{A})_{\text{FF}} = \hat{\boldsymbol{r}}\hat{\boldsymbol{r}} \cdot \frac{\partial^2 \boldsymbol{A}}{\partial r^2}. \tag{1.266}$$

The electric field given by Equation (1.196) can be written as

$$\boldsymbol{E} = -j\omega[\boldsymbol{A} + \frac{1}{k^2} \nabla (\nabla \cdot \boldsymbol{A})] \tag{1.267}$$

where $k^2 = \omega^2 \mu \epsilon$. Now, using Equation (1.266) in Equation (1.267) yields an expression for the far-zone vector electric field

$$\boldsymbol{E}_{\text{FF}} = -j\omega \left[\boldsymbol{A} + \frac{1}{k^2} \hat{\boldsymbol{r}}\hat{\boldsymbol{r}} \cdot \frac{\partial^2 \boldsymbol{A}}{\partial r^2} \right]. \tag{1.268}$$

Repeating Equation (1.201) for the magnetic vector potential

$$\boldsymbol{A}(\boldsymbol{r}) = \mu \int_V \boldsymbol{J}(\boldsymbol{r}') \frac{e^{-jk|\boldsymbol{r} - \boldsymbol{r}'|}}{4\pi|\boldsymbol{r} - \boldsymbol{r}'|} dv' \tag{1.269}$$

and in computing the far field of a radiating source, Equation (1.269) can be simplified by approximating $|\boldsymbol{r} - \boldsymbol{r}'|$ in the exponential argument as

$$|\boldsymbol{r} - \boldsymbol{r}'| \approx r - \hat{\boldsymbol{r}} \cdot \boldsymbol{r}' \tag{1.270}$$

and using

$$kr >> 1 \tag{1.271}$$

such that the factor $|\boldsymbol{r} - \boldsymbol{r}'|$ in the denominator can be replaced by r. Thus, under far-field conditions, Equation (1.269) can be expressed as

$$\boldsymbol{A}_{\text{FF}} = \mu \frac{e^{-jkr}}{4\pi r} \int_V \boldsymbol{J}(\boldsymbol{r}') e^{jk\hat{\boldsymbol{r}}\cdot\boldsymbol{r}'}\, dv'. \tag{1.272}$$

Define a far-field vector current moment $\boldsymbol{f}(\theta, \phi)$ in spherical coordinates as the integral in Equation (1.272) as

$$\boldsymbol{f}(\theta, \phi) = f_r \hat{\boldsymbol{r}} + f_\theta \hat{\boldsymbol{\theta}} + f_\phi \hat{\boldsymbol{\phi}} = \int_V \boldsymbol{J}(\boldsymbol{r}') e^{jk\hat{\boldsymbol{r}}\cdot\boldsymbol{r}'}\, dv'. \tag{1.273}$$

Using Equation (1.273) in Equation (1.272) yields the far-field magnetic vector potential as

$$\boldsymbol{A}_{\text{FF}} = \mu \frac{e^{-jkr}}{4\pi r} \boldsymbol{f}(\theta, \phi). \tag{1.274}$$

Now, comparing Equations (1.267) and (1.268), the contribution of $\boldsymbol{\nabla}(\boldsymbol{\nabla}\cdot\boldsymbol{A})$ to the far-field vector electric field depends on the second partial derivative of the free-space Green's function. Performing the first partial derivative operation, it follows that

$$\frac{\partial}{\partial r}\left[\frac{e^{-jkr}}{4\pi r}\right] = -\frac{jke^{-jkr}}{4\pi r} - \frac{e^{-jkr}}{4\pi r^2}, \tag{1.275}$$

and then the second partial derivative is determined to be

$$\frac{\partial^2}{\partial r^2}\left[\frac{e^{-jkr}}{4\pi r}\right] = -k^2 \frac{e^{-jkr}}{4\pi r} + 2jk\frac{e^{-jkr}}{4\pi r^2} + 2\frac{e^{-jkr}}{4\pi r^3} \tag{1.276}$$

and in the far field, the $1/r^2$ and $1/r^3$ terms go to zero compared to $1/r$, and thus only the $1/r$ term contributes to the far-zone electric field. Thus, in the far field,

$$\frac{\partial^2}{\partial r^2}\left[\frac{e^{-jkr}}{4\pi r}\right]_{\text{FF}} = -k^2 \frac{e^{-jkr}}{4\pi r} \tag{1.277}$$

and it follows that

$$\frac{\partial^2 \boldsymbol{A}_{\text{FF}}}{\partial r^2} = \frac{\partial^2}{\partial r^2}\left[\mu \frac{e^{-jkr}}{4\pi r} \boldsymbol{f}(\theta, \phi)\right] = -\mu k^2 \frac{e^{-jkr}}{4\pi r} \boldsymbol{f}(\theta, \phi). \tag{1.278}$$

Note that another derivative of the scalar Green's function, in this case divided by r, used subsequently in evaluating Equation (1.461), is given by

$$\frac{\partial}{\partial r}\left[\frac{e^{-jkr}}{4\pi r^2}\right] = -\frac{jke^{-jkr}}{4\pi r^2} - \frac{2e^{-jkr}}{4\pi r^3}. \tag{1.279}$$

Substituting Equations (1.274) and (1.278) in Equation (1.268) yields

$$\boldsymbol{E}_{\text{FF}} = -j\omega\left[\mu\frac{e^{-jkr}}{4\pi r}\boldsymbol{f}(\theta,\phi) - \mu\frac{e^{-jkr}}{4\pi r}\hat{\boldsymbol{r}}\hat{\boldsymbol{r}}\cdot\boldsymbol{f}(\theta,\phi)\right], \tag{1.280}$$

which can also be expressed in dyadic/vector notation as

$$\boldsymbol{E}_{\text{FF}} = -j\omega\mu\frac{e^{-jkr}}{4\pi r}(\overline{\boldsymbol{I}} - \hat{\boldsymbol{r}}\hat{\boldsymbol{r}})\cdot\boldsymbol{f}(\theta,\phi) \tag{1.281}$$

where

$$\overline{\boldsymbol{I}} - \hat{\boldsymbol{r}}\hat{\boldsymbol{r}} = \hat{\boldsymbol{\theta}}\hat{\boldsymbol{\theta}} + \hat{\boldsymbol{\phi}}\hat{\boldsymbol{\phi}}. \tag{1.282}$$

Now using Equations (1.282) and (1.239) and performing the dyadic/vector operations in Equation (1.281) yields the vector form of the desired expression for the far-zone electric field as

$$\boldsymbol{E}_{\text{FF}} = -j\omega\mu\frac{e^{-jkr}}{4\pi r}(\hat{\boldsymbol{\theta}}f_\theta + \hat{\boldsymbol{\phi}}f_\phi). \tag{1.283}$$

Noting that

$$\frac{1}{r}\frac{\partial}{\partial r}(r\boldsymbol{E}_{\text{FF}}) = -jk\boldsymbol{E}_{\text{FF}} \tag{1.284}$$

the far-zone magnetic field is then computed using the far-zone electric field (Equation (1.283)), the curl operation (1.98) and vector cross product (1.100) as

$$\boldsymbol{H}_{\text{FF}} = \frac{1}{-j\omega\mu}\boldsymbol{\nabla}\times\boldsymbol{E}_{\text{FF}}(\boldsymbol{r}) = \frac{k\hat{\boldsymbol{r}}}{\omega\mu}\times\boldsymbol{E}_{\text{FF}}(\boldsymbol{r}). \tag{1.285}$$

Performing the vector operation in Equation (1.285) yields

$$\boldsymbol{H}_{\text{FF}} = -jk\frac{e^{-jkr}}{4\pi r}(\hat{\boldsymbol{\phi}}f_\theta - \hat{\boldsymbol{\theta}}f_\phi). \tag{1.286}$$

Referring to Equation (1.409) and using Equations (1.283) and (1.285), the time-average Poynting's power density in the far zone is now computed as

$$<\boldsymbol{S}>= \frac{1}{2}\text{Re}\{\boldsymbol{E}\times\boldsymbol{H}^*\} = \hat{\boldsymbol{r}}\frac{1}{2}\sqrt{\frac{\mu}{\epsilon}}\left(\frac{k}{4\pi r}\right)^2(|f_\theta|^2 + |f_\phi|^2). \tag{1.287}$$

Return now to the general form of the electric field given by Equation (1.209) and express the electric field in terms of the scalar Green's function $g(\boldsymbol{r}, \boldsymbol{r}')$ as

$$\boldsymbol{E}(\boldsymbol{r}) = -j\omega\mu[\overline{\boldsymbol{I}} + \frac{1}{k^2}\boldsymbol{\nabla}\boldsymbol{\nabla}]\cdot \int_V \boldsymbol{J}(\boldsymbol{r}')g(\boldsymbol{r}, \boldsymbol{r}')\, dv'. \tag{1.288}$$

Now the bracketed terms in front of the integral depend only on the unprimed coordinates and can be moved inside the integral in Equation (1.288) and rearranging yields

$$\boldsymbol{E}(\boldsymbol{r}) = -j\omega\mu \int_V \left[[\overline{\boldsymbol{I}} + \frac{1}{k^2}\boldsymbol{\nabla}\boldsymbol{\nabla}]g(\boldsymbol{r}, \boldsymbol{r}')\right] \cdot\boldsymbol{J}(\boldsymbol{r}')\, dv'. \tag{1.289}$$

It is assumed here that the observation point is outside the source volume. In cases where the observation point is within the source volume, additional integration terms must be taken into account [94]. The dyadic Green's function $\overline{\boldsymbol{G}}(\boldsymbol{r}, \boldsymbol{r}')$ can now be expressed in terms of the scalar Green's function $g(\boldsymbol{r}, \boldsymbol{r}')$ as

$$\overline{\boldsymbol{G}}(\boldsymbol{r}, \boldsymbol{r}') = [\overline{\boldsymbol{I}} + \frac{1}{k^2}\boldsymbol{\nabla}\boldsymbol{\nabla}]g(\boldsymbol{r}, \boldsymbol{r}'). \tag{1.290}$$

Thus, Equation (1.289) can be rewritten in the more compact dyadic Green's function representation as

$$\boldsymbol{E}(\boldsymbol{r}) = -j\omega\mu \int_V \overline{\boldsymbol{G}}(\boldsymbol{r}, \boldsymbol{r}')\cdot\boldsymbol{J}(\boldsymbol{r}')\, dv'. \tag{1.291}$$

The dyadic and scalar Green's functions are now examined in more detail. Let us return now to the time-harmonic form of Maxwell's equations corresponding to Equations (1.149) and (1.150), that is,

$$\boldsymbol{\nabla}\times\boldsymbol{E}(\boldsymbol{r}) = -j\omega\mu\boldsymbol{H}(\boldsymbol{r}) - \boldsymbol{M}(\boldsymbol{r}) \tag{1.292}$$

$$\boldsymbol{\nabla}\times\boldsymbol{H}(\boldsymbol{r}) = j\omega\epsilon\boldsymbol{E}(\boldsymbol{r}) + \boldsymbol{J}(\boldsymbol{r}) \tag{1.293}$$

where μ and ϵ are assumed to be complex in general and the current densities $\boldsymbol{J}$ and $\boldsymbol{M}$ are source currents. Assuming that the equivalent magnetic current is zero, that is, $\boldsymbol{M} = 0$, $\boldsymbol{H}(\boldsymbol{r})$ can be eliminated from Equation (1.292) by taking the curl of the curl of $\boldsymbol{E}$ as

$$\boldsymbol{\nabla}\times\boldsymbol{\nabla}\times\boldsymbol{E}(\boldsymbol{r}) = -j\omega\mu\boldsymbol{\nabla}\times\boldsymbol{H}(\boldsymbol{r}) \tag{1.294}$$

and substituting Equation (1.293) yields

$$\boldsymbol{\nabla}\times\boldsymbol{\nabla}\times\boldsymbol{E}(\boldsymbol{r}) = -j\omega\mu j\omega\epsilon\boldsymbol{E}(\boldsymbol{r}) - j\omega\mu\boldsymbol{J}(\boldsymbol{r}) \tag{1.295}$$

and rearranging yields the following inhomogeneous wave equation for $\boldsymbol{E}(\boldsymbol{r})$, that is,

$$\boldsymbol{\nabla}\times\boldsymbol{\nabla}\times\boldsymbol{E}(\boldsymbol{r}) - k^2\boldsymbol{E}(\boldsymbol{r}) = -j\omega\mu\boldsymbol{J}(\boldsymbol{r}). \tag{1.296}$$

To solve Equation (1.296), the dyadic Green's function method can be used for an arbitrary response vector function $\boldsymbol{y}$ and source vector function $\boldsymbol{f}$ as follows.

Let L_{op} be a linear differential operator such that

$$L_{\text{op}}\boldsymbol{y}(\boldsymbol{r}) = \boldsymbol{f}(\boldsymbol{r}). \tag{1.297}$$

It is desired to determine the dyadic Green's function $\overline{\boldsymbol{G}}$ that is given by the solution to

$$L_{\text{op}}\overline{\boldsymbol{G}}(\boldsymbol{r},\boldsymbol{r}') = \overline{\boldsymbol{I}}\delta(\boldsymbol{r}-\boldsymbol{r}') \tag{1.298}$$

and boundary conditions must be satisfied. A solution for $\boldsymbol{y}(\boldsymbol{r})$ is then

$$\boldsymbol{y}(\boldsymbol{r}) = \int \overline{\boldsymbol{G}}(\boldsymbol{r},\boldsymbol{r}')\cdot\boldsymbol{f}(\boldsymbol{r}')dv'. \tag{1.299}$$

Repeating Equation (1.291), the vector electric field is expressed in terms of the dyadic Green's function as

$$\boldsymbol{E}(\boldsymbol{r}) = -j\omega\mu\int_V \overline{\boldsymbol{G}}(\boldsymbol{r},\boldsymbol{r}')\cdot\boldsymbol{J}(\boldsymbol{r}')\,dv' \tag{1.300}$$

where it should be noted that, in addition, any boundary conditions must be enforced. In Equation (1.296), it is desirable to express the source density in terms of the primed coordinate system. One way to achieve this transformation from unprimed to primed coordinates is to make use of the three-dimensional delta function $\delta(\boldsymbol{r}-\boldsymbol{r}')$ as described by Kong [3, p. 496] as

$$\boldsymbol{J}(\boldsymbol{r}) = \int_V \delta(\boldsymbol{r}-\boldsymbol{r}')\overline{\boldsymbol{I}}\cdot\boldsymbol{J}(\boldsymbol{r}')dv'. \tag{1.301}$$

Substituting Equations (1.300) and (1.301) in Equation (1.296) yields

$$\begin{aligned}-j\omega\mu\textstyle\int_V \boldsymbol{\nabla}\times\boldsymbol{\nabla}\times\overline{\boldsymbol{G}}(\boldsymbol{r},\boldsymbol{r}')\cdot\boldsymbol{J}(\boldsymbol{r}')\,dv' + j\omega\mu k^2\textstyle\int_V \overline{\boldsymbol{G}}(\boldsymbol{r},\boldsymbol{r}')\cdot\boldsymbol{J}(\boldsymbol{r}')\,dv'&\\ = -j\omega\mu\textstyle\int_V \delta(\boldsymbol{r}-\boldsymbol{r}')\overline{\boldsymbol{I}}\cdot\boldsymbol{J}(\boldsymbol{r}')dv'.&\end{aligned} \tag{1.302}$$

Since the integral operation involving $\boldsymbol{J}(\boldsymbol{r}')$ appears in each term of Equation (1.302), it follows that the integral operation on $\boldsymbol{J}(\boldsymbol{r}')$ and the $j\omega\mu$ quantity can be eliminated, which gives

$$\boldsymbol{\nabla}\times\boldsymbol{\nabla}\times\overline{\boldsymbol{G}}(\boldsymbol{r},\boldsymbol{r}') - k^2\overline{\boldsymbol{G}}(\boldsymbol{r},\boldsymbol{r}') = \overline{\boldsymbol{I}}\delta(\boldsymbol{r}-\boldsymbol{r}'). \tag{1.303}$$

Substituting Equation (1.290) in Equation (1.303 yields

$$\nabla\times\nabla\times\left[\overline{I}+\frac{1}{k^2}\nabla\nabla\right]g(\boldsymbol{r},\boldsymbol{r}')-k^2\left[\overline{I}+\frac{1}{k^2}\nabla\nabla\right]g(\boldsymbol{r},\boldsymbol{r}')=\overline{I}\delta(\boldsymbol{r}-\boldsymbol{r}'). \tag{1.304}$$

Now Equation (1.304) can be simplified by first observing that the curl of a gradient of an arbitrary vector is zero, that is,

$$\nabla\times\nabla\boldsymbol{V}=0, \tag{1.305}$$

and recognizing that the gradient of the scalar Green's function $g(\boldsymbol{r},\boldsymbol{r}')$ is a vector, it follows that

$$\frac{1}{k^2}\nabla\times\nabla\times\nabla\nabla g(\boldsymbol{r},\boldsymbol{r}')=\frac{1}{k^2}\nabla\times[\nabla\times\nabla(\nabla g(\boldsymbol{r},\boldsymbol{r}'))]=0. \tag{1.306}$$

Thus, Equation (1.304) reduces to

$$\nabla\times\nabla\times\overline{I}g(\boldsymbol{r},\boldsymbol{r}')-k^2\left[\overline{I}+\frac{1}{k^2}\nabla\nabla\right]g(\boldsymbol{r},\boldsymbol{r}')=\overline{I}\delta(\boldsymbol{r}-\boldsymbol{r}'). \tag{1.307}$$

Next, using the dyadic identity

$$\nabla\times\nabla\times\overline{D}=\nabla(\nabla\cdot\overline{D})-\nabla^2\overline{D}, \tag{1.308}$$

it follows that

$$\nabla\times\nabla\times\overline{I}g=\nabla(\nabla\cdot\overline{I}g)-\nabla^2(\overline{I}g); \tag{1.309}$$

but from the identity

$$\nabla\cdot(\phi\overline{I})=\nabla\phi \tag{1.310}$$

it follows that

$$\nabla\cdot\overline{I}g=\nabla g. \tag{1.311}$$

Therefore,

$$\nabla\times\nabla\overline{I}g=\nabla\nabla g-\overline{I}\nabla^2 g. \tag{1.312}$$

Substituting Equation (1.312) in Equation (1.307) yields

$$\nabla\nabla g(\boldsymbol{r},\boldsymbol{r}')-\overline{I}\nabla^2 g(\boldsymbol{r},\boldsymbol{r}')-k^2\left[\overline{I}+\frac{1}{k^2}\nabla\nabla\right]g(\boldsymbol{r},\boldsymbol{r}')=\overline{I}\delta(\boldsymbol{r}-\boldsymbol{r}'), \tag{1.313}$$

and canceling like terms yields the simplified inhomogeneous wave equation

$$\nabla^2 g(\boldsymbol{r},\boldsymbol{r}')+k^2 g(\boldsymbol{r},\boldsymbol{r}')=-\delta(\boldsymbol{r}-\boldsymbol{r}'). \tag{1.314}$$

To determine the functional form of the scalar Green's function $g(\boldsymbol{r},\boldsymbol{r}')$ in Equation (1.314), it is necessary first to solve the homogeneous form of the equation and then invoke the boundary conditions in the inhomogeneous case. To simplify the solution for g, shift the source position to the origin, which means $\boldsymbol{r}' = 0$. Thus, Equation (1.314) can now be expressed as

$$\nabla^2 g(\boldsymbol{r}) + k^2 g(\boldsymbol{r}) = -\delta(\boldsymbol{r}), \tag{1.315}$$

and it is observed that the solution for $g(\boldsymbol{r})$ is independent of the spherical coordinate angles θ and ϕ.

In spherical coordinates, the Laplacian of a scalar g that depends only on r can be expressed as [3, p. 18]

$$\nabla^2 g(r) = \frac{1}{r^2}\frac{\partial}{\partial r}\left[r^2 \frac{\partial g(r)}{\partial r}\right] = \frac{1}{r}\frac{\partial^2}{\partial r^2}(rg(r)). \tag{1.316}$$

Thus, Equation (1.315) can be rewritten as the inhomogeneous form

$$\frac{1}{r}\frac{d^2}{dr^2}(rg(\boldsymbol{r})) + k^2 g(\boldsymbol{r}) = -\delta(\boldsymbol{r}). \tag{1.317}$$

If the observation point is not at the origin, that is, $r \neq 0$, then Equation (1.317) reduces to the homogeneous differential equation

$$\frac{d^2}{dr^2}(rg(\boldsymbol{r})) + k^2 rg(\boldsymbol{r}) = 0. \tag{1.318}$$

The solution to the differential equation expressed by Equation (1.318) is an outgoing wave of the form

$$g(r) = C\frac{e^{-jkr}}{r}. \tag{1.319}$$

To determine the constant C in Equation (1.319), it is necessary to satisfy the boundary conditions at infinity and at the origin where the point source is located. In the limit as $r \to \infty$ in Equation (1.319) $g(r) \to 0$, which satifies the condition that the wave amplitude must attenuate to zero because of field spreading loss. To satisfy the boundary condition at the origin, it is necessary to integrate Equation (1.315) over an infinitesimal sphere of radius r_s centered at the origin and let the sphere radius tend to zero. Expressing the Laplacian operation on the scalar g as

$$\nabla^2 g = \nabla \cdot \nabla g \tag{1.320}$$

and noting that the delta function acts as a sampling function, that is,

$$\int f(x)\delta(x - x_o)\, dx = f(x_o) \tag{1.321}$$

and substituting Equation (1.319) in Equation (1.315), the following integral equation can be written as

$$C\int_V \nabla\cdot\left(\nabla\frac{e^{-jkr}}{r}\right)dv + C\int_V k^2\frac{e^{-jkr}}{r}dv = -\int_V \delta(r)\,dv = -1. \tag{1.322}$$

Now, applying the divergence theorem

$$\int_V \nabla\cdot\boldsymbol{A}\,dv = \oint_S \boldsymbol{A}\cdot\hat{\boldsymbol{n}}\,ds \tag{1.323}$$

to the first term in Equation (1.322) yields

$$C\int_V \nabla\cdot\left(\nabla\frac{e^{-jkr}}{r}\right)dv = C\oint_S \hat{\boldsymbol{n}}\cdot\hat{\boldsymbol{r}}\frac{\partial}{\partial r}\left(\frac{e^{-jkr}}{r}\right)ds. \tag{1.324}$$

The surface integral in Equation (1.324) is evaluated as follows. Noting that for the assumed spherical volume $\hat{\boldsymbol{n}} = \hat{\boldsymbol{r}}$ and $ds = r^2\sin\theta\,d\theta\,d\phi$, the surface integral in the first term of Equation (1.322) becomes

$$\begin{aligned}\oint_S \tfrac{d}{dr}\left(\tfrac{e^{-jkr}}{r}\right)ds &= \int_0^{2\pi}\int_0^{\pi}\left[\tfrac{d}{dr}\left(\tfrac{e^{-jkr}}{r}\right)\right]r^2\sin\theta\,d\theta\,d\phi\\ &= \int_0^{2\pi}\int_0^{\pi}\left[\tfrac{e^{-jkr}(-jkr-1)}{r^2}\right]_{r=r_s}\sin\theta\,d\theta\,d\phi\\ &= \int_0^{2\pi}\int_0^{\pi}\left[e^{-jkr}(-jkr-1)\right]_{r=r_s\to 0}\sin\theta\,d\theta\,d\phi\\ &= \int_0^{2\pi}\int_0^{\pi}[-1]\sin\theta\,d\theta\,d\phi = -4\pi.\end{aligned} \tag{1.325}$$

Now the integral in the second term in Equation (1.324) is evaluated as

$$\begin{aligned}\int_V k^2\tfrac{e^{-jkr}}{r}dv &= k^2\int_0^{r_s}\int_0^{2\pi}\int_0^{\pi}\tfrac{e^{-jkr}}{r}r^2\,\sin\theta\,dr\,d\theta\,d\phi\\ &= 4\pi k^2\int_0^{r_s\to 0} re^{-jkr}dr\\ &= 4\pi k^2\left[\tfrac{e^{-jkr}}{k^2}(-jkr-1)\right]_0^{r_s\to 0} = 0.\end{aligned} \tag{1.326}$$

Substituting the results of Equations (1.325) and (1.326) in Equation (1.322), it follows that

$$C = \frac{1}{4\pi}, \tag{1.327}$$

and substituting the result for the constant in Equation (1.319) yields the scalar Green's function as a function of r

$$g(r) = \frac{e^{-jkr}}{4\pi r}. \tag{1.328}$$

Since the radius r in Equation (1.328) is the distance between the source and the observation point, it is possible to shift the source to the vector position $\boldsymbol{r}'$, and thus the transformed distance between the source and observation point becomes $|\boldsymbol{r} - \boldsymbol{r}'|$. Therefore, the general form of the scalar Green's function is given by

$$g(\boldsymbol{r}, \boldsymbol{r}') = \frac{e^{-jk|\boldsymbol{r}-\boldsymbol{r}'|}}{4\pi|\boldsymbol{r} - \boldsymbol{r}'|}. \tag{1.329}$$

Substituting Equation (1.329) in Equation (1.290) yields the dyadic Green's function in terms of the scalar Green's function as

$$\overline{\boldsymbol{G}}(\boldsymbol{r}, \boldsymbol{r}') = [\overline{\boldsymbol{I}} + \frac{1}{k^2}\boldsymbol{\nabla}\boldsymbol{\nabla}]\frac{e^{-jk|\boldsymbol{r}-\boldsymbol{r}'|}}{4\pi|\boldsymbol{r} - \boldsymbol{r}'|}. \tag{1.330}$$

The electric field is then determined by using Equation (1.330) in Equation (1.300).

1.5 Boundary Conditions

In determining the solution of electromagnetic fields for a general antenna or scattering problem, it is necessary to satisfy boundary conditions between different media for the tangential and normal field components. The boundary conditions depend on the type of media and also whether any charges or currents are present at the interface between the media. To derive the boundary conditions from Maxwell's equations, two vector identities are needed, namely the divergence theorem expressed as

$$\int_V \boldsymbol{\nabla}\cdot\boldsymbol{A}\, dv = \oint_S \boldsymbol{A}\cdot d\boldsymbol{s} \tag{1.331}$$

and Stokes' theorem expressed as

$$\int_S (\boldsymbol{\nabla}\times\boldsymbol{A})\cdot d\boldsymbol{s} = \oint_C \boldsymbol{A}\cdot d\boldsymbol{l}. \tag{1.332}$$

To be completely general and allow for boundary conditions for antenna apertures, it is necessary to include the contribution from the fictitious magnetic current density $\boldsymbol{M}$ in Maxwell's curl of $\boldsymbol{E}$ equation as a mathematical convenience, that is,

$$\boldsymbol{\nabla}\times\boldsymbol{E} = -\boldsymbol{M} - \frac{\partial \boldsymbol{B}}{\partial t}. \tag{1.333}$$

Note, the magnetic current density has units of V/m^2.

The divergence theorem given by Equation (1.331) states that the volume integral of the divergence of $\boldsymbol{A}$ is equal to the outward flux of the vector field $\boldsymbol{A}$ through the surface S enclosing the volume V. Assuming that $\boldsymbol{A}$ and the curl of $\boldsymbol{A}$ are continuous, Stokes' theorem given by Equation (1.332) states that the curl of $\boldsymbol{A}$ over the open surface S bounded by C is equal to the circulation of $\boldsymbol{A}$ around the closed path C.

Consider now Figure 1.22, which shows two media separated by an interface surface that can be of arbitrary shape and can support surface charges ρ_s and surface currents $\boldsymbol{J}_s$ and $\boldsymbol{M}_s$. To determine the boundary conditions for the normal field components, consider Figure 1.23. From Maxwell's divergence of $\boldsymbol{D}$ equation (Equation (1.91)) and the divergence theorem, we can write

$$\int_V \boldsymbol{\nabla}\cdot\boldsymbol{D}dv = \oint_S \boldsymbol{A}\cdot d\boldsymbol{s} = \int_V \rho dv \tag{1.334}$$

where ρ is the volume charge density. In the limit where Δh approaches zero, only the top and bottom of the cylindrical box contribute, such that in vector notation,

$$\hat{\boldsymbol{n}}\cdot(\boldsymbol{D}_2 - \boldsymbol{D}_1) = \rho_s \tag{1.335}$$

where ρ_s is the surface charge density. Similarly, it can be shown that

$$\hat{\boldsymbol{n}}\cdot(\boldsymbol{B}_2 - \boldsymbol{B}_1) = 0. \tag{1.336}$$

Now, referring to Figure 1.24 from Maxwell's curl of $\boldsymbol{E}$ equation (Equation (1.89)) including the fictitious magnetic current density and Stokes' theorem, we can write

$$\int_S (\boldsymbol{\nabla}\times\boldsymbol{E})\cdot d\boldsymbol{s} = \oint_C \boldsymbol{E}\cdot d\boldsymbol{l} = -\int_S \boldsymbol{M}\cdot d\boldsymbol{s} - \int_S \frac{\partial \boldsymbol{B}}{\partial t}\cdot d\boldsymbol{s}. \tag{1.337}$$

As Δh goes to zero, the term involving the integral of $\boldsymbol{B}$ goes to zero, and now in vector form, Equation (1.337) becomes

$$(\boldsymbol{E}_2 - \boldsymbol{E}_1)\times\hat{\boldsymbol{n}} = \boldsymbol{M}_s \tag{1.338}$$

where $\boldsymbol{M}_s$ is the fictitious magnetic surface current. Similarly, it can be shown that

$$\hat{\boldsymbol{n}}\times(\boldsymbol{H}_2 - \boldsymbol{H}_1) = \boldsymbol{J}_s. \tag{1.339}$$

From Equations (1.338) and (1.339) for general media in the absence of sources on the boundary (i.e., $\boldsymbol{J}_s = 0$ and $\boldsymbol{M}_s = 0$), the tangential components of $\boldsymbol{E}$ and $\boldsymbol{H}$ are continuous across the boundary. When medium 1 is a perfect electric conductor ($\sigma = \infty$), which means $\boldsymbol{E}_1 = 0$, from Equation (1.338), the tangential component of $\boldsymbol{E}_2$ must be zero at the perfect electric conductor.

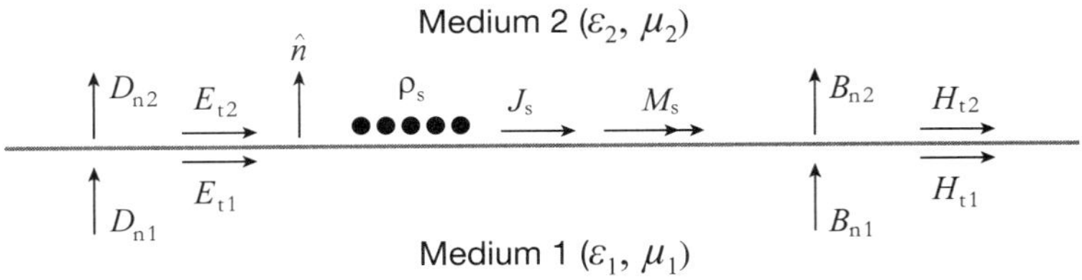

Figure 1.22 Electric charge, currents, and fields at a general interface between two media.

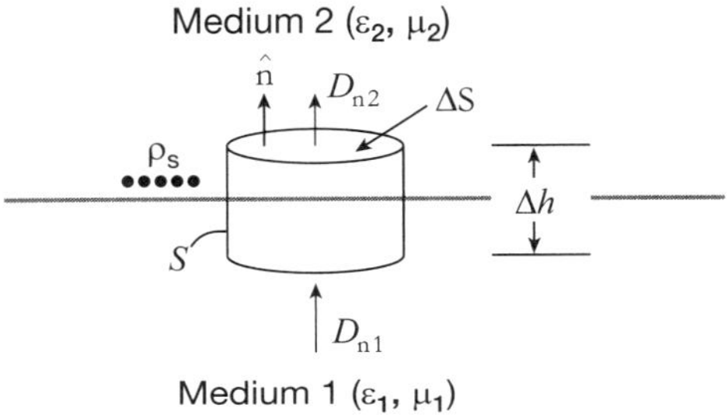

Figure 1.23 Electric charge and the normal components of electric flux density with a closed surface for evaluation of boundary conditions.

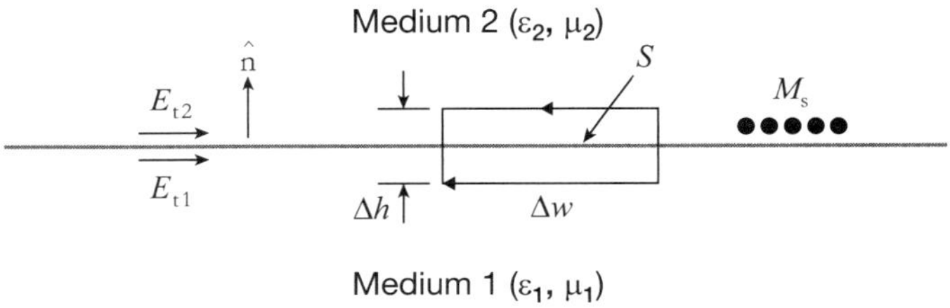

Figure 1.24 Magnetic current and tangential E fields with a closed contour for evaluation of boundary conditions.

1.6 Wave Equation for Conducting Media, Propagation Parameters

Consider Maxwell's equations given by Equations (1.149) and (1.150), but now in a source-free conducting region with $\boldsymbol{J}_{\text{source}} = 0$ and $\boldsymbol{M}_{\text{source}} = 0$, that is,

$$\nabla \times \boldsymbol{E} = -j\omega\mu_c \boldsymbol{H} \tag{1.340}$$

$$\nabla \times \boldsymbol{H} = j\omega\epsilon_c \boldsymbol{E} \tag{1.341}$$

$$\nabla \cdot \boldsymbol{D} = 0 \tag{1.342}$$

$$\nabla \cdot \boldsymbol{B} = 0. \tag{1.343}$$

Taking the curl of Equation (1.340) and substituting Equations (1.153) and (1.341) yields

$$\nabla \times \nabla \times \boldsymbol{E} = -j\omega\mu_c j\omega\epsilon_c \boldsymbol{E} = -j\omega\mu_c(\sigma + j\omega\epsilon')\boldsymbol{E} \tag{1.344}$$

or

$$\nabla \times \nabla \times \boldsymbol{E} = -j\omega\mu_c j\omega\epsilon_c \boldsymbol{E} = \omega^2\mu_c\epsilon_c \boldsymbol{E} \tag{1.345}$$

and defining the complex wavenumber k here as

$$k = \omega\sqrt{\mu_c\epsilon_c} \tag{1.346}$$

it follows that

$$\nabla \times \nabla \times \boldsymbol{E} = k^2 \boldsymbol{E}. \tag{1.347}$$

By using the vector identity

$$\nabla \times \nabla \times \boldsymbol{E} = \nabla(\nabla \cdot \boldsymbol{E}) - \nabla^2 \boldsymbol{E} \tag{1.348}$$

and noting that the divergence of $\boldsymbol{E}$ is zero by Equation (1.342), it follows that

$$\nabla^2 \boldsymbol{E} + k^2 \boldsymbol{E} = 0, \tag{1.349}$$

which is a form of the vector wave equation in complex media. A similar wave equation involving the vector magnetic field is given by

$$\nabla^2 \boldsymbol{H} + k^2 \boldsymbol{H} = 0. \tag{1.350}$$

The vector wave equations expressed by Equations (1.349) and (1.350) are also commonly expressed as

$$\nabla^2 \boldsymbol{E} - \gamma^2 \boldsymbol{E} = 0 \tag{1.351}$$

$$\nabla^2 \boldsymbol{H} - \gamma^2 \boldsymbol{H} = 0 \tag{1.352}$$

where γ is the complex propagation constant. The above vector wave equations allow for spherical, cylindrical, and plane wave solutions.

For example, consider a $\hat{\boldsymbol{z}}$-polarized plane wave propagating in the $\hat{\boldsymbol{x}}$ direction as given by the linearly polarized electric field

$$\boldsymbol{E}(x) = e^{-\gamma x}\hat{\boldsymbol{z}}. \tag{1.353}$$

The plane-wave function given by Equation (1.353) is a solution of Equation (1.351), which can be verified by observing that the wave equation can be expressed in cartesian coordinates as

$$\nabla^2 \boldsymbol{E} - \gamma^2 \boldsymbol{E} = \frac{\partial^2 \boldsymbol{E}}{\partial x^2} - \gamma^2 \boldsymbol{E} = 0 \tag{1.354}$$

and then performing the partial derivative operation.

The relationship between the complex propagation constant γ and the complex wavenumber k is

$$\gamma = jk = j\omega\sqrt{\mu_c \epsilon_c}. \tag{1.355}$$

Thus,

$$\gamma^2 = -k^2 = -\omega^2 \mu_c \epsilon_c. \tag{1.356}$$

By using Equation (1.344), it follows that

$$\gamma^2 = j\omega\mu_c(\sigma + j\omega\epsilon'). \tag{1.357}$$

Assume now an electrically conducting medium, and then it is readily shown that the propagation constant can be expressed as

$$\gamma = \pm\sqrt{j\omega\mu'(\sigma + j\omega\epsilon')} = \pm j\omega\sqrt{\mu'\epsilon'}\sqrt{1 - j\frac{\sigma}{\omega\epsilon'}}. \tag{1.358}$$

Again, the quantity (refer to Equation (1.159))

$$\frac{\sigma}{\omega\epsilon'} = \frac{\epsilon''}{\epsilon'} = \tan\delta \tag{1.359}$$

is referred to as the electric loss tangent or dissipation factor. Both σ and $\tan\delta$ are used in tables of dielectric characteristics of materials. It is common to express the complex propagation constant as

$$\gamma = \alpha + j\beta \tag{1.360}$$

where α is the attenuation constant in nepers/meter and β is the phase constant in radians/meter. Again, the complex wavenumber is expressed as

$$jk = \alpha + j\beta \tag{1.361}$$

and either way of analyzing the wave propagation provides the same results. The units of nepers/meter can be converted to decibels per meter by multiplying by the factor 8.686. The units of radians/meter can be converted to degrees/meter by multiplying by the factor $180/\pi = 57.296$.

The plane wave solution given by Equation (1.353) can now be expressed in terms of α and β as

$$\boldsymbol{E}(x) = e^{-\gamma x}\hat{\boldsymbol{z}} = e^{-\alpha x}e^{-j\beta x}\hat{\boldsymbol{z}}. \tag{1.362}$$

The skin depth, denoted δ_s, refers to the distance for which a plane wave travels in a good conductor such that it attenuates to the value of $e^{-1} = 0.368$ or -8.68 dB. From the exponential term involving the attenuation constant in Equation (1.362) it follows that

$$e^{-\alpha x} = e^{-\alpha \delta_s} = e^{-1} \tag{1.363}$$

Thus, for a good conductor the skin depth is given by

$$\delta_s = \frac{1}{\alpha} \tag{1.364}$$

The constants α and β are found by setting Equation (1.358) equal to Equation (1.360) and then squaring both sides, equating the real and imaginary parts, and solving the pair of simultaneous equations, with the result

$$\alpha = \frac{\omega\sqrt{\mu'\epsilon'}}{\sqrt{2}}\left\{\sqrt{1+\left(\frac{\sigma}{\omega\epsilon'}\right)^2}-1\right\}^{1/2} \tag{1.365}$$

$$\beta = \frac{\omega\sqrt{\mu'\epsilon'}}{\sqrt{2}}\left\{\sqrt{1+\left(\frac{\sigma}{\omega\epsilon'}\right)^2}+1\right\}^{1/2}. \tag{1.366}$$

The wavelength λ in the lossy dielectric is then computed from the phase constant given by Equation (1.366) as

$$\lambda = \frac{2\pi}{\beta}. \tag{1.367}$$

The intrinsic wave impedance η is related to the ratio of the electric and magnetic fields and in lossy dielectric material and using Equation (1.153) is given by [8]

$$\eta = \sqrt{\frac{\mu}{\epsilon}} = \sqrt{\frac{\mu'}{\epsilon_c}} = \sqrt{\frac{j\omega\mu'}{\sigma + j\omega\epsilon'}} = \sqrt{\frac{\mu'}{\epsilon'}}\frac{1}{\sqrt{1-j\frac{\sigma}{\omega\epsilon'}}}. \tag{1.368}$$

In free space, the electrical conductivity σ is equal to zero, and it follows from Equation (1.368) that the free-space intrinsic wave impedance is

$$\eta_{\text{free space}} = \sqrt{\frac{\mu_o}{\epsilon_o}} = 120\pi = 376.99 \approx 377 \text{ ohms}. \tag{1.369}$$

In the case of a good electric conductor where σ has a large value, the quantity

$$\left(\frac{\sigma}{\omega\epsilon'}\right)^2 >> 1 \tag{1.370}$$

Then for a good conductor the attenuation constant given by Equation (1.365) reduces to

$$\alpha = \sqrt{\frac{\omega\mu_o\sigma}{2}} = \sqrt{\pi f \mu_o \sigma} \tag{1.371}$$

and the skin depth is now expressed as

$$\delta_s = \frac{1}{\alpha} = \frac{1}{\sqrt{\pi f \mu_o \sigma}} \tag{1.372}$$

In the general media case, in which both electric and magnetic loss are present, the constants α and β are determined from Equations (1.356) and (1.360) using Equations (1.146) and (1.147), by solving the simultaneous equations for the real and imaginary components with the result

$$\alpha = \frac{\omega\sqrt{\mu'\epsilon'}}{\sqrt{2}}\left\{\sqrt{\left(1+\tan^2\delta_e\right)\left(1+\tan^2\delta_m\right)} - \left(1-\tan\delta_e\tan\delta_m\right)\right\}^{1/2} \tag{1.373}$$

$$\beta = \frac{\omega\sqrt{\mu'\epsilon'}}{\sqrt{2}}\left\{\sqrt{\left(1+\tan^2\delta_e\right)\left(1+\tan^2\delta_m\right)} + \left(1-\tan\delta_e\tan\delta_m\right)\right\}^{1/2} \tag{1.374}$$

where

$$\tan\delta_e = \frac{\epsilon''}{\epsilon'} = \frac{\sigma_e}{\omega\epsilon'} = \frac{\sigma_e}{\omega\epsilon_o\epsilon_r'} \tag{1.375}$$

and

$$\tan\delta_m = \frac{\mu''}{\mu'} = \frac{\sigma_m}{\omega\mu'} = \frac{\sigma_m}{\omega\mu_o\mu_r'} \tag{1.376}$$

are the electric and magnetic loss tangents, respectively. For general lossy media, the wave impedance is given by

$$\eta = \sqrt{\frac{\mu}{\epsilon}} = \sqrt{\frac{\mu' - j\mu''}{\epsilon' - j\epsilon''}} = \sqrt{\frac{\mu'}{\epsilon'}}\sqrt{\frac{1-j\tan\delta_m}{1-j\tan\delta_e}}. \tag{1.377}$$

Taking account of boundary conditions as described in Section 1.7, for normal angle of incidence of a plane wave, the reflection coefficient Γ for a uniform plane wave traveling in medium 1 and reflecting at the boundary of medium 2 is given by

$$\Gamma = \frac{\eta_2 - \eta_1}{\eta_2 + \eta_1}, \tag{1.378}$$

and the transmission coefficient T for a plane wave at normal incidence transmitted into medium 2 is given by

$$T = \frac{2\eta_2}{\eta_2 + \eta_1} \tag{1.379}$$

where η_1 is the impedance of medium 1 and η_2 is the impedance of medium 2. The more general cases of plane-wave incidence, including oblique incidence, polarization effects, and different media, are covered by Balanis [2, pp. 173–257].

It is convenient to have a simple equation for computing the wave propagation between any two points in the near field of an isolated transmitting antenna in conducting media. Consider a time-harmonic source radiating a spherical wave into an infinite homogeneous conducting medium. For an isotropic radiator and suppressing the $e^{j\omega t}$ time dependence where $\omega = 2\pi f$, the electric field as a function of range r can be expressed in phasor notation as a spherical wave

$$E(r) = E_o \frac{e^{-\gamma r}}{r} \tag{1.380}$$

where E_o is a constant and γ is the propagation constant given by Equation (1.358). For some cases in which the $1/r$ dependence can be ignored, a plane wavefront approximation to Equation (1.380) can be used. For example, a plane wavefront propagating along the x-axis can be expressed as

$$E(x) = E_o e^{-\gamma x} = E_o e^{-(\alpha + j\beta)x}. \tag{1.381}$$

Equation (1.380) or Equation (1.381) can be used in a simplified ray-tracing analysis to compute the approximate electric field, generated by one or more transmitting antennas, at any point in a homogeneous lossy dielectric. If we let the electric field at the rectangular coordinates (x, y, z) due to the nth transmitting antenna in an array of N transmitting antennas be denoted as $E_n(x, y, z)$, then the total electric field of the transmitting array can be expressed as the following summation,

$$E_{\text{total}}(x, y, z) = \sum_{n=1}^{N} E_n(x, y, z). \tag{1.382}$$

For a transmitting point source at the origin, the amplitude of the electric field at range r_1 is given by

$$|E(r_1)| = E_o \frac{e^{-\alpha r_1}}{r_1} \tag{1.383}$$

and at range r_2 by

$$|E(r_2)| = E_o \frac{e^{-\alpha r_2}}{r_2}. \tag{1.384}$$

The total propagation loss between ranges r_1 and r_2 is found by taking the ratio of Equations (1.384) and (1.383) as

$$\frac{|E(r_2)|}{|E(r_1)|} = \frac{r_1}{r_2} e^{-\alpha(r_2 - r_1)}. \tag{1.385}$$

The field attenuation A_α in decibels from range r_1 to range r_2 caused by the lossy dielectric is simply

$$A_\alpha = 20 \log_{10}(e^{-\alpha(r_2 - r_1)}). \tag{1.386}$$

If we let $d = r_2 - r_1$, the above equation can be written in a convenient form as

$$A_\alpha = 20 \log_{10}(e^{-\alpha d}), \tag{1.387}$$

which is the relative field attenuation as a function of distance d into a lossy dielectric or transmission loss of a plane wave in decibels per meter. The attenuation constant α has units of nepers per unit length or typically in the literature of nepers per meter (Np/m), where neper is dimensionless. If d is one meter and $\alpha = 1$ Np/m, then the field strength in decibels is $20 \log_{10}(e^{-1}) = 20 \log_{10} |0.368| = -8.686$ dB. Thus, the attenuation in decibels per meter is 8.686 times the attenuation constant in nepers per meter. If the attenuation constant is expressed in nepers/cm, then to convert to attenuation in decibels per cm, the multiplying factor remains 8.686. Similarly, the $1/r$ attenuation loss A_r in decibels is

$$A_r = 20 \log_{10} \frac{r_1}{r_2}. \tag{1.388}$$

1.7 Electromagnetic Energy Flow

To demonstrate that an electromagnetic field can transfer electromagnetic energy into an electrically conducting volume (e.g., soil as in a ground-penetrating radar application), consider the following application of Maxwell's curl equations in the time domain in a source-free region where $\boldsymbol{J}_{\text{source}} = 0$ and $\boldsymbol{M}_{\text{source}} = 0$. Starting with the curl of the instantaneous magnetic field as in Equation (1.90), substitute Equations (1.108) and (1.126) with the result

$$\nabla \times \boldsymbol{H}(\boldsymbol{r}, t) = \sigma \boldsymbol{E} + \epsilon' \frac{\partial \boldsymbol{E}}{\partial t}. \tag{1.389}$$

Next, the curl of the instantaneous electric field from Equation (1.89) is given by

$$\boldsymbol{\nabla}\times\boldsymbol{E}(\boldsymbol{r},t) = -\mu'\frac{\partial \boldsymbol{H}}{\partial t}. \tag{1.390}$$

Fundamentally, it is desired to derive a power-flow expression for the electric and magnetic fields. The electric field has units of volts/meter and the magnetic field has units of amperes/meter. The product of volts/meter and amperes/meter produces watts/meter^2, which is power density. There are two possibilities for manipulating the electric and magnetic field expressions to produce a relation for power density, either a dot product or a cross product of the two field quantities. To produce a meaningful relation here, the cross product is the desired vector operation on the electric and magnetic field vectors. A vector identity will also be needed to convert to the desired field relation. Of the available general vector identities, the one that applies here is

$$\boldsymbol{\nabla}\cdot(\boldsymbol{A}\times\boldsymbol{B}) = \boldsymbol{B}\cdot(\boldsymbol{\nabla}\times\boldsymbol{A}) - \boldsymbol{A}\cdot(\boldsymbol{\nabla}\times\boldsymbol{B}). \tag{1.391}$$

On the basis of the vector identity in Equation (1.391), it is observed that the desired mathematical manipulations are to dot the electric field into the curl of the magnetic field and to dot the magnetic field into the curl of the electric field. Once these manipulations are done, the desired cross product between $\boldsymbol{E}$ and $\boldsymbol{H}$ will appear. Thus, from Equations (1.389), (1.390), and (1.391) and by using $\boldsymbol{J}\cdot\boldsymbol{E} = \boldsymbol{E}\cdot\boldsymbol{J}$, it follows that

$$\boldsymbol{E}\cdot(\boldsymbol{\nabla}\times\boldsymbol{H}) = \boldsymbol{J}\cdot\boldsymbol{E} + \boldsymbol{E}\cdot\epsilon'\frac{\partial \boldsymbol{E}}{\partial t} = \boldsymbol{J}\cdot\boldsymbol{E} + \frac{1}{2}\epsilon'\frac{\partial E^2}{\partial t} \tag{1.392}$$

and

$$\boldsymbol{H}\cdot(\boldsymbol{\nabla}\times\boldsymbol{E}) = -\boldsymbol{H}\cdot\mu'\frac{\partial \boldsymbol{H}}{\partial t} = -\frac{1}{2}\mu'\frac{\partial H^2}{\partial t}. \tag{1.393}$$

Simplifying and subtracting Equation (1.392) from Equation (1.393) yields

$$\boldsymbol{E}\cdot(\boldsymbol{\nabla}\times\boldsymbol{H}) - \boldsymbol{H}\cdot(\boldsymbol{\nabla}\times\boldsymbol{E}) = \boldsymbol{J}\cdot\boldsymbol{E} + \frac{1}{2}\epsilon'\frac{\partial E^2}{\partial t} + \frac{1}{2}\mu'\frac{\partial H^2}{\partial t} \tag{1.394}$$

or applying the vector identity Equation (1.391) in Equation (1.394) yields

$$\boldsymbol{\nabla}\cdot(\boldsymbol{E}\times\boldsymbol{H}) = -\boldsymbol{J}\cdot\boldsymbol{E} - \frac{1}{2}\epsilon'\frac{\partial E^2}{\partial t} - \frac{1}{2}\mu'\frac{\partial H^2}{\partial t} \tag{1.395}$$

and now integrating over a volume of media

$$\int_V \boldsymbol{\nabla}\cdot(\boldsymbol{E}\times\boldsymbol{H})dv = -\int_V \boldsymbol{J}\cdot\boldsymbol{E}dv - \frac{\partial}{\partial t}\int_V [\frac{1}{2}\epsilon' E^2 + \frac{1}{2}\mu' H^2]dv. \tag{1.396}$$

The divergence theorem for any vector $\boldsymbol{A}$ for a closed surface S enclosing volume V is given by

$$\int_V \boldsymbol{\nabla}\boldsymbol{\cdot}\boldsymbol{A}dv = \oint_S \boldsymbol{A}\boldsymbol{\cdot}\hat{\boldsymbol{n}}ds \tag{1.397}$$

where dv is the elemental volume and ds is the elemental area perpendicular to the outward unit normal $\hat{\boldsymbol{n}}$. Thus, applying Equation (1.397) to the left-hand side of Equation (1.396) yields

$$\int_V \boldsymbol{\nabla}\boldsymbol{\cdot}(\boldsymbol{E}\times\boldsymbol{H})dv = \oint_S (\boldsymbol{E}\times\boldsymbol{H})\boldsymbol{\cdot}\hat{\boldsymbol{n}}ds. \tag{1.398}$$

Finally, Equation (1.396) reduces to

$$\oint_S (\boldsymbol{E}\times\boldsymbol{H})\boldsymbol{\cdot}\hat{\boldsymbol{n}}ds = -\int_V \boldsymbol{J}\boldsymbol{\cdot}\boldsymbol{E}dv - \frac{\partial}{\partial t}\int_V [\frac{1}{2}\epsilon' E^2 + \frac{1}{2}\mu' H^2]dv. \tag{1.399}$$

The term on the left side of Equation (1.399) represents the net inward flux of the vector

$$\boldsymbol{P}(\boldsymbol{r},t) = \boldsymbol{E}(\boldsymbol{r},t)\times\boldsymbol{H}(\boldsymbol{r},t). \tag{1.400}$$

In Equations (1.399) and (1.400), the quantity $\boldsymbol{P}(\boldsymbol{r},t) = \boldsymbol{E}(\boldsymbol{r},t)\times\boldsymbol{H}(\boldsymbol{r},t)$ is known as the instantaneous Poynting vector (due to J.H. Poynting [95]) and can be interpreted as the instantaneous power-flow density in watts/meter^2 at each observation point ($\boldsymbol{r}$) on the surface S, or equivalently the power delivered to the volume V by external electromagnetic sources. Equation (1.399) is referred to as the integral form of Poynting's theorem and describes conservation of energy for electromagnetic waves. From Equation (1.108), it follows that

$$\boldsymbol{J}\boldsymbol{\cdot}\boldsymbol{E} = \sigma\boldsymbol{E}\boldsymbol{\cdot}\boldsymbol{E} = \sigma|\boldsymbol{E}|^2 \tag{1.401}$$

where the dot product of the real vector $\boldsymbol{E}(t)$ with itself is just the magnitude of $\boldsymbol{E}$ squared. The first integral term on the right side of Equation (1.399) involving $\boldsymbol{J}\boldsymbol{\cdot}\boldsymbol{E} = \sigma|E|^2$ represents the instantaneous ohmic losses within the volume enclosed by the surface S. The integral in the second term on the right side of Equation (1.399) is the total energy (with units of watt-seconds = joules), due to electric and magnetic fields, stored within the volume V.

To understand the electromagnetic field energy transfer to conductive media such as lossy dielectrics, it is necessary to determine the time-average power delivered to and dissipated within the volume. The time-average Poynting vector or time-average power density given by Equation (1.399) is computed over one period T as

$$\boldsymbol{P}_{\text{ave}}(\boldsymbol{r}) = \frac{1}{T}\int_0^T \boldsymbol{E}(\boldsymbol{r},t)\times\boldsymbol{H}(\boldsymbol{r},t)\, dt \tag{1.402}$$

where the period is equal to

$$T = 2\pi/\omega. \tag{1.403}$$

Substituting the exponential forms given by Equations (1.118) and (1.119) into Equation (1.402) yields

$$\boldsymbol{P}_{\text{ave}}(\boldsymbol{r}) = \frac{1}{T}\int_0^T \left[\text{Re}[\boldsymbol{E}(\boldsymbol{r})e^{j\omega t}]\times\text{Re}[\boldsymbol{H}(\boldsymbol{r})e^{j\omega t}]\right] dt. \tag{1.404}$$

Making use of the relation

$$\text{Re}[\boldsymbol{A}B] = \frac{1}{2}[\boldsymbol{A}B + \boldsymbol{A}^* B^*], \tag{1.405}$$

(with * denoting complex conjugate) it follows that

$$\text{Re}[\boldsymbol{E}(\boldsymbol{r})e^{j\omega t}]\times\text{Re}[\boldsymbol{H}(\boldsymbol{r})e^{j\omega t}] = \frac{1}{2}[\boldsymbol{E}e^{j\omega t} + \boldsymbol{E}^* e^{-j\omega t}]\times\frac{1}{2}[\boldsymbol{H}e^{j\omega t} + \boldsymbol{H}^* e^{-j\omega t}]. \tag{1.406}$$

Expanding Equation (1.406) and using the relation

$$\text{Re}[\boldsymbol{E}\times\boldsymbol{H}^*] = \frac{1}{2}[\boldsymbol{E}\times\boldsymbol{H}^* + \boldsymbol{E}^*\times\boldsymbol{H}] \tag{1.407}$$

it can be shown that Equation (1.404) reduces to

$$\boldsymbol{P}_{\text{ave}} = \frac{1}{T}\int_0^{\frac{2\pi}{\omega}} \left[\frac{1}{2}\text{Re}[\boldsymbol{E}\times\boldsymbol{H}^*] + \frac{1}{4}\boldsymbol{E}\times\boldsymbol{H}e^{j2\omega t} + \frac{1}{4}\boldsymbol{E}^*\times\boldsymbol{H}^* e^{-j2\omega t}\right] dt. \tag{1.408}$$

It can be shown, by integration, that Equation (1.408) reduces to

$$\boldsymbol{P}_{\text{ave}}(\boldsymbol{r}) = \frac{1}{2}\text{Re}[\boldsymbol{E}\times\boldsymbol{H}^*], \tag{1.409}$$

which is the desired expression for the time-average power density of the electromagnetic field, or the time-average Poynting's vector, which has units of watts/meter^2. Note that the factor of 1/2 is used since the electric and magnetic fields are expressed in terms of peak values. If root mean square (rms) fields are used, then the factor of 1/2 would be eliminated.

As an example, for the case of a plane wave propagating in a conducting medium, the electric field is given by

$$\boldsymbol{E}(\boldsymbol{r}) = \eta\, \hat{\boldsymbol{n}} \times \boldsymbol{H}(\boldsymbol{r}) \tag{1.410}$$

and the magnetic field is given by

$$\boldsymbol{H}(\boldsymbol{r}) = \hat{\boldsymbol{n}} \times \boldsymbol{E}(\boldsymbol{r})/\eta \tag{1.411}$$

and thus, by substituting Equation (1.411) in Equation (1.409), it follows from a vector identity that the time-average Poynting vector for a plane wave is equal to

$$\boldsymbol{P}_{\text{ave}}(\boldsymbol{r}) = \hat{\boldsymbol{n}}\frac{1}{2}|\boldsymbol{E}(\boldsymbol{r})|^2/\eta \tag{1.412}$$

where $\hat{\boldsymbol{n}}$ is the direction of propagation.

Next, it is desired to quantify the average power deposition in a lossy dielectric, because this quantity is needed to determine the absorption of electromagnetic energy in any dielectric materials that the antenna might be constructed from, or in the case in which the antenna is radiating in the presence of such lossy materials. Referring to Equation (1.399), for the first term on the right-hand side, the time-average power dissipation per unit volume is computed over one period T as

$$P_d(\boldsymbol{r}) = \frac{1}{T}\int_0^T \boldsymbol{J}(\boldsymbol{r},t)\cdot\boldsymbol{E}(\boldsymbol{r},t)dt. \tag{1.413}$$

Substituting Equations (1.118) and (1.120) into Equation (1.413) yields

$$P_d(\boldsymbol{r}) = \frac{1}{T}\int_0^T \left[\text{Re}[\boldsymbol{J}(\boldsymbol{r})e^{j\omega t}]\cdot\text{Re}[\boldsymbol{E}(\boldsymbol{r})e^{j\omega t}]\right] dt. \tag{1.414}$$

Making use of the relation given in Equation (1.405), it follows that

$$\text{Re}[\boldsymbol{J}(\boldsymbol{r})e^{j\omega t}]\cdot\text{Re}[\boldsymbol{E}(\boldsymbol{r})e^{j\omega t}] = \frac{1}{2}[\boldsymbol{J}e^{j\omega t} + \boldsymbol{J}^*e^{-j\omega t}]\cdot\frac{1}{2}[\boldsymbol{E}e^{j\omega t} + \boldsymbol{E}^*e^{-j\omega t}]. \tag{1.415}$$

Using Equation (1.415) in Equation (1.414) and observing that the current density $\boldsymbol{J}$ and electric field $\boldsymbol{E}$ are in phasor form, it follows that

$$\boldsymbol{J}\cdot\boldsymbol{E}^* = \sigma\boldsymbol{E}\cdot\boldsymbol{E}^* = \sigma|\boldsymbol{E}|^2 \tag{1.416}$$

$$\boldsymbol{J}^*\cdot\boldsymbol{E} = \sigma\boldsymbol{E}^*\cdot\boldsymbol{E} = \sigma|\boldsymbol{E}|^2 \tag{1.417}$$

and it can be shown that

$$P_d = \frac{1}{T}\int_0^{T=\frac{2\pi}{\omega}} \left[\frac{1}{2}\sigma|\boldsymbol{E}|^2 + \frac{1}{4}\boldsymbol{J}\cdot\boldsymbol{E}e^{j2\omega t} + \frac{1}{4}\boldsymbol{J}^*\cdot\boldsymbol{E}^*e^{-j2\omega t}\right] dt. \tag{1.418}$$

It also can be shown that Equation (1.418) reduces to

$$P_d(\boldsymbol{r}) = P_{\text{ave dissipated}} = \frac{1}{2}\sigma|\boldsymbol{E}(\boldsymbol{r})|^2, \tag{1.419}$$

which is the desired expression for the time-average power dissipation per unit volume in the lossy dielectric, which has units of watts/m^3.

The specific absorption rate (SAR) [82, p. 11] in the dielectric is the power dissipated or absorbed per unit mass (W/kg) of the medium (lossy dielectric), so from Equation (1.419), it follows that

$$\mathrm{SAR}(\boldsymbol{r}) = \frac{P_d(\boldsymbol{r})}{\rho} = \frac{\sigma}{2\rho}|\boldsymbol{E}(\boldsymbol{r})|^2 \tag{1.420}$$

where ρ is the density of the medium in kg/m^3. To estimate the initial linear temperature rise ΔT versus time interval Δt in a lossy dielectric that is characterized by the specific heat c_d with units of kJ/(kg-°C), the following calculation can be made:

$$\Delta T = \frac{\mathrm{SAR}\,\Delta t}{c_d} \tag{1.421}$$

where in Equation (1.421), it has been assumed that any thermal conduction or thermal boundary conditions have been ignored.

1.8 Fields of Short Electric and Magnetic Dipoles

1.8.1 Introduction

Arbitrarily shaped antennas can be considered to be composed of the superposition of a number of short electric dipoles and/or magnetic dipoles (electric loops) as depicted in Figure 1.25. The closed-form expressions for the near-zone electromagnetic fields in free space resulting from an electrically short (Hertzian) dipole of length l with current I_e along the z axis, and an electrically small loop or short (small) Hertzian magnetic dipole of length l with current I_m along the z axis are derived in this section.

To derive the near-zone fields, there are two approaches. One approach is to compute the magnetic vector potential $\boldsymbol{A}$ from the source current $\boldsymbol{J}$, and then the magnetic field $\boldsymbol{H}$ can be computed. The electric field $\boldsymbol{E}$ is then

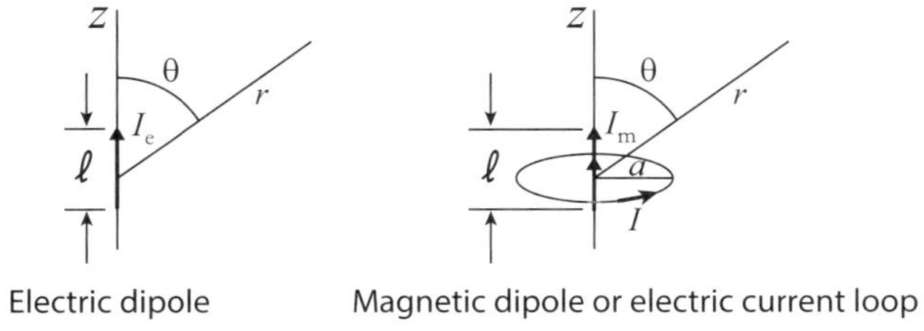

Figure 1.25 Geometry for short electrical dipole with length l and current I_e and short magnetic dipole with length l and magnetic current I_m (small electric current loop with radius a and electric current I).

obtained from the magnetic field $\boldsymbol{H}$ by using Maxwell's curl equation. The other approach is to compute the electric field directly from the source current $\boldsymbol{J}$ using the dyadic Green's function. It is instructive to derive the fields using both approaches as discussed below.

1.8.2 Derivation of Fields for $\hat{z}$ Hertzian Dipole

1.8.2.1 Magnetic Vector Potential Approach

In the first approach, the magnetic vector potential is needed and is given by Equation (1.201), which is repeated below for general isotropic media:

$$\boldsymbol{A}(\boldsymbol{r}) = \mu \int_V \boldsymbol{J}(\boldsymbol{r}') \frac{e^{-jk|\boldsymbol{r}-\boldsymbol{r}'|}}{4\pi|\boldsymbol{r}-\boldsymbol{r}'|} dv'. \tag{1.422}$$

Repeating Equation (1.178), the magnetic field $\boldsymbol{H}$ in terms of $\boldsymbol{A}$ is given by

$$\boldsymbol{H} = \frac{1}{\mu} \boldsymbol{\nabla} \times \boldsymbol{A}. \tag{1.423}$$

In the case of a $\hat{\boldsymbol{z}}$ Hertzian dipole, let the electric current density be given by

$$\boldsymbol{J}(\boldsymbol{r}') = \hat{\boldsymbol{z}} I_e l \delta(r') \tag{1.424}$$

where the delta function $\delta(r')$ is equal to unity when $r' = 0$ and is equal to zero elsewhere. Substituting Equation (1.424) in Equation (1.422) yields

$$\boldsymbol{A}(r) = \hat{\boldsymbol{z}} \frac{\mu I_e l}{4\pi} \frac{e^{-jkr}}{r}. \tag{1.425}$$

Now, use the standard conversion from rectangular components to spherical components as,

$$\begin{bmatrix} A_r \\ A_\theta \\ A_\phi \end{bmatrix} = \begin{bmatrix} \sin\theta\cos\phi & \sin\theta\sin\phi & \cos\theta \\ \cos\theta\cos\phi & \cos\theta\sin\phi & -\sin\theta \\ -\sin\phi & \cos\phi & 0 \end{bmatrix} \begin{bmatrix} A_x \\ A_y \\ A_z \end{bmatrix}. \tag{1.426}$$

If the x and y components of $\boldsymbol{A}$ are zero, the transformation reduces to

$$A_r = A_z \cos\theta \tag{1.427}$$

$$A_\theta = -A_z \sin\theta \tag{1.428}$$

and note that the A_ϕ component is zero. Thus, from Equations (1.425), (1.427), and (1.428), it follows that

$$A_r = \mu I_e l \frac{e^{-jkr}}{4\pi r} \cos\theta \tag{1.429}$$

$$A_\theta = -\mu I_e l \frac{e^{-jkr}}{4\pi r} \sin\theta \tag{1.430}$$

$$A_\phi = 0. \tag{1.431}$$

Since the A_ϕ component is zero and the ϕ variation of A_r and A_θ is zero for a z-directed current, referring to Equation (1.97), the magnetic field $\boldsymbol{H}$ in spherical coordinates is reduced to

$$\boldsymbol{H} = \frac{1}{\mu}\boldsymbol{\nabla}\times\boldsymbol{A} = \hat{\boldsymbol{\phi}}\frac{1}{\mu r}\left[\frac{\partial}{\partial r}(rA_\theta) - \frac{\partial A_r}{\partial \theta}\right]. \tag{1.432}$$

Performing the partial derivatives in Equation (1.432) yields the only nonzero magnetic-field component

$$H_\phi = jkI_e l \sin\theta \left[1 + \frac{1}{jkr}\right] \frac{e^{-jkr}}{4\pi r}. \tag{1.433}$$

In general, the Maxwell curl equation for the magnetic field in time-harmonic form is Equation (1.150), which is repeated here as

$$\boldsymbol{\nabla}\times\boldsymbol{H} = \boldsymbol{J}_{\text{source}} + j\omega\epsilon_c\boldsymbol{E}, \tag{1.434}$$

and when the magnetic field is computed at a point removed from the source, the magnetic field expression is reduced to

$$\boldsymbol{\nabla}\times\boldsymbol{H} = j\omega\epsilon\boldsymbol{E} \tag{1.435}$$

where it is now assumed that the complex permittivity is denoted by ϵ. Furthermore, using Equations (1.196) and (1.435), the electric field as a function of the vector distance $\boldsymbol{r}$ from the source can be computed from either the magnetic vector potential $\boldsymbol{A}$ or magnetic field $\boldsymbol{H}$ as expressed below:

$$\boldsymbol{E}(\boldsymbol{r}) = -j\omega\boldsymbol{A} - \frac{j}{\omega\mu\epsilon}\boldsymbol{\nabla}(\boldsymbol{\nabla}\cdot\boldsymbol{A}) = \frac{1}{j\omega\epsilon}\boldsymbol{\nabla}\times\boldsymbol{H}. \tag{1.436}$$

Substituting either Equations (1.429) and (1.430) or Equation (1.433) into (1.436) yields

$$E_r(r,\theta) = 2\eta I_e l \cos\theta \left[\frac{1}{r} + \frac{1}{jkr^2}\right] \frac{e^{-jkr}}{4\pi r} \tag{1.437}$$

$$E_\theta(r,\theta) = j\eta k I_e l \sin\theta \left[1 + \frac{1}{jkr} - \frac{1}{(kr)^2}\right] \frac{e^{-jkr}}{4\pi r} \tag{1.438}$$

$$E_\phi = 0. \tag{1.439}$$

In the far field, the terms involving $1/r^2$ and $1/r^3$ can be neglected relative to the $1/r$ term. Thus, the only significant far-field radiating components of the short vertical electric dipole are E_θ and H_ϕ. For the small horizontal loop antenna (short vertical magnetic dipole), the only significant far-field radiating components are E_ϕ and H_θ. In each case, in the far field, it can be shown that the ratio of the electric to magnetic field amplitude is equal to the impedance of free space. That is,

$$\frac{E}{H} = \eta_o. \tag{1.440}$$

1.8.2.2 Dyadic Green's Function Approach

The same near fields derived in the previous section can be computed for the $\hat{\boldsymbol{z}}$ Hertzian dipole using the dyadic Green's function approach described below. With this approach, the electric field is computed directly, from which computation the magnetic field can be computed using Maxwell's $\boldsymbol{\nabla}\times\boldsymbol{E}$ equation. Repeating Equation (1.209),

$$\boldsymbol{E}(\boldsymbol{r}) = -j\omega\mu[\overline{\boldsymbol{I}} + \frac{1}{k^2}\boldsymbol{\nabla}\boldsymbol{\nabla}]\cdot \int_V \boldsymbol{J}(\boldsymbol{r}')\frac{e^{-jk|\boldsymbol{r}-\boldsymbol{r}'|}}{4\pi|\boldsymbol{r}-\boldsymbol{r}'|}\, dv' \tag{1.441}$$

and substituting Equation (1.424) in Equation (1.441) yields

$$\boldsymbol{E}(\boldsymbol{r}) = -j\omega\mu[\overline{\boldsymbol{I}} + \frac{1}{k^2}\boldsymbol{\nabla}\boldsymbol{\nabla}]\cdot\hat{\boldsymbol{z}} I_e l\frac{e^{-jkr}}{4\pi r} \tag{1.442}$$

or using spherical components (refer to Equation (1.245)) for $\hat{\boldsymbol{z}}$ as

$$\boldsymbol{E}(\boldsymbol{r}) = -j\omega\mu[\overline{\boldsymbol{I}} + \frac{1}{k^2}\boldsymbol{\nabla}\boldsymbol{\nabla}]\cdot \left[\left[\hat{\boldsymbol{r}} I_e l\cos\theta - \hat{\boldsymbol{\theta}} I_e l\sin\theta\right]\frac{e^{-jkr}}{4\pi r}\right] \tag{1.443}$$

and then define

$$\boldsymbol{V} = \left[\hat{\boldsymbol{r}} I_e l\cos\theta - \hat{\boldsymbol{\theta}} I_e l\sin\theta\right]\frac{e^{-jkr}}{4\pi r} = \hat{\boldsymbol{r}} V_r + \hat{\boldsymbol{\theta}} V_\theta. \tag{1.444}$$

Next, use the following identity

$$(\boldsymbol{\nabla}\boldsymbol{\nabla})\cdot\boldsymbol{V} = \boldsymbol{\nabla}(\boldsymbol{\nabla}\cdot\boldsymbol{V}) \tag{1.445}$$

and it follows that Equation (1.443) becomes

$$\boldsymbol{E}(\boldsymbol{r}) = -j\omega\mu[\hat{\boldsymbol{r}} V_r + \hat{\boldsymbol{\theta}} V_\theta] - j\omega\mu\frac{1}{k^2}\boldsymbol{\nabla}\left[\boldsymbol{\nabla}\cdot\left[\hat{\boldsymbol{r}} V_r + \hat{\boldsymbol{\theta}} V_\theta\right]\right]. \tag{1.446}$$

Carrying out the vector operations in Equation (1.446) using Equation (1.260) yields the near-zone E-field components given previously in Equations (1.437), (1.438), and (1.439), which the reader can verify.

1.8.3 Derivation of Fields for an Arbitrarily Polarized Hertzian Dipole

The electric field produced by a source current density $\boldsymbol{J}$ with arbitrary polarization unit vector $\hat{\boldsymbol{p}}$ in infinite homogeneous space can be determined as follows [3, p. 506]. Assume a Hertzian dipole with vector electric current $\boldsymbol{I}_e l$ located at the origin, that is,

$$\boldsymbol{J}(\boldsymbol{r}') = \boldsymbol{I}_e l \delta(\boldsymbol{r}') \tag{1.447}$$

where

$$\boldsymbol{I}_e l = \hat{\boldsymbol{p}} I_e l \tag{1.448}$$

and I_e is the complex current. Repeating Equation (1.209)

$$\boldsymbol{E}(\boldsymbol{r}) = -j\omega\mu[\overline{\boldsymbol{I}} + \frac{1}{k^2}\boldsymbol{\nabla}\boldsymbol{\nabla}]\cdot \int_V \boldsymbol{J}(\boldsymbol{r}') \frac{e^{-jk|\boldsymbol{r}-\boldsymbol{r}'|}}{4\pi|\boldsymbol{r}-\boldsymbol{r}'|}\, dv' \tag{1.449}$$

and substituting in Equation (1.447) yields

$$\boldsymbol{E}(\boldsymbol{r}) = -j\omega\mu[\overline{\boldsymbol{I}} + \frac{1}{k^2}\boldsymbol{\nabla}\boldsymbol{\nabla}]\cdot \boldsymbol{I}_e l \frac{e^{-jkr}}{4\pi r}. \tag{1.450}$$

Using the vector identity given by Equation (1.445) and

$$\boldsymbol{\nabla}\cdot\psi\boldsymbol{A} = \psi(\boldsymbol{\nabla}\cdot\boldsymbol{A}) + \boldsymbol{A}\cdot(\boldsymbol{\nabla}\psi) \tag{1.451}$$

and from Equation (1.99), it is observed that

$$\boldsymbol{\nabla}\cdot\boldsymbol{I}_e l = 0, \tag{1.452}$$

and it follows that

$$\boldsymbol{\nabla}\boldsymbol{\nabla}\cdot\boldsymbol{I}_e l \frac{e^{-jkr}}{4\pi r} = \boldsymbol{\nabla}\left[\frac{e^{-jkr}}{4\pi r}(\boldsymbol{\nabla}\cdot\boldsymbol{I}_e l) + \boldsymbol{I}_e l\cdot\left(\boldsymbol{\nabla}\frac{e^{-jkr}}{4\pi r}\right)\right] = \boldsymbol{\nabla}\left[\boldsymbol{I}_e l\cdot\left(\boldsymbol{\nabla}\frac{e^{-jkr}}{4\pi r}\right)\right]. \tag{1.453}$$

Next, use the vector identity

$$\boldsymbol{\nabla}(\boldsymbol{A}\cdot\boldsymbol{B}) = (\boldsymbol{A}\cdot\boldsymbol{\nabla})\boldsymbol{B} + (\boldsymbol{B}\cdot\boldsymbol{\nabla})\boldsymbol{A} + \boldsymbol{A}\times(\boldsymbol{\nabla}\times\boldsymbol{B}) + \boldsymbol{B}\times(\boldsymbol{\nabla}\times\boldsymbol{A}). \tag{1.454}$$

Letting $\boldsymbol{A} = \boldsymbol{I}_e l$ and $\boldsymbol{B} = \boldsymbol{\nabla}(e^{-jkr}/4\pi r)$ and using the vector identity

$$\boldsymbol{\nabla}\times(\boldsymbol{\nabla}\psi) = 0 \tag{1.455}$$

and noting that $\nabla \boldsymbol{I}_e l = 0$ and $\nabla \times \boldsymbol{I}_e l = 0$, it follows that the rightmost side of Equation (1.453) reduces to

$$\nabla \left[\boldsymbol{I}_e l \cdot \left(\nabla \frac{e^{-jkr}}{4\pi r} \right) \right] = (\boldsymbol{I}_e l \cdot \nabla) \nabla \frac{e^{-jkr}}{4\pi r}. \tag{1.456}$$

Now the gradient operation in Equation (1.456) is evaluated to be

$$\nabla \left(\frac{e^{-jkr}}{4\pi r} \right) = \hat{\boldsymbol{r}} \frac{\partial}{\partial r} \left(\frac{e^{-jkr}}{4\pi r} \right) = \hat{\boldsymbol{r}} \left(-jk - \frac{1}{r} \right) \frac{e^{-jkr}}{4\pi r}. \tag{1.457}$$

Next, using Equation (1.457) in the right-hand side of Equation (1.456) yields

$$(\boldsymbol{I}_e l \cdot \nabla) \nabla \frac{e^{-jkr}}{4\pi r} = \boldsymbol{I}_e l \cdot \nabla \nabla \frac{e^{-jkr}}{4\pi r} = \boldsymbol{I}_e l \cdot \left(\nabla \hat{\boldsymbol{r}} \left(-jk - \frac{1}{r} \right) \frac{e^{-jkr}}{4\pi r} \right). \tag{1.458}$$

To perform the gradient operation on the right-hand side of Equation (1.458), it is observed that

$$\nabla (\hat{\boldsymbol{r}} \psi(r)) = \hat{\boldsymbol{r}} \nabla \psi(r) + \psi(r) \nabla \hat{\boldsymbol{r}}. \tag{1.459}$$

Now let

$$\psi(r) = \left(-jk - \frac{1}{r} \right) \frac{e^{-jkr}}{4\pi r} \tag{1.460}$$

and compute the gradient of $\psi(r)$ as

$$\nabla \psi(r) = \nabla \left[\left(-jk - \frac{1}{r} \right) \frac{e^{-jkr}}{4\pi r} \right] = \hat{\boldsymbol{r}} \left[-k^2 + \frac{2jk}{r} + \frac{2}{r^2} \right] \frac{e^{-jkr}}{4\pi r}. \tag{1.461}$$

Using Equation (1.256) and the unit vector derivatives given below Equation (1.239), it follows that

$$\nabla \hat{\boldsymbol{r}} = \frac{1}{r} \hat{\boldsymbol{\theta}} \hat{\boldsymbol{\theta}} + \frac{1}{r} \hat{\boldsymbol{\phi}} \hat{\boldsymbol{\phi}}. \tag{1.462}$$

Using the results of Equations (1.459), (1.460), (1.461), and (1.462), it follows that the dyadic operation in Equation (1.458) is evaluated as

$$\nabla \nabla \frac{e^{-jkr}}{4\pi r} = \hat{\boldsymbol{r}} \hat{\boldsymbol{r}} \left[-k^2 + \frac{2jk}{r} + \frac{2}{r^2} \right] \frac{e^{-jkr}}{4\pi r} + \left[\hat{\boldsymbol{\theta}} \hat{\boldsymbol{\theta}} + \hat{\boldsymbol{\phi}} \hat{\boldsymbol{\phi}} \right] \left[-\frac{jk}{r} - \frac{1}{r^2} \right] \frac{e^{-jkr}}{4\pi r}. \tag{1.463}$$

Note that substituting Equation (1.463) in Equation (1.330) yields the explicit dyadic Green's function for a point source at the origin as

$$\overline{\boldsymbol{G}}(\boldsymbol{r}) = \hat{\boldsymbol{r}}\hat{\boldsymbol{r}}\left[\frac{2j}{kr} + \frac{2}{k^2r^2}\right]\frac{e^{-jkr}}{4\pi r} + \left[\hat{\boldsymbol{\theta}}\hat{\boldsymbol{\theta}} + \hat{\boldsymbol{\phi}}\hat{\boldsymbol{\phi}}\right]\left[1 - \frac{j}{kr} - \frac{1}{k^2r^2}\right]\frac{e^{-jkr}}{4\pi r}. \tag{1.464}$$

Using Equations (1.463), (1.453), and (1.246) in Equation (1.450) yields the near-zone electric field for an arbitrary polarized Hertzian dipole at the origin as

$$\boldsymbol{E}(\boldsymbol{r}) = -j\omega\mu\frac{e^{-jkr}}{4\pi r}\left[\left[1 - \frac{j}{kr} - \frac{1}{k^2r^2}\right]\boldsymbol{I}_e l + \hat{\boldsymbol{r}}\left(\hat{\boldsymbol{r}}\cdot\boldsymbol{I}_e l\right)\left[-1 + \frac{3j}{kr} + \frac{3}{k^2r^2}\right]\right] \tag{1.465}$$

The near-zone magnetic field is then determined from Maxwell's curl of $\boldsymbol{E}$ equation as

$$\boldsymbol{H}(\boldsymbol{r}) = -\frac{1}{j\omega\mu}\nabla\times\boldsymbol{E}(\boldsymbol{r}) = \nabla\times\left[\overline{\boldsymbol{I}} + \frac{1}{k^2}\nabla\nabla\right]\cdot\boldsymbol{I}_e l\frac{e^{-jkr}}{4\pi r} = \nabla\times\boldsymbol{I}_e l\frac{e^{-jkr}}{4\pi r} \tag{1.466}$$

which follows from the identity that the curl of the gradient of any scalar is equal to zero (Equation (1.455)). To solve Equation (1.466), the following vector identity is used:

$$\nabla\times(\Psi\boldsymbol{A}) = \Psi\nabla\times\boldsymbol{A} - \boldsymbol{A}\times\nabla\Psi. \tag{1.467}$$

Noting that

$$\nabla\times\boldsymbol{I}_e = 0 \tag{1.468}$$

and using Equation (1.457), it follows that the near-zone magnetic field for an arbitrary polarized Hertzian dipole is expressed as

$$\boldsymbol{H}(\boldsymbol{r}) = \nabla\times\boldsymbol{I}_e l\frac{e^{-jkr}}{4\pi r} = -\hat{\boldsymbol{r}}\times\boldsymbol{I}_e ljk\left[1 - \frac{j}{kr}\right]\frac{e^{-jkr}}{4\pi r}. \tag{1.469}$$

To compute the time-average power density of the electromagnetic field generated by the arbitrary polarized dipole, first substitute Equations (1.465) and (1.469) into the bracketed expression in Equation (1.409) using complex power density denoted $\boldsymbol{P}_c = \boldsymbol{E}\times\boldsymbol{H}^*$ with the result

$$\boldsymbol{P}_c = \eta\left[\frac{k}{4\pi r}\right]^2\left\{\hat{\boldsymbol{r}}(Il)^2(1 - \tfrac{j}{k^3r^3}) - \hat{\boldsymbol{r}}(\hat{\boldsymbol{r}}\cdot\boldsymbol{I}l)^2\left[1 - \tfrac{2j}{kr} - \tfrac{3j}{k^3r^3}\right] - (\hat{\boldsymbol{r}}\cdot\boldsymbol{I}l)\boldsymbol{I}l\left[\tfrac{2j}{kr} + \tfrac{2j}{k^3r^3}\right]\right\} \tag{1.470}$$

from which it follows that the time-average power density of the arbitrary dipole is

$$\boldsymbol{P}_{\text{ave}}^{\text{dipole}}(\boldsymbol{r}) = \frac{1}{2}\text{Re}[\boldsymbol{E}\times\boldsymbol{H}^*] = \frac{1}{2}\text{Re}[\boldsymbol{P}_c] = \eta\left[\frac{k}{4\pi r}\right]^2\left[\hat{\boldsymbol{r}}(Il)^2 - \hat{\boldsymbol{r}}(\hat{\boldsymbol{r}}\cdot\boldsymbol{I}l)^2\right]. \tag{1.471}$$

In the case of a z-directed dipole, referring to Equations (1.479) to (1.482), it can be shown that the complex power density is

$$\boldsymbol{P}_c^{\text{z-dipole}} = \eta \left[\frac{kIl}{4\pi r}\right]^2 \left\{\hat{\boldsymbol{r}} \left[1 + \left(\frac{j}{kr}\right)^3\right] \sin^2\theta + \hat{\boldsymbol{\theta}} \left[\left(\frac{j}{kr}\right) - \left(\frac{j}{kr}\right)^3\right] \sin 2\theta\right\} \tag{1.472}$$

from which it follows that the time-average power density of the z-directed dipole is

$$\boldsymbol{P}_{\text{ave}}^{\text{z-dipole}}(\boldsymbol{r}) = \frac{1}{2}\text{Re}[\boldsymbol{E}\times\boldsymbol{H}^*] = \frac{1}{2}\text{Re}[\boldsymbol{P}_c] = \hat{\boldsymbol{r}}\frac{\eta}{2}\left[\frac{kIl}{4\pi r}\right]^2 \sin^2\theta. \tag{1.473}$$

1.8.4 Duality and Fields for Hertzian Loop and Dipole Antennas

In the case of a Hertzian loop antenna with radius a centered at the origin (Figure 1.25), the electric and magnetic fields can be readily found from the Hertzian electric dipole antenna fields by utilizing the duality principle. For a small circular loop of radius a that has area $A = \pi a^2$, the magnetic dipole moment $m = IA = I\pi a^2$, which has units of amperes meters squared. The vector magnetic dipole moment can be expressed as $\boldsymbol{m} = m\hat{\boldsymbol{p}}$, where $\hat{\boldsymbol{p}}$ is the polarization unit vector for the magnetic dipole moment, which is perpendicular to the plane of the loop. In terms of the electric current I flowing in the small current loop, the vector magnetic dipole moment is expressed as

$$\boldsymbol{m} = IA\hat{\boldsymbol{p}}. \tag{1.474}$$

The relevant duality relationships with the following replacement of quantities have been described by Kong [3, pp. 510, 511], that is, $\boldsymbol{E} \to \boldsymbol{H}$, $\boldsymbol{H} \to -\boldsymbol{E}$, $\mu \to \epsilon$, $\epsilon \to \mu$, $k \to k$, $\eta \to 1/\eta$, $\boldsymbol{D} \to \boldsymbol{B}$, $\boldsymbol{B} \to -\boldsymbol{D}$, and $\boldsymbol{I}_e l \to j\omega\mu\boldsymbol{m}$, $j\omega\mu\boldsymbol{m} \to -\boldsymbol{I}_e l$. Summarizing the near-zone fields of the Hertzian electric dipole (referred to as dipole), we have from Equations (1.465) and (1.469)

$$\boldsymbol{E}_{\text{dipole}}(\boldsymbol{r}) = -j\omega\mu\frac{e^{-jkr}}{4\pi r}\left\{\left[1 - \frac{j}{kr} - \frac{1}{k^2r^2}\right]\boldsymbol{I}_e l + \hat{\boldsymbol{r}}\,(\hat{\boldsymbol{r}}\cdot\boldsymbol{I}_e l)\left[-1 + \frac{3j}{kr} + \frac{3}{k^2r^2}\right]\right\} \tag{1.475}$$

$$\boldsymbol{H}_{\text{dipole}}(\boldsymbol{r}) = -\hat{\boldsymbol{r}}\times\boldsymbol{I}_e ljk\left[1 - \frac{j}{kr}\right]\frac{e^{-jkr}}{4\pi r}. \tag{1.476}$$

Using the above duality relationships, the fields for the Hertzian magnetic dipole (referred to as electric current loop) are found to be

$$\boldsymbol{H}_{\text{loop}}(\boldsymbol{r}) = k^2\frac{e^{-jkr}}{4\pi r}\left\{\left[1 - \frac{j}{kr} - \frac{1}{k^2r^2}\right]\boldsymbol{m} + \hat{\boldsymbol{r}}\,(\hat{\boldsymbol{r}}\cdot\boldsymbol{m})\left[-1 + \frac{3j}{kr} + \frac{3}{k^2r^2}\right]\right\} \tag{1.477}$$

$$\boldsymbol{E}_{\text{loop}}(\boldsymbol{r}) = -\hat{\boldsymbol{r}}\times\boldsymbol{m}\omega\mu k\left[1-\frac{j}{kr}\right]\frac{e^{-jkr}}{4\pi r}. \tag{1.478}$$

From Equations (1.475) and (1.476), the near-field components for a z-directed dipole are determined to be

$$H_{\phi}^{\text{NF, z-dipole}}(r,\theta) = jkI_e l\sin\theta\left[1+\frac{1}{jkr}\right]\frac{e^{-jkr}}{4\pi r} \tag{1.479}$$

$$E_{r}^{\text{NF, z-dipole}}(r,\theta) = 2\eta I_e l\cos\theta\left[\frac{1}{r}+\frac{1}{jkr^2}\right]\frac{e^{-jkr}}{4\pi r} \tag{1.480}$$

$$E_{\theta}^{\text{NF, z-dipole}}(r,\theta) = j\eta kI_e l\sin\theta\left[1+\frac{1}{jkr}-\frac{1}{(kr)^2}\right]\frac{e^{-jkr}}{4\pi r} \tag{1.481}$$

$$E_{\phi}^{\text{NF, z-dipole}} = 0, \tag{1.482}$$

which are the same as Equations (1.433), (1.437), (1.438), and (1.439), respectively. Thus, the dyadic Green's function approach has provided confirmation of the field equations derived by the magnetic vector potential approach. The corresponding near-zone fields for a z-directed loop can be determined by applying the duality relationships $\boldsymbol{E}\rightarrow\boldsymbol{H}$, $\boldsymbol{H}\rightarrow-\boldsymbol{E}$, $\mu\rightarrow\epsilon$, $\epsilon\rightarrow\mu$, $k\rightarrow k$, $\eta\rightarrow 1/\eta$, and $I_e l\rightarrow j\omega\mu m = j\omega\mu IA$, to the fields of the z-directed Hertzian dipole given by Equations (1.479), (1.480), (1.481), and (1.482), and noting that

$$\eta = \omega\mu/k, \tag{1.483}$$

leads to the simplified results as

$$E_{\phi}^{\text{NF, z-loop}}(r,\theta) = \eta IA\sin\theta\left[k^2-\frac{jk}{r}\right]\frac{e^{-jkr}}{4\pi r} \tag{1.484}$$

$$H_{r}^{\text{NF, z-loop}}(r,\theta) = 2IA\cos\theta\left[\frac{jk}{r}+\frac{1}{r^2}\right]\frac{e^{-jkr}}{4\pi r} \tag{1.485}$$

$$H_{\theta}^{\text{NF, z-loop}}(r,\theta) = IA\sin\theta\left[-k^2+\frac{jk}{r}+\frac{1}{r^2}\right]\frac{e^{-jkr}}{4\pi r} \tag{1.486}$$

$$H_{\phi}^{\text{NF, z-loop}} = 0. \tag{1.487}$$

As an example of the amplitude of the magnetic flux density $B_\theta = \mu_o H_\theta$ in teslas at the angle $\theta = 90^\circ$ versus range distance r, consider a horizontal z-loop (Figure 1.25(b)) with loop diameter 1 meter and loop current I =0.1A radiating at 100 kHz. Then using Equation (1.486), Figure 1.26 compares on a log-log scale the individual terms ($1/r$, $1/r^2$, and $1/r^3$) with B_θ calculated

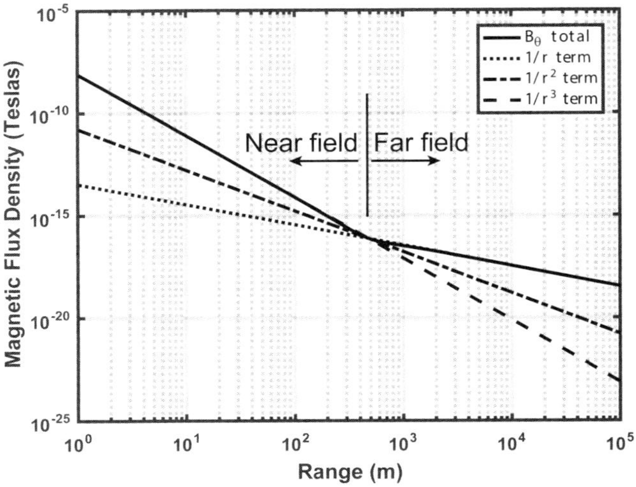

Figure 1.26 Magnetic flux density $B_\theta = \mu_o H_\theta$ versus near-field range distance for a 1m diameter z-loop radiating at 100 kHz with 0.1A electric current.

using all three terms (total). At 100 kHz, the wavelength is 3,000 meters. The bend in the solid curve occurs where the range distance is given by

$$r = \frac{\lambda}{2\pi} = \frac{3000\text{m}}{2\pi} = 477.5\text{m} \tag{1.488}$$

in which case all three terms have equal amplitude, which the reader can verify from Equation (1.486). For range distances less than $\lambda/2\pi$, the $1/r^3$ term dominates. For range distance greater than $\lambda/2\pi$, the $1/r$ term dominates. Thus, the *far-field region* is defined to begin when the $1/r$ term dominates and the range distance is greater than $\lambda/2\pi$. The *near-field region* is defined for when the $1/r^3$ term dominates and the range distance is less than $\lambda/2\pi$. By duality, the shape of the curves in Figure 1.26 and the near-field / far-field definitions apply to Hertzian dipoles as well. For aperture antennas, the far-field condition given by Equation (1.213) must also be satisfied.

1.9 Far-Zone Fields of Arbitrary Dipoles and Loops

1.9.1 Introduction

The far-zone fields can be computed by ignoring the higher-order terms $1/r^2$ and $1/r^3$ as mentioned earlier. Thus, in the far field for arbitrary Hertzian dipole current orientation, Equations (1.475) and (1.476) reduce to

$$\boldsymbol{E}_{\text{dipole}}^{FF}(\boldsymbol{r}) = -j\omega\mu\frac{e^{-jkr}}{4\pi r}\left[\boldsymbol{I}_e l - \hat{\boldsymbol{r}}\left(\hat{\boldsymbol{r}}\cdot\boldsymbol{I}_e l\right)\right] \tag{1.489}$$

$$\boldsymbol{H}_{\text{dipole}}^{FF}(\boldsymbol{r}) = -\hat{\boldsymbol{r}}\times\boldsymbol{I}_e l jk\frac{e^{-jkr}}{4\pi r}. \tag{1.490}$$

Similarly, using Equations (1.477), and (1.478), the far-zone fields for the arbitrary Hertzian magnetic dipole (referred to as electric current loop) are found to be

$$\boldsymbol{H}_{\text{loop}}^{FF}(\boldsymbol{r}) = k^2\frac{e^{-jkr}}{4\pi r}\left[\boldsymbol{m} - \hat{\boldsymbol{r}}\left(\hat{\boldsymbol{r}}\cdot\boldsymbol{m}\right)\right] \tag{1.491}$$

$$\boldsymbol{E}_{\text{loop}}^{FF}(\boldsymbol{r}) = -\hat{\boldsymbol{r}}\times\boldsymbol{m}\omega\mu k\frac{e^{-jkr}}{4\pi r}. \tag{1.492}$$

Equations (1.489) to (1.492) are evaluated below for $\hat{\boldsymbol{x}}$, $\hat{\boldsymbol{y}}$, and $\hat{\boldsymbol{x}}$ polarizations, which are utilized for the vector sensor antennas described in Chapter 6.

1.9.2 Far-Zone Fields for $\hat{z}$ Oriented Dipoles and Loops

For example, when the electric dipole moment has only a z orientation, that is, $\boldsymbol{I}_e l = \hat{\boldsymbol{z}} I_e l$, and referring to Equation (1.245) $\hat{\boldsymbol{z}} = \hat{\boldsymbol{r}}\cos\theta - \hat{\boldsymbol{\theta}}\sin\theta$, Equations (1.489) and (1.490) reduce to the far-field results

$$\boldsymbol{E}_{\text{z-dipole}}^{FF}(\boldsymbol{r}) = \hat{\boldsymbol{\theta}} j\omega\mu I_e l\frac{e^{-jkr}}{4\pi r}\sin\theta \tag{1.493}$$

$$\boldsymbol{H}_{\text{z-dipole}}^{FF}(\boldsymbol{r}) = \hat{\boldsymbol{\phi}} I_e l jk\frac{e^{-jkr}}{4\pi r}\sin\theta. \tag{1.494}$$

The far-zone fields for the Hertzian magnetic dipole (electric current loop) oriented with the axis in the z direction, that is, $\boldsymbol{m} = m\hat{\boldsymbol{z}}$, where $m = IA$, are found from Equations (1.493) and (1.494) using the duality relations given in the previous section, and noting $\hat{\boldsymbol{\phi}}\times\hat{\boldsymbol{\theta}} = -\hat{\boldsymbol{r}}$, to be

$$\boldsymbol{H}_{\text{z-loop}}^{FF}(\boldsymbol{r}) = -\hat{\boldsymbol{\theta}} k^2 IA\frac{e^{-jkr}}{4\pi r}\sin\theta \tag{1.495}$$

$$\boldsymbol{E}_{\text{z-loop}}^{FF}(\boldsymbol{r}) = \hat{\boldsymbol{\phi}} IA\omega\mu k\frac{e^{-jkr}}{4\pi r}\sin\theta. \tag{1.496}$$

As discussed earlier, by taking the ratio of the magnitudes of the electric field and magnetic fields, it is readily shown that the far-zone fields of the dipole and loop in free space satisfy $|E|/|H| = \eta_o$, where η_o is the impedance of free space. In the case of a multiturn small loop antenna with N turns, the field expressions given in this section are generalized by simply replacing the quantity IA with NIA such that the vector magnetic dipole moment is now given by

$$\boldsymbol{m} = NIA\hat{\boldsymbol{p}}. \tag{1.497}$$

Consider now an example of the near-zone electric field dependence with range distance from a z directed Hertzian dipole antenna. For an observation point in the xy-plane, using Equation (1.481) it can be shown that the magnitude of the z component of the electric field is proportional to

$$|E_z(r)| \propto \sqrt{\left[\frac{1}{r} - \frac{1}{k^2 r^3}\right]^2 + \left[\frac{1}{kr^2}\right]^2} \tag{1.498}$$

Assume that the radiating frequency is 300 MHz, so that the wavelength is $\lambda = 1$m and the corresponding wavenumber is $k = 2\pi$ radians/m. The normalized electric-field magnitude (envelope) versus near-field range distance is shown in Figure 1.27. From Equation (1.116), the time variation of the electric field is given by

$$\boldsymbol{E}(\boldsymbol{r}, t) = |\boldsymbol{E}(\boldsymbol{r})| \cos(\omega t - kr) \tag{1.499}$$

and taking snapshots of just the cosine factor at fixed instants in time, say $t = 0, t = 1$ns with $k = 2\pi$ radians/m, Figure 1.28 shows that the wave moves to the right at the speed of light (0.3m per nanosecond). The wavelength 1m is observed between two consecutive positive (or negative) peaks. Now, taking

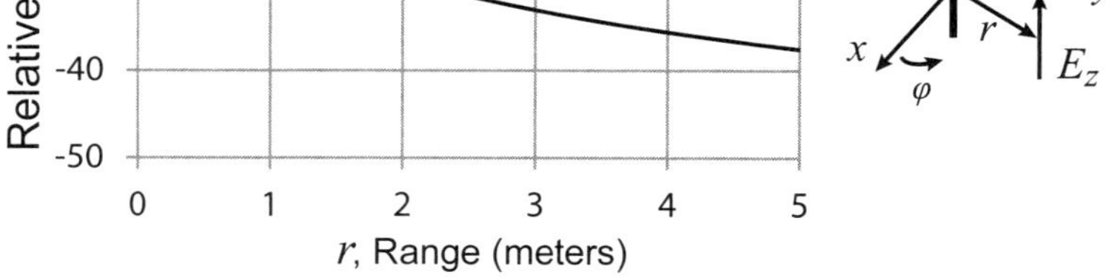

Figure 1.27 Normalized electric-field magnitude (envelope) versus near-field range distance for a z-polarized Hertzian dipole antenna radiating at 300 MHz.

a snapshot of the electric-field amplitude at $t = 0$, from Equation (1.499), the spatial variation of the electric field is given by

$$\boldsymbol{E}(\boldsymbol{r}, 0) = |\boldsymbol{E}(\boldsymbol{r})| \cos(kr) \tag{1.500}$$

and Figure 1.29 shows the decrease in the amplitude of the peaks and troughs as the range increases. Next, the magnitude in dB of the instantaneous E-field at $t = 0$ versus range distance is shown in Figure 1.30 Finally, the instantaneous phase versus range distance is shown in Figure 1.31, and the phase is observed to rotate through 360° every 1m (one wavelength in this case) distance.

1.9.3 Far-Zone Fields for $\hat{x}$ Oriented Dipoles and Loops

When the electric dipole moment has only an x orientation, that is, $\boldsymbol{I}_e l = \hat{\boldsymbol{x}} I_e l$, and referring to Equation (1.243) then Equation (1.490) reduces to

$$\boldsymbol{H}^{FF}_{\text{x-dipole}}(\boldsymbol{r}) = -\left[\hat{\boldsymbol{\theta}} \sin\phi + \hat{\boldsymbol{\phi}} \cos\theta \cos\phi\right] jkI_e l \frac{e^{-jkr}}{4\pi r}. \tag{1.501}$$

Then, applying Equation (1.410), $\hat{\boldsymbol{\phi}} \times \hat{\boldsymbol{r}} = \hat{\boldsymbol{\theta}}$, and $\hat{\boldsymbol{\theta}} \times \hat{\boldsymbol{r}} = -\hat{\boldsymbol{\phi}}$ yields

$$\boldsymbol{E}^{FF}_{\text{x-dipole}}(\boldsymbol{r}) = \left[-\hat{\boldsymbol{\theta}} \cos\theta \cos\phi + \hat{\boldsymbol{\phi}} \sin\phi\right] jkI_e l\eta \frac{e^{-jkr}}{4\pi r}. \tag{1.502}$$

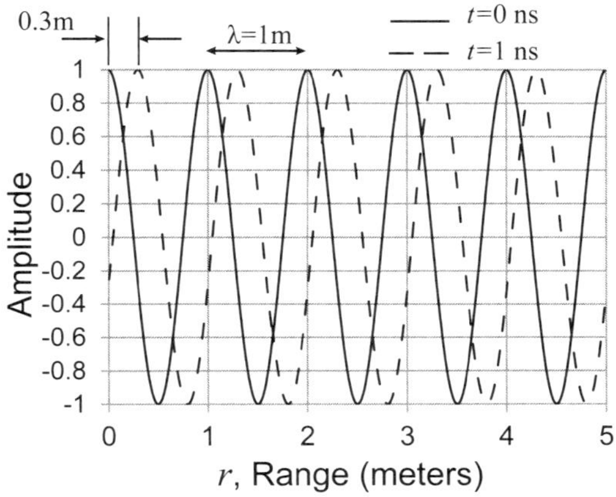

Figure 1.28 Instantaneous electric-field amplitude (ignoring the envelope factor) versus near-field range distance for a z-polarized Hertzian dipole antenna radiating at 300 MHz.

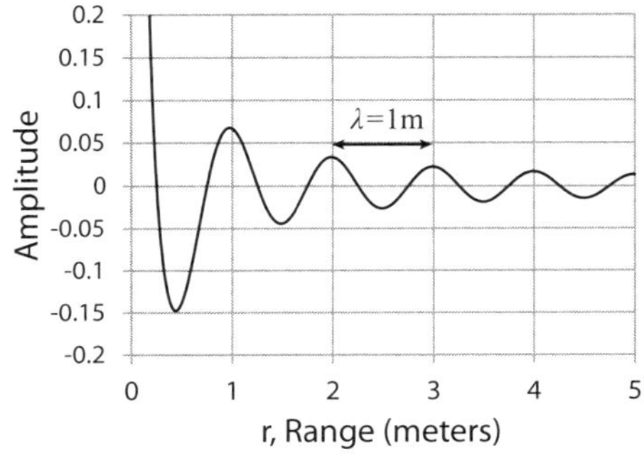

Figure 1.29 Instantaneous electric-field amplitude (including the envelope factor) $|E_z(r)|\cos(kr)$ versus near-field range distance for a z-polarized Hertzian dipole antenna radiating at 300 MHz.

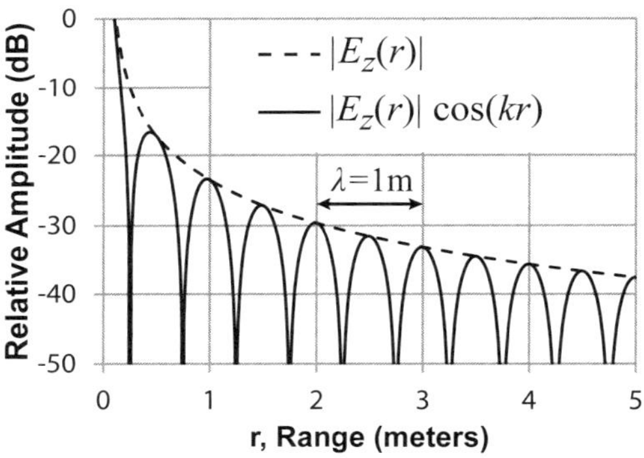

Figure 1.30 Comparison of the electric-field envelope (dashed curve) and instantaneous electric-field magnitude (solid curve) in dB (including the envelope factor) versus near-field range distance for a z-polarized Hertzian dipole antenna radiating at 300 MHz.

By applying duality, the loop fields are given by

$$\boldsymbol{H}^{FF}_{\text{x-loop}}(\boldsymbol{r}) = \left[\hat{\boldsymbol{\theta}}\cos\theta\cos\phi - \hat{\boldsymbol{\phi}}\sin\phi\right] k^2 IA\frac{e^{-jkr}}{4\pi r}. \tag{1.503}$$

$$\boldsymbol{E}^{FF}_{\text{x-loop}}(\boldsymbol{r}) = -\left[\hat{\boldsymbol{\theta}}\sin\phi + \hat{\boldsymbol{\phi}}\cos\theta\cos\phi\right] \eta k^2 IA\frac{e^{-jkr}}{4\pi r}. \tag{1.504}$$

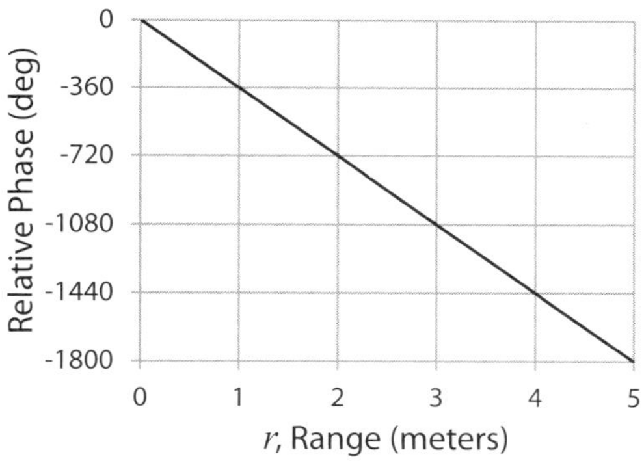

Figure 1.31 Electric-field phase versus near-field range distance for a z-polarized Hertzian dipole antenna radiating at 300 MHz.

1.9.4 Far-Zone Fields for $\hat{y}$ Oriented Dipoles and Loops

When the electric dipole moment has only a y orientation, that is, $\boldsymbol{I}_e l = \hat{\boldsymbol{y}} I_e l$, and referring to Equation (1.244) then Equation (1.490) reduces to

$$\boldsymbol{H}^{FF}_{\text{y-dipole}}(\boldsymbol{r}) = \left[\hat{\boldsymbol{\theta}} \cos\phi - \hat{\boldsymbol{\phi}} \cos\theta \sin\phi\right] jkI_e l \frac{e^{-jkr}}{4\pi r}. \tag{1.505}$$

Then, applying Equation (1.410), $\hat{\boldsymbol{\phi}} \times \hat{\boldsymbol{r}} = \hat{\boldsymbol{\theta}}$, and $\hat{\boldsymbol{\theta}} \times \hat{\boldsymbol{r}} = -\hat{\boldsymbol{\phi}}$ yields

$$\boldsymbol{E}^{FF}_{\text{y-dipole}}(\boldsymbol{r}) = -\left[\hat{\boldsymbol{\theta}} \cos\theta \sin\phi + \hat{\boldsymbol{\phi}} \cos\phi\right] jk\eta I_e l\eta \frac{e^{-jkr}}{4\pi r}. \tag{1.506}$$

By applying duality, the loop fields are given by

$$\boldsymbol{H}^{FF}_{\text{y-loop}}(\boldsymbol{r}) = \left[\hat{\boldsymbol{\theta}} \cos\theta \sin\phi + \hat{\boldsymbol{\phi}} \cos\phi\right] k^2 IA \frac{e^{-jkr}}{4\pi r}. \tag{1.507}$$

$$\boldsymbol{E}^{FF}_{\text{y-loop}}(\boldsymbol{r}) = \left[\hat{\boldsymbol{\theta}} \cos\phi - \hat{\boldsymbol{\phi}} \cos\theta \sin\phi\right] \eta k^2 IA \frac{e^{-jkr}}{4\pi r}. \tag{1.508}$$

1.9.5 Image Theory for Electric and Magnetic Dipole Antennas

Image theory can be used to compute the fields caused by an electric dipole or a magnetic dipole (electric current loop) over a perfect electric conductor as depicted in Figure 1.32. The sign of the image can be determined by using the exact electric field expresssions (as given in the previous section) for the electric dipole and magnetic dipole and by enforcing boundary conditions that

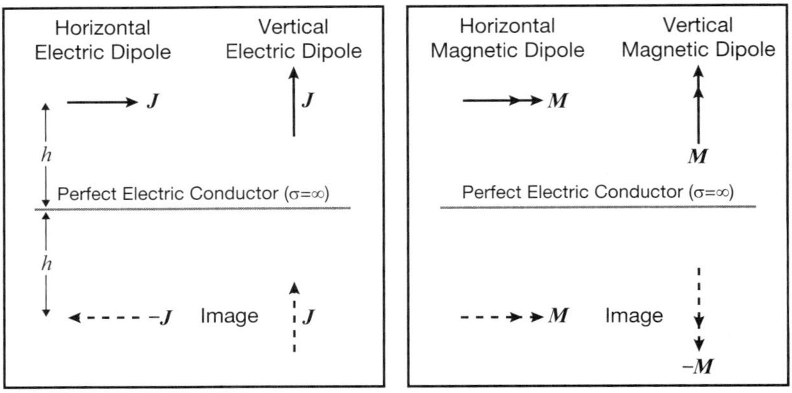

Figure 1.32 Image theory applied to electric and magnetic dipoles.

the total tangential electric field be zero at the surface of the perfect electric conductor. A horizontal electric dipole will have a negative image, whereas a horizontal magnetic dipole will have a positive image. Additionally, a vertical electric dipole has a positive image, whereas a vertical magnetic dipole has a negative image. For these cases, the dipole over a perfect electric conductor is replaced by the equivalent situation of the original dipole and the dipole image with the perfect electric conductor removed. Thus, the electromagetic fields above the perfect electric conductor can be computed by analyzing a two-element array with the proper phasing, as described for general arrays in the next chapter.

1.10 Electromagnetic Wave Polarization and Receive Antennas

1.10.1 Wave Polarization Theory

In this section, a brief discussion of electromagnetic wave polarization in the context of antennas is given based on standard IEEE definitions and electrical engineering notation [97, 98]. The polarization of an electromagnetic wave is defined with the antenna transmitting, by the movement of the tip of the electric field vector as a function of time at a fixed position in space. When the E-field tip follows a straight line, the wave is linearly polarized. When the tip follows a circle, the wave is said to be circularly polarized. In the general case, the tip of the electric field will follow the shape of an ellipse. Two orthogonal electric field components can be superimposed to form the total electric field vector. The polarization state of the total electric field depends on the amplitude and phase relations of these orthogonal field components.

Consider Figure 1.33, which shows examples of linear, circular, and elliptical polarizations. In this figure, the wave is viewed as approaching the reader (moving in the positive z direction) at $z = 0$. To characterize the polarization state of a general electromagnetic wave propagating along the positive z axis as depicted now in Figure 1.34(a), we can write the complex electric field in terms of the complex x component E_x, complex y component E_y, E-field amplitude E_1 in the x direction and E_2 in the y direction in phasor form as

$$\boldsymbol{E}(z) = E_x\hat{\boldsymbol{x}} + E_y\hat{\boldsymbol{y}} = E_1 e^{-jkz}\hat{\boldsymbol{x}} + E_2 e^{-jkz} e^{j\delta_p}\hat{\boldsymbol{y}}, \tag{1.509}$$

where δ_p is the relative phase shift between the two electric-field components. Note that a related diagram known as the Poincaré sphere is shown in Figure 1.34(b), which is discussed later. Equation (1.515) can now be written in instantaneous form as

$$\boldsymbol{E}(z,t) = \mathrm{Re}[E_1 e^{j(\omega t - kz)}\hat{\boldsymbol{x}}] + \mathrm{Re}[E_2 e^{j(\omega t - kz + \delta_p)}\hat{\boldsymbol{y}}]. \tag{1.510}$$

In Equation (1.510), expressing the exponentials in terms of cosines and sines and taking the real part yields

$$\boldsymbol{E}(z,t) = E_x\hat{\boldsymbol{x}} + E_y\hat{\boldsymbol{y}} = E_1\cos(\omega t - kz)\hat{\boldsymbol{x}} + E_2\cos(\omega t - kz + \delta_p)\hat{\boldsymbol{y}}. \tag{1.511}$$

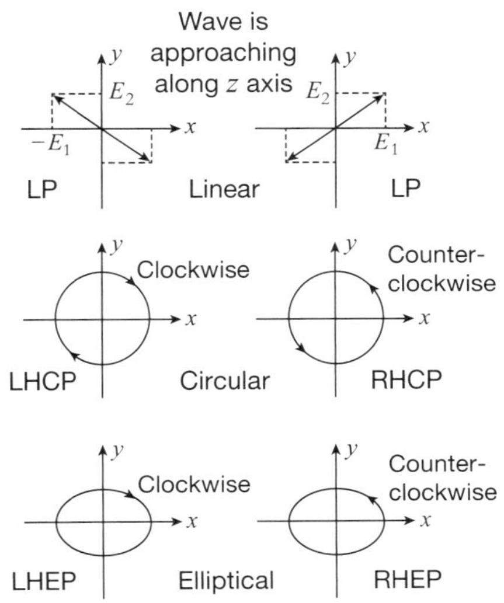

Figure 1.33 Depiction of linear, circular, and elliptical polarizations.

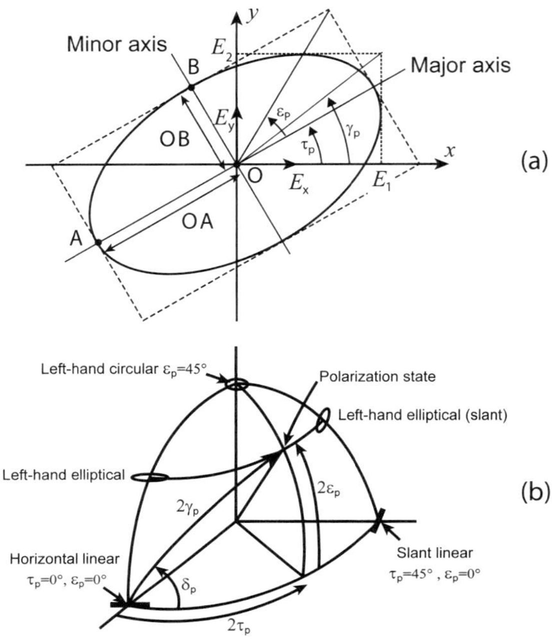

Figure 1.34 Depictions of electromagnetic wave polarization. (a) General tilted polarization ellipse, and (b) one quadrant of the Poincaré sphere.

Note that in Equation (1.511), the amplitudes $E_x = E_1 \cos(\omega t - kz)$ and $E_y = E_2 \cos(\omega t - kz + \delta_p)$ follow the tip of the electric-field vector as a function of time and spatial position. For the case when $z = 0$, Equation (1.511) reduces to

$$\boldsymbol{E}(t) = E_x \hat{\boldsymbol{x}} + E_y \hat{\boldsymbol{y}} = E_1 \cos(\omega t) \hat{\boldsymbol{x}} + E_2 \cos(\omega t + \delta_p) \hat{\boldsymbol{y}}. \tag{1.512}$$

Referring to Figure 1.34(a), the angle γ_p is expressed in terms of E_1 and E_2 as

$$\gamma_p = \tan^{-1} \frac{E_2}{E_1}. \tag{1.513}$$

In Equation (1.513) we note that E_1 and E_2 are peak values and thus $0° \leq \gamma_p \leq 90°$. Furthermore, note that

$$|\boldsymbol{E}| = \boldsymbol{E} \cdot \boldsymbol{E}^* = |E_x|^2 + |E_y|^2 = |E_1|^2 + |E_2|^2. \tag{1.514}$$

At $z = 0$, Equation (1.515) reduces to

$$\boldsymbol{E}(z = 0) = E_x \hat{\boldsymbol{x}} + E_y \hat{\boldsymbol{y}} = E_1 \hat{\boldsymbol{x}} + E_2 e^{j\delta_p} \hat{\boldsymbol{y}}, \tag{1.515}$$

Further, note that the amplitudes $E_1 = |\boldsymbol{E}|\cos\gamma_p$ and $E_2 = |\boldsymbol{E}|\sin\gamma_p$, and then the electric field unit vector can be expressed as

$$\hat{\boldsymbol{e}} = \frac{\boldsymbol{E}}{|\boldsymbol{E}|} = \frac{E_1}{|\boldsymbol{E}|}\hat{\boldsymbol{x}} + \frac{E_2}{|\boldsymbol{E}|}\hat{\boldsymbol{y}} \tag{1.516}$$

which reduces to

$$\hat{\boldsymbol{e}} = \cos\gamma_p\hat{\boldsymbol{x}} + \sin\gamma_p e^{j\delta_p}\hat{\boldsymbol{y}}. \tag{1.517}$$

For the case with $z = 0$, it is observed that Equation (1.515) can also be expressed as

$$\boldsymbol{E}(\gamma_p, \delta_p) = \sqrt{E_1^2 + E_2^2}(\cos\gamma_p\hat{\boldsymbol{x}} + \sin\gamma_p e^{j\delta_p}\hat{\boldsymbol{y}}) = |\boldsymbol{E}|\hat{\boldsymbol{e}}. \tag{1.518}$$

In Equation (1.518), $\hat{\boldsymbol{e}}$ is the complex unit vector polarization state of the electromagnetic wave.

It is worth noting that the more general case of spherical coordinates for a plane wave propagating in the radial direction r can be expressed as

$$\boldsymbol{E}(r) = E_\theta\hat{\boldsymbol{\theta}} + E_\phi\hat{\boldsymbol{\phi}} = E_1 e^{-jkr}\hat{\boldsymbol{\theta}} + E_2 e^{-jkr}e^{j\delta_p}\hat{\boldsymbol{\phi}}, \tag{1.519}$$

For the case with $r = 0$, it is observed that Equation (1.519) can be expressed as

$$\boldsymbol{E}(\gamma_p, \delta_p) = \sqrt{E_1^2 + E_2^2}(\cos\gamma_p\hat{\boldsymbol{\theta}} + \sin\gamma_p e^{j\delta_p}\hat{\boldsymbol{\phi}}) = |\boldsymbol{E}|\hat{\boldsymbol{e}}. \tag{1.520}$$

Now for an incident plane wave propagating in the negative radial direction, note that the cross product of the unit vectors for phi and theta gives

$$\hat{\boldsymbol{\phi}} \times \hat{\boldsymbol{\theta}} = -\hat{\boldsymbol{r}} \tag{1.521}$$

For a wave traveling in the $-\hat{\boldsymbol{r}}$ direction, if we compare $\hat{\boldsymbol{\phi}}$ with $\hat{\boldsymbol{x}}$ and $\hat{\boldsymbol{\theta}}$ with $\hat{\boldsymbol{y}}$ in Figure 1.34(a) and with Equation (1.517) then it follows that the incident wave electric-field unit vector $\hat{\boldsymbol{e}}^{\text{inc}}$ or incident polarization unit vector $\hat{\boldsymbol{p}}^{\text{inc}}$ in spherical coordinates is given by

$$\hat{\boldsymbol{e}}^{\text{inc}} = \hat{\boldsymbol{p}}^{\text{inc}} = \frac{E_\theta}{|\boldsymbol{E}|}\hat{\boldsymbol{\theta}} + \frac{E_\phi}{|\boldsymbol{E}|}\hat{\boldsymbol{\phi}} = \sin\gamma_p e^{j\delta_p}\hat{\boldsymbol{\theta}} + \cos\gamma_p\hat{\boldsymbol{\phi}}. \tag{1.522}$$

Referring now to the upper left diagram of Figure 1.33, the wave shown is slanted to the left, and it is linearly polarized. In contrast, in the upper right diagram of Figure 1.33, the linearly polarized wave is slanted to the right. For circularly polarized (CP) waves, the amplitude of the two orthogonal electric field components will be equal, that is, $E_1 = E_2$, and the relative phase will be

$\delta_p = \pm 90°$. By the IEEE definition [98], for a left-hand circularly polarized (LHCP) wave approaching, the relative phase will be $\delta_p = 90°$ and the tip of the electric field vector will rotate clockwise as in the middle-left diagram in Figure 1.33. For a right-hand circularly polarized (RHCP) wave approaching, $\delta_p = -90°$, and the tip will rotate counter-clockwise as in the middle-right diagram in Figure 1.33.

In the case where E_x and E_y are known, the left-hand and right-hand circular polarization fields can be computed for a wave traveling in the $\hat{z}$ direction as [96, p. 132]

$$E_{\text{LHCP}} = \frac{1}{\sqrt{2}}(E_x - jE_y) \tag{1.523}$$

$$E_{\text{RHCP}} = \frac{1}{\sqrt{2}}(E_x + jE_y) \tag{1.524}$$

In the case where horizontally (H) and vertically (V) polarized fields are known, the left-hand and right-hand circular polarization fields can be computed for a wave traveling in the positive range direction as [97, p. 87]

$$E_{\text{LHCP}} = \frac{1}{\sqrt{2}}(E_H - jE_V) \tag{1.525}$$

$$E_{\text{RHCP}} = \frac{1}{\sqrt{2}}(E_H + jE_V) \tag{1.526}$$

In the case where E_θ and E_ϕ are known, the left-hand and right-hand circular polarization fields can be computed for a wave traveling in the $\hat{r}$ direction as

$$E_{\text{LHCP}} = \frac{1}{\sqrt{2}}(E_\theta - jE_\phi) \tag{1.527}$$

$$E_{\text{RHCP}} = \frac{1}{\sqrt{2}}(E_\theta + jE_\phi) \tag{1.528}$$

For elliptically polarized (EP) waves, the amplitudes E_1 and E_2 will not be equal. If the relative phase is $\delta_p = 90°$, then the wave will be left-hand elliptically polarized (LHEP) with the major axis along the x-axis (lower left diagram in Figure 1.33). If the relative phase is $\delta_p = -90°$, then the wave will be right-hand elliptically polarized (RHEP) with the major axis also along the x-axis (lower right diagram in Figure 1.33). For general elliptically polarized waves, the relative phase δ_p will vary over $\pm 180°$. The elliptical polarization will be right-handed for $-180° < \delta_p < 0$ and will be left-handed for $0 < \delta_p < 180°$.

For the general case of an elliptically polarized wave [95], the polarization ellipse major axis will tilt away at the angle τ_p from the x axis as depicted in Figure 1.34(a). In this diagram, the ratio of the major axis to the minor axis is defined as the axial ratio (AR), that is,

$$\mathrm{AR} = \pm\frac{\mathrm{OA}}{\mathrm{OB}}. \tag{1.529}$$

The sign of the axial ratio is positive for right-hand polarization and negative for left-hand polarization [106, p. 38]. The magnitude of the axial ratio varies from 1 to infinity, that is, $1 \leq |\mathrm{AR}| \leq \infty$. The polarization ellipticity angle, denoted ϵ_p, is expressed as

$$\epsilon_p = \cot^{-1}(-\mathrm{AR}). \tag{1.530}$$

In decibels, the axial ratio is expressed as

$$\mathrm{AR}_{\mathrm{dB}} = 20\log_{10}|\mathrm{AR}|. \tag{1.531}$$

The polarization state of a general electromagnetic wave can be completely specified by either the pair of angles (ϵ_p, τ_p) or (γ_p, δ_p). The transformation of the angles (ϵ_p, τ_p) to (γ_p, δ_p) is given by

$$\gamma_p = \frac{1}{2}\cos^{-1}(\cos 2\epsilon_p \cos 2\tau_p) \tag{1.532}$$

$$\delta_p = \tan^{-1}\frac{\tan 2\epsilon_p}{\sin 2\tau_p}. \tag{1.533}$$

In Equation (1.533), the angle $2\tau_p$ represents longitude and $2\epsilon_p$ represents latitude for a point on a sphere known as the Poincaré sphere [97, pp. 36–38] as depicted for one quadrant in Figure 1.34(b). In the Poincaré sphere, linear polarization is depicted along the equator. Left-hand circular polarization is located at the north pole. Right-hand circular polarization (not shown) would be located at the south pole. Elliptical polarization is located, in general, between the equator and the poles.

Consider now the polarization efficiency (or antenna-wave coupling) between transmit and receive linearly polarized antennas as depicted in Figure 1.35. The transmit antenna produces the incident linear polarization denoted $\hat{\boldsymbol{p}}_{\mathrm{inc}}$. The polarization efficiency η_p is expressed as the magnitude squared of the dot product of the unit vectors for the incident linear polarization and complex conjugate of the receiving antenna's unit vector polarization defined under transmitting conditions (denoted $\hat{\boldsymbol{p}}_{\mathrm{a}}$), that is,

$$\eta_p = |\hat{\boldsymbol{p}}_{\mathrm{inc}}\cdot\hat{\boldsymbol{p}}_{\mathrm{a}}^*|^2 = \cos^2\psi \tag{1.534}$$

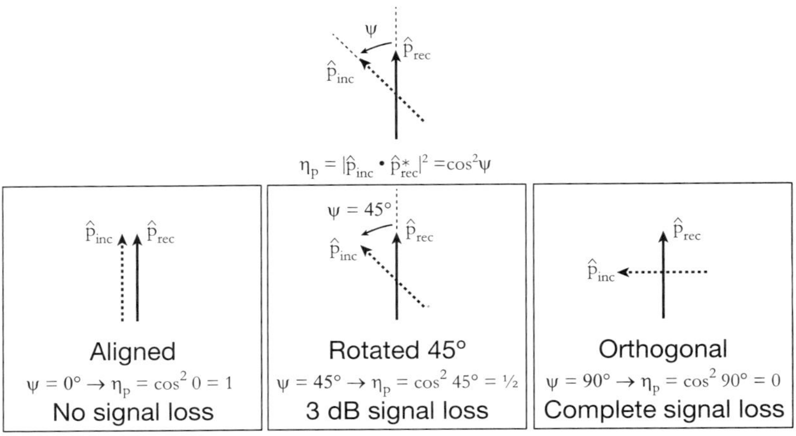

Figure 1.35 Depiction of polarization efficiency for an incident wave and a receive antenna.

where ψ is the projected angle between the associated transmit and receive linear-polarized antenna unit vectors, and the complex conjugate is required to take account of the relative coordinate systems for the incident field and the receive antenna (the propagation directions are opposite). That is, the polarization of the receive antenna $\hat{\boldsymbol{p}}_{\text{rec}}$ is the complex conjugate of the receive antenna's polarization when it is used to transmit, that is,

$$\hat{\boldsymbol{p}}_{\text{rec}} = \hat{\boldsymbol{p}}_{\text{a}}^{*} \tag{1.535}$$

For linearly polarized antennas the conjugate operation in Equation (1.535) is not used since the polarization unit vector is real. In the case of circular polarization the conjugate operation must used as can be seen in the following equations for the polarization of a transmit antenna. For the case where the antenna in Cartesian coordinates is transmitting a wave along the positive z axis the transmit antenna LHCP, RHCP polarization unit vectors are complex and given by

$$\hat{\boldsymbol{p}}_{\text{LHCP}}^{t} = \frac{\hat{\boldsymbol{x}} + j\hat{\boldsymbol{y}}}{\sqrt{2}} \tag{1.536}$$

$$\hat{\boldsymbol{p}}_{\text{RHCP}}^{t} = \frac{\hat{\boldsymbol{x}} - j\hat{\boldsymbol{y}}}{\sqrt{2}} \tag{1.537}$$

Similarly, if the antenna is now analyzed in spherical coordinates and it is transmitting a circularly polarized wave along the outgoing radius r the transmit antenna polarization unit vectors are complex and given by [25, p.

79]

$$\hat{\boldsymbol{p}}_{\text{LHCP}}^{t} = \frac{\hat{\boldsymbol{\theta}} + j\hat{\boldsymbol{\phi}}}{\sqrt{2}} \tag{1.538}$$

$$\hat{\boldsymbol{p}}_{\text{RHCP}}^{t} = \frac{\hat{\boldsymbol{\theta}} - j\hat{\boldsymbol{\phi}}}{\sqrt{2}} \tag{1.539}$$

An incoming wave traveling in the $-\hat{\boldsymbol{r}}$ direction for LHCP and RHCP will have the following wave polarization unit vectors

$$\hat{\boldsymbol{p}}_{\text{LHCP}}^{\text{wave}} = \frac{\hat{\boldsymbol{\theta}} - j\hat{\boldsymbol{\phi}}}{\sqrt{2}} \tag{1.540}$$

$$\hat{\boldsymbol{p}}_{\text{RHCP}}^{\text{wave}} = \frac{\hat{\boldsymbol{\theta}} + j\hat{\boldsymbol{\phi}}}{\sqrt{2}} \tag{1.541}$$

When two linearly polarized antennas are polarization aligned (co-polarized), as in the left diagram in Figure 1.34, the angle ψ is zero, the polarization efficiency is 100% ($\eta_p = 1$), and there is no signal loss. When the antennas are rotated 45° with respect to each other (middle diagram in Figure 1.35), the polarization efficiency is 0.5, and there is a 3 dB signal loss. For the case in which the linear polarized antennas are orthogonal (right diagram in Figure 1.35), there is complete signal loss. For the case where a circularly polarized wave is incident on a linearly polarized antenna, the signal loss is 3 dB. More general cases of polarization efficiency between antennas with circular and elliptical polarizations are considered by Stutzman [97]. It should be mentioned here that, in general, electromagnetic waves can consist of completely polarized, partially polarized, and unpolarized waves. Completely polarized waves will have single or orthogonal electric-field components that are constant with respect to time in terms of amplitude and relative phase shift. The amplitude and phase of orthogonal electric field components will vary with time for partially polarized waves.

For example, partially polarized and unpolarized waves can occur from celestial sources. To quantify the general polarization case, Stokes parameters (I, Q, U, V) are often used [97, pp. 43–46, 107-115]. A partially polarized wave will consist, mathematically, of a completely polarized wave and a completely unpolarized wave.

Now, define time average over an interval of time T as as

$$< \cdot >= \frac{1}{T}\int_0^T \cdot \, dt \tag{1.542}$$

For partially polarized waves, the orthogonal electric field components E_x and E_y are random variables in terms of amplitude and phase, that is,

$$E_x(t) = E_1(t)e^{j(\omega t+\Phi_1(t))} \tag{1.543}$$

$$E_y(t) = E_2(t)e^{j(\omega t+\Phi_2(t))} \tag{1.544}$$

and the Stokes parameters are expressed in terms of wave power densities as

$$I = S_{\text{ave}} = \frac{< |E_x(t)|^2 >}{2\eta} + \frac{< |E_y(t)|^2 >}{2\eta} =< S_x(t) > + < S_y(t) > \tag{1.545}$$

$$Q = \frac{< |E_x(t)|^2 >}{2\eta} - \frac{< |E_y(t)|^2 >}{2\eta} =< S_x(t) > - < S_y(t) > \tag{1.546}$$

or

$$Q = S_{\text{ave}} < \cos 2\epsilon_p \cos 2\tau_p > \tag{1.547}$$

$$U = \frac{1}{2\eta} 2 \text{ Re} < E_x E_y^* >= S_{\text{ave}} < \cos 2\epsilon_p \sin 2\tau_p > \tag{1.548}$$

$$V = \frac{1}{2\eta} 2 \text{ Im} < E_x E_y^* >= S_{\text{ave}} < \sin 2\epsilon_p > \tag{1.549}$$

If the wave is completely polarized then the time averaging is not required and the Stokes parameters reduce to

$$I = S = \frac{|E_x(t)|^2}{2\eta} + \frac{|E_y(t)|^2}{2\eta} = S_x(t) + S_y(t) \tag{1.550}$$

$$Q = \frac{|E_x(t)|^2}{2\eta} - \frac{|E_y(t)|^2}{2\eta} = S_x(t) - S_y(t) = S \cos 2\epsilon_p \cos 2\tau_p \tag{1.551}$$

$$U = S \cos 2\epsilon_p \sin 2\tau_p \tag{1.552}$$

$$V = S \sin 2\epsilon_p \tag{1.553}$$

The Stokes parameters can be normalized by the wave intensity S and then be expressed as a column vector as

$$\boldsymbol{s} = \begin{bmatrix} s_o \\ s_1 \\ s_2 \\ s_3 \end{bmatrix} = \begin{bmatrix} 1 \\ \cos 2\epsilon_p \cos 2\tau_p \\ \cos 2\epsilon_p \sin 2\tau_p \\ \sin 2\epsilon_p \end{bmatrix} \tag{1.554}$$

The normalized Stokes parameters represent projections into the Poincaré sphere's rectangular coordinate system. The Stokes parameters for horizontal, vertical and slant polarizations are:

$$\text{HPol:} \begin{bmatrix} 1 \\ 1 \\ 0 \\ 0 \end{bmatrix} \quad \text{VPol:} \begin{bmatrix} 1 \\ -1 \\ 0 \\ 0 \end{bmatrix} \quad \text{S}(45^\circ): \begin{bmatrix} 1 \\ 0 \\ 1 \\ 0 \end{bmatrix} \quad \text{S}(-45^\circ): \begin{bmatrix} 1 \\ 0 \\ -1 \\ 0 \end{bmatrix} \tag{1.555}$$

for circular polarization:

$$\text{LHCP:} \begin{bmatrix} 1 \\ 0 \\ 0 \\ 1 \end{bmatrix} \quad \text{RHCP:} \begin{bmatrix} 1 \\ 0 \\ 0 \\ -1 \end{bmatrix} \tag{1.556}$$

and for unpolarized waves:

$$\text{Stokes unpolarized:} \begin{bmatrix} 1 \\ 0 \\ 0 \\ 0 \end{bmatrix} \tag{1.557}$$

It is useful to determine the open-circuit terminal voltage (denoted V_{oc}) induced when a polarized electromagnetic wave impinges on a receiving antenna. By using a Thévinen equivalent circuit, the open-circuit induced voltage can then be used as a generator together with the antenna input impedance $Z_A = R_A + jX_A$ to determine the current I flowing through a load impedance $Z_L = R_L + jX_L$ at a receiver. That is,

$$I = \frac{V_{\text{oc}}}{Z_A + Z_L}. \tag{1.558}$$

Note that the real part of the antenna input impedance is equal to the antenna radiation resistance (denoted R_r), that is, $R_A = R_r$. The time-average power P_L delivered to the receiver, in terms of the resistive component R_L of the receiver load impedance, is given by

$$P_L = \frac{1}{2}|I|^2 R_L = \frac{1}{2}|V_{\text{oc}}|^2 \frac{R_L}{|Z_A + Z_L|^2} \tag{1.559}$$

and the complex voltage across the receiver load is given by

$$V_{\text{rec}} = V_{\text{oc}} \frac{Z_L}{Z_A + Z_L}. \tag{1.560}$$

Utilizing the generalized antenna vector effective height ($\boldsymbol{h}$), which has units of meters, the induced open-circuit voltage across the terminals of an antenna is given by [97, pp. 137–144]

$$V_{\text{oc}} = \boldsymbol{E}_{\text{inc}} \cdot \boldsymbol{h}^* \tag{1.561}$$

where

$$\boldsymbol{h} = h\hat{\boldsymbol{p}}_{\text{a}} \tag{1.562}$$

and where h is the magnitude of the vector effective height. Now express the incident electric field $\boldsymbol{E}_{\text{inc}}$ with magnitiude E_o, phase ψ_o, with polarization $\hat{\boldsymbol{p}}_{\text{inc}}$ as

$$\boldsymbol{E}_{\text{inc}} = E_o e^{\psi_o} \hat{\boldsymbol{p}}_{\text{inc}}. \tag{1.563}$$

An example of the induced open-circuit voltage given by Equation (1.561) will now be considered. Assume 45° slant-left and 45° slant-right linearly-polarized short dipoles oriented in the yz-plane such that

$$\hat{\boldsymbol{p}}_{\text{slant left}} = \frac{-\hat{\boldsymbol{y}} + \hat{\boldsymbol{z}}}{\sqrt{2}} \tag{1.564}$$

$$\hat{\boldsymbol{p}}_{\text{slant right}} = \frac{\hat{\boldsymbol{y}} + \hat{\boldsymbol{z}}}{\sqrt{2}} \tag{1.565}$$

Assume plane wave incidence at the arbitrary position (x, y, z) with vertical and horizontal polarization, repectively, as

$$\boldsymbol{E}^{v}_{\text{inc}} = \hat{\boldsymbol{\theta}} E^{i}_{\theta}(x, y, z) \tag{1.566}$$

$$\boldsymbol{E}^{h}_{\text{inc}} = \hat{\boldsymbol{\phi}} E^{i}_{\phi}(x, y, z) \tag{1.567}$$

Substituting Equations (1.564), (1.565), (1.566), and (1.567) in Equation (1.561) yields the induced open-circuit voltages

$$V^{v}_{\text{slant left}} = E^{i}_{\theta}(x, y, z)\hat{\boldsymbol{\theta}} \cdot \left[\frac{-\hat{\boldsymbol{y}} + \hat{\boldsymbol{z}}}{\sqrt{2}}\right] \tag{1.568}$$

$$V^{v}_{\text{slant right}} = E^{i}_{\theta}(x, y, z)\hat{\boldsymbol{\theta}} \cdot \left[\frac{\hat{\boldsymbol{y}} + \hat{\boldsymbol{z}}}{\sqrt{2}}\right] \tag{1.569}$$

$$V^{h}_{\text{slant left}} = E^{i}_{\phi}(x, y, z)\hat{\boldsymbol{\phi}} \cdot \left[\frac{-\hat{\boldsymbol{y}} + \hat{\boldsymbol{z}}}{\sqrt{2}}\right] \tag{1.570}$$

$$V^{h}_{\text{slant right}} = E^{i}_{\phi}(x, y, z)\hat{\boldsymbol{\phi}} \cdot \left[\frac{\hat{\boldsymbol{y}} + \hat{\boldsymbol{z}}}{\sqrt{2}}\right] \tag{1.571}$$

From Equations (1.241) and (1.242) and referring to Figure 1.21 it follows for source incidence angles $\theta = \theta^i$ and $\phi = \phi^i$ that

$$\hat{\boldsymbol{\theta}} \cdot \hat{\boldsymbol{y}} = \cos\theta^i \sin\phi^i \tag{1.572}$$

$$\hat{\boldsymbol{\theta}} \cdot \hat{\boldsymbol{z}} = -\sin\theta^i \tag{1.573}$$

$$\hat{\boldsymbol{\phi}} \cdot \hat{\boldsymbol{y}} = \cos\phi^i \tag{1.574}$$

$$\hat{\boldsymbol{\phi}} \cdot \hat{\boldsymbol{z}} = 0 \tag{1.575}$$

It follows then that the induced open-circuit voltages are

$$V^v_{\text{slant left}} = -E^i_\theta(x, y, z)\left[\frac{\cos\theta^i \sin\phi^i + \sin\theta^i}{\sqrt{2}}\right] \tag{1.576}$$

$$V^v_{\text{slant right}} = -E^i_\theta(x, y, z)\left[\frac{\cos\theta^i \sin\phi^i - \sin\theta^i}{\sqrt{2}}\right] \tag{1.577}$$

$$V^h_{\text{slant left}} = -E^i_\phi(x, y, z)\frac{\cos\phi^i}{\sqrt{2}} \tag{1.578}$$

$$V^h_{\text{slant right}} = E^i_\phi(x, y, z)\frac{\cos\phi^i}{\sqrt{2}} \tag{1.579}$$

For plane-wave incidence with wavenumber $k = 2\pi/\lambda$ at an arbitrary nth position (x_n, y_n, z_n), the electric field has a constant amplitude and a phase Ψ^i that varies with angle of incidence as

$$\Psi^i = k\sin\theta^i(x_n\cos\phi^i + y_n\sin\phi^i) + kz_n\cos\theta^i \tag{1.580}$$

and the complex electric field is then expressed as

$$E^i(k, x_n, y_n, z_n, \theta^i, \phi^i) = E_o e^{jk[\sin\theta^i(x_n\cos\phi^i + y_n\sin\phi^i) + z_n\cos\theta^i]} \tag{1.581}$$

Next, it is desired to determine the power transferred to the receiver load from the induced open-circuit voltage. It follows from Equations (1.534) and (1.561) that

$$|V_{\text{oc}}| = E_o h|\hat{\boldsymbol{p}}_{\text{inc}} \cdot \hat{\boldsymbol{p}}^*_{\text{a}}| = E_o h\sqrt{\eta_p}. \tag{1.582}$$

Substituting Equation (1.582) in Equation (1.559) yields

$$P_L = \frac{1}{2}|V_{\text{oc}}|^2 \frac{R_L}{|Z_A + Z_L|^2} = \frac{1}{2}E_o^2 h^2 \eta_p \frac{R_L}{|Z_A + Z_L|^2}. \tag{1.583}$$

In the case where the receive antenna is connected to a conjugate-matched load impedance ($Z_L = R_A - jX_A$), it follows from Equation (1.583) that the time-average power transferred to the resistive part (R_A) of the receiver load is given by

$$P_{\mathrm{L}} = \frac{1}{2}\frac{|V_{\mathrm{oc}}|^2}{4R_A} = \frac{1}{8R_A}E_o^2 h^2 \eta_p. \tag{1.584}$$

Now the magnitude of the time-average power density (denoted S) incident on an antenna in free space is given by (refer to Equation (1.412))

$$S = \frac{1}{2}\frac{E_o^2}{\eta_o} \tag{1.585}$$

where η_o is the impedance of free space. Next, in terms of effective receive aperture (denoted A_{eff}), the time-average received power at the conjugate-matched load can be expressed as

$$P_{\mathrm{L}} = \eta_p S A_{\mathrm{eff}} = \eta_p \frac{1}{2}\frac{E_o^2}{\eta_o} A_{\mathrm{eff}}. \tag{1.586}$$

Now, setting Equation (1.584) equal to Equation (1.586) yields

$$h = 2\sqrt{\frac{R_A}{\eta_o} A_{\mathrm{eff}}} \tag{1.587}$$

which is a general expression for the effective height of an antenna assuming a conjugate matched load. Solving Equation (1.587) for the effective aperture gives

$$A_{\mathrm{eff}} = \frac{h^2 \eta_o}{4R_A}. \tag{1.588}$$

Note, in the literature, the electric antenna factor (denoted AF_e) with units of 1/m is often used to convert the incident electric field magnitude (E_o) to received voltage magnitude, that is,

$$|V_{\mathrm{rec}}| = \frac{E_o}{\mathrm{AF}_e}, \tag{1.589}$$

and it follows that the electric antenna factor and effective antenna height are related as

$$\mathrm{AF}_e = \frac{1}{h}. \tag{1.590}$$

Additionally, in the literature, the magnetic antenna factor (denoted AF_m) with units of 1/m is often used to convert the incident magnetic field magnitude (H_o) to received voltage magnitude, that is,

$$|V_{\mathrm{rec}}| = \frac{H_o}{\mathrm{AF}_m}, \tag{1.591}$$

and it follows that the magnetic antenna factor and magnetic effective antenna height h_m are related as

$$\mathrm{AF}_m = \frac{1}{h_m}. \tag{1.592}$$

To quantify the maximum effective area given by Equation (1.588), it is necessary to determine the radiation resistance R_r, or equivalently the input resistance R_A, for the antenna of interest, as discussed below for the short dipole and small loop antenna. Since an antenna can be treated as a load, the radiation resistance can be determined by quantifying the time-average radiated power (denoted P_{ave}) or equivalently the time-average power delivered to the antenna, and using the relation

$$P_{\text{rad}} = \frac{1}{2} I_o^2 R_r \tag{1.593}$$

and solving for R_r as

$$R_r = \frac{2P_{\text{rad}}}{I_o^2}. \tag{1.594}$$

Now the time-average radiated power can be found by integrating the dot product of the radial unit vector with the time-average power density over the volume surrounding the antenna as the range distance tends to infinity, that is,

$$P_{\text{rad}} = \int_0^{2\pi} \int_0^{\pi} \hat{\boldsymbol{r}} \cdot \frac{1}{2} \mathrm{Re}[\boldsymbol{E} \times \boldsymbol{H}^*] r^2 \sin\theta d\theta d\phi. \tag{1.595}$$

Repeating the general far-field Equations (1.283) and (1.286), we have

$$\boldsymbol{E}_{\text{FF}} = -j\omega\mu \frac{e^{-jkr}}{4\pi r} (\hat{\boldsymbol{\theta}} f_\theta + \hat{\boldsymbol{\phi}} f_\phi) \tag{1.596}$$

$$\boldsymbol{H}_{\text{FF}} = -jk \frac{e^{-jkr}}{4\pi r} (\hat{\boldsymbol{\phi}} f_\theta - \hat{\boldsymbol{\theta}} f_\phi). \tag{1.597}$$

To compute the far-zone electric ($\boldsymbol{E}_{\text{FF}}$) and magnetic ($\boldsymbol{H}_{\text{FF}}$) fields, it is necessary to determine the vector current moment $\boldsymbol{f}(\theta, \phi)$ for the antenna of interest.

1.10.2 Short Dipole Receive Characteristics

For an electrically short dipole with total length l in free space, the current distribution has a triangular shape [23]. The maximum effective height is determined by integrating the current over the dipole physical length divided by the terminal current I_o, and for symmetrical currents,

$$h = \frac{1}{I_o}\int_{-\frac{l}{2}}^{\frac{l}{2}} I(z')dz' = \frac{2}{I_o}\int_{0}^{\frac{l}{2}} I(z')dz'. \tag{1.598}$$

A triangular current distribution is a good approximation for short dipoles of total length l, that is,

$$I(z') = \frac{2I_o}{l}\left[\frac{l}{2} - |z'|\right]. \tag{1.599}$$

Substituting Equation (1.599) in Equation (1.598) and integrating, the effective height is found to be equal to one-half the physical length of the short dipole with triangular current, that is,

$$h_{\text{dipole}} = \frac{1}{2}l. \tag{1.600}$$

Thus, the open-circuit voltage induced in a short dipole is determined by substituting Equation (1.600) into Equation (1.582) with the result (ignoring the radiation pattern)

$$|V_{\text{dipole}}| = \frac{1}{2}E_o l\sqrt{\eta_p}. \tag{1.601}$$

For the short dipole, the far-zone electric and magnetic fields can be determined from the vector current moment given by Equation (1.273), and using Equation (1.599) for the triangular current distribution, it follows that

$$\boldsymbol{J}(\boldsymbol{r}') = \hat{\boldsymbol{z}}I(z')\delta(x')\delta(y') = \hat{\boldsymbol{z}}\frac{2I_o}{l}\left[\frac{l}{2} - |z'|\right]\delta(x')\delta(y'). \tag{1.602}$$

Now, using Equations (1.240) and (1.245), it follows from Equation (1.273) that for the triangular current distribution that the vector current moment is

$$\boldsymbol{f}(\theta,\phi) = \int_V \boldsymbol{J}(\boldsymbol{r}')e^{jk\hat{\boldsymbol{r}}\cdot\boldsymbol{r}'}dv' = -\hat{\boldsymbol{\theta}}\frac{2I_o}{l}\sin\theta\int_{-\frac{l}{2}}^{\frac{l}{2}}\left[\frac{l}{2} - |z'|\right]e^{jkz'\cos\theta}dz' \tag{1.603}$$

where the only nonradial component in this case is the $\hat{\boldsymbol{\theta}}$ component. For an electrically short dipole, the phase term in Equation (1.603) can be ignored, and thus the vector current moment for the triangular current distribution is determined to be

$$f_\theta(\theta) = -\frac{l}{2}I_o\sin\theta. \tag{1.604}$$

Using the above expressions given by Equations (1.595), (2.35), (1.597), and (1.604) the radiated power is now determined as

$$P_{\text{rad}} = 2\pi \left[\frac{\eta k^2}{(4\pi)^2} \right] \frac{1}{2} \int_0^{\pi} |f_\theta|^2 \sin\theta d\theta = \pi \left[\frac{120\pi k^2}{(4\pi)^2} \right] \frac{l^2 I_o^2}{4} \int_0^{\pi} \sin^3\theta d\theta \tag{1.605}$$

and noting that

$$\int_0^{\pi} \sin^3\theta d\theta = \frac{4}{3}, \tag{1.606}$$

it follows that the power radiated in watts is

$$P_{\text{rad}}^{\text{dipole, triangular current}} = \frac{5}{2}(kl)^2 I_o^2. \tag{1.607}$$

Thus, the radiation resistance of a short dipole with triangular current distribution is

$$R_r^{\text{dipole, triangular current}} = \frac{2P_{\text{rad}}}{I_o^2} = 5(kl)^2. \tag{1.608}$$

Substituting Equations (1.600) and (1.608) in Equation (1.588) yields the effective area of a short dipole with triangular current as

$$A_{\text{eff}}^{\text{dipole, triangular current}} = \frac{h^2\eta_o}{4R_A} = \frac{\frac{l^2}{4}120\pi}{4\cdot 5(kl)^2} = \frac{3\lambda^2}{8\pi} \approx \frac{\lambda^2}{8}. \tag{1.609}$$

1.10.3 Small Current Loop Receive Characteristics

Consider now an N-turn loop antenna of arbitrary shape with physical area A illuminated by a uniform magnetic flux density with magnitude $B_o = \mu_o H_o$, where μ_o is the permeability of free space and H_o is the magnitude of the magnetic field. The magnitude of the open-circuit voltage is readily determined, according to the time rate of change of magnetic flux density through the loop, using Faraday's law in time-harmonic form as

$$|V_{\text{loop}}| = |-j\omega A N B_o \sqrt{\eta_p}| = \omega A N \mu_o H_o \sqrt{\eta_p}. \tag{1.610}$$

For a plane wave, $H_o = E_o/\eta_o$, where η_o is the impedance of free space, and noting $\omega\mu_o/\eta_o = k = 2\pi/\lambda$, it follows that the open-circuit voltage of the loop is

$$|V_{\text{loop}}| = AN\frac{\omega\mu_o}{\eta_o}E_o\sqrt{\eta_p} = ANkE_o\sqrt{\eta_p} \tag{1.611}$$

and so setting Equation (1.611) equal to Equation (1.582), the maximum effective height of an N-turn loop antenna with physical area A is

$$h_{\text{loop}} = ANk. \tag{1.612}$$

In the case of an N-turn small loop antenna with uniform current, from Equations (1.495) and (1.496), the far-zone fields are

$$\boldsymbol{H}^{FF}_{\text{z-loop, N turns}}(\boldsymbol{r}) = -\hat{\boldsymbol{\theta}} k^2 NIA \frac{e^{-jkr}}{4\pi r} \sin\theta \tag{1.613}$$

$$\boldsymbol{E}^{FF}_{\text{z-loop, N turns}}(\boldsymbol{r}) = \hat{\boldsymbol{\phi}} NIA\omega\mu k \frac{e^{-jkr}}{4\pi r} \sin\theta. \tag{1.614}$$

Substituting Equations (5.12) and (5.11) in Equation (1.595) and evaluating the integral yields the radiated power

$$P_{\text{rad}}^{\text{loop, uniform current}} = 10(k^2 NA)^2 I_o^2. \tag{1.615}$$

Thus, the radiation resistance of an N-turn loop antenna with uniform current distribution and physical area A is

$$R_r^{\text{loop, uniform current}} = \frac{2P_{\text{rad}}}{I_o^2} = 20(k^2 NA)^2. \tag{1.616}$$

Substituting Equations (1.622) and (1.616) in Equation (1.588) yields the effective area of an N-turn small loop antenna with uniform current as

$$A_{\text{eff}}^{\text{loop, uniform current}} = \frac{h^2 \eta_o}{4R_A} = \frac{(ANk)^2 120\pi}{4 \cdot 20(k^2 NA)^2} = \frac{3\lambda^2}{8\pi} \approx \frac{\lambda^2}{8}, \tag{1.617}$$

and this result is identical to the result obtained for the short dipole (see Equation (1.609).

1.10.4 Ferrite-Loaded Small Current Loop Receive Characteristics

Consider now a ferrite-loaded N-turn loop antenna [20] with physical area A illuminated by a uniform magnetic flux density with magnitude $B_o = \mu_o H_o$, where μ_o is the permeability of free space and H_o is the magnitude of the magnetic field in free space. The N-turn loop is assumed to be wound around a uniform ferrite core with relative intrinsic permeability of the unbounded ferrite material denoted as $\mu_{ri\ \text{ferrite core}}$. The relative effective permeability of the ferrite core is denoted as $\mu_{re\ \text{ferrite core}}$, which is related to $\mu_{ri\ \text{ferrite core}}$ as

$$\mu_{re\ \text{ferrite core}} = \frac{\mu_{ri\ \text{ferrite core}}}{1 + D(\mu_{ri\ \text{ferrite core}} - 1)} \tag{1.618}$$

where D is the static demagnetization factor. The static demagnetization factor is geometry dependent, and for an ellipsoidal core with length L and radius a (where $L >> a$), D is given by

$$D = \left(\frac{2a}{L}\right)^2 \left[\ln\left(\frac{L}{a}\right) - 1\right] \tag{1.619}$$

The presence of the ferrite material increases the amplitude of the incident magnetic field and open-circuit voltage linearly by the relative effective permeability of the ferrite core as [20]

$$H_{o \text{ ferrite core}} = \mu_{re \text{ ferrite core}} H_o \tag{1.620}$$

and from Equation (1.611)

$$\left|V_{\text{loop}}\right| = ANk\mu_{re \text{ ferrite core}} E_o \sqrt{\eta_p} \tag{1.621}$$

and so setting Equation (1.621) equal to Equation (1.582), the maximum effective height of an N-turn ferrite loop antenna with physical area A is

$$h_{\text{loop}} = ANk\mu_{re \text{ ferrite core}}. \tag{1.622}$$

The radiation resistance of an N-turn ferrite-loaded loop antenna, with uniform current distribution and physical area A, increases by the square of the relative effective permeability of the ferrite and is given by

$$R_r^{\text{ferrite loop, uniform current}} = \frac{2P_{\text{rad}}}{I_o^2} = 20(k^2NA)^2\mu_{re \text{ ferrite core}}^2 \tag{1.623}$$

or in terms of loop circumference C using $A = C^2/(4\pi)$ as

$$R_r^{\text{ferrite loop, uniform current}} = 20\pi^2 \left(\frac{C}{\lambda}\right)^4 N^2 \mu_{re \text{ ferrite core}}^2. \tag{1.624}$$

1.11 Antenna Bandwidth and Quality Factor

1.11.1 Introduction

Bandwidth (B) is defined as the difference between the highest f_2 and lowest f_1 operating frequencies, that is,

$$B = f_2 - f_1 = \Delta f. \tag{1.625}$$

The terms highest and lowest operating frequencies usually refer to the radar or communications system meeting some desired performance metric, such as radiation pattern, gain, beamwidth, sidelobe level, reflection coefficient or VSWR, transmit power level, sensitivity, and so on. In specifying a bandwidth value, sometimes the -3 dB point is used, and sometimes a different level, say the -10 dB point, is used; hence, it is important to define what is meant by bandwidth. Fractional bandwidth (FBW) is defined as the ratio of the bandwidth to the arithmetic mean center frequency f_c, that is,

$$\text{FBW} = \frac{f_2 - f_1}{f_c} = \frac{\Delta f}{f_c}. \tag{1.626}$$

The percent bandwidth (PBW) is computed as

$$\text{PBW} = \text{FBW} \times 100\% = \frac{\Delta f}{f_c} \times 100\%. \tag{1.627}$$

The ratio bandwidth (RBW) is defined as the ratio of the highest operating frequency to the lowest operating frequency, or

$$\text{RBW} = \frac{f_2}{f_1}. \tag{1.628}$$

Antennas that have ultrawideband (or wideband) characteristics have numerous applications [99–112]. To quantify the bandwidth, consider an ultrawideband system that requires an antenna to operate from 30 MHz to 3,000 MHz. It follows that the bandwidth $B = 2,700$ MHz, the FBW = $(3000 - 30)/1515 = 1.96$, the PBW = 196%, and the RBW = 3000 / 30 = 100. In the limit as the bandwidth grows linearly, the FBW approaches 2, the PBW approaches 200%, and the RBW grows linearly. Each of the bandwidth definitions are useful and are applied as appropriate to the desired application.

Bandwidth is important in radar systems requiring fine range resolution. The time t it takes for an electromagnetic wave to travel a range distance r between two points in free space is fundamentally related to the speed of light c (measured value $c = 2.99792458 \times 10^8$ m/s $\approx 3.0 \times 10^8$ m/s) as

$$t = \frac{r}{c}. \tag{1.629}$$

The wavelength of an electromagnetic wave in free space is expressed as

$$\lambda = \frac{c}{f} \tag{1.630}$$

where f is the frequency of the wave. In a radar system application, the range resolution Δr in free space is given in terms of the radar pulse width τ as [47]

$$\Delta r = \frac{c\tau}{2}. \tag{1.631}$$

Since the bandwidth of the radar pulse is inversely proportional to the pulse width, that is,

$$B = \frac{1}{\tau} \tag{1.632}$$

it follows that the range resolution can also be expressed as

$$\Delta r = \frac{c}{2B}. \tag{1.633}$$

For example, a 1 μs pulse width has a bandwidth of 1 MHz; thus, the range resolution is computed from Equation (1.633) to be 150m. Similarly, a 10 ns pulse width has a bandwidth of 100 MHz, which corresponds to a range resolution of 1.5m. In the case of an ultrawideband pulse with a bandwidth on the order of gigahertz, the range resolution is on the order of centimeters and Equation (1.633) can be written conveniently as

$$\Delta r\text{cm} = \frac{30}{2B_{\text{GHz}}}. \tag{1.634}$$

Thus, a radar with a pulse width of 1 ns, which has a bandwidth of 1 GHz, will have a range resolution of

$$\Delta r = \frac{30 \text{ cm}}{2} = 15 \text{ cm} \approx 6 \text{ inches}. \tag{1.635}$$

The terms ultrawideband [99] or wideband are used to refer to an antenna that covers a large frequency range. As there is no standard definition for ultrawideband antennas, we simply define that an ultrawideband antenna will have a PBW typically 25% or more or a RBW of 1.25 or more as described in the 2007 *IEEE Standard for Ultrawideband Radar Definitions* [99]. Over this bandwidth, the antenna will have desired radiation characteristics and will not have any significant amplitude degradation for the application of interest.

Ultrawideband array performance requirements are highly dependent on the application [100]. For example, in a receive-only application with a strong signal, it might only be necessary to provide an antenna with a reflection coefficient on the order of −6 dB or better with sufficient pattern gain to close the link, but for a transmit application, the reflection coefficient requirement might be on the order of −10 dB or better to avoid significant reflected signal toward the transmitter and significant loss in effective isotropic radiated power. For ground-penetrating radar or impulse radar applications [100, 101], since targets are at very close range (1 to 2m), it is important to design the ultrawideband antenna with sufficient resistive attenuation to avoid signal ringing effects [102–103]. A wide variety of ultrawideband or frequency independent antennas are addressed in the literature [104–110].

The above definitions of bandwidth do not take into account the size of the antenna aperture, which affects the time delay for transmitted and received signals. For a transmitted or incident wavefront with either a pulsed, noise, or other wideband spectrum, the antenna design and analysis must take into account the time delay as will be described for phased arrays in the next chapter.

A quantity that is related to antenna bandwidth is the quality factor or Q, which has been investigated for electrically small antennas by a number of researchers including Chu [111], Wheeler [112], Harrington [113], Hansen [114], McLean [115], Balanis [25, pp. 637–641], and Yaghjian and Best [116]. For an electrically small antenna with maximum radius a, the minimum possible Q has been derived by McLean [115] as

$$Q_{\text{min}} = \frac{1}{k^3 a^3} + \frac{1}{ka} \tag{1.636}$$

where $k = 2\pi/\lambda$ is the wavenumber. The next section quantifies bandwidth in terms of the antenna input impedance and quality factor Q.

1.11.2 Derivation of Q Factor from Input Impedance

The impedance bandwidth for an allowed value of VSWR of an antenna can be computed from a quantity known as the quality factor that depends on the reactance and radian frequency derivatives of the resistance and reactance. The so-called quality factor (denoted $Q(\omega)$ for an antenna tuned to have zero reactance at the radian frequency ω is defined for example by Yaghjian and Best [116] as

$$Q(\omega) = \frac{\omega |W(\omega)|}{P_A(\omega)} \tag{1.637}$$

where $W(\omega)$ is the internal energy and P_A is the total power accepted by the antenna. Assume that the antenna has an input impedance expressed as

$$Z_A(\omega) = R_A(\omega) + jX_A(\omega) \tag{1.638}$$

where $R_A(\omega)$ and $X_A(\omega)$ are the antenna input resistance and input reactance, respectively. Tuning can be achieved at radian frequency ω_o with a series reactance, denoted X_s, for either a series inductor L_s or series capacitor C_s to make the total reactance of an antenna equal to zero at a radian frequency ω_o. That is, the total reactance of the antenna plus series reactance is expressed as,

$$X_T(\omega) = X_A(\omega) + X_s(\omega) \tag{1.639}$$

where

$$X_s(\omega) = \begin{cases} \omega L_s & \text{if } X(\omega_o) < 0 \\ -\frac{1}{\omega C_s} & \text{if } X(\omega_o) > 0 \end{cases} \tag{1.640}$$

Yaghjian and Best assume that after the antenna has been tuned at frequency ω_o, then the antenna is connected to a transmission line with characteristic impedance equal to the antenna's tuned input resistance (denoted R_o). Taking the derivative of $X_T(\omega)$ with respect to radian frequency yields

$$\frac{dX_T(\omega)}{d\omega} = \begin{cases} \frac{dX_A(\omega)}{d\omega} + L_s & \text{if } X(\omega_o) < 0 \\ \frac{dX_A(\omega)}{d\omega} + \frac{1}{\omega^2 C_s} & \text{if } X(\omega_o) > 0 \end{cases} \tag{1.641}$$

which can be written in simplified form as

$$\frac{dX_T(\omega)}{d\omega} = \frac{dX_A(\omega)}{d\omega} + \frac{|X(\omega)|}{\omega} \tag{1.642}$$

Now, the power accepted by the antenna is given by

$$P_A(\omega) = \frac{1}{2}\text{Re}[V_o(\omega)I_o^*(\omega)] \tag{1.643}$$

where * means complex conjugate. The power accepted by the antenna is also expressed as the sum of the radiated power and any power lost due to the material of the antenna as

$$P_A(\omega) = P_{\text{radiated}}(\omega) + P_{\text{loss}}(\omega) \tag{1.644}$$

The radiated power is quantified by

$$P_{\text{radiated}}(\omega) = \frac{1}{2}|I_o(\omega)|^2 R_r(\omega) \tag{1.645}$$

where $R_r(\omega)$ is defined as the radiation resistance. The power loss due to any lossy material of the antenna is determined as

$$P_{\text{loss}}(\omega) = \frac{1}{2}|I_o(\omega)|^2 R_{\text{loss}}(\omega) \tag{1.646}$$

where R_{loss} is defined as the loss resistance. Then, applying Maxwell's equations in a lengthy derivation by Yaghjian and Best [116, Equation 96], an accurate approximate expression for the antenna quality factor (denoted by the authors as Q_z) is given by

$$Q_z = \frac{\omega}{2R(\omega)}\sqrt{\left[\frac{dR(\omega)}{d\omega}\right]^2 + \left[\frac{dX(\omega)}{d\omega} + \left|\frac{X(\omega)}{\omega}\right|\right]^2} \tag{1.647}$$

The following MATLAB code is used to compute the derivatives of the given input impedance $Z = R + jX$ and values of Q_z given by Equation (1.647):

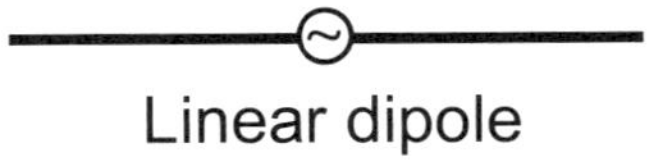

Figure 1.36 Center-fed linear dipole antenna.

```
R_prime=diff(Z_real)./delta_omega;
X_prime=diff(Z_imag)./delta_omega;
QZ=(omega(1:N)./(2.*R(1:N))).*...
      sqrt(R_prime.^2+(X_prime+...
      (abs(X(1:N)./omega(1:N)))).^2);
```

Furthermore, Yaghjian and Best have shown that another accurate approximate quality factor Q_B is expressed in terms of allowed VSWR and fractional bandwidth (FBW) as

$$Q_B = \frac{\text{VSWR-1}}{\text{FBW}\sqrt{\text{VSWR}}} \tag{1.648}$$

Yaghjian and Best have shown that the approximate relations given by Q_z and Q_B are in very good agreement with the exact calculation of Q, that is $Q_z = Q_B = Q$ for engineering purposes, thus, solving for the fractional bandwidth in terms of Q and allowed VSWR yields

$$\text{FBW} = \frac{f_2 - f_1}{f_c} = \frac{\text{VSWR-1}}{Q\sqrt{\text{VSWR}}} \tag{1.649}$$

1.11.3 Example: Q Factor for a Dipole Antenna

As in the paper by Yaghjian and Best [116], consider now an electromagnetic simulation of a center-fed linear dipole antenna (Figure 1.36) having total wire length 1m with a 1 mm wire diameter. This simulation is performed here using the commercial FEKO software with a method of moments solver — the moment method technique is described in the next chapter. Figure 1.37 shows the simulated complex input impedance (real and imaginary components) over the frequency band 50 MHz to 2,000 MHz. This antenna is one wavelength long at 300 MHz and is thus one-half wavelength long at 150 MHz. Resonance occurs at 13 frequencies where the imaginary component is zero. The first resonance occurs at 144 MHz, which can be seen more clearly in Figure 1.38, and this resonance is referred to as a natural resonance where the slope of the imaginary component is positive. The second resonance occurs at 272 MHz where the slope of the imaginary component is negative and this type of resonance is referred to as antiresonant. At each frequency where the natural resonance with positive slope in the imaginary component occurs,

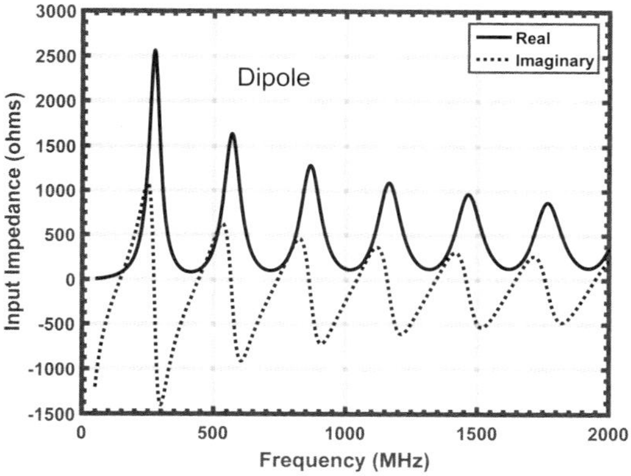

Figure 1.37 Moment method simulated input impedance versus frequency (50 MHz to 2,000 MHz) for a linear dipole with overall length 1 meter and wire diameter 1 mm.

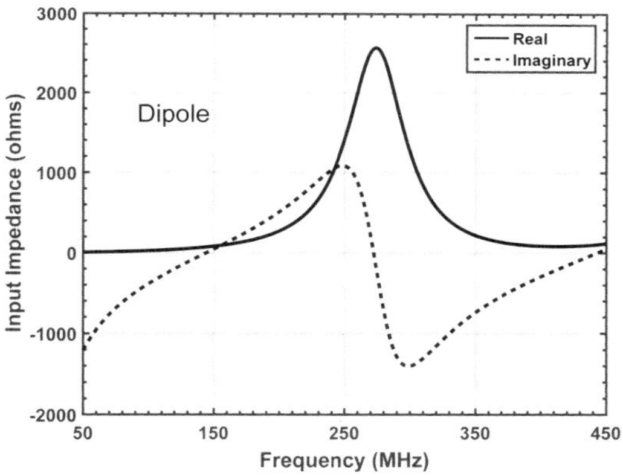

Figure 1.38 Moment method simulated input impedance versus frequency (50 MHz to 450 MHz) for a linear dipole with overall length 1 meter and wire diameter 1 mm.

the corresponding real (resistive) component is low (starting at 72 ohms at the first natural resonance and increasing gradually to about 200 ohms at high frequencies). At each frequency where the antiresonance with negative slope in the imaginary component occurs, the corresponding real (resistive) component is very high (starting at about 2,500) ohms at the first antiresonance and reducing gradually to about 1,000 ohms for the antiresonance just below 1,500 MHz. Taking 72 ohms as the reference impedance, the

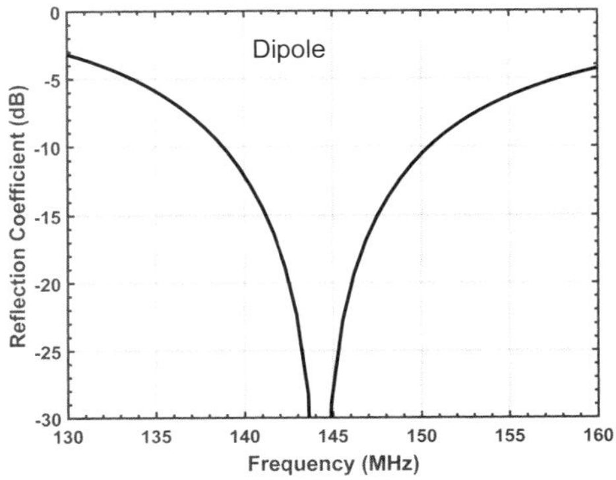

Figure 1.39 Moment method simulated reflection coefficient versus frequency (130 MHz to 160 MHz) for a linear dipole with overall length 1 meter and wire diameter 1 mm. The reference impedance is 72 ohms.

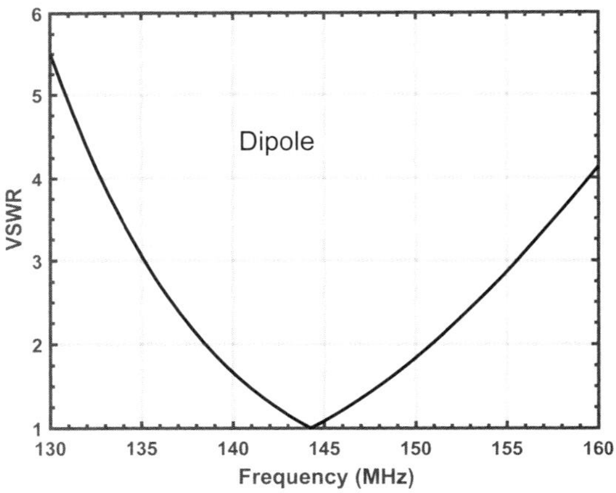

Figure 1.40 Moment method simulated VSWR versus frequency (130 MHz to 160 MHz) for a linear dipole with overall length 1 meter and wire diameter 1 mm. The reference impedance is 72 ohms.

magnitude of the reflection coefficient in dB is plotted versus frequency in Figure 1.39. The reflection coefficient has a minimum at approximately 144 MHz. Since the reflection coefficient magnitude is very low, the corresponding VSWR at 144 MHz is approximately equal to 1 (Figure 1.40). Figure 1.41 shows the quality factor Q_Z computed using Equation (1.647) over the band 50 MHz to 450 MHz. Using percent bandwidth PBW equal to $100 \times$ FBW, Figure 1.42 shows the percent bandwidth achieved by Equation (1.649)

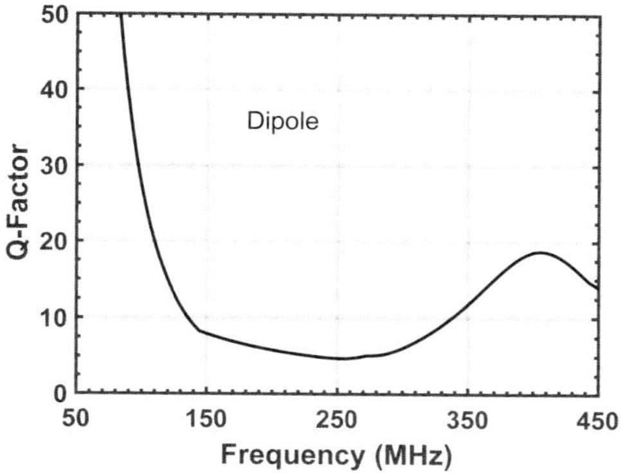

Figure 1.41 Quality factor Q_Z calculated from moment method simulated input impedance versus frequency (50 MHz to 450 MHz) for a linear dipole with overall length 1 meter and wire diameter 1 mm.

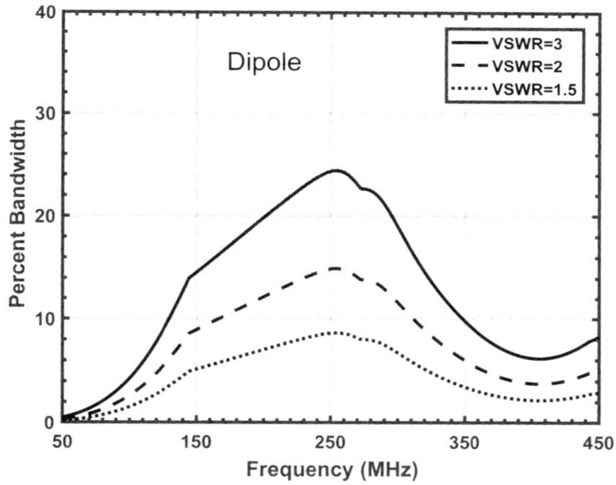

Figure 1.42 Calculated percent bandwidth based on quality factor Q_Z and assumed values of VSWR (1.5, 2, and 3) versus operating frequency (50 MHz to 450 MHz) for a linear dipole with overall length 1 meter and wire diameter 1 mm. The quality factor was calculated from the moment method simulated input impedance.

assuming allowed VSWR values equal to 1.5, 2, and 3. The peak value of PBW occurs near the first antiresonance, where the corresponding resistive component of input impedance is very high. If we assume that the antenna would be driven with a transmission line with a low characteristic impedance, say 72 ohms, then the frequency at the first natural resonance would be used as the center frequency. We can now compare the percent bandwidth observed in the method of moments simulated VSWR versus frequency plot in Figure 1.40 with the percent bandwidth computed using Equation (1.649). The results are tabulated in Table 1.2 and good agreement between the bandwidth calculated from the simulated input impedance data using Equation (1.647) and directly from the reflection coefficient.

Table 1.2
Comparison of Percent Bandwidth Computed from Equations (1.647) and (1.649) and from the Method of Moments Simulation for a 1m Linear Wire Dipole with 1 mm Wire Diameter Operating at a 144 MHz Center Frequency with $Q_z = 8.28$

	Percent Bandwidth for Wire Dipole	
Allowed VSWR	Eqs. (1.647), (1.649)	FEKO (Method of Moments)
1.5	4.9%	4.9%
2.0	8.5%	9.0%
3.0	14.0%	14.6%

1.11.4 Example: Q Factor for a Circular Loop Antenna

Consider now an electromagnetic simulation of a circular loop antenna with a single feed point as shown in Figure 1.43. The loop antenna is assumed to have a circumference equal to 1m and has a 1 mm wire diameter as in the article by Yaghjian and Best [116]. Again, the simulation is performed here using the commercial FEKO software with a method of moments solver. Figure 1.44 shows the simulated complex input impedance up to 2 GHz. Similar to the dipole, resonance occurs at 13 frequencies where the imaginary component is zero. The first resonance occurs at 142 MHz, which can be seen more clearly in the expanded scale plot in Figure 1.45, but unlike the dipole this first resonance is an antiresonance where the slope of the imaginary

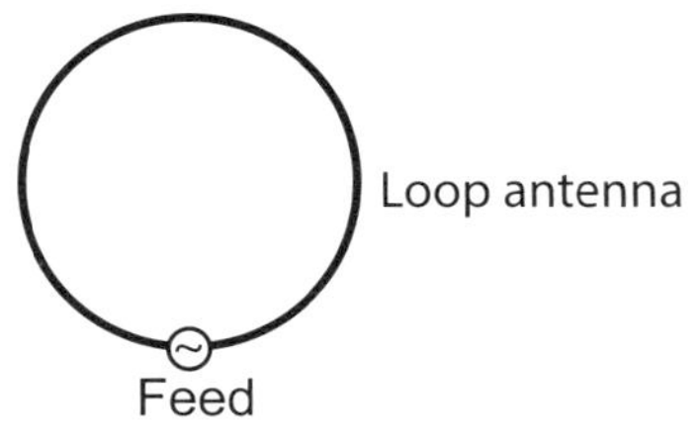

Figure 1.43 Circular loop antenna with single feed point.

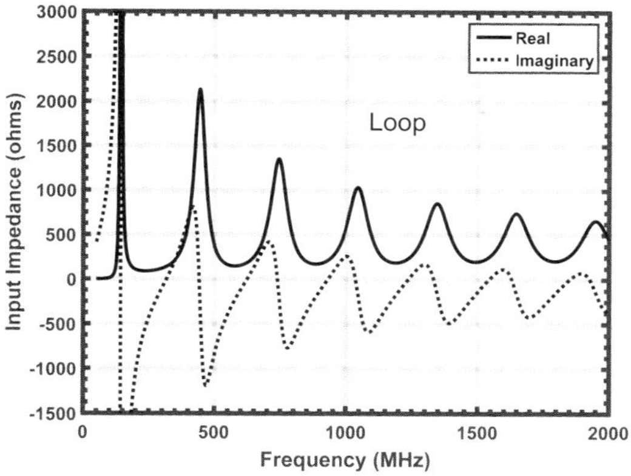

Figure 1.44 Moment method simulated input impedance versus frequency (50 MHz to 2,000 MHz) for a circular wire loop antenna with circumference 1 meter and wire diameter 1 mm.

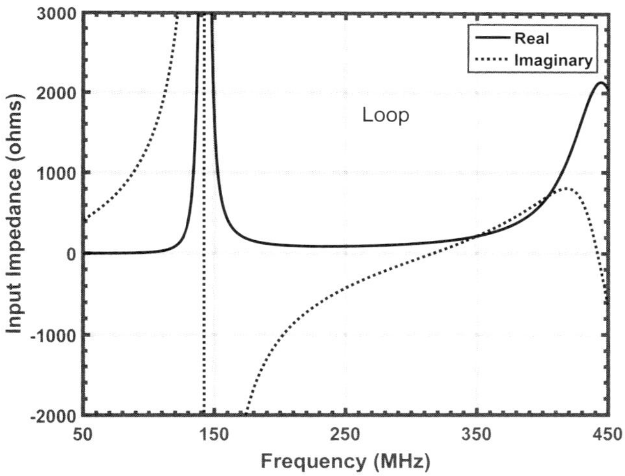

Figure 1.45 Moment method simulated input impedance versus frequency (50 MHz to 450 MHz) for a circular wire loop antenna with circumference 1 meter and wire diameter 1 mm.

component is negative. The second resonance occurs at 317 MHz and it is a natural resonance where the slope of the imaginary component is positive. At each frequency where the natural resonance occurs, the corresponding real (resistive) component is low (starting at 140 ohms) at the first natural resonance and increasing gradually to about 266 ohms at the high frequencies

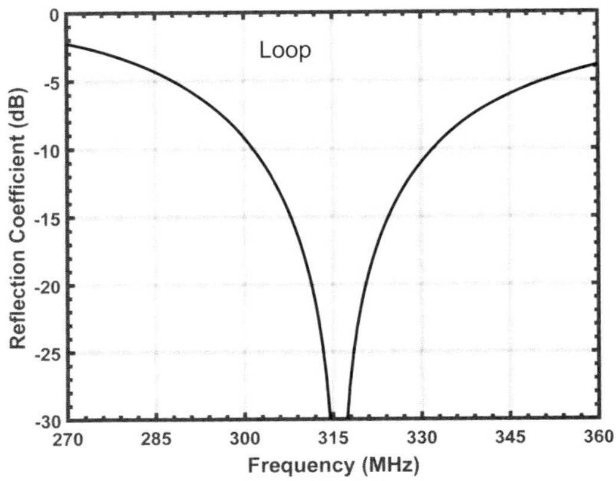

Figure 1.46 Moment method simulated reflection coefficient versus frequency (270 MHz to 360 MHz) for a circular wire loop antenna with circumference 1 meter and wire diameter 1 mm. The reference impedance is 140 ohms.

shown. At each frequency where the antiresonance occurs, the corresponding real (resistive) component is very high (starting on the order of 40,000 ohms) at the first anti-resonance (142 MHz) and reducing quickly down to the range of about 600 to 2,100 ohms for the antiresonances below 2,000 MHz. Also, at each frequency where the antiresonance occurs, the corresponding imaginary (inductive) component is very high (starting on the order of 20,000 ohms) at the first anti-resonance (142 MHz) and reducing quickly down to the range of about 800 to 100 ohms for the antiresonances below 2,000 MHz. Assume now that the loop antenna is driven with a transmission line with a characteristic impedance equal to 140 ohms corresponding to the input resistance at the first natural resonance. Taking 140 ohms as the reference impedance, the magnitude of the reflection coefficient in dB is plotted versus frequency in Figure 1.46. The reflection coefficient has a minimum at approximately 317 MHz. The corresponding VSWR is shown in Figure 1.47. Figure 1.48 shows the quality factor Q_Z computed using Equation (1.647) over the band 50 MHz to 450 MHz. Figure 1.49 shows the percent bandwidth achieved by Equation (1.649) assuming allowed VSWR values equal to 1.5, 2, and 3. For this loop antenna, the peak value of PBW occurs near the first natural resonance. The results are tabulated for the loop in Table 1.3 and good agreement between the bandwidth calculated from the simulated input impedance data using Equation (1.647) and directly from the reflection coefficient.

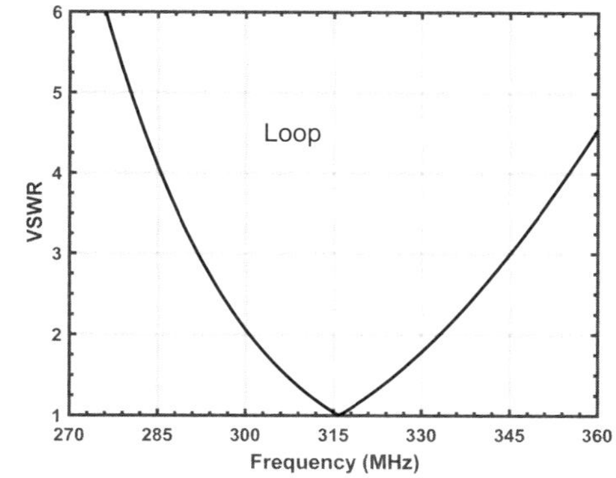

Figure 1.47 Moment method simulated VSWR versus frequency (270 MHz to 360 MHz) for a circular wire loop antenna with circumference 1 meter and wire diameter 1 mm. The reference impedance is 140 ohms.

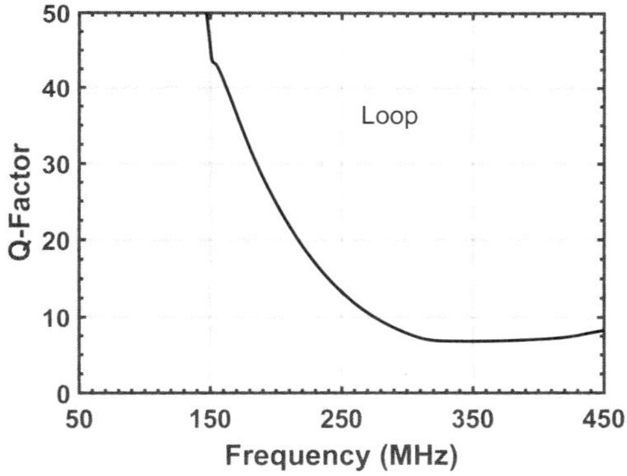

Figure 1.48 Quality factor Q_Z calculated from moment method simulated input impedance versus frequency (50 MHz to 450 MHz) for a circular wire loop antenna with circumference 1 meter and wire diameter 1 mm.

1.12 Radiation Intensity

Antenna radiation intensity $U(\theta, \phi)$ refers to the power radiated from an antenna per unit solid angle per square radians or per square degrees. The term *steradians* or sr is often used to represent the units of square radians. Consider Figure 1.50 that depicts the differential solid angle in spherical coordinates. The surface area of a sphere with radius r is equal to $4\pi r^2$, and normalizing the area with respect to r^2 there are 4π steradians in the unit closed sphere, as derived below. The elemental or differential area on the surface of the sphere

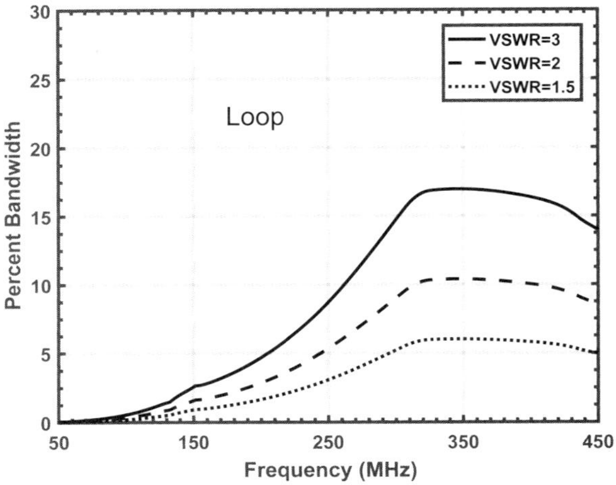

Figure 1.49 Calculated percent bandwidth based on quality factor Q_Z and assumed values of VSWR (1.5, 2, and 3) versus operating frequency (50 MHz to 450 MHz) for a circular wire loop antenna with circumference 1 meter and wire diameter 1 mm. The quality factor was calculated from the moment method simulated input impedance.

Table 1.3
Comparison of Percent Bandwidth Computed from Equations (1.647) and (1.649) and from the Method of Moments Simulation for a Circular Wire Loop Antenna with 1m Circumference and 1 mm wire Diameter Operating at a 317 MHz Center Frequency with $Q_z = 6.94$

	Percent Bandwidth for Wire Loop	
Allowed VSWR	Eqs. (1.647), (1.649)	FEKO (Method of Moments)
1.5	6.0%	6.0%
2.0	10.3%	10.4%
3.0	16.9%	17.7%

is expressed as

$$dA = r^2 \sin\theta d\theta d\phi \tag{1.650}$$

Normalizing the differential area dA by the distance r squared yields the differential solid angle $d\Omega$ as

$$d\Omega = \frac{dA}{r^2} = \sin\theta d\theta d\phi \tag{1.651}$$

Integrating Equation (1.651) over all angles yields the solid angle Ω as

$$\Omega = \int_0^\pi \int_0^{2\pi} d\Omega = \int_0^\pi \int_0^{2\pi} \sin\theta d\theta d\phi = 4\pi \tag{1.652}$$

Equivalently, the radiation intensity can be computed from the ratio of the power density generated by the antenna in a particular direction and at a particular range distance r to the average radiated power density at the same distance.

Repeating Equation (1.409), the time-average power density of the electromagnetic field, or the time-average Poynting's vector, which has units of watts/meter^2 is given by

$$\boldsymbol{P}_d^{\mathrm{ave}}(\boldsymbol{r}) = \frac{1}{2}\mathrm{Re}[\boldsymbol{E}\times\boldsymbol{H}^*] \tag{1.653}$$

The time-average radiated power in watts is computed by integrating the time-average power density over the closed sphere as

$$P_{\mathrm{rad}}^{\mathrm{ave}} = \int_0^\pi \int_0^{2\pi} \frac{1}{2}\mathrm{Re}[\boldsymbol{E}\times\boldsymbol{H}^*] \cdot r^2\hat{\boldsymbol{r}} \sin\theta d\theta d\phi \tag{1.654}$$

The radiation intensity, denoted $U(\theta, \phi)$, with units of watts per steradian is now defined as independent of range by taking the dot product of the time-average power density (watts/meter^2) with the range squared (meter^2) times the radial unit vector as

$$U(\theta, \phi) = \boldsymbol{P}_d^{\mathrm{ave}}(\boldsymbol{r}) \cdot r^2\hat{\boldsymbol{r}} = \frac{1}{2}\mathrm{Re}[\boldsymbol{E}\times\boldsymbol{H}^*] \cdot r^2\hat{\boldsymbol{r}} \tag{1.655}$$

Next, let the normalized antenna far-field pattern be expressed, for example, in terms of the electric field component (E_θ or E_ϕ) as

$$F(\theta, \phi) = E(\theta, \phi)/|E(\theta, \phi)| \tag{1.656}$$

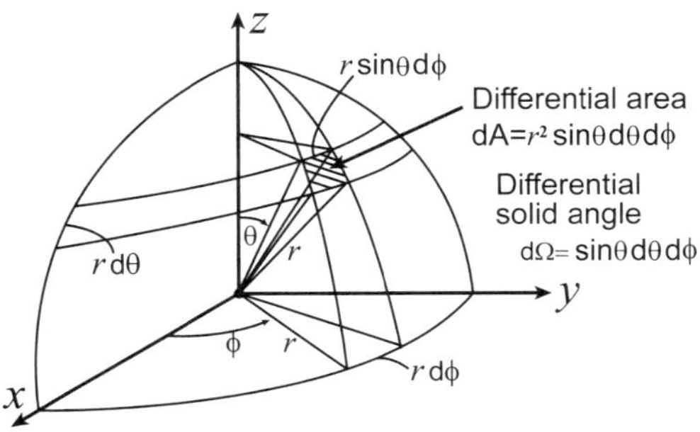

Figure 1.50 Diagram depicting the differential solid angle in spherical coordinates.

which has a maximum value of unity.

The radiation intensity of an antenna can now be computed from the maximum radiation intensity $U_{\max}$ and the normalized antenna radiation pattern $F(\theta, \phi)$ as

$$U(\theta, \phi) = U_{\max}|F(\theta, \phi)|^2 \tag{1.657}$$

where * denotes conjugate.

The total power radiated from the antenna is computed by integrating the radiation intensity over all θ and ϕ angles as

$$P_r^{\text{total}} = \iint U(\theta, \phi) d\Omega = \iint U_{\max}|F(\theta, \phi)|^2 d\Omega \tag{1.658}$$

The average radiation intensity is equal to the total power radiated P_r^{total} divided by 4π steradians as

$$U_{\text{ave}} = \frac{P_r^{\text{total}}}{4\pi} = \frac{1}{4\pi} \int_{\theta=0}^{\pi} \int_{\phi=0}^{2\pi} U(\theta, \phi) d\Omega = \frac{1}{4\pi} \int_{\theta=0}^{\pi} \int_{\phi=0}^{2\pi} U_{\max}|F(\theta, \phi)|^2 d\Omega \tag{1.659}$$

1.13 Antenna Directivity

The directivity of an antenna is a function of the radiation intensity normalized by the average radiation intensity as

$$D(\theta, \phi) = \frac{U(\theta, \phi)}{U_{\text{ave}}} = \frac{4\pi U(\theta, \phi)}{P_r^{\text{total}}} \tag{1.660}$$

Substituting Equations (1.657) and (1.659) in (1.660) yields

$$D(\theta, \phi) = \frac{U(\theta, \phi)}{\frac{1}{4\pi} \iint U(\theta, \phi) d\Omega} = \frac{U_{\max}|F(\theta, \phi)|^2}{\frac{1}{4\pi} \iint U_{\max}|F(\theta, \phi)|^2 d\Omega} = \frac{|F(\theta, \phi)|^2}{\frac{1}{4\pi} \iint |F(\theta, \phi)|^2 d\Omega} \tag{1.661}$$

In the case of an isotropic radiating antenna the radiation pattern is unity, that is,

$$F^{\text{isotropic}}(\theta, \phi) = 1 \tag{1.662}$$

Substituting Equation (1.662) in (1.661) and using Equation (1.652) the directivity of the isotropic antenna is unity, that is,

$$D^{\text{isotropic}}(\theta, \phi) = 1 \tag{1.663}$$

In Equation (1.661), it is observed that the antenna directivity, $D(\theta, \phi)$, relative to an isotropic radiator, depends only on the shape of the radiation pattern $F(\theta, \phi)$ and does not depend on the power delivered to the antenna or the maximum radiation intensity. The antenna directivity in decibels relative to an isotropic radiator (dBi) is computed by evaluating $10 \log_{10} D(\theta, \phi)$.

A general expression for the maximum directivity of a uniformly illuminated aperture antenna can be obtained by noting that the total far-field power must be equal to the total power passing through the aperture [23, pp. 292–293]. Integrating the power density over the surface (denoted S') of the aperture gives the total radiated power as

$$P_r^{\text{total}} = \frac{1}{2\eta_o} \iint |\boldsymbol{E}|^2 dS' \tag{1.664}$$

where η_o is the impedance of free space. The maximum radiation intensity is computed at broadside from

$$U_{\max} = \frac{1}{2\lambda^2 \eta_o} \left| \iint \boldsymbol{E}\, dS' \right|^2 \tag{1.665}$$

Then substituting Equations (1.664) and (1.665) into Equation (1.660) yields

$$D_{\max} = \frac{4\pi}{\lambda^2} \frac{|\iint \boldsymbol{E}\, dS'|^2}{\iint |\boldsymbol{E}|^2\, dS'} \tag{1.666}$$

In the case of a uniformly illuminated aperture with area A, the electric field is equal to unity and Equation (1.666) simplifies to

$$D_{\max} = \frac{4\pi A}{\lambda^2} \tag{1.667}$$

The projected physical aperture will decrease as a cosine function as the observation angle moves from broadside, in which case a general expression for the directivity of a uniformly illuminated aperture is given by

$$D(\theta) = \frac{4\pi A}{\lambda^2} \cos\theta \tag{1.668}$$

For a uniformly illuminated rectangular aperture with dimensions L_x and L_y the directivity is given by

$$D^{\text{rectangular}}(\theta) = \frac{4\pi L_x L_y}{\lambda^2} \cos\theta \tag{1.669}$$

In the case of a uniformly illuminated circular aperture with diameter d, the directivity is expressed as

$$D^{\text{circular}}(\theta) = \left(\frac{\pi d}{\lambda}\right)^2 \cos\theta \tag{1.670}$$

If an amplitude taper is applied to the aperture, then an aperture illumination effciency factor, denoted η_a, is used to reduce the directivity relative to uniform illumination, that is,

$$D(\theta) = \eta_a \frac{4\pi A}{\lambda^2} \cos\theta \tag{1.671}$$

1.13.1 Hertzian Dipole or Electrically Small Loop: Directivity Pattern and Radiation Intensity

Consider now the far-field radiation pattern of a z-directed Hertzian (electrically short) dipole antenna (refer to Figure 1.25), which has a θ electric field component that varies in observation angle only as $\sin\theta$ according to Equation (1.493). That is, the field pattern function $F(\theta)$ is equal to

$$F(\theta) = \sin\theta \tag{1.672}$$

$$D^{\text{z-dipole}}(\theta,\phi) = \frac{U(\theta,\phi)}{U_{\text{ave}}} = \frac{|F(\theta,\phi)|^2}{\frac{1}{4\pi}\iint |F(\theta,\phi)|^2 d\Omega} \tag{1.673}$$

Substituting Equations (1.672) and (1.651) in (1.673) yields

$$D^{\text{z-dipole}}(\theta,\phi) = \frac{\sin^2\theta}{\frac{1}{4\pi}\iint \sin^2\theta \sin\theta d\theta d\phi} \tag{1.674}$$

Noting that the definite integral for sine cubed is evaluated as

$$\int_0^\pi \sin^3\theta d\theta = \left[-\cos\theta + \frac{1}{3}\cos^3\theta\right]_0^\pi = \frac{4}{3} \tag{1.675}$$

and the integral with respect to ϕ is just 2π it follows that Hertzian dipole directivity pattern is

$$D^{\text{z-dipole}}(\theta,\phi) = \frac{\sin^2\theta}{\frac{1}{4\pi}\left(\frac{4}{3}\right)2\pi} = \frac{3}{2}\sin^2\theta \tag{1.676}$$

Repeating Equation (1.473), the time-average power density of the z-directed dipole is

$$\boldsymbol{P}_{\text{ave}}^{\text{z-dipole}}(\boldsymbol{r}) = \frac{1}{2}\text{Re}[\boldsymbol{E}\times\boldsymbol{H}^*] = \hat{\boldsymbol{r}}\frac{\eta}{2}\left[\frac{kIl}{4\pi r}\right]^2 \sin^2\theta. \tag{1.677}$$

Then, applying Equation (1.655) yields the radiation intensity for the Hertzian dipole

$$U^{\text{z-dipole}}(\theta,\phi) = \boldsymbol{P}_{\text{ave}}^{\text{z-dipole}}(\boldsymbol{r}) \cdot r^2\hat{\boldsymbol{r}} = \frac{\eta}{2}\left[\frac{kIl}{4\pi}\right]^2 \sin^2\theta \tag{1.678}$$

So it follows that the maximum radiation intensity of the Hertzian dipole is

$$U_{\text{max}}^{\text{z-dipole}} = \frac{\eta}{2}\left[\frac{kIl}{4\pi}\right]^2 \tag{1.679}$$

Now since $F(\theta) = \sin^2\theta$, the average radiation intensity for the dipole is determined as

$$U_{\text{ave}}^{\text{z-dipole}} = \frac{1}{4\pi}\int_{\theta=0}^{\pi}\int_{\phi=0}^{2\pi} U_{\text{max}}|F(\theta,\phi)|^2 d\Omega = \frac{2}{3}U_{\text{max}} \tag{1.680}$$

Substituting Equations (1.678), (1.679), and (1.680) into Equation (1.661), it follows that the angular dependence of the Hertzian dipole directivity is expressed again as

$$D^{\text{z-dipole}}(\theta,\phi) = \frac{3}{2}\sin^2\theta \tag{1.681}$$

The electrically short z-oriented dipole normalized 3D radiation pattern in decibels is shown in Figure 1.51, and it is observed that the pattern is a figure of revolution about the z axis, with a null along the z axis. From Equation (1.681) the half-power (-3 dB) beamwidth is 90°. A polar diagram of the absolute directivity pattern (θ component) as a function of θ is shown in Figure 1.52, where the peak directivity of 1.76 dBi is observed to occur at $\theta = 90°$. A z-oriented electrically small loop antenna (Figure 1.25) has the same $\sin\theta$ angular dependence (refer to Equation (1.496)), so it has the same 3D radiation and directivity patterns as in Figures 1.51 and 1.52, respectively. Thus, the directivity patterns for a z-directed electrically small loop and a z-directed electrically short dipole are equal, that is,

$$D^{\text{z-loop}}(\theta,\phi) = D^{\text{z-dipole}}(\theta,\phi) = 1.5\sin^2\theta. \tag{1.682}$$

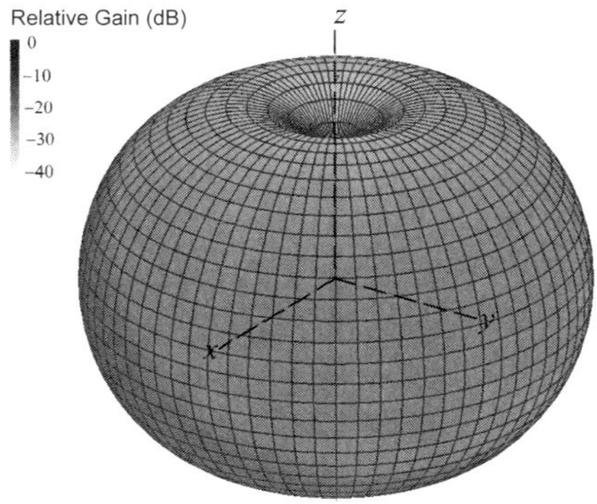

Figure 1.51 3D radiation pattern for a z-directed Hertzian dipole in free space.

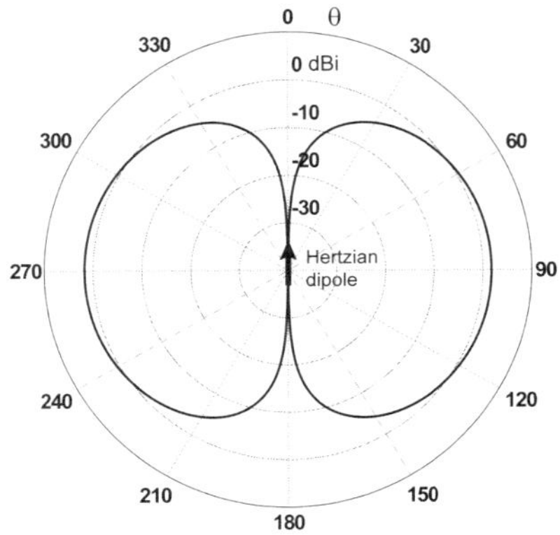

Figure 1.52 Polar directivity pattern D_θ for a z-directed Hertzian dipole antenna.

1.13.2 Crossed Hertzian Dipoles (Turnstile Antenna): Directivity Pattern

Crossed dipole antennas have a number of applications including operation for independent dual linear polarization, circular polarization, and nearly omnidirectional coverage with horizontal polarization depending on the phase relation between the dipoles. Consider the crossed Hertzian (short) dipoles in the xy-plane in free space as shown in Figure 1.53. As an example of circular polarization on transmit, let the y-polarized dipole be transmitting with a $-90°$ phase shift with respect to the x-polarized dipole. This antenna configuration is referred to as a turnstile antenna [117]. By the right-hand rule, this quadrature phase relation between the two dipoles generates peak RHCP gain at $\theta = 0°$ and peak LHCP gain at $\theta = 180°$. Figure 1.54 shows a polar diagram of the moment method simulated circularly polarized elevation radiation patterns for these crossed-dipole antennas in free space. These radiation patterns are a figure of revolution about the z axis. For example, the 3D LHCP directivity pattern is shown in Figure 1.55. Peak RHCP directivity 1.76 dBi and half-power beamwidth 131° occurs in the forward hemisphere at $\theta = 0°$. Figure 1.56 shows a polar diagram of the corresponding linearly

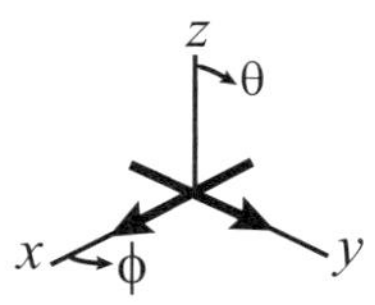

Figure 1.53 Crossed Hertzian (short) dipole antennas in free space (turnstile).

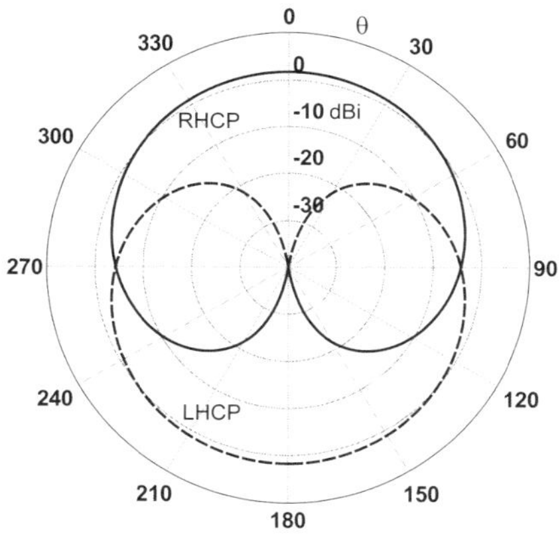

Figure 1.54 Simulated circularly polarized elevation radiation patterns for crossed-dipole antennas (turnstile) phased for RHCP in the upper hemisphere.

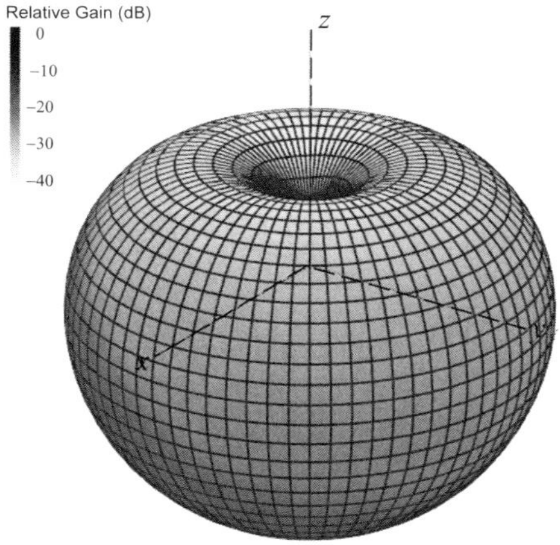

Figure 1.55 Simulated 3D LHCP radiation pattern for crossed-dipole antennas (turnstile) in free space.

polarized elevation radiation patterns, and again these patterns are a figure of revolution. The $\hat{\phi}$ polarized component has a constant, omnidirectional, gain at the level of -1.24 dBi as a function of elevation angle θ. In fact, as shown in Figure 1.57 for these turnstile Hertzian dipoles the $\hat{\phi}$ polarized component is isotropic with respect to all θ and ϕ angles. As expected for the linearly polarized phi component, this constant value of -1.24 dBi is 3 dB below the peak circularly polarized directivity. The linearly polarized $\hat{\theta}$ component has the same radiation pattern (same peak directivity and same beamwidth) as in Figure 1.52.

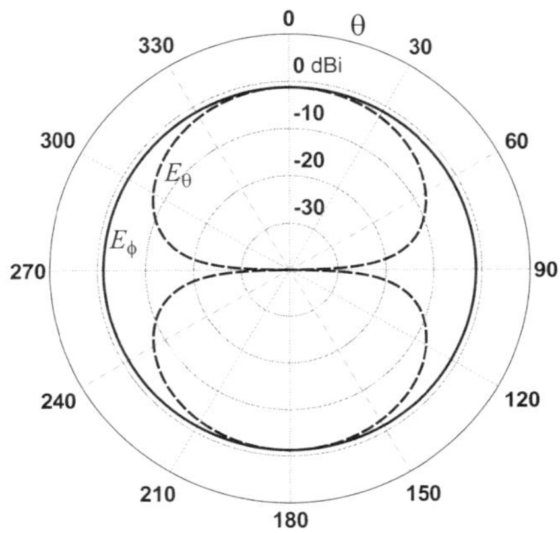

Figure 1.56 Simulated linearly polarized elevation radiation patterns for circularly polarized crossed-dipole antennas (turnstile) in free space.

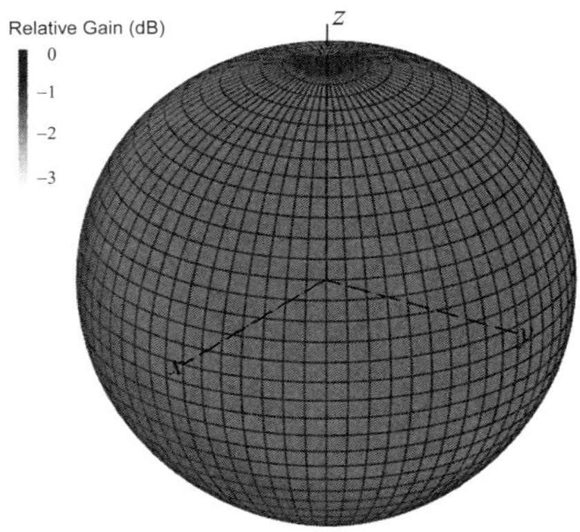

Figure 1.57 Simulated 3D horizontally polarized ($\hat{\phi}$ component) *isotropic* radiation pattern for crossed-dipole antennas in free space (turnstile).

1.13.3 Circularly Polarized Spiral Antenna

Planar spiral antennas are an example of a frequency independent circularly polarized antenna [24, pp. 13-7–13-34] with the geometry described by angles. A center-fed two-arm planar spiral antenna operating in the mode 1 (sum beam) is simulated in this section. In practice, the two arms of the spiral are fed with a balun to maintain a balanced differential phase excitation. The active radiating region of the spiral moves with frequency and occurs where the circumference is approximately one wavelength. Spiral antennas will typically be used over a ground plane or over an electrically conducting cavity that is sometimes loaded with microwave absorber. The phase center of a planar spiral will be located very close to the spiral surface over a wide bandwidth and, thus, can be used as a wideband feed for a parabolic reflector.

Consider Figure 1.58(a) which shows a right-hand wound two-arm Archimedean spiral antenna over a circular ground plane. Here, the two spiral arms are wound according to the following equations

$$r_1 = r_o + a\phi \quad \text{and} \quad r_2 = r_o + a(\phi - \pi) \tag{1.683}$$

where in this example $r_o = 0.254$ cm is the feed radius, $a = 0.097$ is the growth rate, and ϕ is the angle in the xy plane. The simulated spiral has 10 turns of perfectly conducting 0.5 mm diameter wire, and the outer diameter of the spiral is 12.7 cm. The radiating center frequency in this example is chosen to be 2.5 GHz and a 15.24 cm diameter ground plane is placed one-quarter wavelength from the planar spiral. The FEKO multilevel fast multipole method (MLFMM) simulated 3D relative gain radiation pattern is shown in Figure 1.58(b), and in this example the peak directivity is 9 dBi (RHCP). The simulated input impedance of this planar spiral antenna over ground plane is $Z_{\text{in}} = 253+j36.5$ ohms. Next, as depicted in Figure 1.59(a) the spiral

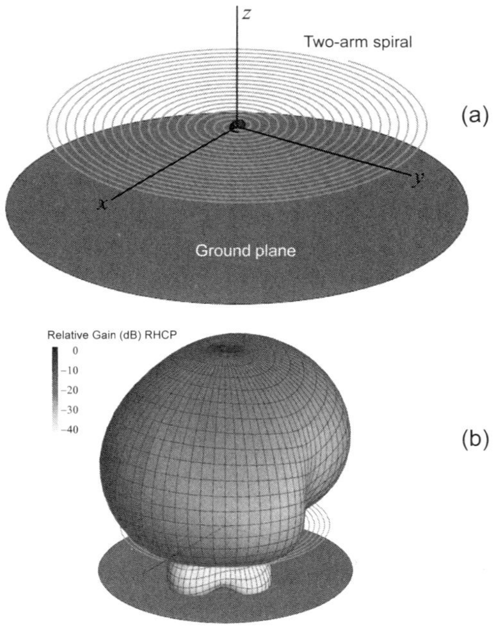

Figure 1.58 Circularly polarized spiral antenna example. (a) Spiral antenna over a ground plane geometry, and (b) simulated RHCP 3D radiation pattern.

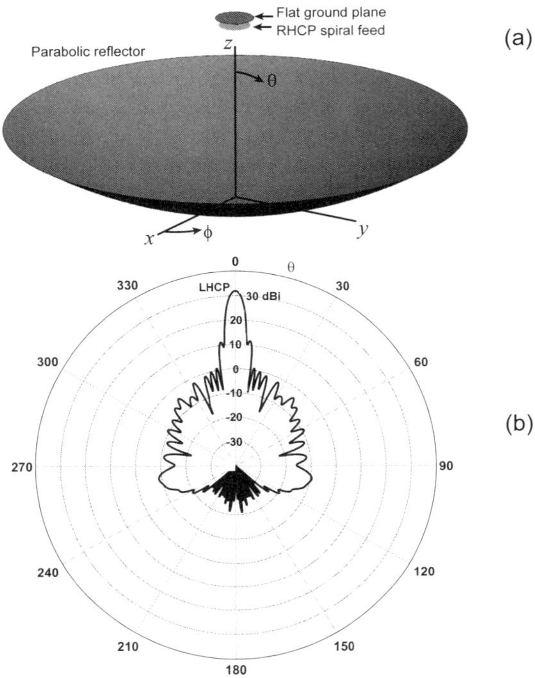

Figure 1.59 Circularly polarized spiral antenna with parabolic reflector. (a) Antenna geometry, and (b) simulated LHCP polar directivity radiation pattern.

antenna with ground plane is used as a feed for a 1.83m (6 ft) diameter parabolic reflector with a focal distance to diameter ratio (f/D) equal to 0.4. The RHCP illumination of the parabolic reflector yields a LHCP radiation pattern as shown in Figure 1.59(b) with 32 dBi peak directivity. For a circular aperture with diameter $d = 1.8$m and uniform illumination, the calculated ideal directivity using Equation (1.669) is 33.6 dBi. The aperture illumination efficiency of this spiral antenna fed reflector system is the difference between the ideal directivity and the MLFMM simulated directivity, which is -1.6 dB or $\eta_a = 0.69$.

1.14 Gain, Realized Gain, Transmit Power, and Receive Power

The antenna directivity discussed in Section 1.13 does not take into account any losses in the antenna and its surroundings. The terms antenna gain and antenna realized gain are used to take into account both the radiation pattern (directivity as in Equation (1.661)) as well as certain losses as summarized in Figure 1.60 and is discussed below.

Terminology in the literature regarding antenna gain and power can be vague in some instances, so a brief discussion here is appropriate. Transmitter power, transmit power, available power, power accepted by the antenna, input power, power input to the antenna, and power radiated by the antenna are terms found in the literature. The terms directive gain (replaced by the term directivity), power gain, effective gain, gain loss due to impedance mismatch, gain loss due to polarization mismatch, gain loss due to mutual coupling, and gain loss due to power absorbed by the antenna or surrounding structures are found in the literature. In antenna electromagnetic modeling analysis or measurements, it is important to clearly define these terms.

Consider now a transmission line with a transmitter source of a given power level attached at one end of the transmission line. At the other end of this transmission line, which will be the attachment point to the transmitting antenna, a power meter (matched to the transmission line) is attached and registers a power level of P_o watts. Thus, the amount of incident power that is connected to the antenna is accurately known. Now the transmitting antenna element can be mismatched to the transmission line, resulting in reflected power into the transmission line (assumed to be terminated in a resistive load matched to the characteristic impedance of the transmission line) that will be completely absorbed back at the transmitting source location. This reflected power reduces the realized antenna gain and can be quantified and referred to as impedance mismatch loss, transmission mismatch loss, or gain loss. In the case where there are antennas that surround the transmitting antenna (forming an array of antennas), there will be mutual coupling between array elements and, as discussed in Chapter 2, some of the power radiated by the transmit element will be received and consumed (additional power loss or additional gain loss) by the surrounding array elements. The net effect of the

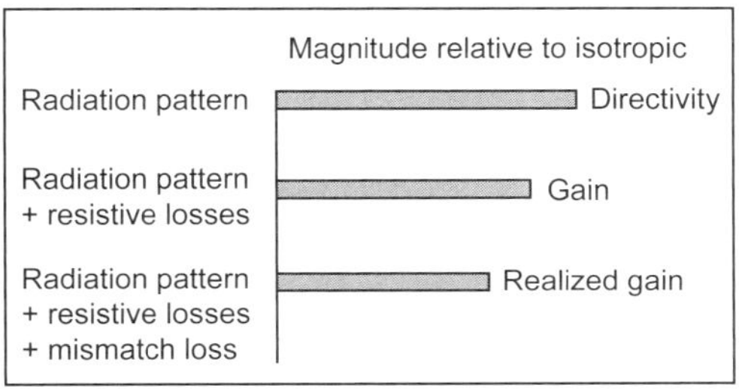

Figure 1.60 Comparisons of the relative magnitude for antenna directivity, gain, and realized gain with respect to an isotropic antenna.

gain loss mechanisms will be a reduced transmitted power density (watts per meter squared) at a near-field or far-field position from the array.

The antenna gain, relative to an isotropic radiator with the same input power, can be expressed in terms of the radiation intensity U and input power as

$$G(\theta, \phi) = \frac{4\pi U(\theta, \phi)}{P_{\text{in}}} \tag{1.684}$$

where P_{in} is the input power, that is, the power at the port of the transmission line that will be connected to the antenna. As a result of losses from antenna impedance mismatch and losses attributed to components between the transmitter and the antenna terminals, the power transmitted by the antenna is less than the transmitter power. For electrically small antenna elements, the finite resistive loss of the antenna element itself can reduce both the efficiency and gain of the antenna. The antenna gain in dBi (decibels relative to isotropic) is computed by evaluating $10 \log_{10} G(\theta, \phi)$.

Assuming that the antenna, with input impedance Z_{in}, is connected to a transmission line with characteristic impedance Z_o, the voltage reflection coefficient Γ is given by

$$\Gamma = \frac{Z_{\text{in}} - Z_o}{Z_{\text{in}} + Z_o}. \tag{1.685}$$

The reflection coefficient in dB is computed as

$$\Gamma_{\text{dB}} = 20 \log_{10} |\Gamma|. \tag{1.686}$$

The reflection coefficient in dB is also referred to in the literature as return loss (a reflection coefficient of zero is the same as a minus infinity dB return loss or a perfect match usually to 50 ohms). In antenna designs, it is often desirable to provide an impedance matching network to improve the reflection coefficient [71]. The loss in antenna gain resulting from antenna input impedance mismatch alone is computed from the impedance mismatch loss efficiency, denoted η_{ML} (same as transmission mismatch loss efficiency given by Equation (1.21)),

$$\eta_{\text{ML}} = 1 - |\Gamma|^2. \tag{1.687}$$

By using Equation (1.687), the mismatch loss ML (or gain loss) in dB as a positive value is computed as

$$\text{ML}_{\text{dB}} = -10 \log_{10} \eta_{\text{ML}} = -10 \log_{10} \left[1 - |\Gamma|^2\right]. \tag{1.688}$$

The radiation loss efficiency factor η_L of an antenna can be calculated as follows [23, pp. 43-48]. The radiation loss efficiency of an antenna is expressed as

$$\eta_L = R_r / R_A \tag{1.689}$$

where R_r is the radiation resistance and

$$R_A = R_r + R_{\text{ohmic}} \tag{1.690}$$

is the total input resistance of the antenna. For example, the ohmic resistance for an electrically short dipole of length L and diameter d is given by

$$R_{\text{ohmic}} = \frac{LR_s}{3\pi d} \tag{1.691}$$

where

$$R_s = \sqrt{\pi f \mu / \sigma} \tag{1.692}$$

is the surface resistance of the dipole, which depends on the frequency f, permeability μ, and electrical conductivity σ of the dipole material. Thus, since the ohmic resistance is inversely proportional to the dipole diameter, the efficiency can be increased by using a larger-diameter dipole.

The realized antenna gain, denoted as G_{realized}, takes account of impedance mismatch and efficiency loss. The realized antenna gain and directivity are related by the following expression

$$G_{\text{realized}}(\theta, \phi) = D(\theta, \phi)\eta_{\text{L}}\eta_{\text{ML}} \tag{1.693}$$

where η_L is the antenna efficiency loss factor, which includes resistive losses in the antenna. For a 100% efficient (lossless) antenna with $\eta_{\text{L}} = 1$, and with no mismatch loss ($\eta_{\text{ML}} = 1$), the gain is equal to the directivity. For a transmitting antenna, the effective isotropic radiated power (EIRP(θ, ϕ)) is equal to the product of the transmission line power incident on the antenna P_o and the realized gain $G_{\text{realized}}(\theta, \phi)$, that is

$$\text{EIRP}(\theta, \phi) = P_o G_{\text{realized}}(\theta, \phi). \tag{1.694}$$

Alternately, if the impedance mismatch loss is included in the power budget, then the transmit power accepted by the antenna is denoted as P_t and then the EIRP is equal to the following product

$$\text{EIRP}(\theta, \phi) = P_t G_t(\theta, \phi) \tag{1.695}$$

where $G_t(\theta, \phi)$ is the gain of the transmit antenna, which includes any resistive losses associated with the antenna and its surroundings.

The far-zone power density P_d (with units of W/m^2) at a distance r from the aperture (one-way path) is given by

$$P_d = \frac{P_t G_t}{4\pi r^2}. \tag{1.696}$$

Based on a receive antenna's maximum effective aperture, denoted A_{em}, the aperture collects incident power density and converts it to an output power. Thus, the received power P_r of an antenna aperture can be expressed as

$$P_r = P_d A_{\text{em}}. \tag{1.697}$$

Furthermore, the receive antenna gain is expressed as

$$G_r = 4\pi A_{\text{em}}/\lambda^2, \tag{1.698}$$

and solving for A_{em} yields

$$A_{\text{em}} = G_r \lambda^2/(4\pi). \tag{1.699}$$

Thus, it follows that

$$P_r = \frac{P_t G_t G_r \lambda^2}{(4\pi r)^2}. \tag{1.700}$$

Equation (1.700) can be rearranged to express the mutual coupling power ratio (P_r/P_t) between two antennas as

$$\frac{P_r}{P_t} = \frac{G_t G_r \lambda^2}{(4\pi r)^2}. \tag{1.701}$$

which is known as the Friis transmission equation [118].

1.15 Types of Antennas

The antennas discussed in this chapter have concentrated on dipoles and loops, which are the fundamental building blocks for antenna technology. There are numerous types of antennas that can be considered as listed in Table 1.4, and Figure 1.61 depicts several of these antennas. Array antennas can consist of groups of any of the antennas listed in Table 1.4 or shown in Figure 1.61. Monopoles, dipoles, and loops can take on various forms and orientations, for example the planar, cylindrical, and conical shapes as shown in Figures 1.61(a–d). Dipoles can be arranged in an orthogonal relationship (Figure 1.61(c)) for independent dual polarization (discussed in Chapters 2 and 3). Vertical monopoles provide omnidirectional vertical polarization (discussed in Chapters 2, 4, and 5). An omnidirectional discone (discussed in Chapter 4) or biconical dipole antenna either in wire form (Figure 1.61(d)) or solid form, or a log-periodic dipole array (Figure 1.61(e)) can provide wide bandwidth performance. Horizontal loops (Figure 1.61(f)) or slotted cylinder antennas (Figure 1.61(j)) can provide omnidirectional

Table 1.4
Some Common Antenna Types

Arrays (phased, broadside, endfire, corporate-fed, lens, reflect)
Biconical (symmetric, asymmetric, solid surface, wire frame)
Dielectric rod
Dipole (linear, folded, vee, ultrawideband)
Discone (solid, wire frame)
Helix [axial mode (endfire), normal mode (broadside)]
Horn (pyramidal, conical, corrugated, stepped)
Lens (dielectric, waveguide)
Log periodic (dipole, monopole)
Long wire
Loop (circular, square)
Monocone (solid surface, wire frame)
Monopole (wire, sleeve, top loaded, blade)
Notch (flared Vivaldi, stepped)
Parasitic (Yagi-Uda array)
Patch (circular, rectangular, triangular, microstrip, suspended)
Reflector (corner, parabolic, shaped, dual)
Resistively loaded
Slot
Spiral (planar: Archimedean, equiangular; conical)
Subreflector (parabolic, elliptical, hyperbola, shaped)
Waveguide (open-ended: rectangular, circular)

horizontal polarization (discussed in Chapter 4). Dipoles and loops are discussed in the context of a vector sensor array in Chapter 5. Thin monopoles tend to be narrowband, but in Chapter 6 impedance matching is used to extend the bandwidth. Horizontally polarized dipole arrays are discussed in Chapters 2 and 7. Open-end circular waveguide antennas are described in Chapter 8. Microstrip patch antennas typically have a circular (Figure 1.61(g)), rectangular (Figure 1.61(h)), or square (Figure 1.61(i)) shape. Circular patch antennas are described in Chapter 9. Antennas such as the Vivaldi flared notch shown in an array in Figure 1.61(k) and the dual-polarized flared quad-ridge horn in Figure 1.61(l) have wideband performance. Reflector antennas can be offset fed (Figure 1.61(m)) or symmetrically fed (Chapters 3 and 10) and can have rolled edged (Chapter 12) or serrated edges (Figure 1.61(n)) to reduce edge diffraction. Multiple

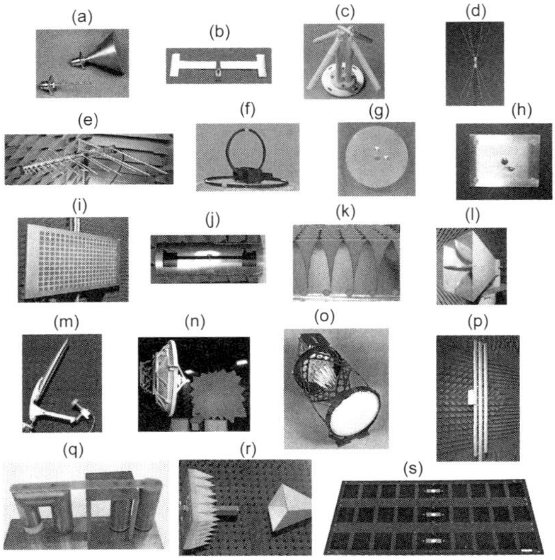

Figure 1.61 Some examples of antennas. (a) Cylindrical and conical monopoles, (b) end-loaded strip dipole on flexible circuit, (c) dual-polarized swept-back dipole, (d) wire biconical, (e) log periodic, (f) loops, (g) circular patch, dual polarized, (h) rectangular patch, dual band, (i) phased array of square patches, (j) slotted cylinder, (k) Vivaldi array, (l) quad-ridge horn, (m) offset-fed reflector, (n) parabolic reflector and serrated-edge compact range reflector, (o) multiple beam array of conical horns with dielectric tubes feeding a dielectric lens, (p) parasitic dipole array, (q) parasitic monopole array, (r) rectangular waveguide probe and standard gain pyramidal horn, (s) resistively loaded dipole array.

beams can be generated with multiple feeds displaced from the focal point of a lens antennas (Figure 1.61(o)) [39, pp. 177–189]. Parasitic dipole arrays (Figure 1.61(p)) [47, pp. 214–222] and parasitic monopole arrays (Figure 1.61(q)) [39, pp. 227–238] can achieve large bandwidths. Open end rectangular waveguide probes surrounded with absorber are used in near field testing of pyramidal horns and other antennas (Figure 1.61(r)) (Chapter 12). Resistively loaded dipole arrays (Figure 1.61(s)) are used in ultrawideband ground penetrating radar antenna applications [47, pp. 181–196]. Advances in materials and fabrication techniques such as 3D printing (Chapter 9), metamaterials, and graphene for antenna development are being investigated by a number of researchers, and may lead to future improvements in antenna technology [119–122].

1.16 Summary

The theory given in this chapter can be used to gain some basic insight into the fundamental radiation properties of antennas. In the case of an array of antennas, the radiation properties of one array antenna element have an effect on the surrounding array antenna elements, and thus the interaction between the array antenna elements must be taken into account in any detailed analysis. Some methods for computing the performance of phased array antennas taking account of mutual coupling effects, along with examples, are discussed in Chapter 2.

1.17 PROBLEM SET

1.1 An antenna has a reflection coefficient magnitude of 0.33. Calculate the reflection coefficient (return loss) in dB and mismatch loss in dB. Answer: Return loss is -9.6 dB, mismatch loss is 0.5 dB.

1.2 Show that the input impedance given by Equation (1.29) at a point $z = -l$ from an antenna with reflection coefficient Γ simplifies to Equation (1.31).

1.3 An antenna is connected to a transmission line with characteristic impedance 50 ohms. The reflection coefficient has magnitude of 0.33 and phase angle $40°$ referenced to the terminals of the antenna. Calculate the input impedance at the terminals of the antenna and at a distance $\lambda/4$ from the terminals. Answer: 73.1 $+j33.0$ ohms at the terminals, and 28.4 $-j12.8$ ohms at a distance $\lambda/4$ from the terminals.

1.4 A one-half wavelength electrically thin dipole has an input impedance equal to $73 + j42.5$ ohms at 300 MHz. The imaginary part is positive so the input impedance is inductive. What value of a lumped element series capacitor can be used used to bring the antenna to resonance? Answer: 22.6 pF.

1.5 Using Maxwell's curl equation for the magnetic field, derive the continuity equation given by Equation (1.105).

1.6 Consider an electrically conducting medium. Derive expressions for the attenuation and phase constants as given by Equations (1.365) and (1.366), respectively.

1.7 Using Equation (1.372), compute the skin depth δ at 300 MHz for good conductors with the following electrical conductivities: $\sigma_{\text{aluminum}} = 3.5 \times 10^7$ S/m, $\sigma_{\text{brass}} = 1.1 \times 10^7$ S/m, $\sigma_{\text{copper}} = 5.7 \times 10^7$ S/m, $\sigma_{\text{gold}} = 4.1 \times 10^7$ S/m, and $\sigma_{\text{silver}} = 6.1 \times 10^7$ S/m. Answer: $\delta_{\text{aluminum}} = 4.9\ \mu$m, $\delta_{\text{brass}} = 8.8\ \mu$m, $\delta_{\text{copper}} = 3.9\ \mu$m, $\delta_{\text{gold}} = 4.5\ \mu$m, $\delta_{\text{silver}} = 3.7\ \mu$m.

1.8 A 3D printed antenna uses an electrically conducting layer of silver with thickness 200 μm. At 300 MHz, using Equation (1.372) how many skin depths does this thickness represent? Answer: 54.

1.9 Consider a 1m diameter circular loop antenna radiating at 1 MHz with 1A electric current. Using Equation (1.486), compute and plot the total magnetic flux density (B-field) and the $1/r$, $1/r^2$, and $1/r^3$ terms versus range distance from 1m to 10^5m for $\theta = 90°$. The plot should look similar to Figure 1.26. At what value of range distance r are the $1/r$, $1/r^2$, and $1/r^3$ terms equal? Answer: $r = 47.8$m, which is the boundary between the near field and far field for this antenna design.

1.10 The far-zone electric field of a z-polarized Hertzian electric dipole with electric current I and length l is given by Equation (1.494), that is, $E_\theta = j\omega\mu Il\frac{e^{-jkr}}{4\pi r}\sin\theta$. Suppose that the dipole length is 1m and the dipole current is 1A with a radiating frequency 1 MHz. Assume that the permeability of free space is $\mu = 1.257 \times 10^{-6}$ Henries/m. (a) What is the magnitude of the electric field (volts/m) in free space when $\theta = 90°$ and the range distance is 1 km? (b) What is the magnitude of the magnetic field (A/m) in free space when $\theta = 90°$ and the range distance is 1 km? Hint: Refer to Equation (1.440). (c) What is the time-average Poynting's power density (W/m^2) in free space when $\theta = 90°$ and the range distance is 1 km? Hint: use Equation (1.585) $S = \frac{1}{2}|E|^2/\eta$ Answers: (a) 0.63 mV/m, (b) 1.67 μA/m, (c) 0.53 nW/m^2.

References

[1] T.K. Sarkar, R.J. Mailloux, A.A. Oliner, M. Salazar-Palma, D.L. Sengupta, *History of Wireless,* Hoboken, NJ: Wiley, 2006.

[2] C.A. Balanis, *Advanced Engineering Electromagnetics, 2nd ed.* New Jersey: Wiley, 2012.

[3] J.A. Kong, *Electromagnetic Wave Theory.* Cambridge, Mass.: EMW Publishing, 2008.

[4] D.T. Paris and F.K. Hurd, *Basic Electromagnetic Theory.* New York: McGraw-Hill, 1969.

[5] J.A. Stratton, *Electromagnetic Theory.* New York: McGraw-Hill, 1941.

[6] J.D. Kraus, *Electromagnetics.* New York: McGraw-Hill, 1953.

[7] R.F. Harrington, *Time-Harmonic Electromagnetic Fields.* New York: McGraw-Hill, 1961.

[8] W.H. Hayt and J.A. Buck, *Engineering Electromagnetics, 6th ed.* New York: McGraw-Hill, 2001.

[9] C.H. Durney and C.C. Johnson, *Introduction to Modern Electromagnetics.* New York: McGraw-Hill, 1969.

[10] J.D. Kraus and K.R. Carver, *Electromagnetics, 2nd ed.* New York: McGraw-Hill, 1973, pp. 612–613.

[11] D.K. Cheng, *Field and Wave Electromagnetics.* Reading, Mass.: Addison-Wesley, 1983.

[12] S. Ramo, J.R. Whinnery, and T. van Duzer, *Fields and Waves in Communication Electronics.* New York: Wiley, 1984.

[13] M.N.O. Sadiku, *Elements of Electromagnetics, 2nd ed.* New York: Oxford University Press, 1995.

[14] G.S. Smith, *An Introduction to Classical Electromagnetic Radiation.* Cambridge, UK: Cambridge University Press, 1997.

[15] J.D. Kraus and R.J. Marhefka, *Antennas for All Applications, 3rd ed.* New York: McGraw-Hill, 2002.

[16] R.W.P. King, *The Theory of Linear Antennas,* Cambridge, MA: Harvard University Press, 1956.

[17] S.A. Schelkunoff and H.T. Friis, *Antennas: Theory and Practice.* New York: Wiley, 1952.

[18] S. Silver, ed., *Microwave Antenna Theory and Design,* MIT Radiation Lab Series, vol. 12, New York: McGraw-Hill, 1949.

[19] C.H. Walter, *Traveling Wave Antennas*, New York: McGraw-Hill, 1968.

[20] W.L. Weeks, *Antenna Engineering*, New York: McGraw-Hill, 1968.

[21] R.S. Elliott, *Antenna Theory and Design*, New York: Prentice-Hall, 1981.

[22] A.W. Rudge, K. Milne, A.D. Olver, and P. Knight, *The Handbook of Antenna Design, vols. 1 and 2* London: Peter Perigrinus, 1992.

[23] W.L. Stutzman and G.A. Thiele, *Antenna Theory and Design, 2nd ed.* New York: Wiley, 1998.

[24] J. Volakis, ed., *Antenna Engineering Handbook, 4th ed.* New York: McGraw-Hill, 2007.

[25] C.A. Balanis, *Antenna Theory and Design, 3rd ed.* New York: Wiley, 2005.

[26] R.A. Burberry, *VHF and UHF Antennas* London: Peter Peregrinus, 1992.

[27] T.A. Milligan, *Modern Antenna Design, 2nd ed.* Hoboken, NJ: Wiley-IEEE Press, 2005

[28] L.V. Blake and M.W. Long, *Antennas: Fundamentals, Design, Measurement*, Raleigh, NC: SciTech Publishing, 2009.

[29] W. Stutzman, "Bibliography for Antennas: A List of Every English-Language Antenna Book Ever Written," *IEEE Antennas and Propagation Magazine*, vol. 50, no. 4, August 2008, pp. 128–143.

[30] N. Amitay, V. Galindo, and C.P. Wu, *Theory and Analysis of Phased Array Antennas.* New York: Wiley, 1972, ch. 1.

[31] A.A. Oliner and G.H. Knittel, eds., *Phased Array Antennas.* Dedham, Mass.: Artech House, 1972.

[32] R.C. Hansen, ed., *Significant Phased Array Papers*. Dedham, Mass.: Artech House, 1973.

[33] R.J. Mailloux, *Phased Array Handbook, 2nd ed.* Norwood, Mass.: Artech House, 2005.

[34] E. Brookner, ed., *Practical Phased-Array Antenna Systems*. Norwood, Mass.: Artech House, 1991.

[35] R.C. Hansen, *Microwave Scanning Antennas, vol. II: Array Theory and Practice.* Los Altos, Calif.: Peninsula Publishing, 1985.

[36] R.C. Hansen, *Phased Array Antennas, 2nd ed.* New York: Wiley, 2009.

[37] A.K. Bhattacharyya, *Phased Array Antennas.* New York: Wiley, 2006.

[38] T. Jeffrey, *Phased-Array Radar Design: Application of Radar Fundamentals.* Raleigh, North Carolina: Scitech Pub Inc., 2009.

[39] A.J. Fenn, *Adaptive Antennas and Phased Arrays for Radar and Communications.* Norwood, Mass.: Artech House, 2008.

[40] S.P. Skobelev, *Phased Array Antennas with Optimized Element Patterns.* Norwood, Mass.: Artech House, 2011.

[41] H.J. Visser, *Array and Phased Array Antenna Basics.* New York: Wiley, 2005.

[42] R.L. Haupt, *Antenna Arrays: A Computational Approach.* New York: Wiley-IEEE Press, 2010.

[43] D. Ehyale, *Novel Approaches to the Design of Phased Array Antennas: Low Cost, Low Complexity Phased Array Antenna Design.* LAP Lambert Academic Publishing, 2012.

[44] E. Brookner, "Phased-Array Radars," *Scientific American,* vol. 252, no. 2, 1985, pp. 94–102.

[45] A.J. Fenn, D.H. Temme, W.P. Delaney, and W.E. Courtney, "The Development of Phased Array Radar Technology," *Lincoln Laboratory Journal*, vol. 12, no. 2, 2000, pp. 321–340.

[46] R.J. Mailloux, "Phased Array Theory and Technology," *Proc. IEEE*, vol. 70, no. 3, 1982, pp. 246–291.

[47] A.J. Fenn and P.T. Hurst, *Ultrawideband Phased Array Antenna Technology for Sensing and Communications Systems*, Cambridge, MA: The MIT Press, 2015.

[48] P.W. Howells, "Explorations in Fixed and Adaptive Resolution at GE and SURC," *IEEE Trans. Antennas and Propagation*, vol. 24, no. 5, 1976, pp. 575–584.

[49] S.P. Applebaum, "Adaptive Arrays," *IEEE Trans. Antennas and Propagation*, vol. 24, no. 5, 1976, pp. 585–598.

[50] W.F. Gabriel, "Adaptive Arrays – An Introduction," *Proc. IEEE,* vol. 64, 1976, pp. 239–271.

[51] J.T. Mayhan, "Some Techniques for Evaluating the Bandwidth Characteristics of Adaptive Nulling Systems," *IEEE Trans. Antennas and Propagation,* vol. 27, no. 3,

1979, pp. 363–373.

[52] J.T. Mayhan, A.J. Simmons, and W.C. Cummings, "Wide-Band Adaptive Antenna Nulling Using Tapped Delay Lines," *IEEE Trans. Antennas and Propagation,* vol. 29, no. 6, 1981, pp. 923–936.

[53] A.J. Fenn, "Evaluation of Adaptive Phased Array Far-Field Nulling Performance in the Near-Field Region," *IEEE Trans. Antennas and Propagation*, vol. 38, no. 2, 1990, pp. 173–185.

[54] R.A. Monzingo and T.W. Miller, *Introduction to Adaptive Arrays*. New York: Wiley, 1980.

[55] R.T. Compton, Jr., *Adaptive Antennas, Concepts and Performance*. Upper Saddle River, New Jersey: Prentice-Hall, 1988.

[56] M.M. Weiner, *Adaptive Antennas and Receivers*. Boca Raton, Florida: CRC Press, 2006.

[57] S. Chandran, ed., *Adaptive Antenna Arrays: Trends and Applications*. Berlin: Springer-Verlag, 2004.

[58] D.G. Manolakis, V.K. Ingle, and S.M. Kogon, *Statistical and Adaptive Signal Processing: Spectral Estimation, Signal Modeling, Adaptive Filtering, and Array Processing*. Norwood, Mass.: Artech House, 2005.

[59] R. Nitzberg, *Adaptive Signal Processing for Radar*. Norwood, Mass.: Artech House, 1992.

[60] A. Farina, *Antenna-Based Signal Processing Techniques for Radar Systems*. Norwood, Mass.: Artech House, 1992.

[61] J.E. Hudson, *Adaptive Array Principles*. New York: Peter Peregrinus, 1981.

[62] D.M. Pozar, *Microwave Engineering, 4th ed.*, Hoboken, NJ: Wiley, 2012.

[63] P. H. Smith, "Transmission line calculator," *Electronics*, vol. 12, no. 1, pp. 29–31, Jan. 1939.

[64] P. H. Smith, "An Improved Transmission Line Calculator," *Electronics*, vol. 17, no. 1, pp. 130–133, p. 318, p. 320, p. 322, and pp. 324–325, Jan. 1944.

[65] P. H. Smith, "Electronic Applications of the Smith Chart," – In *Waveguide, Circuit, and Component Analysis*, 2nd ed. Raleigh, NC: SciTech Publishing, 1995.

[66] T. Lee, "The Smith Chart Comes Home," IEEE Microwave Magazine, p. 10, 12, and 25, Nov. 2015, Digital Object Identifier: 10.1109/MMM.2015.2467953, Date of Publication: 14 Oct. 2015.

[67] H.J. Reich, ed., Harvard Radio Research Laboratories, *Very High Frequency Techniques*, New York: Mc-Graw Hill, Vol. 1, Chapter 3, 1947.

[68] R.M. Fano, "Theoretical Limitations on the Broadband Matching of Arbitrary Impedances," Massachusetts Institute of Technology, Research Laboratory of Electronics, Technical Report 41, Jan. 2, 1948.

[69] H.A. Wheeler, "Wideband Impedance Matching," Wheeler Laboratories Report 418, Great Neck, New York, 1950.

[70] V.H. Rumsey, "The Design of Frequency-Compensating Matching Sections," *Proc. IRE*, October 1950, pp. 1191–1196.

[71] R.L. Thomas, *A Practical Introduction to Impedance Matching*, Dedham, MA: Artech House, 1976.

[72] W.N. Caron, *Antenna Impedance Matching*, Newington, CT: American Radio Relay League, 1989.]

[73] G.L. Matthaei, L. Young, E.M.T. Jones, *Microwave Filters, Impedance-Matching Networks, and Coupling Structures*, New York, McGraw-Hill, 1964.

[74] T.R. Cuthbert, *Circuit Design Using Personal Computers*, New York: John Wiley, 1983.

[75] T.R. Cuthbert, *Broadband Direct-Coupled and Matching RF Networks*, Greenwood, Arkansas: TRCPEP Publications, 1999.

[76] C. Bowick, *RF Circuit Design*, Burlington, MA: Newnes, 2008.

[77] R. Mavaddat, *Network Scattering Parameters,* Advanced Series in Circuits and Systems, vol. 2, Singapore, World Scientific Publishing, 1996.

[78] D.A. Frickey, "Conversions Between S, Z, Y, h, ABCD, and T Parameters Which are Valid for Complex Source and Load Impedances," *IEEE Trans. Microwave Theory and Techniques*, vol. 42, no. 2, February 1994, pp. 205-211.

[79] W.F. Egan, *Practical RF System Design*, New York: IEEE, Hoboken, New Jersey: Wiley.

[80] M. Lazebnik, E.L. Madsen, G.R. Frank, and S.C. Hagness, "Tissue-Mimicking Phantom Materials for Narrowband and Ultrawideband Microwave Applications," *Physics in Medicine and Biology*, vol. 50, 2005, pp. 4245–4258.

[81] S. Gabriel, R.W. Lau, and C. Gabriel, "The Dielectric Properties of Biological Tissues: Part III. Parametric Models for the Dielectric Spectrum of Tissues," *Physics in Medicine and Biology,* vol. 41, 1996, pp. 2271–2293, online http://niremf.ifac.cnr.it/tissprop.

[82] C-K Chou and J. D'Andrea, *IEEE Standard for Safety Levels with Respect to Human Exposure to Radio Frequency Electromagnetic Fields, 3 kHz to 300 GHz*, IEEE Std C95.1-2005. New York: Institute of Electrical and Electronics Engineers, Inc.

[83] J.F. Lehmann, ed., *Therapeutic Heat and Cold, 3rd ed.* Baltimore, Maryland: Williams & Wilkins, 1982.

[84] A.R. von Hippel, *Dielectrics and Waves.* New York: John Wiley, 1954.

[85] C. Cohen-Tannoudji, J. Dupont-Roc, and G. Grynberg, *Photons and Atoms: Introduction to Quantum Electrodynamics.* Weinheim: Wiley-VCH, Verlag GmbH & Co. KGaA, 2004.

[86] R. Chrapkiewicz, M. Jachura, K. Banaszek, W. Wasilewski, "Hologram of a Single Photon," *Nature Photonics*, vol. 10, 2016, pp. 576–579.

[87] J.D. Jackson, *Classical Electrodynamics, 3rd ed.* New York: Wiley, 1999, p. 3.

[88] J.C. Maxwell, *A Treatise on Electricity and Magnetism, vols. 1 and 2.* Oxford, UK: Clarendon Press, 1873.

[89] C.T. Tai, *Dyadic Green's Functions in Electromagnetic Theory.* London: Intext Educational Publishers, 1971.

[90] C.T. Tai, *Generalized Vector and Dyadic Analysis.* New York: IEEE Press, 1992.

[91] C.T. Tai, "A Survey of the Improper Uses of ∇ in Vector Analysis," Radiation Laboratory, The University of Michigan, Technical Report RL 909, November 1994.

[92] C.T. Tai, "A Historical Study of Vector Analysis," Radiation Laboratory, The University of Michigan, Technical Report RL 915, May 1995.

[93] G. Green, "An Essay on the Application of Mathematical Analysis to the Theories of Electricity and Magnetism," Nottingham, England, printed for the author by T. Wheelhouse, 1828.

[94] R. Nevels and J. Jeong, "The Time Domain Green's Function and Propagator for Maxwell's equations," *IEEE Trans. Antennas and Propagat.*, vol. 52, no. 11, 2004, pp. 3012–3018.

[95] J.H. Poynting, "On the Transfer of Energy in the Electromagnetic Field," *Philosophical Transactions of the Royal Society of London*, vol. 175, 1884, pp. 343–361.

[96] H. Mott, *Antennas for Radar and Communications: A Polarimetric Approach,* New York: John Wiley, 1992.

[97] W.L. Stutzman, *Polarization in Electromagnetic Systems.* Norwood, Mass.: Artech House, 1993.

[98] W.H. Kummer, (Chair), "IEEE Standard Test Procedures for Antennas," IEEE Std 149-1979 (R2008), ISBN 0-471-08032-2, 2008.

[99] *IEEE Standard for Ultrawideband Radar Definitions*, IEEE Std 1672-2006, 4 May 2007 (2nd printing, 11 Dec. 2008), doi: 10.1109/IEEESTD.2007.359972.

[100] J.D. Taylor, *Introduction to Ultra-Wideband Radar Systems.* Boca Raton, Florida: CRC Press, 1995.

[101] L. Peters, Jr., J.J. Daniels, and J.D. Young, "Ground Penetrating Radar as a Subsurface Environmental Sensing Tool," *Proc. IEEE*, vol. 82, no. 12, 1994, pp. 1802–1822.

[102] D. Uduwawala et al., "A Deep Parametric Study of Resistor-Loaded Bow-Tie Antennas for Ground-Penetrating Radar Applications Using FDTD," *IEEE Trans Geoscience and Remote Sensing*, vol. 42, no. 4, 2004, pp. 732–742.

[103] T.P. Montoya and G.S. Smith, "Land Mine Detection Using a Ground Penetrating Radar Based on Resistively Loaded Vee Dipoles," *IEEE Trans. Antennas and Propagation,* vol. 47, no. 12, 1999, pp. 1795–1806.

[104] V.H. Rumsey, *Frequency Independent Antennas.* New York: Academic Press, Inc., 1966.

[105] H. Schantz, *The Art and Science of Ultrawideband Antennas.* Norwood, Mass.: Artech House, 2005.

[106] D. Valderas, J.I. Sancho, D. Puente, C. Ling, and X. Chen, *Ultrawideband Antennas,*

Design and Applications. London: Imperial College Press, 2011.

[107] T.S. Ralston, G.L. Charvat, and J.E. Peabody, "Real-Time Through-Wall Imaging Using an Ultrawideband Multiple-Input Multiple-Output (MIMO) Phased Array Radar System," *IEEE Int. Symp. on Phased Array Systems and Technology*, 2010, pp. 551–558.

[108] A. Rajagopalan, G. Gupta, A.S. Konanur, B. Hughes, and G. Lazzi, "Increasing Channel Capacity of an Ultrawideband MIMO System Using Vector Antennas," *IEEE Trans. on Antennas and Propagation*, vol. 55, no. 10, 2007, pp. 2880–2887.

[109] C.M. Keller, and D.P. Young, "Ultra-Wideband (UWB) Signal Localization Using a Vehicle-Sized Array," *2005 IEEE International Conference on Ultra-Wideband*, 2005, pp. 290–295.

[110] E.W. Reid, L. Ortiz-Balbuena, A. Ghadiri, and K. Moez, "A 324-Element Vivaldi Antenna Array for Radio Astronomy Instrumentation," *IEEE Trans. Instrumentation and Measurement*, vol. 61, no. 1, 2012, pp. 241–250.

[111] L.J. Chu, "Physical Limitations of Omnidirectional Antennas," *Journal of Applied Physics*, 1948, vol. 19, pp. 1163–1175.

[112] H.A. Wheeler, "Fundamental Limitations of Small Antennas," *Proc. Institute of Radio Engineers*, 1947, 35, pp. 1479–1484.

[113] R. F. Harrington, "Effect of Antenna Size on Gain, Bandwidth and Efficiency," *J. Research National Bureau of Standards*, vol. 64-D, Jan./Feb. 1960, pp. 1–12.

[114] R. C. Hansen, "Fundamental Limitations in Antennas," *Proc. of the IEEE*, vol. 69, no. 2, Feb. 1981, pp. 170–182.

[115] J.S. McLean, "A Re-Examination of the Fundamental Limits on the Radiation Q of Electrically Small Antennas," *IEEE Trans. Antennas Propag.*, vol. AP-44, 1996, pp. 672–676.

[116] A.D. Yaghjian and S.R. Best, "Impedance, Bandwidth, and Q of Antennas," *IEEE Trans. Antennas and Propagat.*, vol. 53, no. 4, April 2005, pp. 1298–1324.

[117] G. Brown, "The Turnstile Antenna," *Electronics*, 1936.

[118] H. T. Friis, "A Note on a Simple Transmission Formula," *Proc. Institute of Radio Engineers*, vol. 34, no. 5, May 1946, pp. 254–256.

[119] M. Liang and H. Xin, "Three-Dimensionally Printed/Additive Manufactured Antennas," in *Handbook of Antenna Technologies*, Z.N. Chen, (ed.), Singapore: Springer Science, 2016, pp. 661–697.

[120] E. Brookner, "Metamaterial Advances for Radar and Communications," *2017 IEEE Radar Conference*,2017 pp. 1614–1621.

[121] B.A. Munk, *Metamaterials, Critique and Alternatives,* Hoboken, New Jersey: Wiley, 2009.

[122] X. Huang, A. Alburaikan, T. Leng, Z. Hu, Y. Qin, and P. Liu, Graphene Metamaterials Array Based Reconfigurable Antenna, *2016 Int. Symposium on Antennas and Propagation (ISAP)*, 24–28 Oct. 2016, pp. 106–107.

2

Phased Array Antennas

2.1 Introduction

In Chapter 1, electromagnetic theory for antennas was reviewed in detail. In this chapter, the subject of phased array antennas [1–62] is briefly reviewed. Development of phased array radar technology is described by Fenn et al. [22]. The concept of a phased array and early experiments were presented over 100 years ago in a 1909 Nobel Prize lecture by Braun [54]. A history of phased arrays has been reviewed by Mailloux [55]. Delaney [56] has reviewed the development and deployment of operational phased array radars, with an emphasis on achieving a goal of a solid state phased array. Solid-state phased arrays are now being utilized in many radar applications as described by Brookner [57]. Herd and Conway [58] have described the evolution of low-cost solid-state phased arrays, for radar weather sensing and aircraft surveillance, that utilize high-volume commercial manufacturing and packaging technologies that implement a modern tile array architecture. Millimeter-wave phased arrays have been mass produced for automotive radar systems for collision avoidance [59]. Millimeter-wave phased arrays are being investigated for fifth generation (5G) mobile and wireless communications systems for high-data-rate applications [60-62].

This chapter begins with basic phased array theory in Section 2.2. To take into account the simulated effects of array mutual coupling, equivalence principles (Section 2.3), the reciprocity theorem (Section 2.4) and reaction integral equation (Section 2.5) are reviewed, from which the method of moments (Section 2.6) is derived. The method of moments is used in many of the electromagnetic simulations of antennas presented in this book. Section 2.7 provides examples of broadside and endfire linear arrays of Hertzian (short) dipoles. Examples of mutual coupling effects on the embedded element pattern in a two-dimensional array of linear dipoles in free space and linear monopoles over a ground plane are shown in Section 2.8. For wide-angle scanning, swept-back dipole antenna elements are used in a number of phased array radar applications. Swept-back dipole array measurements and simulations are presented in Section 2.9. Simulations of the scanned radiation patterns for a linear array of rectangular waveguides are presented in Section 2.10. Section 2.11 provides a summary and Section 2.12 provides a problem set.

2.2 Phased Array Basics

2.2.1 Introduction

Antenna systems are often required to have apertures with significant radiated power levels, sensitive receive capability, rapid beam scanning, wide bandwidth, and durability. Phased array antennas have been deployed on many ground based systems, automobiles, ships, aircraft, and satellites. Phased array antenna systems with just a few antenna elements or with hundreds or thousands of individual radiating antenna elements can provide electronically controlled beam agility and graceful degradation. Adaptive arrays can be used in a number of applications such as to suppress interference (jamming) and to suppress radar clutter (background scatter) [63–73], and for cancer treatment with focused microwaves [74, 75]. Antennas operating with desired performance characteristics over very wide bandwidths are desirable in many applications. An antenna design that performs well over very wide bandwidths [76–85] avoids the situation of having to develop multiple, different antenna designs to cover multiple bands spanning a large frequency range.

Figure 2.1 depicts a phased array aperture with multiple antenna elements (represented as dots) that produces a narrow-beam gain radiation pattern that can be electronically scanned over some field of view. In general, the phased array aperture can be composed of radiating and/or receiving antenna elements that are located on a planar or conformal surface. Figure 2.2 shows an N-element two-dimensional planar phased array with a beamforming network that produces a phase coherent combined main channel. In this diagram the antenna array is located on a rectangular ground plane. Depending on required gain, beamwidth, sidelobe levels and other requirements, the antenna elements can be located in a periodic lattice, either rectangular or triangular, or they could be sparsely or randomly positioned on the ground plane. The amplitude and phase of RF signals to and from the antenna elements are electronically adjusted using complex element weighting as $\boldsymbol{W} = (W_1, W_2, \cdots, W_N)^T$, where the superscript T means matrix transpose. The combiner is used to provide a phase coherent summing junction with either uniform amplitude combining or with a fixed taper. The combiner is typically an RF component, although if the element signals are digitized the combiner can be accomplished by digital summation of the data. When the antenna element signals are coherently combined, the resulting radiation pattern will have a main lobe scanned to some desired angle and there will be a number of sidelobes formed away from the main beam. An adaptive beamformer as shown can be added to this architecture if desired

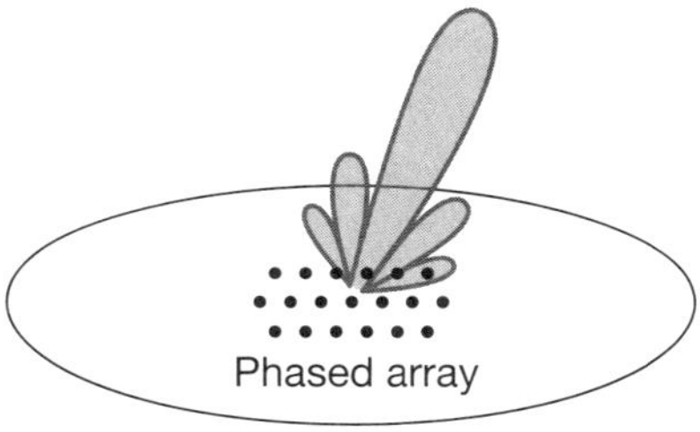

Figure 2.1 Phased array aperture radiating from a general surface that produces an electronically steerable narrow-beam radiation pattern.

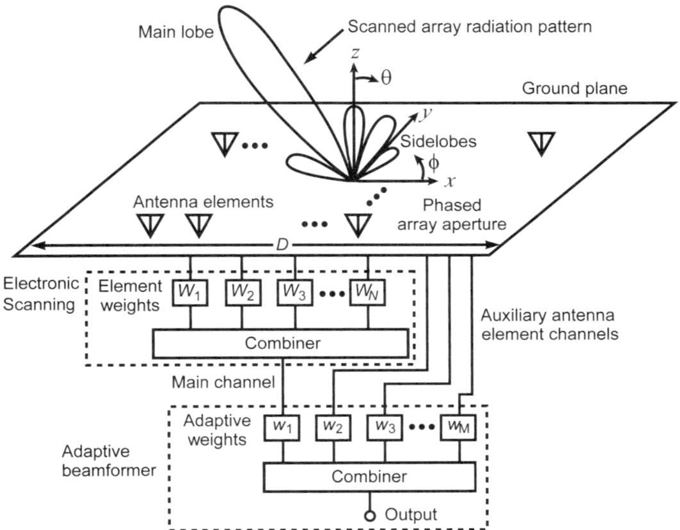

Figure 2.2 Two-dimensional phased array antenna aperture with a beamforming network that electronically scans a narrow-beam radiation pattern. Auxiliary antenna elements and adaptive weights provide adaptive nulling capability.

for providing adaptive sidelobe cancellation of interference. In this case, M auxiliary antenna element channels are provided with complex adaptive nulling weights denoted as $\boldsymbol{w} = (w_1, w_2, \cdots, w_M)^T$. These auxiliary antenna elements can be separate antennas or they could use signals coupled from the antenna elements used in the phased array aperture.

A sketch of a phased array antenna aperture in spherical coordinates is shown in Figure 2.3. In this diagram, the aperture of the array is oriented parallel to the yz plane. As such, the positive z axis could point to zenith and the angle ϕ could represent azimuth. The angle $\theta = 90°$ would be taken as the horizontal plane, and the angle $90° - \theta$ would then be the elevation angle with respect to the horizon. In the far field, depending on the array element design and antenna orientation, one or both of two orthogonal electric-field components (E_θ, E_ϕ) can be produced, and in the near field, a radial component of the electric field (denoted E_r) can also be generated or received. If the array elements are, for example, linear dipoles oriented in the $\hat{z}$ direction, then only an E_θ far-field component would be transmitted or received. If the linear dipole elements are tilted away from the $\hat{z}$ direction, then in general both E_θ and E_ϕ far-field components will be produced. Besides dipoles, many other types of antennas such as monopoles, waveguides, slots, horns, patches, log periodics, spirals, helix antennas, Yagi-Uda antennas, biconical antennas, reflector antennas, and lenses have been used in phased array applications [4].

Some of the key parameters that are involved in phased array antenna design are bandwidth, polarization, scan sector, beamwidth, antenna gain, reflection coefficient, and peak and average sidelobe levels. Additional design

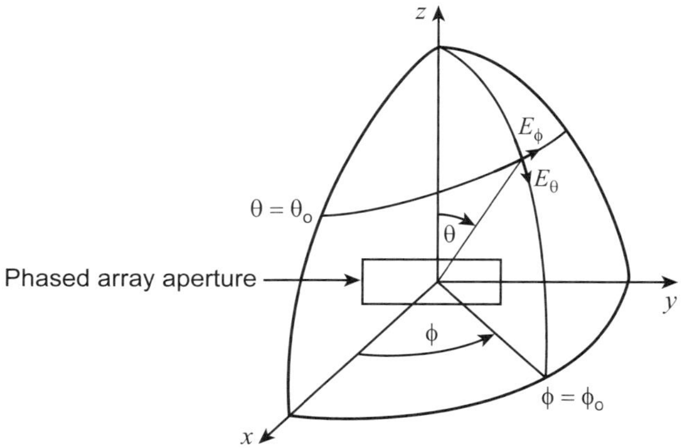

Figure 2.3 Phased array antenna aperture in the spherical coordinate system.

parameters involve volume, mass, power, thermal/cooling, complexity, and cost for the antenna array system, which need to be addressed for each unique application. Transmit/receive (T/R) modules containing phase shifters to steer the main beam, amplifiers (power amplifiers for transmit and low-noise amplifiers for receive) to provide the desired signal level and noise figure, and attenuators for generating low sidelobe radiation patterns are used in many phased arrays.

Electronic scanning of the array antenna main beam is effected by means of phase shifters and/or time delays connected to individual array elements or subarrays (groups) of elements. A phased array can be used to transmit or receive a focused electromagnetic wavefront for plane waves or spherical waves as discussed in the next section.

2.2.2 Wavefront Basics

Consider now an electromagnetic wavefront incident on an array of antennas. The source generating the wavefront can be located at close range, producing a spherical wave, or the source could be located sufficiently far from the antenna so that a plane wave is incident upon the array. In this book, we will be dealing primarily with the plane wave, far-field source case, such as for a radar antenna with a target at a range distance of many kilometers. An example of a source producing a spherical wave impinging on an array could be the case of a close-range target in a ground-penetrating radar application or a source of near-field scattering or interference.

The contrast between plane wave incidence and spherical wave incidence [15, 68] is depicted in Figure 2.4. The amount of wavefront dispersion (time delay multiplied by bandwidth) observed by the antenna is a function of the antenna length, angle of incidence, and bandwidth [42]. Dispersion can also be introduced by reflections and other frequency-dependent characteristics within the antenna beamformer or by dispersive materials used in the antenna construction. Wavefront dispersion can cause significant signal distortion unless time-delay weighting across the array antenna is used. Interference wavefront dispersion is an effect that can limit the depth of null (or cancellation) achieved by an adaptive antenna [65–67]. To compensate for wavefront dispersion, the adaptive array process can include multiple auxiliary antenna element channels and tapped-delay line weighting. In the case of a pulsed radar system, if the pulse width is narrower than the amount of time delay across the aperture, then to properly transmit or receive the radar waveform, time delay must be provided in the beamformer. Basic time delay or dispersion models for spherical wave incidence and plane wave incidence

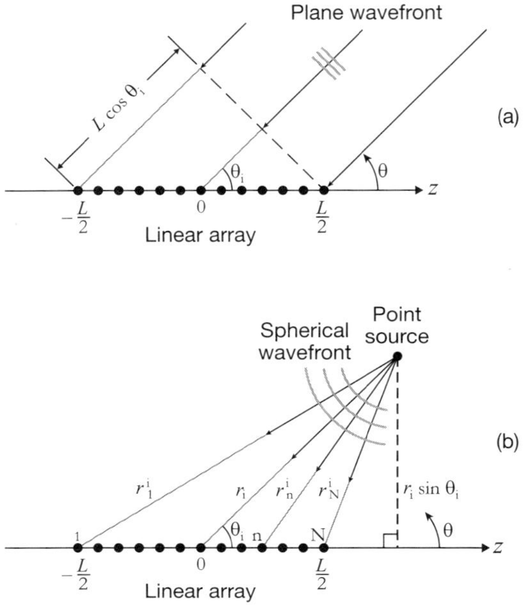

Figure 2.4 Geometry for far-field and near-field wave incidence for a linear array. (a) Plane wavefront, and (b) spherical wavefront.

can be made by considering only the wavefront time delay or dispersion observed by the end points of an array [68].

Consider first a plane wave arriving from infinity and a linear array of length L as shown in Figure 2.4(a). The far-field time delay $\Delta\tau_{FF}$ of the wavefront across the array is given by

$$\Delta\tau_{FF} = \frac{L}{c}\cos\theta_i = \frac{L}{c}\beta_{FF} \tag{2.1}$$

where

$$\beta_{FF} = \cos\theta_i \tag{2.2}$$

will be referred to as the far-field dispersion multiplier, c is the speed of light, and θ_i is the angle of incidence with respect to the axis of the array. Note that the time delay is maximum for endfire incidence ($\theta_i = 0$) and is zero for broadside incidence ($\theta_i = \pi/2$). The dispersion for this case is denoted γ_{FF} and is computed according to the product of bandwidth and time delay as

$$\gamma_{FF} = B\Delta\tau_{FF} = \frac{BL}{c}\cos\theta_i = \frac{BL}{c}\beta_{FF} \tag{2.3}$$

where B is the waveform bandwidth.

Next, consider the same array and now a point source, at range $r = r_i$ and angle $\theta = \theta_i$, which produces a spherical wavefront as depicted in Figure 2.4(b). The near-field time delay $\Delta\tau_{NF}$ is given by

$$\Delta\tau_{NF} = \frac{L}{c}\beta_{NF} \tag{2.4}$$

where the quantity β_{NF} denotes the near-field dispersion multiplier, which, from the difference between the path lengths r_1^i and r_N^i, is expressed as

$$\beta_{NF} = \sqrt{\alpha^2 + \alpha\,\cos\theta_i + \frac{1}{4}} - \sqrt{\alpha^2 - \alpha\,\cos\theta_i + \frac{1}{4}} \tag{2.5}$$

where

$$\alpha = \frac{F}{L} \tag{2.6}$$

and where F is the focal length of the array such that the source range is equal to the focal distance, that is, $r_i = F$. The near-field dispersion, denoted γ_{NF}, is given by

$$\gamma_{NF} = B\Delta\tau_{NF} = \frac{BL}{c}\beta_{NF} \tag{2.7}$$

where, again, the quantity β_{NF} denotes the near-field time-delay or dispersion multiplier in this case. In comparing Equation (2.3) and Equation (2.7), it is seen that the far-field and near-field dispersions differ only by their respective multipliers β_{FF} and β_{NF}. Figure 2.5 shows a plot of β_{FF} and β_{NF} versus angle of incidence for values of $\alpha = 0.2$ to 2 (i.e., focal lengths $0.2L$ to $2L$). From this figure, it is seen that the near-field dispersion approaches the value of the far-field dispersion for source range distances greater than approximately one aperture diameter ($\alpha \geq 1$). Clearly, at source range distances such that $\alpha \leq 0.5$ (one-half aperture diameter), the near-field dispersion is significantly different from the far-field dispersion. Thus, for this simple dispersion model, it is expected that near-field dispersion will be similar to far-field dispersion at source range distances greater than one aperture diameter.

For a phased array system with a large instantaneous bandwidth, to keep the scanned beam pointing angle fixed, it is necessary to use true time-delay phase shifters. Under far-field conditions, for a linear array with element spacing d and scan angle θ_s as measured from the axis of the array, the electrical phase shift between elements is given by

$$\Delta\Phi_{\text{elements}} = \frac{2\pi}{\lambda}d\cos\theta_s = \frac{2\pi f}{c}d\cos\theta_s. \tag{2.8}$$

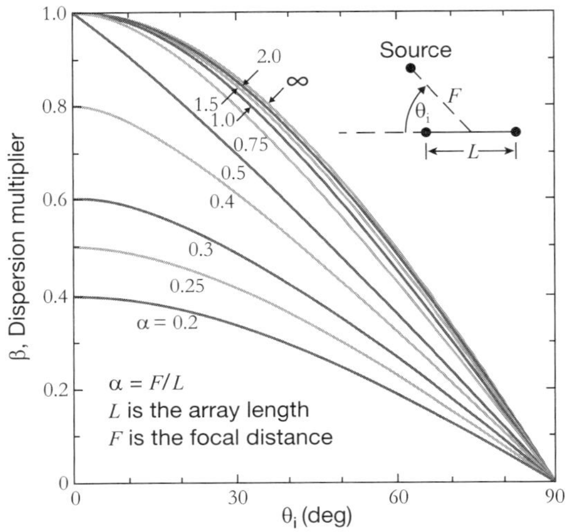

Figure 2.5 Dispersion multiplier as a function of source incidence angle for various source range distances.

To compensate for the linear phase shift versus frequency that occurs from the element spacing and scan angle, the time-delay unit must provide a linear phase shift versus frequency. A time-delay phase shifter can be designed using coaxial lines controlled by diodes. For example a six-bit time-delay phase shifter provides the following phase states for each of the bits: $(0°, 180°)$, $(0°, 90°)$, $(0°, 45°)$, $(0°, 22.5°)$, $(0°, 11.25°)$, and $(0°, 5.63°)$. It is desired to provide an interelement time delay, denoted ΔL as computed below. In a coaxial transmission line, the effective electrical phase shift of the line $\Delta\Phi_{\text{line}}$ is expressed in terms of the physical length and the relative dielectric constant ϵ_r of the material filling the coaxial region as

$$\Delta\Phi_{\text{coaxial line}} = \frac{2\pi f}{c}\Delta L_{\text{electrical}} = \frac{2\pi f}{c}\Delta L_{\text{physical}}\sqrt{\epsilon_r}. \qquad (2.9)$$

Setting Equation (2.9) equal to Equation (2.8) yields

$$\frac{2\pi f}{c}\Delta L_{\text{physical}}\sqrt{\epsilon_r} = \frac{2\pi f}{c} d\cos\theta_s \qquad (2.10)$$

It follows that the required physical length of true-time-delay coaxial line to compensate for the element spacing d and scan angle θ_s from the axis of the linear array is given by

$$\Delta L_{\text{physical}} = \frac{d}{\sqrt{\epsilon_r}}\cos\theta_s. \qquad (2.11)$$

The next section describes a number of phased array beamformer architectures.

2.2.3 Beamformer Architectures

Consider Figure 2.6, which shows a linear phased array antenna that generates a radiation main beam at an angle θ_s that can be scanned over a field of view. Depending on the spacing in wavelengths between the antenna elements and the phase progression along the array, the scan angle θ_s can vary from $0°$ (broadside direction) to $\pm 90°$ (endfire direction). The broadside array will typically use one-half wavelength spacing, whereas the endfire array will typically use one-quarter wavelength spacing. The challenge in phased array antenna development is to design the array so that main beam scanning can be achieved, with desired radiation characteristics such as gain, power density on transmit, sensitivity on receive, beamwidth, sidelobe levels, polarization, and reflection coefficient. A block diagram showing an active beamformer arrangement for a transmitting linear phased array antenna is depicted in Figure 2.7. An RF source has its signal divided into a number of paths by using a power divider circuit. Each output port from the power divider is connected to a phase shifter device that applies a linear phase shift from element to element such that the array main beam is scanned to a desired angle. In this transmitting phased array example, prior to the divided and phase-shifted signal reaching each of the antenna radiating elements, amplification is applied such that a desired power level is achieved at each array element. With this beamforming architecture where an RF amplifier is connected to each antenna element, the antenna is referred to as an active electronically scanned array (AESA). If an amplifier at each element is not present, the antenna architecture is referred to then as a passive electronically

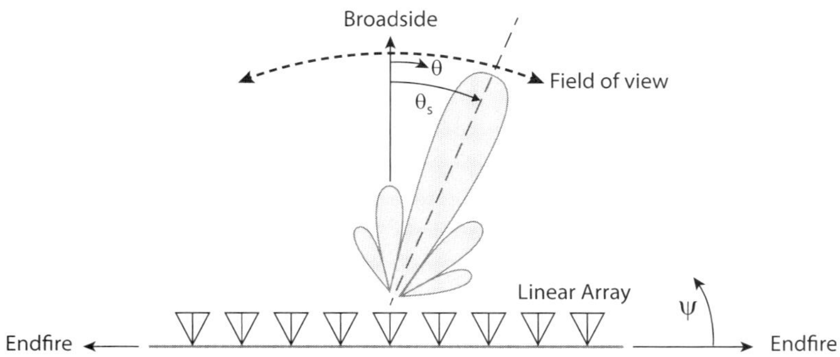

Figure 2.6 Example linear array antenna with the main beam steered to the angle θ_s.

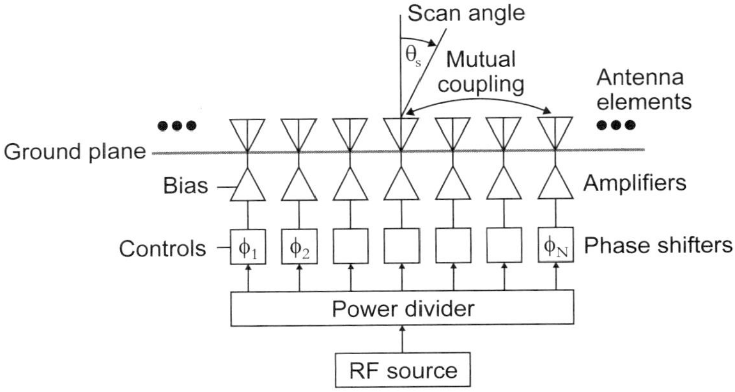

Figure 2.7 Block diagram for a transmitting active electronically scanned array (AESA) antenna. The main beam is steered using phase shifters.

scanned array (PESA). The phase-steered and amplified RF signal from each antenna element is coherent and additive in the scan angle direction. However, as a consequence of typical one-half-wavelength spacing between the array antenna elements, a portion of the signal from each array element is electromagnetically coupled (curved arrow in Figure 2.7) into the surrounding array elements, and the coupled signal generally becomes stronger/weaker as the distance to the coupled element decreases/increases, respectively. This coupled signal is referred to as array mutual coupling, which is a complex quantity.

If phase shifters with constant phase versus frequency are used to steer the main beam, then the main beam pointing angle will be frequency dependent. In a phased array application in which wideband waveforms are involved, time delays with linear phase delay versus frequency need to be used as depicted in Figure 2.8. Time delays can be implemented at intermediate frequencies (IF), radio frequencies (RF), or optical frequencies (OF).

The contrast between an active electronically scanned array (AESA) and a passive electronically scanned array (PESA) [47] is shown in Figure 2.9. In Figure 2.9(a), a transmit/receive (T/R) module is connected to each antenna element. The T/R module contains amplifiers, phase shifters, and other components including switches, circulators, and filters. In a large phased array aperture with thousands of antenna elements, if a T/R module (or a small percentage of modules) fails to operate properly, the array experiences only a minor degradation in performance. In Figure 2.9(b), only a phase shifter is connected to the antenna elements. The PESA requires a high-power transmitter that represents the possibility of a single-point failure. The

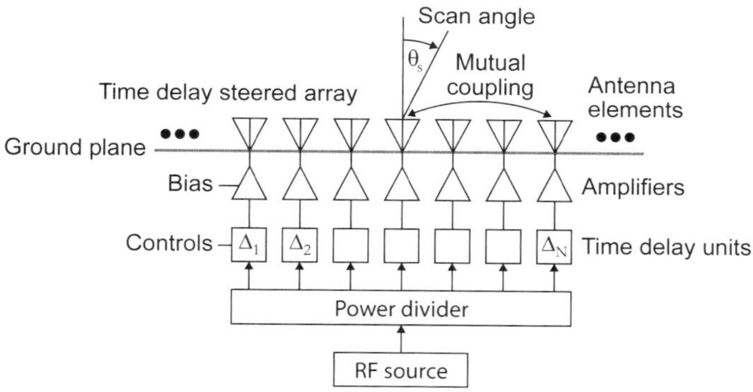

Figure 2.8 Example beamformer for time-delay steered linear array.

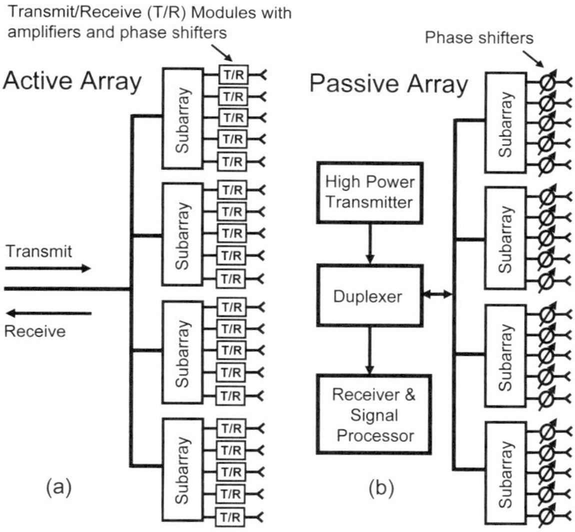

Figure 2.9 Example array beamforming architectures. (a) Active electronically scanned array (AESA). (b) passive electronically scanned array (PESA).

duplexer can be implemented by either a high-power switch or a circulator and represents another possible single-point failure for a PESA. In Figure 2.9, both types of arrays shown are corporate fed with transmission lines that can be, for example, coaxial cables, microstrip lines, or waveguides. In some phased arrays, to eliminate the RF transmission lines a space-fed beamformer approach is used.

Consider Figure 2.10 that shows two types of space-fed phased

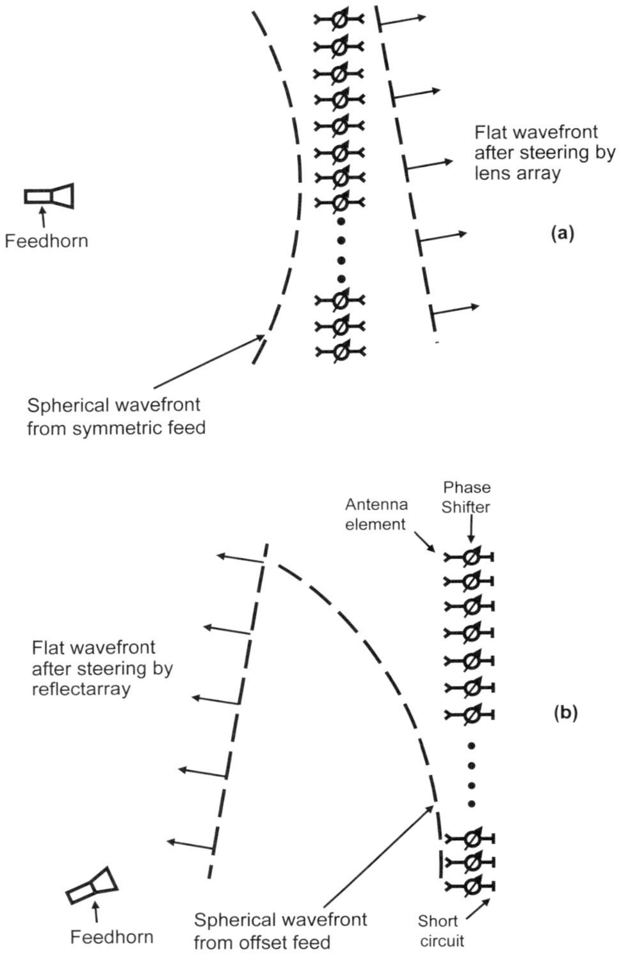

Figure 2.10 Beamforming architectures. (a) Space-fed lens array, (b) reflectarray.

array architectures [47, p.13.47], namely, a space-fed lens and a space-fed reflectarray. In Figure 2.10(a), a horn antenna generates a spherical wavefront that illuminates a passive lens phased array aperture with phase shifters located between the front and back antenna elements. The phase shifters are calibrated and adjusted to remove the spherical wavefront curvature and generate a flat wavefront to steer the radiation to a desired scan angle. In Figure 2.10(b), the reflect array has a single group of antenna elements facing the feedhorn. In this case, the spherical wavefront illuminates the antenna elements and the received signals pass through a calibrated phase shifter, and then these signals are reflected by a short circuit. The reflected signal passes backward through the phase shifter and the resultant wavefront is now flat and steered to a desired scan angle.

A fully digital beamforming receive array architecture is shown in Figure 2.11. With this digital architecture, the RF signals received by the antenna elements are downverted to a lower intermediate frequency (IF) (and possibly to a lower baseband frequency) where analog to digital (A/D) converters are used to digitize the received signals. Digital signal processing is used to detect and analyze received signals as well as to provide adaptive nulling to reduce jamming effects. A hybrid analog/digital beamforming phased array architecture is shown in Figure 2.12. In this case, RF phase shifters are used as part of several RF subarrays where the subarray RF output is then downconverted and digitized.

To achieve a large aperture phased array that can electronically scan a beam over a large field of view, say $\pm 60^\circ$, normally a large number (many thousands) of antenna elements is often required. An alternate geometry is to have a smaller phased array feed a single reflector or multiple reflector antennas [86–89], which can provide limited electronic scanning of a few degrees. For example, Figure 2.13 shows an example of an offset Gregorian reflector antenna system with confocal paraboloids and a planar phased array feed, which has been investigated by Fitzgerald [87]. This dual-reflector geometry allows the aperture of the phased array to be magnified by the ratio of the primary to secondary reflector focal lengths. As the phased array scans from broadside, to avoid subreflector spillover the subreflector diameter must

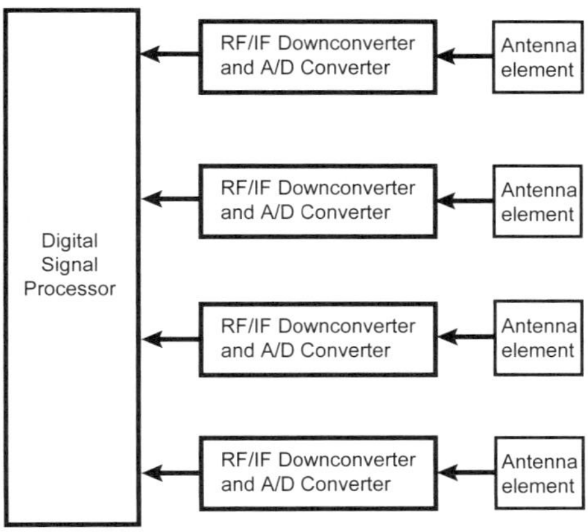

Figure 2.11 Example fully-digital array beamforming architecture.

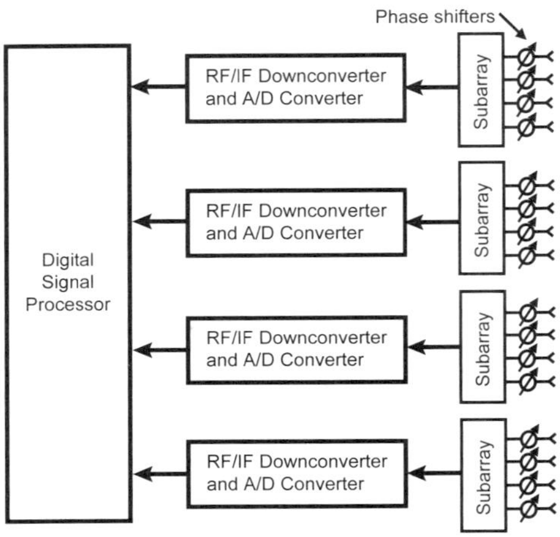

Figure 2.12 Example hybrid analog/digital phased array beamforming architecture.

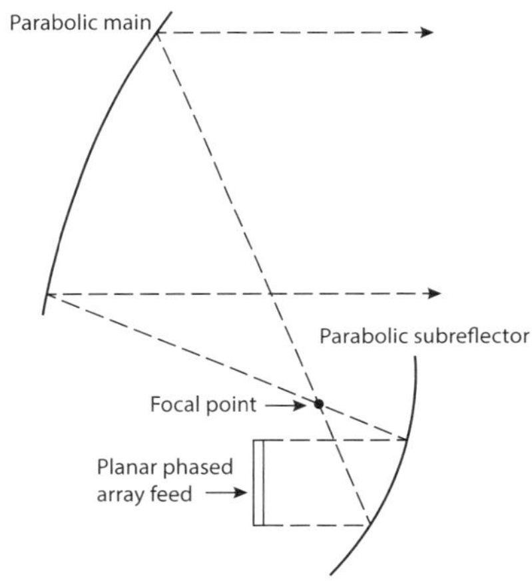

Figure 2.13 Example offset Gregorian reflector system with confocal paraboloids and planar phased array feed. Redrawn from Fitzgerald [83].

be oversized. In Chapter 10, an axisymmetric Gregorian reflector system with a planar array (non-steered) feed is analyzed.

In communications or radar applications where a wide instantaneous field of view is required with high-gain beams, phased array antennas with multiple beams can be generated, for example, by a Butler matrix [9, 37, 47] or by a Rotman lens [9, 47]. In some airborne radar applications, multiple phase centers are desired for example in displaced phased center antenna (DPCA) designs for cancelling stationary ground clutter [43–47] as in Figure 2.14. In

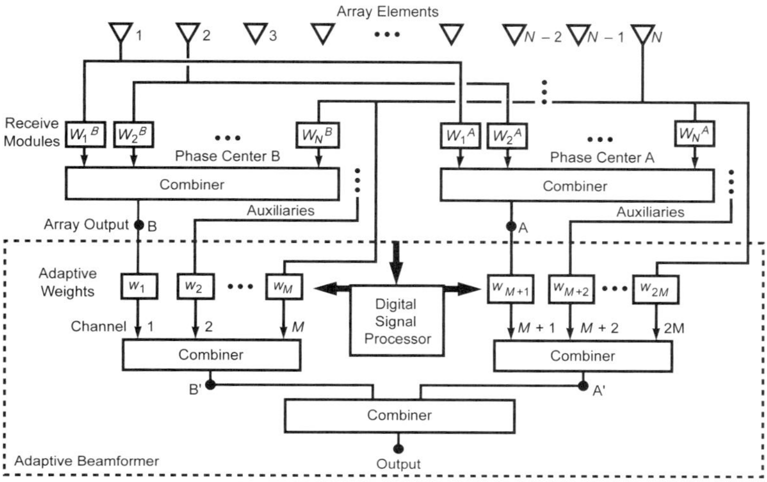

Figure 2.14 Example beamformer design for a displaced phased array antenna with adaptive signal processing to cancel both radar clutter and interference.

this DPCA architecture, there are two independent beamformers that generate two movable phase centers [15]. The phase center spacing is varied by using attenuators in the receive modules to shut off received signals from groups of antenna elements. In general, the DPCA array will contain a total of N antenna elements that are used to form the receive main channel. In addition to these N elements, a guard band of passively terminated elements can be used to provide impedance matching to the active elements and/or isolation from ground plane edges to improve radiation pattern match between the displaced phase centers. Signals received from auxiliary elements in the array are used in adaptively nulling interference and jamming. For this two phase center example, the received RF signal output from each of the array antenna elements is first split into two paths that are weighted and then summed in separate power combiners to form two independent subarray main channels (or moveable phase centers). In each element channel is a transmit/receive module, which has amplitude and phase control. The amplitude control can be used to provide a desired low-sidelobe array illumination function as well as to achieve phase-center displacement. The modules utilize phase shifters, which are used to steer the main beam to a desired angle. Let $\boldsymbol{W}_A = (W_1^A, W_2^A, \cdots, W_N^A)^T$ and $\boldsymbol{W}_B = (W_1^B, W_2^B, \cdots, W_N^B)^T$, where T is transpose, denote the array element weight vectors (including quantization and random errors) of phase centers A and B, respectively (superscript T means transpose). To effect phase center displacement, a portion of each subarray is turned off by applying a large value of attenuation for a group of antenna elements. This amplitude weighting will essentially move the electrical phase center to the center of gravity for the remaining elements. Thus, there is an effective number of elements actually used in receiving signals in phase centers A and B and are denoted as N_A and N_B, respectively. Let a wavefront (either planar or spherical) due to the jth source (either clutter or jammer), be impressed across the array, which results in a set of array element received voltages denoted as the complex vector $\boldsymbol{v} = v_1^j, v_2^j, \cdots, v_N^j$. The channel covariance matrix $\boldsymbol{R}$ consisting of clutter and jamming signals is formed from the correlation of the received voltage vector as $\boldsymbol{R} = \boldsymbol{v}^\dagger \boldsymbol{v}$, where † is complex conjugate transpose, from which adaptive array weights (amplitude and phase) are generated. The number of adaptive channels per phase center is denoted M. For a sidelobe canceller $M = 1 + N_{aux}$ where N_{aux} is the number of auxiliary channels in each phase center. This adaptive system has M degrees of freedom in each phase center and, thus, a total of $2M$ degrees of freedom. Letting a quiescent weight vector be denoted as $\boldsymbol{w}_q$, then the adaptive weight vector $\boldsymbol{w}_a$ to form nulls is given by matrix inversion as [69]

$$\boldsymbol{w}_a = \boldsymbol{R}^{-1} \boldsymbol{w}_q \tag{2.12}$$

A simplified block diagram for an example transmit/receive (T/R) module [22] is shown in Figure 2.15. By means of switches and proper timing, the phase shifter shown is used for both transmit and receive. In this diagram, the switches are adjusted for the transmit mode. To protect the low noise amplifier (LNA), the combination of a circulator and a switched-in load provides high isolation from the high-power transmit signal. The command

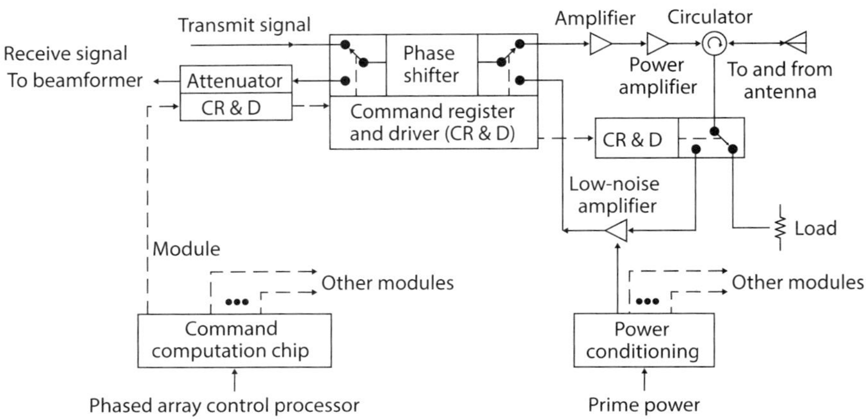

Figure 2.15 Example transmit/receive (T/R) module block diagram.

computation chip provides the desired phase shift and attenuation values to the phase shifter and attenuators. Many different architectures exist for T/R modules, which are designed based on the system requirements. For example, in some radar applications the T/R modules must allow for dual polarization and multiple phase center operation.

Recently, there has been considerable research in the area of multiple-input multiple-output (MIMO) arrays [90-92]. Figure 2.16 shows an artist's concept where a MIMO antenna system uses multiple transmit antennas and multiple receive antennas to help mitigate multipath effects in communications and radar applications.

Now that a number of beamforming architectures have been described, the next section describes basic array theory, which considers phased arrays composed of isotropic antenna elements.

2.2.4 Arrays of Isotropic Antenna Elements

Examples of planar and linear arrays of isotropic radiating antennas are depicted in Figure 2.17. Consider a two-dimensional N-element phased array with the $e^{j\omega t}$ time-harmonic dependence of the currents suppressed, where $\omega = 2\pi f$ is the radian frequency, t is time, and f is the frequency. Referring to Figure 2.17(a), the far-field array radiation pattern, or array factor (AF) in this case, in terms of observation spherical coordinates (θ, ϕ), can be expressed by superposition of the N radiating currents as

$$\text{AF}(\theta, \phi) = \sum_{n=1}^{N} i_n e^{j\Psi_n} \tag{2.13}$$

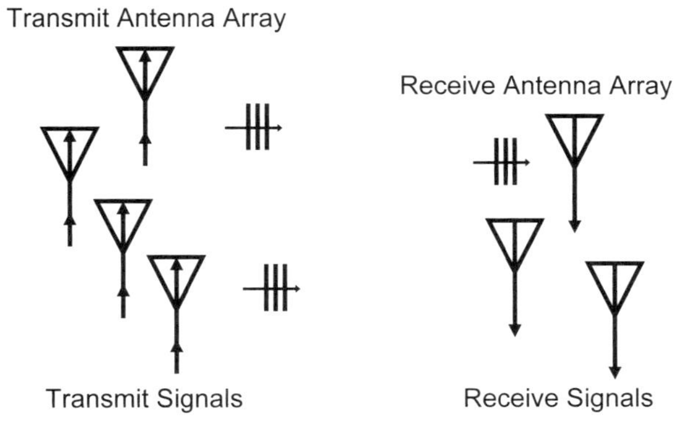

Figure 2.16 Artist's concept of a MIMO array system that consists of multiple transmit antennas and multiple receive antennas.

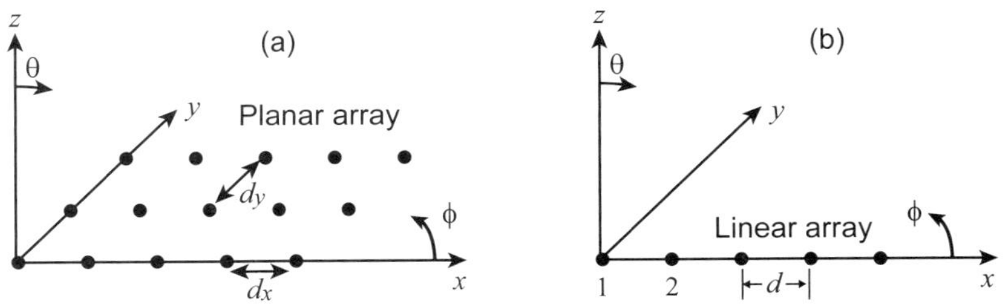

Figure 2.17 Arrays of isotropic elements (point sources). (a) Planar array, and (b) linear array spaced along the x axis.

where i_n is the complex current induced in the nth radiating element, (x_n, y_n, z_n) are the coordinates of the nth array element, and

$$\Psi_n = k \sin\theta(x_n \cos\phi + y_n \sin\phi) + kz_n \cos\theta \tag{2.14}$$

where Ψ_n is the observation phase with respect to the nth element and $k = 2\pi/\lambda$ is the wavenumber with λ denoting the wavelength. In order to electrically scan the main beam to a particular scan angle (θ_s, ϕ_s), the phasing Ψ_{ns} of the element currents is set to be the negative value (phase conjugate) of Ψ_n evaluated at (θ_s, ϕ_s), that is,

$$\Psi_{ns} = -k \sin\theta_s(x_n \cos\phi_s + y_n \sin\phi_s) - kz_n \cos\theta_s. \tag{2.15}$$

The array currents i_n are expressed in complex form as

$$i_n = A_n e^{j\Psi_{ns}} \tag{2.16}$$

where A_n is the current amplitude illumination of the nth array element. Equation (2.16) is general in the sense that amplitude and phase quantization and random errors of the transmit/receive modules can be included. Furthermore, Equations (2.14) and (2.15) allow errors in element positioning $(\Delta x_n, \Delta y_n, \Delta z_n)$ to be taken into account. The array factor for the electrically scanned array can be written as

$$\mathrm{AF}(\theta, \phi) = \sum_{n=1}^{N} A_n e^{j(\Psi_n + \Psi_{ns})}. \tag{2.17}$$

If the current amplitudes are assumed to be uniform with unity amplitude, that is, $A_n = 1$, then when the observation angles (θ, ϕ) are equal to the scan angles (θ_s, ϕ_s), it follows from Equations (2.14) and (2.15) that $\Psi_n + \Psi_{ns} = 0$ and the array factor is evaluated as

$$\mathrm{AF}(\theta_s, \phi_s) = \sum_{n=1}^{N} 1 = N \tag{2.18}$$

That is, for uniform illumination at scan-angle direction (θ_s, ϕ_s) the amplitude of the array factor is equal to the number of array elements.

For a linear array oriented along the x axis, both y_n and z_n are zero. Then with $\phi = 0°$ using Equations (2.14) and (2.15), Equation (2.17) simplifies to

$$\mathrm{AF}(\theta, \theta_s) = \sum_{n=1}^{N} A_n e^{jkx_n(\sin\theta - \sin\theta_s)}. \tag{2.19}$$

It should be noted that for the array factor given by Equation (2.19), since ϕ was set to zero, the angles θ and θ_s are allowed to take on positive and negative values.

A number of phased array illumination functions, or tapers, [5, 33] can be used to control the sidelobe levels and beamwidth of phased array antennas. Some common phased array illumination functions include uniform, binomial, triangular, cosine, Dolph-Chebyshev [93–95], and Taylor [96]. Density tapered or thinned arrays are used to effectively produce a tapered array illumination function for low sidelobes. Uniformly illuminated arrays are typically used for transmit applications and have a narrow beam with high first sidelobes (on the order of -13.2 dB for a linear array or -17.6 dB for a circular filled array, relative to the main beam maximum value). Compared to uniform illumination, the binomial illumination has a broader beam with no sidelobes. The Dolph-Chebyshev illumination has low sidelobes at a constant, prescribed level. The Taylor illumination generates a narrow beam with controlled low sidelobes. Amplitude and phase quantization and random errors in the illumination function can cause the sidelobes to degrade. Generally, as the array sidelobe levels are lowered the main beam broadens and the peak gain is reduced.

Consider now a linear array with isotropic elements (point sources) that have spacing d along the x axis as shown in Figure 2.17(b). For the uniform illumination case $A_n = 1$, let the element position be quantified by $x_n = (n-1)d$ and then let

$$u = kd(\sin\theta - \sin\theta_s). \tag{2.20}$$

Equation (2.19) can now be simplified and then expanded as

$$\mathrm{AF}(u) = \sum_{n=1}^{N} e^{jnu} = 1 + e^{ju} + e^{j2u} + \cdots + e^{j(N-1)u} \tag{2.21}$$

Now multiply both sides of Equation (2.21) by e^{ju} which yields

$$\mathrm{AF}e^{ju} = e^{ju} + e^{j2u} + e^{j3u} + \cdots + e^{jNu} \tag{2.22}$$

Subtracting Equation (2.22) from (2.21) and rearranging yields

$$\text{AF} = \frac{e^{jNu} - 1}{e^{ju} - 1} \tag{2.23}$$

Factoring the exponentials in the numerator and denominator and converting from exponentials to sines yields

$$\text{AF} = e^{j(N-1)u/2} \frac{\sin(Nu/2)}{\sin(u/2)} \tag{2.24}$$

The maximum value of the array factor occurs at the main beam peak, which occurs when the denominator goes to zero, and this condition exists when $u = 2m\pi$ for $m = 0, \pm 1, \pm 2, \cdots$. Applying L'Hôpital's rule to the ratio of the sine terms in Equation (2.24), at $u = 2m\pi$, yields the maximum value of N for the array factor, which can be used to normalize the AF to unity. The minimum (null) values of the array factor occur when the numerator goes to zero, and this condition exists when $u = 2m\pi/N$ for $m = \pm 1, \pm 2, \cdots$.

In designing phased array antennas, the maximum electrical spacing d between the elements of a phased array is usually chosen to avoid grating lobes (multiple peaks) according to the following condition [9]

$$\frac{d}{\lambda_h} < \frac{1}{1 + \sin|\theta_s|} \tag{2.25}$$

where λ_h is the wavelength at the highest frequency and θ_s is the scan angle from broadside. This equation is readily derived in four steps by (1) setting the exponential argument of the array factor in Equation (2.19) equal to -2π, (2) using a positive scan angle, (3) taking the observation angle as the endfire angle $-90°$, where the grating lobe first occurs, and (4) solving for d/λ. Typically, the maximum array-element spacing d is more tightly constrained to keep the first null of the endfire grating lobe at the edge of visible space. In practice, the endfire grating lobe should be kept at least one half-power beamwidth away from visible space. For many wide-angle scanning phased arrays, the electrical spacing is often chosen to be close to one-half wavelength for elements on a rectangular or triangular grid.

Calculated radiation patterns for linear arrays with uniform, Dolph-Chebyshev, and Taylor illumination functions are shown below. Note that the MATLAB Phased Array System Toolbox (http://www.mathworks.com.) provides tapered array illumination functions for one and two-dimensional arrays. For example, the MATLAB statement <taper=chebwin(N,SLLdB)> generates the Dolph-Chebyshev illumination for an N-element linear array with sidelobe levels equal to SLLdB. Similarly, in MATLAB the Taylor illumination statement for N elements is <taper=taylorwin(N,nbar,SLLdB)>, where nbar is a parameter that affects the number of transition sidelobes adjacent to the main beam. A typical value of nbar is equal to five (5) that provides an illumination that is readily implemented in practice.

As an example of the array factor for an ideal uniformly illuminated array, consider an eight-element linear array of isotropic array elements with $\lambda/2$ element spacing, as depicted in Figure 2.18. For this eight-element array, using Equation (2.19), the array factors for uniform illumination for scan angles of $\theta_s = 0°$ and $\theta_s = 45°$ are shown in Figure 2.19. Next, the normalized array factor magnitude for a larger, 16-element linear array of isotropic antenna elements with uniform amplitude illumination scanned to $\theta = 30°$ is shown in Figure 2.20, and the corresponding phase of the array factor is shown in Figure 2.21. It is observed that the phase alternates between 0° and 180° as the angle varies across the sidelobes. Since the phase center of the array is located at the origin $x = 0$, the far-field phase is flat over the

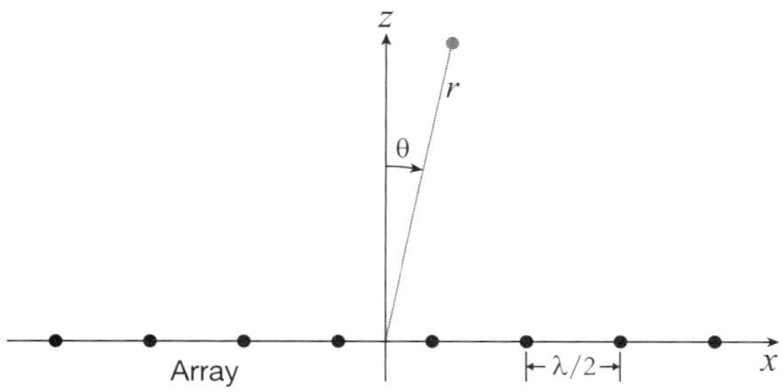

Figure 2.18 Example of an eight-element linear array antenna with isotropic elements.

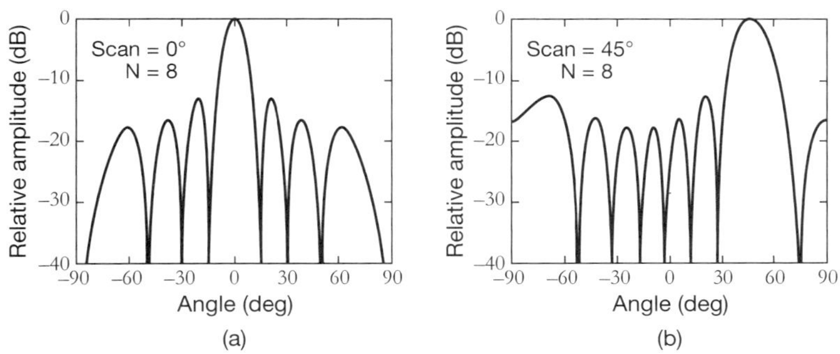

Figure 2.19 Calculated array factor for an eight-element linear array antenna with isotropic elements. (a) Scan angle 0°, and (b) scan angle 45°.

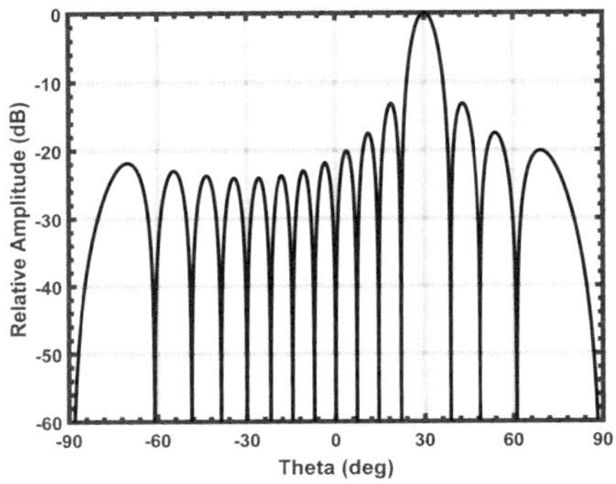

Figure 2.20 Calculated array factor magnitude for a 16-element linear array antenna with isotropic elements. Uniform amplitude illumination with $\theta_s = 30°$.

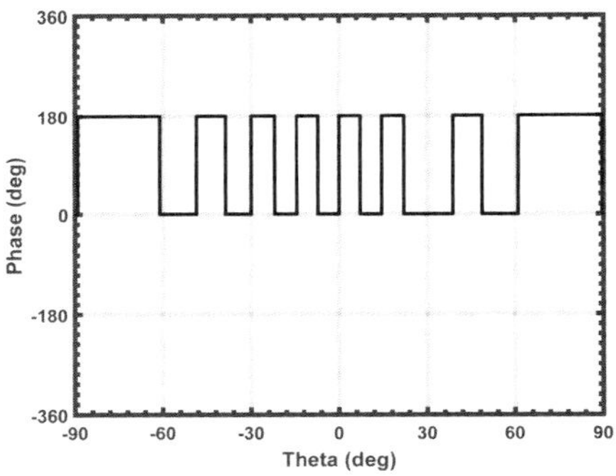

Figure 2.21 Calculated array factor phase for a 16-element linear array antenna with isotropic elements. Uniform amplitude illumination with $\theta_s = 30°$.

main beam. Note that there are $N-2$ number of sidelobes for the N element linear array. The uniformly illuminated array has high sidelobe magnitudes, starting with the first sidelobe which is at the -13 dB level. To reduce the sidelobes to a constant level, the Dolph-Chebshev taper for -40 dB uniform sidelobes was applied to the 16-element linear array and the radiation pattern is shown in Figure 2.22. In this case, to achieve lower sidelobes with the Dolph-Chebyshev taper, the far-field main beam has broadened compared to the radiation pattern for the uniformly illuminated array in Figure 2.20. Again for a 16-element array, the case of a Taylor taper (-40 dB, nbar=5) is shown in Figure 2.23, which is very similar to the Dolph-Chebyshev pattern except for the first few sidelobe levels. If the array elements are not isotropic radiators, but instead have a radiation pattern that depends on angle, then the

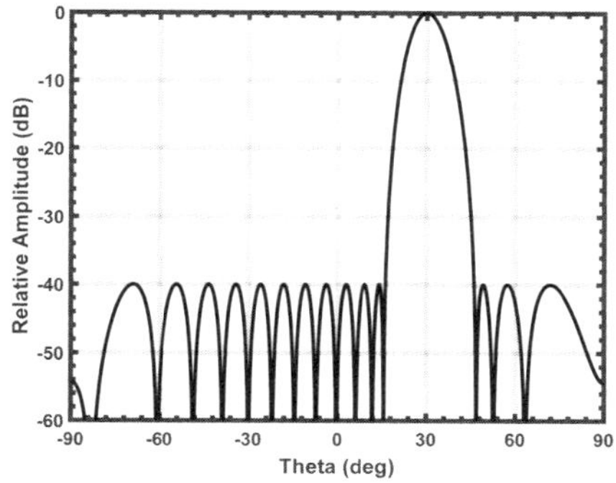

Figure 2.22 Calculated array factor magnitude for a sixteen-element linear array antenna. Dolph-Chebyshev taper (40 dB) with scan angle $\theta_s = 30°$.

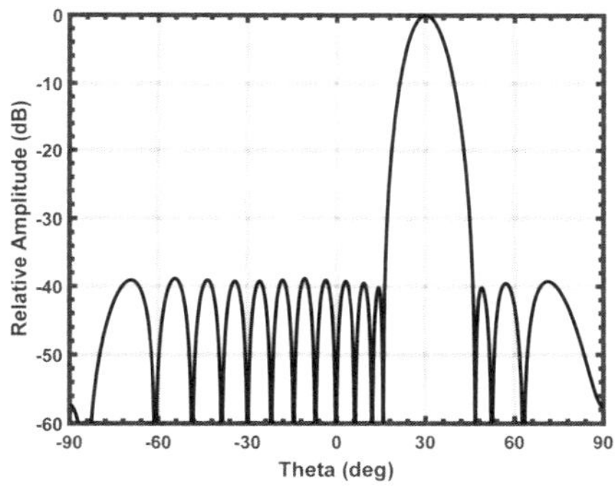

Figure 2.23 Calculated array factor magnitude for a sixteen-element linear array antenna. Taylor taper (40 dB, nbar=5) with scan angle $30°$.

array radiation pattern $P(\theta, \phi)$ can be expressed as the product of the array factor and the element radiation pattern (denoted $p_e(\theta, \phi)$) as

$$P(\theta, \phi) = p_e(\theta, \phi)\mathrm{AF} \tag{2.26}$$

where it is assumed that all array elements have the same radiation pattern.

2.2.5 Polarized Array Far-Zone Electromagnetic Fields

The previous section considered the far-field radiation for isotropic elements and the theory ignored polarization. In this section, the effect of polarization of the elements is included in the theoretical formulation. In Equation (1.273), the far-field vector current moment $\boldsymbol{f}(\theta, \phi)$ was defined in spherical coordinates in terms of a volume integral of the source current $\boldsymbol{J}(\boldsymbol{r}')$ as

$$\boldsymbol{f}(\theta, \phi) = f_r \hat{\boldsymbol{r}} + f_\theta \hat{\boldsymbol{\theta}} + f_\phi \hat{\boldsymbol{\phi}} = \int_V \boldsymbol{J}(\boldsymbol{r}') e^{jk\hat{\boldsymbol{r}} \cdot \boldsymbol{r}'} dv', \tag{2.27}$$

Referring to Figure 2.17(a), consider now a two-dimensional periodic array case, in which the array antenna elements are now Hertzian (short) electric dipoles of length l and arbitrary polarization $\hat{\boldsymbol{p}}$ that are located in a rectangular lattice parallel to the xy plane at $z = z_o$. Assume that the array has N_x elements in the x direction with interelement spacing d_x and N_y elements in the y direction with interelement spacing d_y. The total number of elements is equal to $N = N_x N_y$. Referring to Equation (2.15), let the electrical phasing of the nth Hertzian dipole array element with position (x_n, y_n, z_o) for the scan direction (θ_s, ϕ_s) be given by

$$\Psi_{ns} = -kx_n \sin\theta_s \cos\phi_s - ky_n \sin\theta_s \sin\phi_s - kz_o \cos\theta_s. \tag{2.28}$$

In this case, the current density function $\boldsymbol{J}(\boldsymbol{r}')$ can be expressed in terms of the array element amplitude $A_n = I_n l$ and phase shift Ψ_{ns} as

$$\boldsymbol{J}(\boldsymbol{r}') = \hat{\boldsymbol{p}} \sum_{n=1}^{N} I_n l e^{j\Psi_{ns}} \delta(x' - x_n)\delta(y' - y_n)\delta(z' - z_o). \tag{2.29}$$

The propagation unit vector in terms of the spherical coordinate angles (θ, ϕ) is given by

$$\hat{\boldsymbol{r}} = \hat{\boldsymbol{x}} \sin\theta \cos\phi + \hat{\boldsymbol{y}} \sin\theta \sin\phi + \hat{\boldsymbol{z}} \cos\theta \tag{2.30}$$

and the source vector $\boldsymbol{r}'$ is given by

$$\boldsymbol{r}' = \hat{\boldsymbol{x}}x' + \hat{\boldsymbol{y}}y' + \hat{\boldsymbol{z}}z'. \tag{2.31}$$

The dot product of the propagation direction unit vector and the source vector is equal to

$$\hat{\boldsymbol{r}} \cdot \boldsymbol{r}' = x' \sin\theta \cos\phi + y' \sin\theta \sin\phi + z' \cos\theta \tag{2.32}$$

Substituting Equation (2.29) into Equation (2.27) yields

$$\boldsymbol{f}(\theta, \phi) = \int_V \hat{\boldsymbol{p}} \sum_{n=1}^{N} I_n l e^{j\Psi_{ns}} \delta(x' - x_n)\delta(y' - y_n)\delta(z' - z_o) e^{jk\hat{\boldsymbol{r}} \cdot \boldsymbol{r}'} dv' \tag{2.33}$$

and using Equation (2.32) yields

$$\boldsymbol{f} = \hat{\boldsymbol{p}} \sum_{n=1}^{N} A_n e^{jk[x_n(\sin\theta\cos\phi - \sin\theta_s\cos\phi_s) + y_n(\sin\theta\sin\phi - \sin\theta_s\sin\phi_s) + z_o(\cos\theta - \cos\theta_s)]} \tag{2.34}$$

The far-zone electric and magnetic fields in terms of the far-field vector current moment $\boldsymbol{f}$ are determined from Equations (1.283) and (1.286) as

$$\boldsymbol{E}_{\text{FF}} = -j\omega\mu\frac{e^{-jkr}}{4\pi r}(\hat{\boldsymbol{\theta}} f_\theta + \hat{\boldsymbol{\phi}} f_\phi) \tag{2.35}$$

$$\boldsymbol{H}_{\text{FF}} = -jk\frac{e^{-jkr}}{4\pi r}(\hat{\boldsymbol{\phi}} f_\theta - \hat{\boldsymbol{\theta}} f_\phi). \tag{2.36}$$

The electric and magnetic fields are then computed for elements with a particular polarization $\hat{\boldsymbol{p}}$ by evaluating Equation (2.34) and substituting the result in Equations (2.35) and (2.36), respectively. To obtain the desired spherical components of $\boldsymbol{f}$, the unit vector $\hat{\boldsymbol{p}}$ is dotted into the spherical unit vectors $\hat{\boldsymbol{\theta}}$ and $\hat{\boldsymbol{\phi}}$. In the general case, where the nth element current density has an arbitrary distribution, the volume integral in Equation (2.27) can be evaluated either in closed form or by numerical integration to yield the desired far-field vector current moment $\boldsymbol{f}$, from which the electric and magnetic fields can be computed. When the array is a linear array along the x axis at $z_o = 0$, Equation (2.34) simplifies to

$$\boldsymbol{f}(\theta, \phi, \theta_s) = \hat{\boldsymbol{p}} \sum_{n=1}^{N} A_n e^{jkx_n(\sin\theta\cos\phi - \sin\theta_s)}. \tag{2.37}$$

2.2.6 Array Mutual Coupling Effects

Array mutual coupling, or electromagnetic field coupling between array elements, as depicted in Figure 2.24, is an important effect that can strongly influence the radiation pattern of the individual array elements and the overall array patterns. In this diagram, an input signal is transmitted by one antenna element and this signal is radiated to free space and it is received by the

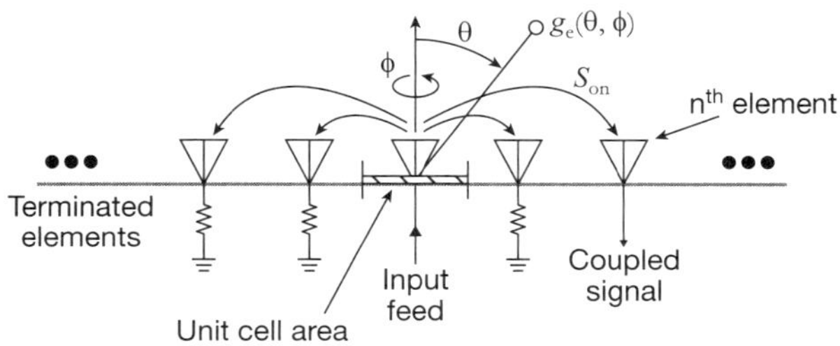

Figure 2.24 Diagram depicting an array with one element driven, with coupling to the surrounding elements.

surrounding antenna elements. In a large array with periodic element spacing, each antenna element will have a gain radiation pattern that is related to the element unit cell area. Early developments of phased array radiating-element technology were conducted from 1959 to 1967. Beginning in 1959, there were numerous contributions to the theoretical understanding of phased arrays, particularly the effects of array mutual coupling on the performance of various configurations of dipole and waveguide arrays; for example, the reports and articles by Allen et al. [23–35]. Array mutual coupling is defined as the ratio of the coupled signal voltage in a surrounding array element relative to the input signal voltage in a reference element, when the surrounding elements are terminated in loads matched to the impedance of the transmission line. Let S_{on} denote the complex mutual coupling between the reference element and the nth array element. The scan reflection coefficient for a reference element in a planar phased array is calculated as [6, p. 22, equation 21]

$$\Gamma_o(\theta_s, \phi_s) = \sum_n S_{on} e^{\Psi_{ns}(x_n, y_n, 0)} = \sum_n S_{on} e^{-jk \sin\theta_s (x_n \cos\phi_s + y_n \sin\phi_s)} \tag{2.38}$$

where (x_n, y_n) are the coordinates of the nth array element.

The scanned array impedance mismatch loss (ML) efficiency denoted η_{ML} is expressed in terms of the scan reflection coefficient as

$$\eta_{\text{ML}}(\theta_s, \phi_s) = 1 - |\Gamma_o(\theta_s, \phi_s)|^2. \tag{2.39}$$

The scan mismatch loss efficiency in dB can be computed as

$$\text{ML}(\theta_s, \phi_s) = -10\log_{10} \eta_{\text{ML}} = -10\log_{10}\left[1 - |\Gamma_o(\theta_s, \phi_s)|^2\right]. \tag{2.40}$$

When a single antenna element is driven in a passively terminated array (also referred to as an embedded element), the fractional dissipated power (denoted F_{pe} relative to the input power to the element) caused by array mutual coupling is computed as the summation of the magnitude squared of the mutual coupling coefficients, that is [6, p. 22, equation 24],

$$F_{pe} = \sum_n |S_{on}|^2. \tag{2.41}$$

This fractional dissipated power is the fractional power lost to the terminating loads in the transmission lines connected to the array elements. The array element gain $g_e(\theta, \phi)$ is then the product of the element directivity $D_e(\theta, \phi)$ and the quantity $(1 - F_{pe})$, that is,

$$g_e(\theta, \phi) = D_e(\theta, \phi)(1 - \sum_n |S_{on}|^2). \tag{2.42}$$

The maximum gain (relative to an isotropic radiator) of an antenna aperture of arbitrary shape is given by the following expression:

$$G_{\text{max}} = 4\pi A_{\text{em}}/\lambda^2 \tag{2.43}$$

where A_{em} is the antenna maximum effective aperture area in square meters and λ is the wavelength in meters. As the frequency increases, the wavelength is reduced according to the following expression,

$$\lambda = c/f \tag{2.44}$$

where c is the speed of light and f is the frequency. Thus, for a fixed aperture dimension in meters, ideally the gain will increase as the frequency increases. However, for a practical antenna aperture, the surface tolerances and, in the case of a phased array, T/R module errors will introduce phase errors and will cause a reduction in antenna gain and an increase in sidelobe levels. In decibels, the maximum gain of an antenna aperture is given by

$$G_{\text{max, dB}} = 10\log_{10} G_{\text{max}}. \tag{2.45}$$

The element gain pattern for an infinite array (or very large finite array) is a function of the unit-cell area, the scanned array impedance mismatch efficiency, and the projected aperture [6, pp. 23-25]. The projected aperture will vary as $\cos\theta$, where θ is the angle from broadside. For a general array in the xy plane, the element gain pattern is given by

$$g_e(\theta,\phi) = \frac{4\pi A_e}{\lambda^2}\left[1 - |\Gamma(\theta,\phi)|^2\right]\cos\theta \tag{2.46}$$

where A_e is the unit-cell area occupied by the element. For a rectangular grid array, the element gain pattern is given by

$$g_{er}(\theta,\phi) = \frac{4\pi d_x d_y}{\lambda^2}\left[1 - |\Gamma(\theta,\phi)|^2\right]\cos\theta \tag{2.47}$$

where d_x and d_y are the element spacings in the x and y directions, respectively. For an equilateral-triangle (hexagonal) array lattice, the element gain is expressed as

$$g_{eh}(\theta,\phi) = \frac{2\pi\sqrt{3}b^2}{\lambda^2}\left[1 - |\Gamma(\theta,\phi)|^2\right]\cos\theta \tag{2.48}$$

where b is the side of the equilateral triangle for the element spacing. The scanned radiation pattern, including mutual coupling effects of a large array of thousands of array elements, can be calculated in an approximate manner by first computing the element gain pattern in a large or infinite array and then multiplying it by the array factor given by Equation (2.19).

One of the fundamental challenges in designing a phased array is that significant portions of the microwave field transmitted by one element of the array can be received by the surrounding array antenna elements. This mutual coupling effect, can sometimes lead to surface waves and result in a substantial or total loss of transmitted or received signal in a desired scan direction, which depends on the coherent combination of all of the mutual-coupling signals in the array. This loss of antenna gain is referred to as a blind spot, as is depicted conceptually in Figure 2.25. In this figure, assume that the phased array is designed for maximum gain at broadside (0° scan angle); however, in this artist's concept element gain pattern nulls (minima) are formed at ±45° (Figure 2.25(a)). Thus, the full scan sector ±60° cannot be covered. The ideal shape of the element gain pattern is a cosine function of the scan angle, when it is assumed that the reflection coefficient amplitude is zero (refer to Equation (2.46) with $|\Gamma| = 0$). When these element gain pattern nulls occur, there will be corresponding peaks occurring in the scan reflection coefficient amplitude (Figure 2.25(b)). A phased array blind spot due to induced surface waves or mode cutoff can be described as either a null in the element gain

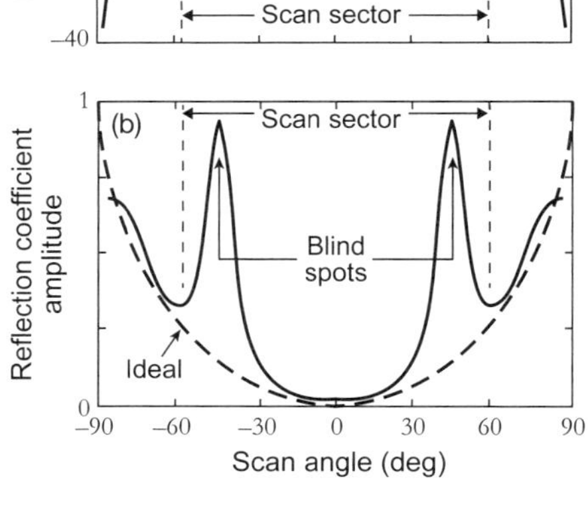

Figure 2.25 Conceptual diagram demonstrating blind spots for a phased array antenna. (a) Element gain pattern, and (b) scan reflection coefficient.

pattern at the blind spot angles θ_b, ϕ_b such that

$$g_e(\theta_b, \phi_b) = 0 \tag{2.49}$$

or as a peak tending toward unity in the scanned array reflection coefficient magnitude, that is,

$$|\Gamma(\theta_b, \phi_b)| = 1. \tag{2.50}$$

The amplitudes and phases of the array mutual-coupling signals depend primarily on the type (shape) of the radiating antenna elements, the electrical spacing between the array elements, and the number of surrounding radiating elements. Numerous studies have investigated many different array element designs, taking into account mutual-coupling effects. In many phased array designs, it is desirable to provide linear-polarized, dual-polarized, or circularly polarized scan coverage with peak gain at broadside and constant gain for any azimuthal (ϕ) angle, as depicted in Figure 2.26.

When all of the phase shifters of an array are properly aligned, the array produces a main beam in the desired pointing direction. Generally, the corporate feed (for example, the power divider network in Figure 2.7) is designed with minimal crosstalk between channels. However, once the signals have reached the radiating antenna elements, a significant amount of crosstalk in the form of array mutual coupling occurs. As discussed above, the amplitudes and phases of these mutual coupling signals can significantly impact the performance of the phased array. If the array element spacing is around one-half wavelength, substantial amounts of mutual coupling can occur. The mutual coupling amplitude tends to increase as the electrical spacing is reduced. This coupling manifests itself in changes in the element's gain radiation pattern and its scan reflection coefficient. Unless care is taken in the design of the array, undesired blind spots in the scan sector can occur. These blind spots occur at angles where the element gain pattern has a null

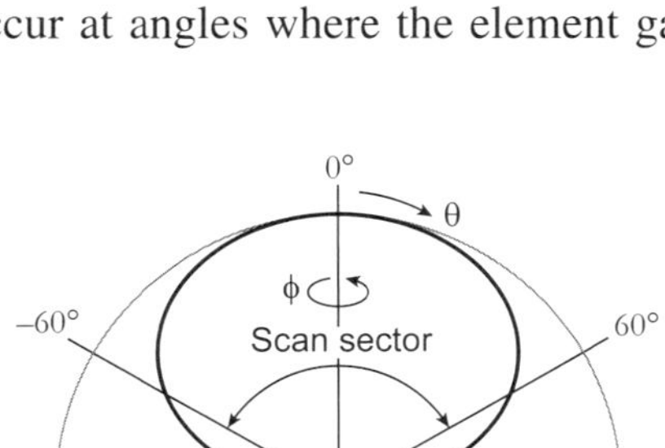

Figure 2.26 Conceptual polar diagram showing an element gain pattern covering a scan sector, with peak radiation at broadside to the plane of the element.

and the reflection coefficient of the array has a peak close to unity, as was depicted in Figure 2.25. At these blind spots, the transmitted or received signal is significantly reduced in amplitude.

In some applications, it is desirable to place an intentional blind spot or null in directions where it is undesirable to transmit or receive energy [36]. For example, Figure 2.27 shows an example of a broadside-null radiation pattern that could be generated by either a monopole or a uniform-current loop antenna element in an array environment over a ground plane. These types of elements could be useful when broadside radiation is undesirable, such as in reducing broadside clutter and jamming in the case of a spaceborne radar [45, 46]. As the radar beam is steered away from $\theta = 0^\circ$ (broadside) toward $\theta = 60^\circ$, the conventional broadside-peak-type element radiation pattern drops off, but the broadside-null-type element radiation pattern increases to a peak at about $\theta = 45^\circ$ to 50°.

2.2.7 Power Density and Array Gain

The effective isotropic radiated power, denoted EIRP, of an aperture is given by the product of the transmitter power P_t delivered to the antenna and the antenna transmit gain G_t as

$$\text{EIRP} = P_t G_t. \tag{2.51}$$

Note that the power transmitted by the antenna is usually less than the power delivered by the generator feeding the antenna. This difference in power is attributed to any RF losses between the generator and the antenna, as well as any mismatch loss resulting from the reflection coefficient at the antenna port.

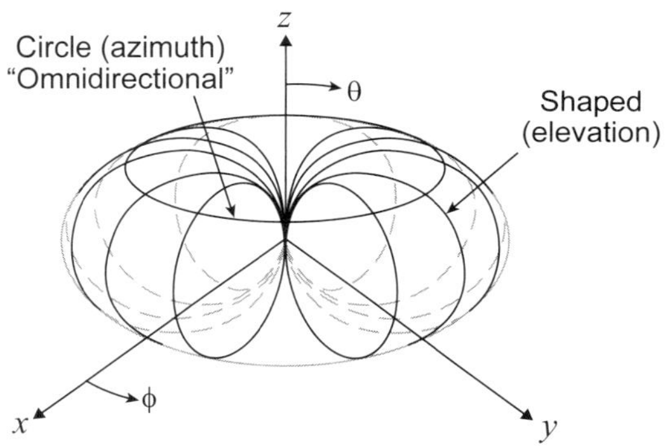

Figure 2.27 Artist depiction of a monopole-type radiation pattern for an element in a phased array antenna.

If the overall antenna aperture is assumed to be an array of identical elements, the aperture can be divided into element apertures, which are also referred to as element unit cells (see Figure 2.24). For a large N-element array with an effective aperture A_{em}, the unit cell area is equal to A_{em}/N. If the embedded element gain patterns are either computed or measured, the array gain can be computed as follows. At range distance r, the embedded-element radiated power density, denoted P_{de}, resulting from the element transmit power (input power to the element) and embedded-element gain is given by (refer to Equation (1.696))

$$P_{de} = \frac{P_{te} g_e(\theta, \phi)}{4\pi r^2} \tag{2.52}$$

and solving for g_e,

$$g_e = 4\pi r^2 \frac{P_{de}}{P_{te}}. \tag{2.53}$$

In the far field, the instantaneous plane-wave power density of the element is given in terms of the array-element electric field intensity $E_e(\theta, \phi)$ as

$$P_{de} = \frac{|E_e(\theta, \phi)|^2}{\eta_o} = \frac{|E_e(\theta, \phi)|^2}{120\pi}. \tag{2.54}$$

Substituting Equation (2.54) in Equation (2.53) and simplifying yields

$$g_e(\theta, \phi) = \frac{r^2}{30} \frac{|E_e(\theta, \phi)|^2}{P_{te}}. \tag{2.55}$$

Thus, it follows that

$$|E_e(\theta, \phi)| = \frac{\sqrt{30}}{r} \sqrt{g_e(\theta, \phi)} \sqrt{P_{te}}. \tag{2.56}$$

The element gain pattern, $g_e(\theta, \phi)$ includes the effect of array mutual coupling and mismatch losses.

The transmit power of the nth element is proportional to A_n^2, and using this relation in Equation (2.56) yields

$$|E_{en}(\theta, \phi)| = \frac{\sqrt{30}}{r} \sqrt{g_{en}(\theta, \phi)} A_n. \tag{2.57}$$

To compute the maximum gain of an array at a particular scan angle (θ_s, ϕ_s), by superposition the electric-field contributions for the embedded elements can be summed as

$$E_{\text{array, max}}(\theta_s, \phi_s) = \frac{\sqrt{30}}{r} \sum_n \sqrt{g_{en}(\theta_s, \phi_s)} A_n. \tag{2.58}$$

Now, following along the lines of Equation (2.55), the array gain can be expressed as

$$G_{\text{array, max}}(\theta_s, \phi_s) = \frac{r^2}{30} \frac{|E_{\text{array, max}}(\theta_s, \phi_s)|^2}{P_{\text{t, array}}} \tag{2.59}$$

where

$$P_{\text{t, array}} = \sum_{n=1}^{N} A_n^2. \tag{2.60}$$

Substituting Equations (2.58) and (2.60) in Equation (2.59) and simplifying yields

$$G_{\text{array, max}}(\theta_s, \phi_s) = \frac{|\sum_{n=1}^{N} \sqrt{g_{en}(\theta_s, \phi_s)} A_n|^2}{\sum_{n=1}^{N} A_n^2}. \tag{2.61}$$

For the case of a large array with uniform array illumination (with unity amplitude at each array element), it can be assumed that all of the elements have the same element gain pattern (i.e., $g_{en} = g_e$) and Equation (2.61) reduces to

$$G_{\text{array, max}}(\theta_s, \phi_s) = g_e(\theta_s, \phi_s) N. \tag{2.62}$$

For uniform illumination, the array gain is equal to the element gain times the number of array elements. If the array illumination is not uniform, then a taper efficiency must be included in computing the array gain from the element gain. To compute the array gain pattern from the element electric field versus angle (θ, ϕ), the element phase pattern $\Psi_{en}(\theta, \phi)$ must be used, that is,

$$E_{en}(\theta, \phi) = |E_{en}(\theta, \phi)| e^{j\Psi_{en}(\theta, \phi)}. \tag{2.63}$$

The array pattern E_{array} is then obtained by the superposition of the individual array element patterns with the desired array amplitude taper and with proper phasing of the array to steer the main beam in a desired direction. Thus,

$$E_{\text{array}}(\theta, \phi) = \sum_n E_{en}(\theta, \phi) e^{j\Psi_{ns}} \tag{2.64}$$

or from Equation (2.63)

$$E_{\text{array}}(\theta, \phi) = \sum_n |E_{en}(\theta, \phi)| e^{j(\Psi_{en} + \Psi_{ns})} \tag{2.65}$$

and using Equation (2.57) yields

$$E_{\text{array}}(\theta, \phi) = \frac{\sqrt{30}}{r} \sum_n \sqrt{g_{en}(\theta, \phi)} A_n e^{j(\Psi_{en} + \Psi_{ns})}. \tag{2.66}$$

The array gain as a function of angle can then be computed from the electric field as

$$G_{\text{array}}(\theta, \phi) = \frac{r^2}{30} \frac{|E_{\text{array}}(\theta, \phi)|^2}{P_{\text{t, array}}}. \tag{2.67}$$

Substituting Equations (2.60) and (2.66) in Equation (2.67) yields the array gain in terms of the element gain and element phase functions and the array illumination, that is,

$$G_{\text{array}}(\theta, \phi) = \frac{|\sum_n \sqrt{g_{en}(\theta, \phi)} A_n e^{j(\Psi_{en} + \Psi_{ns})}|^2}{\sum_n A_n^2}. \tag{2.68}$$

The next four sections describe methods that are used in computing electromagnetic fields of antennas, including array mutual coupling effects and scattering from obstacles. The discussion begins with a description of electromagnetic field equivalence principles in Section 2.3. In Section 2.4, the reciprocity theorem for antennas is derived. To determine the mutual coupling between an antenna and surrounding antennas in an array or interaction with surrounding structures, a generalized integral equation known as the reaction integral equation is then developed in Section 2.5, which is based on the equivalence principle and the reciprocity theorem for antennas. The reaction integral equation is used in analyzing antennas and phased arrays by the method of moments, which is discussed in Section 2.6.

2.3 Equivalence Principles

To describe the various forms of equivalence principles, assume an antenna with arbitrary impressed electric and magnetic currents, $\boldsymbol{J}_i$ and $\boldsymbol{M}_i$, respectively, in free space (region 1) radiating in the presence of a homogeneous object (region 2), such as depicted in Figure 2.28(a). For convenience, assume that all currents and fields are time harmonic and the $e^{j\omega t}$ time dependence is suppressed. The surface of the homogeneous object is denoted S and the unit normal to the surface is denoted $\hat{\boldsymbol{n}}$, which points from region 2 (internal region) to region 1 (external region). Region 2 is assumed in general to have permittivity ϵ and permeability μ, or the surface of region 2 can be a electric conductor with conductivity σ. These impressed currents ($\boldsymbol{J}_i$ and $\boldsymbol{M}_i$) are assumed to radiate known electric and magnetic fields $\boldsymbol{E}^i$ and $\boldsymbol{H}^i$, respectively, in free space without the homogeneous object present. Also, in the presence of the object, the total electric and magnetic fields generated by the impressed currents are $\boldsymbol{E}$ and $\boldsymbol{H}$, respectively. In the exterior region (region 1), the total electric and magnetic fields are equal to the summation of the incident fields and the fields $\boldsymbol{E}^s$ and $\boldsymbol{H}^s$ scattered by the object. In the

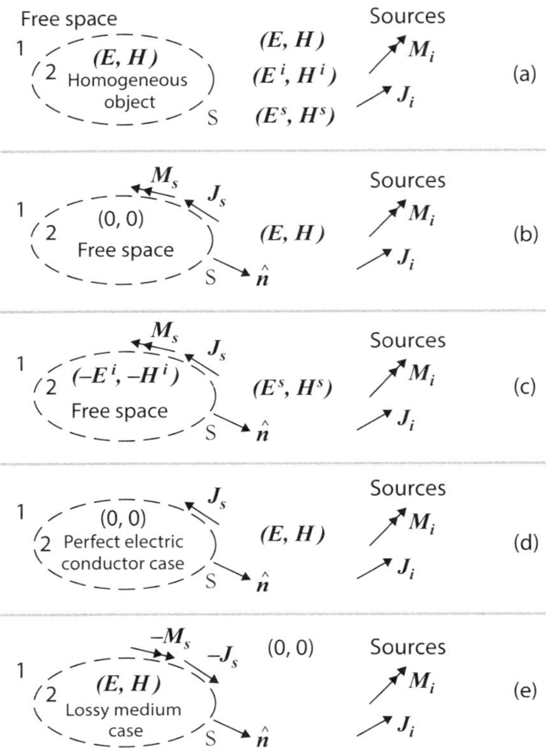

Figure 2.28 Depiction of the surface equivalence principle.

interior region (region 2), if the object is not a perfect electric conductor then internal electric and magnetic fields will exist as well.

Equivalent situations can be developed by using surface equivalence principles such as described by Love [97], Schelkunoff [98], Harrington [99], Mautz [100], Arvas [101], and Jakobus [102]. If the electric and magnetic fields are known over the entire surface of a homogeneous object, the object can be replaced by electric and magnetic surface currents radiating in free space, which will produce the correct fields external to the object. For example, if only the fields external to the object are desired, the homogeneous object is equivalently modeled as in Figure 2.28(b) with equivalent electric and magnetic surface currents $\boldsymbol{J}_s$ and $\boldsymbol{M}_s$ placed on the surface S, that is,

$$\boldsymbol{J}_s = \hat{\boldsymbol{n}} \times [\boldsymbol{H}^1 - \boldsymbol{H}^2] \tag{2.69}$$

$$\boldsymbol{M}_s = [\boldsymbol{E}^1 - \boldsymbol{E}^2] \times \hat{\boldsymbol{n}} \tag{2.70}$$

where the superscripts 1 and 2 on $\boldsymbol{E}$ and $\boldsymbol{H}$ indicate the side of S in which the fields are evaluated. In Figure 2.28(c), the same surface currents from Figure 2.27(b) produce the desired scattered fields $\boldsymbol{E}^s$ and $\boldsymbol{H}^s$ in the exterior region and the negative of the incident fields in the interior region.

In the case of a perfect electrically conducting object, the equivalent magnetic surface current is zero ($\boldsymbol{M}_s = 0$), since the total electric field is zero at the surface by the boundary condition (Section 1.5). As depicted in Figure 2.28(d), the equivalent electric surface current $\boldsymbol{J}_s$ is assumed to generate scattered fields $\boldsymbol{E}^s$ and $\boldsymbol{H}^s$ as

$$\boldsymbol{E}^s = \boldsymbol{E} - \boldsymbol{E}^i \tag{2.71}$$

$$\boldsymbol{H}^s = \boldsymbol{H} - \boldsymbol{H}^i. \tag{2.72}$$

In the interior region of a perfect conductor, the total vector electric and magnetic fields must be zero; thus, the following relation must hold:

$$\boldsymbol{E}^s = -\boldsymbol{E}^i \tag{2.73}$$

$$\boldsymbol{H}^s = -\boldsymbol{H}^i. \tag{2.74}$$

In this case, where the scatterer is a perfect electric conductor, the total tangential electric field on the outer surface is zero ($\boldsymbol{E}_{\tan} = 0$), and the total tangential magnetic field on the outer surface is equal to the induced current density $\boldsymbol{H}_{\tan} = \boldsymbol{J}_s$. On the outer surface of the perfect conductor, the tangential components of the incident and scattered magnetic fields are equal ($\boldsymbol{H}^s_{\tan} = \boldsymbol{H}^i_{\tan}$). In the technique known as physical optics, the induced electric surface current density is determined from Equation (2.69) as $\boldsymbol{J}_s = 2\hat{\boldsymbol{n}} \times \boldsymbol{H}^i$ [5, pp. 337–338]. The scattered electric field can then be computed from the surface currents $\boldsymbol{J}_s$ over the illuminated region of the surface by using Equation (1.209) as a surface integral. Finally, consider the case where the object is a homogeneous lossy medium as depicted in Figure 2.28(e). In this case, the negative of the equivalent currents produce the desired fields inside the object in which the dielectric and magnetic properties of the material must be used in computing the internal fields. Now that the necessary equivalence conditions have been described, the next step in developing a method of moments formulation is to derive the reciprocity theorem.

2.4 Reciprocity Theorem

Let there be two sets of electric and magnetic sources $\boldsymbol{J}_a$, $\boldsymbol{M}_a$ and $\boldsymbol{J}_b$, $\boldsymbol{M}_b$, which produce the fields $\boldsymbol{E}_a$, $\boldsymbol{H}_a$ and $\boldsymbol{E}_b$, $\boldsymbol{H}_b$, respectively. The Lorentz reciprocity theorem [5] is derived in the following manner starting with Maxwell's equations in time-harmonic form as:

$$\nabla \times \boldsymbol{E}_a = -\boldsymbol{M}_a - j\omega\mu' \boldsymbol{H}_a \tag{2.75}$$

$$\nabla \times \boldsymbol{H}_a = \boldsymbol{J}_a + j\omega\epsilon' \boldsymbol{E}_a \tag{2.76}$$

$$\nabla \times \boldsymbol{E}_b = -\boldsymbol{M}_b - j\omega\mu' \boldsymbol{H}_b \tag{2.77}$$

$$\nabla \times \boldsymbol{H}_b = \boldsymbol{J}_b + j\omega\epsilon' \boldsymbol{E}_b \tag{2.78}$$

Now, take the divergence of the difference between the vector cross products of the fields from the a and b sources and apply the vector identity

$$\nabla \cdot (\boldsymbol{A} \times \boldsymbol{B}) = \boldsymbol{B} \cdot (\nabla \times \boldsymbol{A}) - \boldsymbol{A} \cdot (\nabla \times \boldsymbol{B}). \tag{2.79}$$

which yields after eliminating the terms involving the dot products of the vector fields

$$-\nabla \cdot (\boldsymbol{E}_a \times \boldsymbol{H}_b - \boldsymbol{E}_b \times \boldsymbol{H}_a) = \boldsymbol{E}_a \cdot \boldsymbol{J}_b - \boldsymbol{H}_a \cdot \boldsymbol{M}_b - \boldsymbol{E}_b \cdot \boldsymbol{J}_a + \boldsymbol{H}_b \cdot \boldsymbol{M}_a \tag{2.80}$$

which is the Lorentz reciprocity theorem in differential form. Now, taking the volume integral of both sides of Equation (2.80) and applying the divergence theorem

$$\int_V \nabla \cdot \boldsymbol{A} dv = \oint_S \boldsymbol{A} \cdot d\boldsymbol{s} \tag{2.81}$$

to the left-hand side yields

$$\begin{aligned} &- \oint_S \nabla \cdot (\boldsymbol{E}_a \times \boldsymbol{H}_b - \boldsymbol{E}_b \times \boldsymbol{H}_a) \cdot d\boldsymbol{s} = \\ &\int_V (\boldsymbol{E}_a \cdot \boldsymbol{J}_b - \boldsymbol{H}_a \cdot \boldsymbol{M}_b - \boldsymbol{E}_b \cdot \boldsymbol{J}_a + \boldsymbol{H}_b \cdot \boldsymbol{M}_a) dv \end{aligned} \tag{2.82}$$

which is the integral form of the Lorentz reciprocity theorem.

The integral form of the Lorentz reciprocity theorem in a source-free region is expressed as

$$\iint_S (\boldsymbol{E}_a \times \boldsymbol{H}_b - \boldsymbol{E}_b \times \boldsymbol{H}_a) \cdot d\boldsymbol{s} = 0. \tag{2.83}$$

That is, if the surface S does not enclose sources, the volume integral on the right side of Equation (2.82) is zero. Next, consider the case of a radiating antenna with source currents contained within a finite volume. In this case, if the surface S is allowed to extend to infinity, the antenna radiated fields at infinity in the surface integral are attenuated to zero and now the left side of Equation (2.82) is zero. Thus, the reciprocity theorem for antennas can be expressed as

$$\iiint_V (\boldsymbol{E}_a \cdot \boldsymbol{J}_b - \boldsymbol{H}_a \cdot \boldsymbol{M}_b - \boldsymbol{E}_b \cdot \boldsymbol{J}_a + \boldsymbol{H}_b \cdot \boldsymbol{M}_a) dv = 0 \tag{2.84}$$

where the volume integration extends over all space including the antenna source currents in the constrained finite volume.

An example of the application of the reciprocity theorem to antennas is now considered. Assume that there are no magnetic sources present, so that the volume integral of Equation (2.84) is expressed as

$$\iiint_V \boldsymbol{E}_b \cdot \boldsymbol{J}_a dv = \iiint_V \boldsymbol{E}_a \cdot \boldsymbol{J}_b dv \tag{2.85}$$

A depiction of reciprocity for two antennas is shown in Figure 2.29. It is assumed that the antennas, feed networks, and media surrounding the antennas are linear and isotropic. One antenna can be assumed to be located at the origin and the second antenna is located at an arbitrary point with coordinates (r, θ, ϕ). The relative orientations (polarizations) of the two antennas are assumed to be the same for Figures 2.29(a) and 2.29(b). That is, the unit vector polarizations are $\hat{\boldsymbol{p}}_a$ and $\hat{\boldsymbol{p}}_b$ in both cases. In Figure 2.29(a), antenna "a" is transmitting with complex current I_a and antenna "b" is receiving with open-circuit complex voltage V_b. In Figure 2.29(b), antenna "b" is transmitting with complex current I_b and antenna "a" is receiving with voltage V_a. In general, the voltage across the terminals of the two antennas is expressed as

$$V_a = Z_{aa} I_a + Z_{ab} I_b \tag{2.86}$$

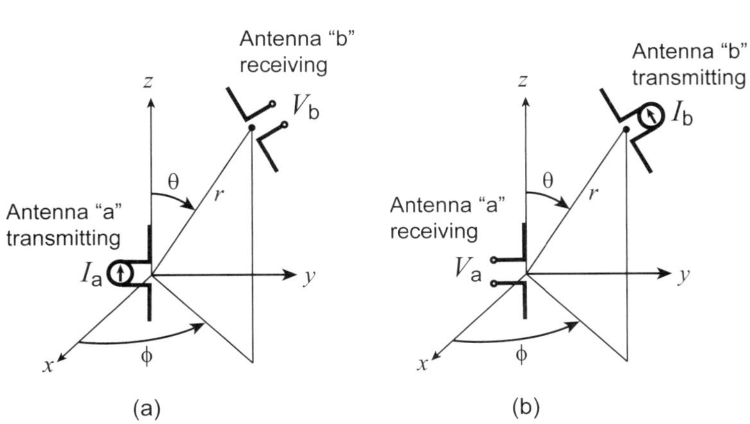

Figure 2.29 Depiction of reciprocity for two antennas.

$$V_b = Z_{ba}I_a + Z_{bb}I_b \tag{2.87}$$

where Z_{aa} and Z_{bb} are the self impedances of antennas "a" and "b", respectively, and Z_{ab} and Z_{ba} are the mutual impedances between antennas "a" and "b". Furthermore, assume that the complex currents I_a and I_b are equal (same amplitude and phase). Note that open-circuit conditions are enforced at antennas "a" and "b" when the currents I_a and I_b, respectively, are set equal to zero. Assume now that each antenna has terminals with a feed region that can be represented as a Hertzian (short) dipole with terminal current I, infinitesimal length dl and polarized in the $\hat{\boldsymbol{p}}_a$ and $\hat{\boldsymbol{p}}_b$ directions. It follows then that the volume integration becomes a line integral evaluated as

$$\iiint_V \boldsymbol{E}_b \cdot \boldsymbol{J}_a dv = \int \boldsymbol{E}_b \cdot \hat{\boldsymbol{p}}_a I_a dl = V_a^{\text{o.c.}} I_a \tag{2.88}$$

$$\iiint_V \boldsymbol{E}_a \cdot \boldsymbol{J}_b dv = \int \boldsymbol{E}_a \cdot \hat{\boldsymbol{p}}_b I_b dl = V_b^{\text{o.c.}} I_b \tag{2.89}$$

In Equations (2.88) and (2.89) the interpretation is that the terminal current I_b generates an electric field $\boldsymbol{E}_b$ that induces an open-circuit voltage $V_a^{\text{o.c.}}$ across the terminals of antenna "a", and terminal current I_a generates an electric field $\boldsymbol{E}_a$ that induces an open-circuit voltage $V_b^{\text{o.c.}}$ across the terminals of antenna "b". Then from Equations (2.88) and (2.89), Equation (2.85) reduces to

$$V_a^{\text{o.c.}} I_a = V_b^{\text{o.c.}} I_b \tag{2.90}$$

which is a statement of reciprocity in terms of currents and voltages. Now in the case where the driving currents for antennas "a" and "b" are assumed to be equal, that is, $I_a = I_b$, it follows then that

$$V_a^{\text{o.c.}} = V_b^{\text{o.c.}} \tag{2.91}$$

Equation (2.90) can expressed in terms of an equality between ratios as

$$\frac{V_a^{\text{o.c.}}}{I_b} = \frac{V_b^{\text{o.c.}}}{I_a} \tag{2.92}$$

The ratios in Equation (2.92) are open-circuit mutual impedances and can be rewritten then as the following reciprocity relation

$$Z_{ab}^{\text{o.c.}} = Z_{ba}^{\text{o.c.}} \tag{2.93}$$

Referring to Figure 2.29, the reciprocity relation given by Equation (2.93) applies at all r, θ, and ϕ coordinates and that means that the transmit and receive vector gain radiation patterns of an antenna are equivalent. This reciprocity relation allows radiation pattern measurements to be performed with the antenna under test in either a transmit or receive mode.

2.5 Reaction Integral Equation

The generalized reaction integral equation is derived from the reciprocity theorem for antennas expressed by Equation (2.84). By definition, the reaction of field a on source current b is given by [104, 105, p. 118]

$$<a,b>=\iiint_V (\boldsymbol{E}_a \cdot \boldsymbol{J}_b - \boldsymbol{H}_a \cdot \boldsymbol{M}_b) dv. \tag{2.94}$$

By reciprocity, the reaction of field a on source current b is the same as the reaction of field b on source current a, that is,

$$<a,b>=<b,a>. \tag{2.95}$$

Let a test source with currents $\boldsymbol{J}_m$ and $\boldsymbol{M}_m$ be placed within the scatterer volume, which has now been replaced by the original ambient medium. These currents radiate the fields $\boldsymbol{E}_m$ and $\boldsymbol{H}_m$ in the ambient medium. It is desired to compute the reaction between the test source currents with the total electric and magnetic fields within the volume within the surface S. Since the total electric and magnetic fields are equal to zero within the scatterer, the reaction between the test source and total fields is zero. Thus, we can write

$$\iiint_m \boldsymbol{J}_m \cdot (\boldsymbol{E}^s + \boldsymbol{E}^i) dv = 0 \tag{2.96}$$

$$\iiint_m \boldsymbol{M}_m \cdot (\boldsymbol{H}^s + \boldsymbol{H}^i) dv = 0. \tag{2.97}$$

To be consistent with the minus sign in the reciprocity theorem, the difference between Equations (2.96) and (2.97) is taken, which results in

$$\iiint_m (\boldsymbol{J}_m \cdot \boldsymbol{E}^s - \boldsymbol{M}_m \cdot \boldsymbol{H}^s) dv + \iiint_m (\boldsymbol{J}_m \cdot \boldsymbol{E}^i - \boldsymbol{M}_m \cdot \boldsymbol{H}^i) dv = 0 \tag{2.98}$$

where the volume integration is over the volume of the test source. In terms of surface integrals and surface currents, Equation (2.98) becomes

$$\iint_S (\boldsymbol{J}_{ms} \cdot \boldsymbol{E}^s - \boldsymbol{M}_{ms} \cdot \boldsymbol{H}^s) ds + \iint_S (\boldsymbol{J}_{ms} \cdot \boldsymbol{E}^i - \boldsymbol{M}_{ms} \cdot \boldsymbol{H}^i) ds = 0 \tag{2.99}$$

where $m = 1, 2, 3, \cdots, N$, $\boldsymbol{J}_{ms}$ is the mth electric surface current, and $\boldsymbol{M}_{ms}$ is the mth magnetic surface current. Equation (2.99) is the desired reaction integral equation. The next section shows how Equation (2.99) can be solved using the method of moments.

2.6 Method of Moments

The reaction integral equation expressed by Equation (2.99) can be solved using the method of moments for the vector currents, from which quantities such as the antenna input impedance, mutual coupling, and gain radiation patterns can be computed. In applying the method of moments, each antenna analysis geometry requires special considerations [105–116]. By using the equivalence principle as shown in Figure 2.28, the moment method solution is simplified for antenna problems involving only electric conductors and, thus, only vector electric currents $\boldsymbol{J}$ are assumed nonzero, and $\boldsymbol{M} = 0$.

In the method of moments, the unknown vector electric surface current density $\boldsymbol{J}_s$ is expressed as a summation of n surface currents $\boldsymbol{J}_n$ where

$$\boldsymbol{J}_n = i_n \boldsymbol{b}_n. \tag{2.100}$$

The current coefficients i_n are complex in general and they are the N unknown coefficients to be determined, and there are N vector basis functions $\boldsymbol{b}_n$ (also referred to as expansion or interpolation functions). Thus, the unknown vector surface current density can be expressed as

$$\boldsymbol{J}_s = \sum_{n=1}^{N} \boldsymbol{J}_n = \sum_{n=1}^{N} i_n \boldsymbol{b}_n. \tag{2.101}$$

It is desired to react an mth vector testing function $\boldsymbol{t}_m$ (also referred to as a vector weighting function) with the radiated field from the nth basis function $\boldsymbol{b}_n$. The electric field from the nth basis function can be expressed as the scattered field $\boldsymbol{E}_n^s(\boldsymbol{J}_n)$, where it is understood that the electric field is determined by an integral operating on the nth electric-current function $\boldsymbol{J}_n$. This integral operation can be expressed as

$$L\text{op}(\boldsymbol{J}_n) = \boldsymbol{E}_n^s(\boldsymbol{J}_n) \tag{2.102}$$

where $L\text{op}$ is the integral operator. For example, the electric field $\boldsymbol{E}_n^s(\boldsymbol{J}_n)$ can be determined using a two-step approach in which the integral in Equation (1.201) is evaluated and is then substituted into Equation (1.196). The integral operation on the current is a linear operator, that is, superposition applies and scalar quantities within the integral can be factored out of the integral. It is noted that the total scattered field is given by

$$\boldsymbol{E}^s = \sum_{n=1}^{N} \boldsymbol{E}_n^s(\boldsymbol{J}_n). \tag{2.103}$$

Thus, it follows from Equation (2.99) that

$$\sum_{n=1}^{N} \iint_S \boldsymbol{t}_m \cdot \boldsymbol{E}_n^s(\boldsymbol{J}_n) ds + \iint_S \boldsymbol{t}_m \cdot \boldsymbol{E}^i ds = 0, \tag{2.104}$$

which is the electric-field integral equation (EFIE) that is valid for open and closed surfaces. Using Equations (2.101) and (2.103) and the linearity of the integral operator, it follows that Equation (2.104) can be rewritten as

$$\iint_S \boldsymbol{t}_m \cdot \boldsymbol{E}^i ds = -\sum_{n=1}^{N} i_n \iint_S \boldsymbol{t}_m \cdot \boldsymbol{E}_n^s(\boldsymbol{b}_n) ds \tag{2.105}$$

where the nth complex current coefficient i_n has been factored out of the double integration. The basis functions $\boldsymbol{b}_n$ can be chosen typically to have an overlapping piecewise sinusoidal or triangular amplitude distribution, and these basis functions are known to provide an accurate representation of currents that flow on antennas and surrounding structures. In the case of a piecewise-sinusoidal basis function of arbitrary unit vector orientation $\hat{\boldsymbol{p}}$ along an arbitrary p axis (for example, p could be the z axis), the basis function takes the following form [108–110] with k the wavenumber,

$$\boldsymbol{b}_n(p) = \hat{\boldsymbol{p}_n} \frac{\sin k(p - p_{n-1})}{\sin k(p_n - p_{n-1})} \qquad p_{n-1} \leq p \leq p_n \tag{2.106}$$

$$\boldsymbol{b}_n(p) = \hat{\boldsymbol{p}_n} \frac{\sin k(p_{n+1} - p)}{\sin k(p_{n+1} - p_n)} \qquad p_n \leq p \leq p_{n+1} \tag{2.107}$$

Triangular basis functions have the following form

$$\boldsymbol{b}_n(p) = \hat{\boldsymbol{p}_n} \frac{p - p_{n-1}}{p_n - p_{n-1}} \qquad p_{n-1} \leq p \leq p_n \tag{2.108}$$

$$\boldsymbol{b}_n(p) = \hat{\boldsymbol{p}_n} \frac{p_{n+1} - p}{p_{n+1} - p_n} \qquad p_n \leq p \leq p_{n+1} \tag{2.109}$$

Figures 2.30(a) and 2.30(b) show examples for the case where there are 5 nodes and 3 overlapping (piecewise) sinusoidal and triangular basis functions along the p axis, respectively. The coefficient a_n that multiplies each basis function b_n is determined by solving a set of simultaneous equations as described below. For modeling antenna and scattering general surfaces, planar [106], curvilinear, and higher-order basis functions [112–114] are used. Now, define the general mth element of the voltage matrix as

$$V_m = \iint_S \boldsymbol{t}_m \cdot \boldsymbol{E}^i ds \tag{2.110}$$

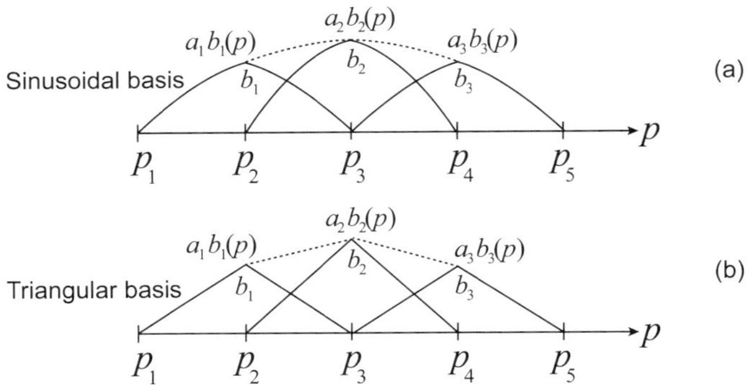

Figure 2.30 Depiction of basis functions for the case of 5 nodes. (a) Piecewise sinusoidal, and (b) triangular.

and the general mnth element of the mutual impedance matrix as

$$Z_{mn} = -\iint_S \boldsymbol{t}_m \cdot \boldsymbol{E}_n^s(\boldsymbol{b}_n) ds. \tag{2.111}$$

The testing and basis functions are typically normalized by dividing by the terminal current, so the units of the voltage matrix and impedance matrix are in volts and ohms, respectively. A Galerkin's formulation is obtained by using testing functions $\boldsymbol{t}$ that are equal to the basis functions $\boldsymbol{b}$. It now follows that

$$V_m = \sum_{n=1}^{N} i_n Z_{mn} \quad m = 1, 2, \ldots, N \tag{2.112}$$

where V_m is the voltage excitation and Z_{mn} is the open-circuit mutual impedance between current elements m and n, expressed as the matrix $\boldsymbol{Z}$. These expressions can be combined in terms of the applied voltage excitation vector $\boldsymbol{V}$, the impedance matrix $\boldsymbol{Z}$, and unknown current vector $\boldsymbol{i}$ to give the matrix equation

$$\boldsymbol{V} = \boldsymbol{Z}\boldsymbol{i}; \tag{2.113}$$

that is,

$$\begin{aligned} V_1 &= Z_{11}i_1 + Z_{12}i_2 + \cdots + Z_{1N}i_N \\ V_2 &= Z_{21}i_1 + Z_{22}i_2 + \cdots + Z_{2N}i_N \\ &\ \ . \\ &\ \ . \\ &\ \ . \\ V_N &= Z_{N1}i_1 + Z_{N2}i_2 + \cdots + Z_{NN}i_N. \end{aligned} \tag{2.114}$$

The unknown current coefficients $i_n, n = 1, 2, \ldots, N$ are found by matrix inversion from

$$i = Z^{-1}V \tag{2.115}$$

or by lower-triangular/upper-triangular (LU) matrix decomposition, for example. For phased array antenna analysis, the current responses for multiple beam-steering voltage vectors are readily computed since the impedance matrix inversion or LU decomposition needs to be performed only once for a given antenna geometry. The voltage source model chosen at the antenna terminals is typically a delta gap generator. This model assumes that the impressed electric field exists only at the antenna element terminals. In this case, the voltage generators can be defined to be one volt in amplitude at the terminals of each array element. If a general skew-symmetric lattice is assumed for the example dipole array depicted in Figure 2.31, then the required steering phase for an element in the pth row and qth column of the array is related to the main beam scan direction (θ_s, ϕ_s) by

$$\psi_x = kqd_x \sin\theta_s \cos\phi_s \tag{2.116}$$

$$\psi_y = k(pd_y + q\Delta y) \sin\theta_s \sin\phi_s \tag{2.117}$$

so

$$V_{pq}^{g} = e^{-j(\psi_x + \psi_y)} \tag{2.118}$$

defines the generator voltages. Note that in Figure 2.31, an artist's concept of an example of three overlapping dashed piecewise triangles, superimposed on the upper right dipole, represent the basis functions that can be used to determine the unknown current coefficients.

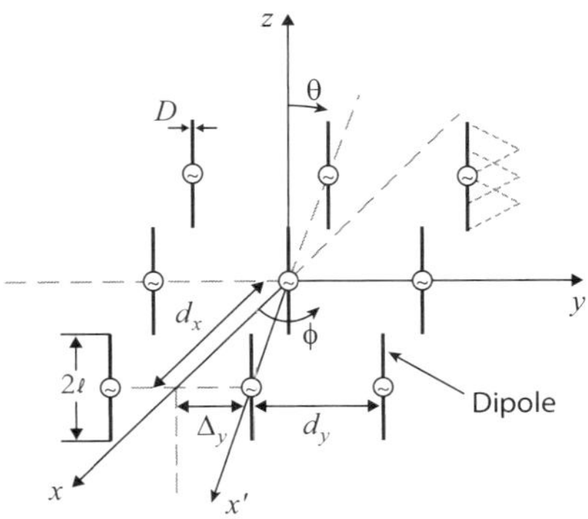

Figure 2.31 Dipole phased array antenna arranged in a two-dimensional skew-symmetric lattice.

In the present form, Equation (2.118) assumes a uniform amplitude excitation. It should be noted, however, that any amplitude taper can be used. To generate low sidelobes, some common array amplitude tapers include Taylor, Chebyshev, cosine, and triangular [5]. The various tapers would be implemented by multiplying the pqth array element terminal voltage of the voltage matrix by the appropriate amplitude coefficient A_{pq}.

The array elements are assumed to be excited by constant incident-power sources. This type of feeding is referred to as free excitation because the complex voltage across the terminals of the array element can vary according to the scan conditions [11]. The equivalent circuit for the mth element port ($m = 1, 2, \ldots, N$) is shown in Figure 2.32. The effect of the generator (or load) impedance (denoted Z_m^g) is included in the driven elements in the voltage excitation matrix, that is,

$$V_m = V_m^g - i_m Z_m^g. \tag{2.119}$$

For an electrically scanned phased array, all generators are on; hence, every V_m^g is nonzero. For a single-element excited array, only one voltage generator is nonzero, which can be assumed equal to one volt. For both types of arrays, the value of Z_m^g can be chosen to be 50 ohms, corresponding to the generator impedance, which is the same as the characteristic impedance of the feedline (usually 50 ohms). The unknown currents $i_n, n = 1, 2, \ldots, N$ are determined from

$$\boldsymbol{i} = [\boldsymbol{Z}']^{-1} \boldsymbol{V}^g \tag{2.120}$$

where

$$\boldsymbol{Z}' = \boldsymbol{Z} + \boldsymbol{I} Z_m^g \tag{2.121}$$

where $\boldsymbol{I}$ represents the identity matrix. The current coefficients i_n determined from the method of moments solution are used in the computation of the array radiation patterns. Improvements in the speed and reduced memory requirements for moment method analysis has been achieved using the multilevel fast multipole method (MLFMM) [112, pp. 209–253, 113].

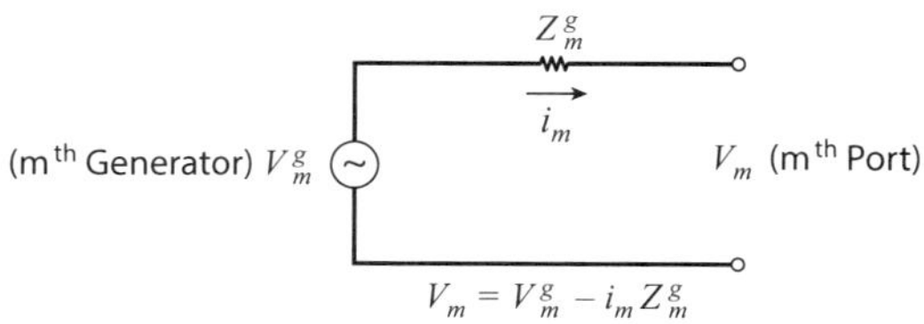

Figure 2.32 Equivalent circuit model for the mth array element port.

In the case of the mth array element, the input impedance, denoted Z_{in}^{m}, is computed from

$$Z_{\text{in}}^{m} = \frac{V_m}{i_m} \tag{2.122}$$

where V_m is the voltage across the terminals of the array antenna element and i_m is the current at the terminals of the mth element determined by the method of moments. Equations (2.27), (2.35), and (2.36) can be used to determine the radiated fields from the moment method basis functions. The far-zone electric field can be computed from the superposition of the radiated electric fields from the moment-method-determined currents computed from Equation (2.120). The scanned array reflection coefficient is obtained from the scanned array input impedance by the use of the following equation,

$$\Gamma(\theta_s, \phi_s) = \frac{Z_{\text{in}}(\theta_s, \phi_s) - Z_c}{Z_{\text{in}}(\theta_s, \phi_s) + Z_c} \tag{2.123}$$

where (θ_s, ϕ_s) is the scan direction, and Z_c is the characteristic impedance of the feed line. For large finite periodic phased arrays, a unit cell can be defined and the finite array is replaced with an infinite array where the array illumination periodic phase shift between array elements satisfies Floquet's theorem [6, pp. 37-42, 9, 52].

2.7 Broadside and Endfire Linear Arrays of Hertzian Dipoles

The radiation pattern of a single z-directed Hertzian dipole in free space was shown in Figures 1.51 and 1.52. Referring now to Figure 2.33, consider two 5-element array cases in which the antenna elements are z-directed Hertzian dipoles where the arrays are designed for either broadside beam direction or endfire beam direction. Figure 2.34 shows the FEKO moment method

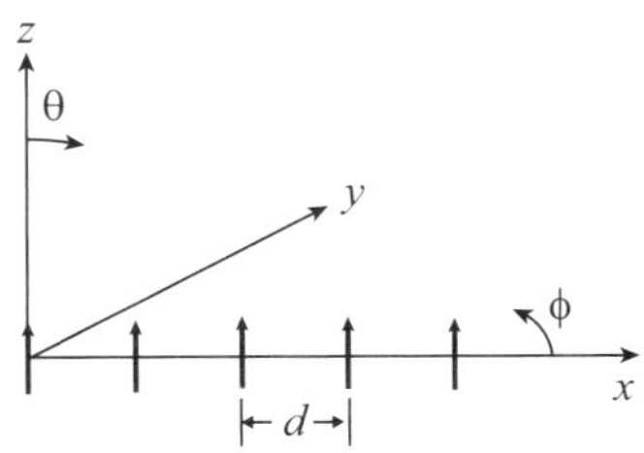

Figure 2.33 Linear array of Hertzian dipole antennas with spacing d along the x axis.

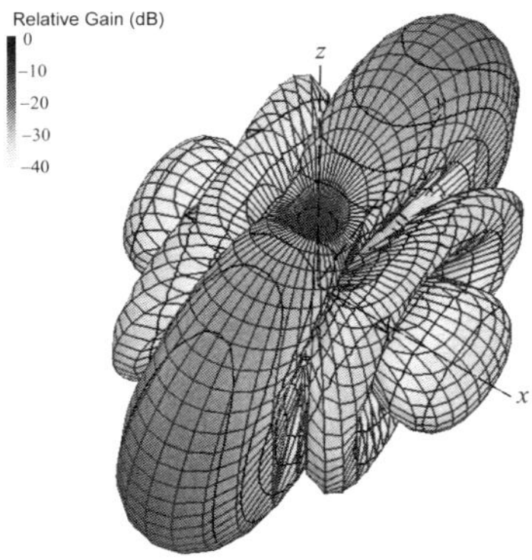

Figure 2.34 Moment method simulated 3D relative gain pattern (G_θ component) for a 5-element broadside array of z-directed Hertzian dipoles with $\lambda/2$ element spacing along the x axis.

simulated 3D relative gain pattern for a 5-element broadside array of z-directed Hertzian dipoles with $\lambda/2$ element spacing along the x axis, and Figure 2.35 is a polar diagram of the radiation pattern in the xy plane. For this simulated broadside array, the half-power beamwidth (HPBW) is 20.7° and the beamwidth between first nulls (BWFN) is 40°. Figure 2.36 is the moment method simulated 3D relative gain pattern for a 5-element endfire array of z-directed Hertzian dipoles with $\lambda/4$ element spacing along the x axis, and Figure 2.37 is a polar diagram of the radiation pattern in the xy plane. For this simulated endfire array, the half-power beamwidth (HPBW) is 75.5° and the beamwidth between first nulls (BWFN) is 156°.

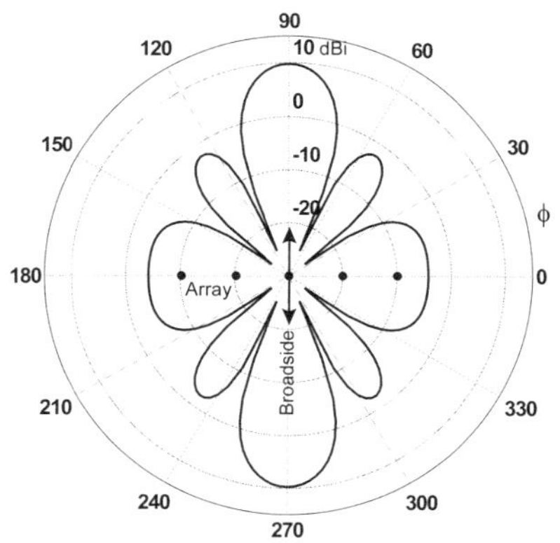

Figure 2.35 Moment method simulated polar gain pattern (G_θ) in the xy plane for a 5-element broadside array of z-directed Hertzian dipoles with $\lambda/2$ spacing.

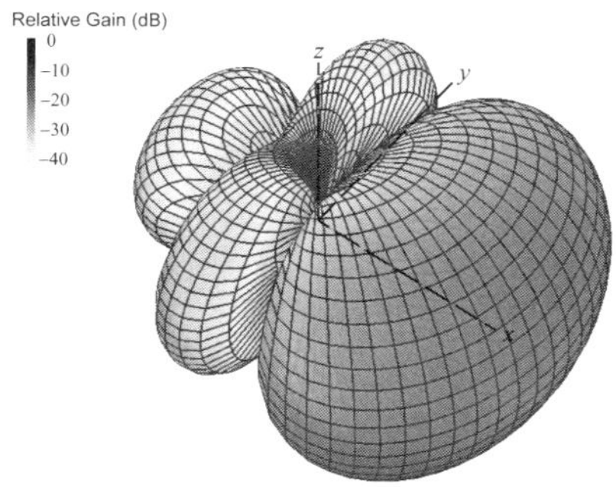

Figure 2.36 Moment method simulated 3D relative gain pattern (G_θ component) for a 5-element endfire array of z-directed Hertzian dipoles with $\lambda/4$ element spacing along the x axis.

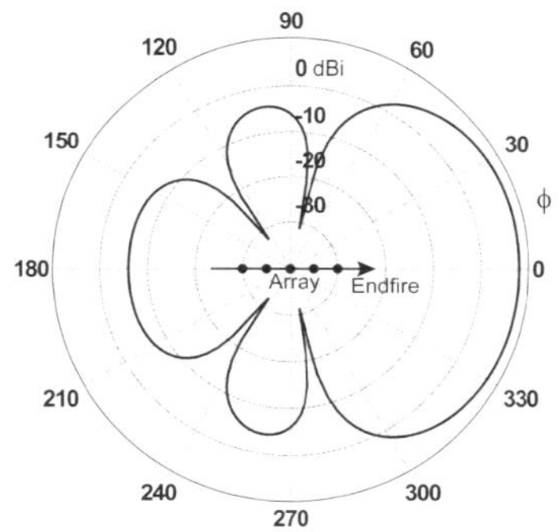

Figure 2.37 Moment method simulated polar absolute gain pattern (G_θ component) in the xy plane for a 5-element endfire array of z-directed Hertzian dipoles with $\lambda/4$ element spacing.

2.8 Example of 2D Array Mutual Coupling Effects

2.8.1 Dipole Element

Consider now a single $\lambda/2$ dipole oriented along the z axis in free space as shown in Figure 2.38(a). The moment method simulated normalized 3D gain pattern (vertically polarized, θ component) for this dipole antenna is shown in Figure 2.38(b). The peak gain is in the xy plane and the pattern is omnidirectional in azimuth with a null along the z axis. To show the effect of array mutual coupling on the element gain pattern, let the same dipole now be driven with a voltage source (having a 100-ohm reference impedance)

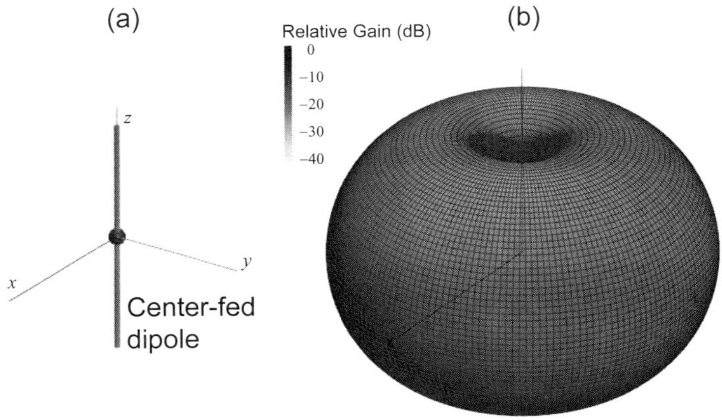

Figure 2.38 (a) Moment method model for a center-fed $\lambda/2$ dipole in free space, and (b) simulated normalized 3D gain pattern (G_θ component) for a single $\lambda/2$ dipole in free space.

and embedded in an 11×11 100-ohm resistively terminated array with $\lambda/2$ element spacing as depicted in the FEKO model shown in Figure 2.39(a). Due to array mutual coupling, in Figure 2.39(b) the moment method simulated normalized gain pattern (vertically polarized, θ component) of the center dipole element now has the peak gain shifted symmetrically upward and downward from the xy plane, and thus has reduced relative gain in the xy plane. There is a null in the direction of the z axis, which means the dipole element in this orientation has a blind spot in that direction.

2.8.2 Monopole Element

Consider now a simulation of a single $\lambda/4$ monopole oriented along the z axis located at the center of a square ground plane as shown in Figure 2.40(a). The perfectly conducting square ground plane is 5 wavelengths on a side. The moment method simulated normalized 3D gain pattern (vertically polarized, θ component) for this monopole antenna is shown in Figure 2.40(b). Figure 2.40(c) shows a polar plot of the gain pattern where the peak gain is 5.3 dBi and the peak angle is at $\theta = 61.5°$. The peak gain is elevated above the xy plane and the pattern is approximately omnidirectional in azimuth, with some ripple due to the square ground plane, and has a null along the z axis. To show the effect of array mutual coupling on the element gain pattern, let the same monopole now be driven with a voltage source (having a 50-ohm reference impedance) and embedded in an 11×11 50-ohm resistively terminated array with $\lambda/2$ element spacing as depicted in the FEKO model

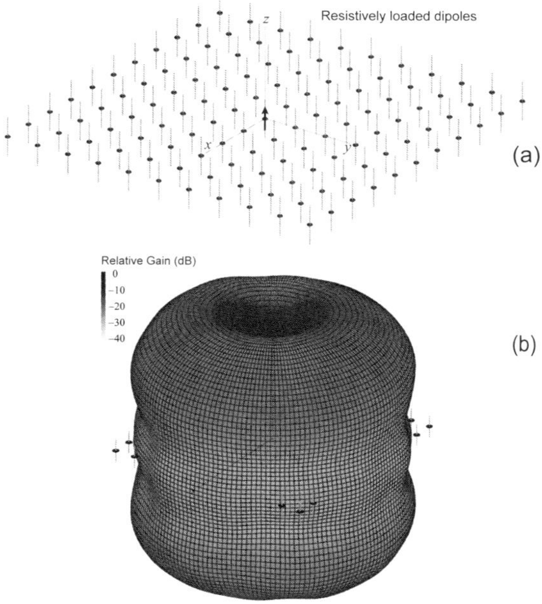

Figure 2.39 (a) Moment method model for an 11 $\times$ 11 array of $\lambda/2$ dipoles with the center element driven and the surrounding elements passively terminated in 100Ω resistive loads. (b) 3D normalized gain pattern (G_θ component) of the center dipole element.

shown in Figure 2.41(a). Due to array mutual coupling, in Figure 2.41(b) the moment method simulated normalized gain pattern (vertically polarized, θ component) of the center dipole element now has the peak gain shifted upward above the xy plane, and thus has reduced relative gain in the xy plane. There is a null in the direction of the z axis, which means the monopole element in this vertical orientation has a natural scanning blind spot in that direction. Figure 2.41(c) shows conical gain patterns at $\theta = 55°$ for the G_θ (co-polarized) and G_ϕ (cross-polarized) components.

2.9 Swept-Back Dipole Array Measurements and Simulations

2.9.1 Introduction

Dipole antennas are commonly used in phased arrays [9, 11, 23–26, 32–34, 48]. Linear dipoles can be designed to radiate or receive linear polarization. Crossed-dipoles can operate with dual polarization or circular polarization,

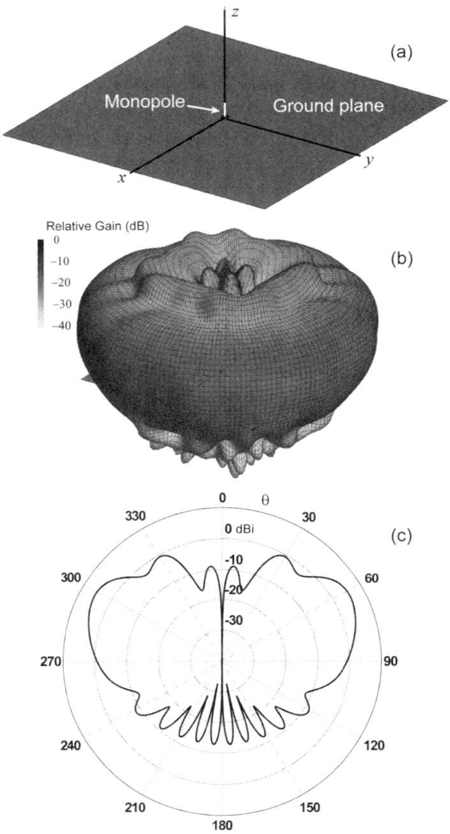

Figure 2.40 (a) Moment method model for a $\lambda/4$ monopole on a square ground plane 5λ on a side. (b) Simulated normalized 3D gain pattern (G_θ component). (c) Simulated polar gain pattern (G_θ component).

which can be advantageous for communications or radar links. For example, the polarization of communications users or radar targets can be arbitrary, so a single linearly polarized system might not close the link. In some cases, the polarization sense can be altered at microwave frequencies due to propagation effects such as Faraday rotation that occurs in the ionosphere [47]. Dipoles typically have straight arms that are parallel to a conducting ground plane; however, for wide-angle pattern coverage for a single element, or wide-angle scanning for an array, the dipole arms can be swept downward toward the ground plane forming a V-shaped antenna [7, 43, 116]. The element gain pattern for finite phased arrays of dual-polarized crossed V-dipole (pronounced *Vee* dipole) antennas above a ground plane is addressed in this

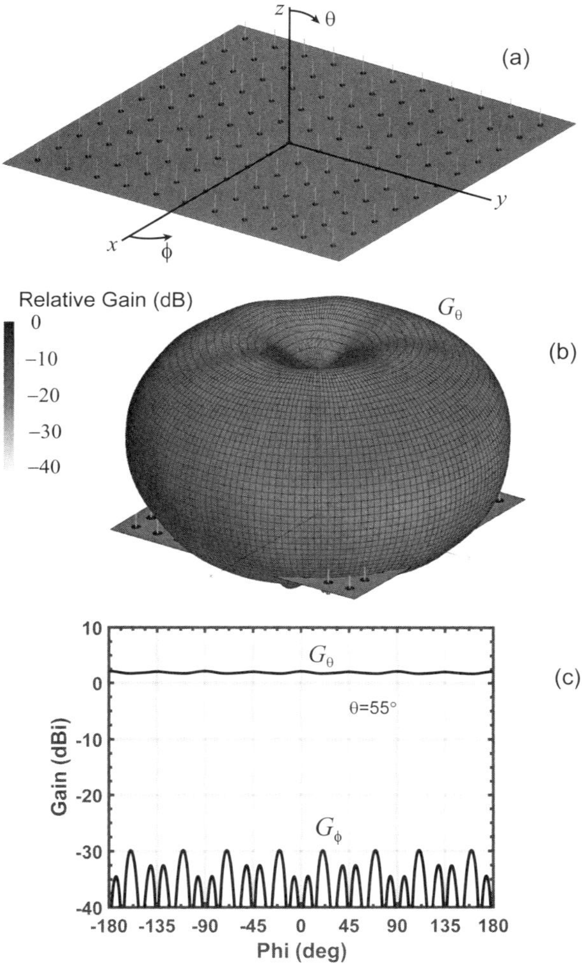

Figure 2.41 (a) Moment method simulation model for an 11×11 array of $\lambda/4$ monopoles with the center element driven and the surrounding elements passively terminated in 50Ω resistive loads. (b) 3D normalized gain pattern (G_θ component) of the center monopole element. (c) Conical gain pattern at $\theta = 55°$ for the G_θ and G_ϕ components.

chapter, both with computer simulation and by measurements. The method of moments is used to compute the center element gain pattern of a finite array of V-dipoles. An experimental 19-element passively terminated planar array is described, and mutual coupling, center element scanned array reflection coefficient, and center-element gain pattern measurements are described here.

It is known that for certain radiating elements scan blindness (blind spots) [6, 7, 9] can occur in large planar phased-array antennas. In phased-array antenna applications, a blind spot is identified with an array element pattern null (minimum) or a scanned array reflection coefficient with amplitude equal to unity. For example, a blind spot has been observed in a balun-fed straight-arm dipole phased array [7, 9]. A center-fed straight dipole with balun is shown in Figure 2.42(a). The dipole has an arm length denoted l, with strip width w, and is located a distance h above a conducting ground plane. The balun in this case is a balanced two-wire line that is shorted at the ground plane, which for a one-quarter wavelength transmission line, appears as an open-circuit at the feed terminals to restrict current flow on the balun. When this element is situated in a large phased array with the main beam steered away from broadside, the illumination of the balun from the array elements is not symmetrical, and unbalanced currents can flow on the balun. The presence of induced currents flowing on the balun can radiate and in some cases cancel the primary field of the dipole in a certain direction, creating the blind spot. This effect can be reduced or eliminated by tilting the dipole arms toward the ground plane at an angle α forming a V-shaped element [7, 9] (see Figure 2.42(b)).

It can be inferred that the tilted arms of the V-dipole act to reduce the amplitude of currents that flow on the balun stubs. If this inference is true, then it can be assumed that the balun and feedline scattering can be ignored in a theoretical analysis of the element gain pattern.

The radiation pattern of a single V-dipole antenna in free space can be computed as follows. Schelkunoff and Friis have given an expression for the radiation pattern of a dipole arm with sinusoidal current distribution [2, p. 501]. If it is assumed that the dipole arm is oriented in the z' direction, the radiation field pattern is expressed as

$$E_\theta = \frac{j30i_t}{\sin kl}\left[\frac{e^{jkl\cos\theta'} - j\cos\theta'\sin kl - \cos kl}{\sin\theta'}\right] \tag{2.124}$$

where $k = 2\pi/\lambda$, and λ is the wavelength. From the diagram shown in Figure 2.43 for a V-dipole in free space, the pertinent angles θ'_L for the left half dipole and θ'_R for the right half of the V-dipole, in terms of the angles θ and α, are given by

$$\theta'_L = \frac{\pi}{2} - (\theta + \alpha) \tag{2.125}$$

$$\theta'_R = \frac{\pi}{2} - (\theta - \alpha) \tag{2.126}$$

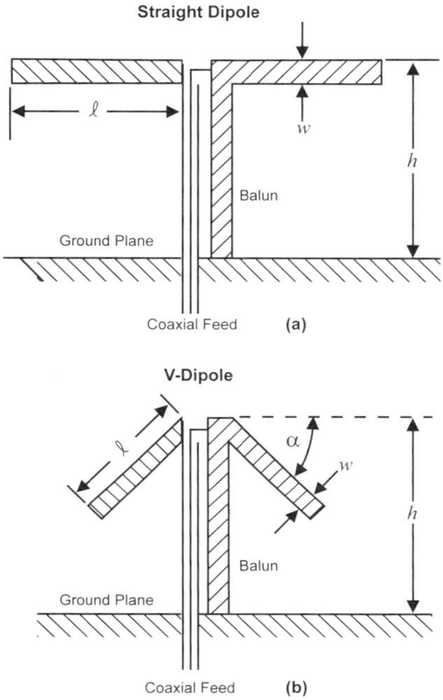

Figure 2.42 Balun-fed dipoles above a ground plane. (a) Straight arm, and (b) V-dipole.

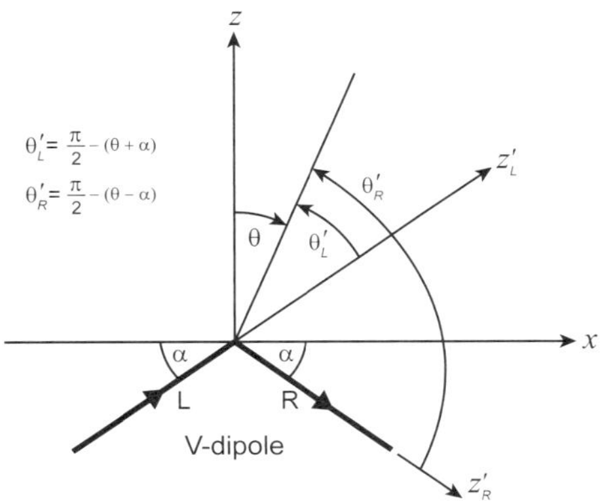

Figure 2.43 Geometry for V-dipole element in free space.

The total electric field due to the left and right arms is given by

$$E_\theta = E_{\theta_L} + E_{\theta_R} \tag{2.127}$$

Using (2.125) and (2.126) for the left and right halves of the V-dipole in (2.124) and using Equation (2.127) yields the total E-field for the V-dipole in free space as

$$E_\theta = \frac{j30i_t}{\sin(kl)}\left(\frac{e^{jkl\sin(\theta+\alpha)} - j\sin(\theta+\alpha)\sin(kl) - \cos(kl)}{\cos(\theta+\alpha)}\right. \\ \left. + \frac{e^{-jkl\sin(\theta-\alpha)} + j\sin(\theta-\alpha)\sin(kl) - \cos(kl)}{\cos(\theta-\alpha)}\right) \quad (2.128)$$

If the V-dipole is located above a conducting ground plane, the method of images can be used to include an image V-dipole forming a two-element array in free space, from which the combined primary plus image V-dipole radiation pattern can be determined. The radiation patterns of an array of V-dipole antennas can be computed by using the appropriate array factor.

To include the complete effects of the balun feed and dipole arms a general-purpose moment method solver such as FEKO can be used to compute the V-dipole input impedance and element gain patterns, which is addressed in Chapter 3.

Infinite array analyses for dipole radiators with feedlines have been investigated [48, 115]. In [115], straight-arm dipole arrays with feedlines were analyzed in terms of the transverse magnetic (TM) feed region modes. It was shown that the dominant TM mode can go into cut-off, producing a blind spot, prior to the onset of grating lobes. This mode cut-off effect was presented in terms of the propagation constant of the TM mode. The theoretical results presented in [48] indicate the presence of blind spots in the active scan impedance for straight dipole elements with feedlines. For swept-back dipoles with feedlines it has been demonstrated with measurements in a finite array that a blind spot does not tend to occur [116]. It is desirable to be able to predict the V-dipole element gain pattern in a finite array, including the effects of mutual coupling. Many phased-array antennas are designed to scan up to about 60° from broadside. The purpose of the next section is to show that over this $\pm 60°$ scan sector embedded element gain pattern simulations are accurate for V-dipole arrays using a simplified method of moments model. Only the current flowing on the dipole arms is included in the moment method model. Balun or feedline effects are not considered here, but the effects of a finite ground plane are included. Crossed dipole elements are useful in generating circular or dual-linear polarization. Dual linearly polarized center-element gain patterns are considered here. Passively terminated arrays are convenient for determining the element gain pattern [6]. In this situation a single element is driven, usually the center element, and the surrounding elements are terminated in resistive loads.

2.9.2 Dipole Element Prototypes

A dipole array element design for straight and V-dipoles is discussed in this section. The desired operating band of interest here is 1.2–1.4 GHz (15.4% bandwidth). At the center frequency (1.3 GHz, λ =23.08 cm), the dipole terminals are located approximately $\lambda/4$ above the ground plane. Both straight-arm and V-dipole elements were fabricated [116], for comparing their measured radiation patterns when each element is mounted above a conducting ground plane. A photograph of the crossed straight-arm dipole antenna used in these measurements is shown in Figure 2.44. The dipole arms ($l = 5.461$ cm, $w = 0.356$ cm, $h = 5.588$ cm) were etched on a double-sided printed circuit (PC) board (1.5 mm FR4 glass-reinforced plastic laminate, dielectric constant 4.7). Two of these PC boards are notched and interleaved, forming the desired crossed dipole antenna. The crossed dipoles are fed by two separate RG-141 (0.358 cm diameter) coaxial cables (with 50-ohm characteristic impedance), which provide independent orthogonally polarized signals. The balun/feedline is made from the RG-141 semi-rigid coaxial cable and a solid brass rod (0.358 cm diameter) that is referred to as a dummy post. Electrical connections and crossover at the pair of dipole terminals are facilitated by the use of a double-sided printed circuit board. The balun coaxial feedline and dummy post effectively forms a two-wire balanced line with length 5.588 cm and center-to-center spacing 0.754 cm. Since the balun length is approximately $\lambda/4$ at the center frequency and is short circuited at the ground plane, the input impedance to the two-wire line is approximately an open circuit and prevents currents from flowing on the outer jacket of both the coaxial cable and dummy rod. For the given balanced transmission line wire diameter and spacing, the balun characteristic impedance is computed to be 165 ohms (refer to Chapter 3, Equation (3.1)). This large value for the balun characteristic impedance (compared to the 50-ohm coaxial cable characteristic impedance), helps to prevent current flow on the outside surfaces of the balun section including the coaxial cable and dummy post over the operating bandwidth. It should be noted that other balun feed designs for dipole

Figure 2.44 Photograph of a crossed straight-arm dipole antenna.

antennas using printed circuit techniques have been developed by a number of researchers [117–122]

The crossed V-dipole element used in these measurements was fabricated in the same manner, but with the arms swept down toward the ground plane at a 45° angle, as shown in Figure 2.45 ($l = 5.461$ cm, $w = 0.356$ cm, $h = 5.588$ cm, $\alpha = 45°$). Nineteen such V-dipoles were fabricated for array testing. The ground plane used for the single element and V-dipole array measurements was a 3.175 mm sheet of aluminum with dimensions 1.22m × 1.22m square.

2.9.3 Measured and Simulated Results

To compare dipole designs, straight-arm and crossed V-dipole antennas were mounted at the center of a 1.22m square aluminum ground plane, and linearly polarized radiation patterns were measured in an anechoic chamber. The measured E-plane and H-plane patterns are shown in Figures 2.46 and 2.47, respectively. In the measured E-plane patterns (E_θ component) shown in Figure 2.46, as expected the V-dipole has a wider angular pattern coverage compared to the straight-arm dipole. For angles greater than about 45° from broadside, the E-plane radiation patterns of the straight and V-dipoles differ significantly. For example, at an angle 60° from broadside, the V-dipole gain is about 5 dB higher than the straight dipole gain. This wider pattern coverage should allow wider scanning coverage compared to the straight-arm dipole, which is a motivating factor for tilting the dipole arms. In the measured H-plane patterns (E_ϕ component) shown in Figure 2.47, the radiation patterns of the straight-arm and V-dipole single elements are approximately the same.

A layout for the finite array of 19 V-dipole antenna elements on a 1.22m square ground plane is shown in Figure 2.48. For wide-angle scanning the V-dipole element spacing in the array is chosen as 12.7 cm (corresponding to 0.55λ at the center frequency of 1.3 GHz) on a hexagonal lattice. This

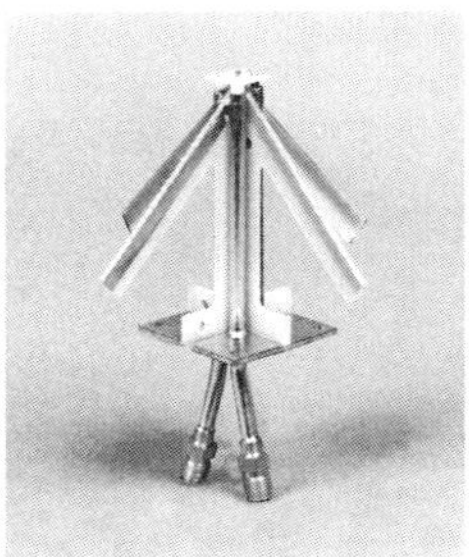

Figure 2.45 Photograph of a crossed V-dipole antenna.

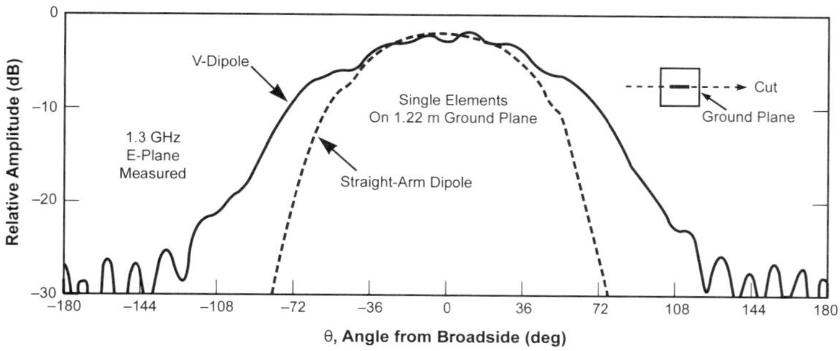

Figure 2.46 Measured E-plane radiation patterns for single crossed straight-arm and crossed V-dipole antennas.

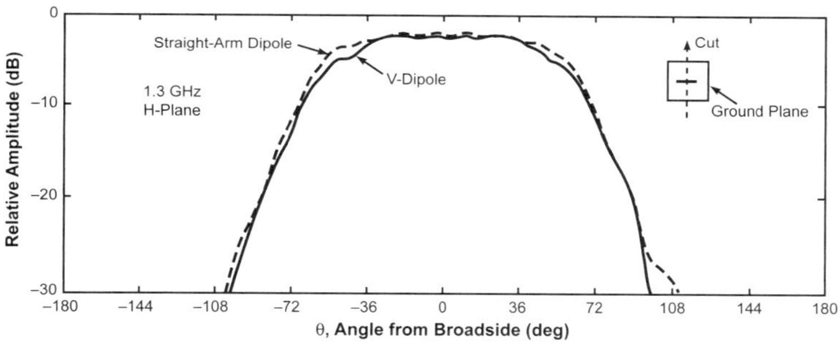

Figure 2.47 Measured H-plane radiation patterns for single crossed straight-arm and crossed V-dipole antennas.

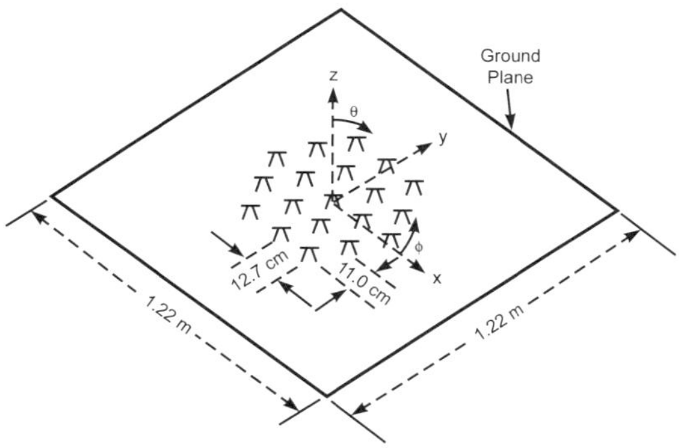

Figure 2.48 Layout for a 19-element crossed V-dipole array.

lattice spacing allows scanning without the formation of grating lobes in any plane out to 60° from broadside. A photograph of the array of 19 V-dipoles is shown in Figure 2.49. For center element gain pattern measurements and moment method simulations, the center element is driven and 50-ohm resistive loads are assumed at the terminals of each surrounding passively terminated dipole element. For array mutual coupling (S-parameter) measurements, the center element is driven and all elements except the receive dipole element are terminated in 50-ohm resistive loads. Center element reflection coefficient and mutual coupling data were measured for the 19-element V-dipole array, and these data are given in Figure 2.50. At 1.3 GHz (center frequency) the measured return loss ($20 \log_{10} |S_{11}|$) was -8.5 dB. In the E-plane ($\phi = 0°$), the measured coupling to the nearest neighbor was -25.5 dB. In the H-plane, the measured coupling to the nearest neighbor was -19.7 dB. These measured complex mutual coupling data can be summed according to Equation (2.38) to compute the array scan reflection coefficient for the center element versus scan angle in both E- and H-planes, as depicted in Figure 2.51. Based on these calculations, the array scan mismatch loss ranges from about 0.5 dB to 1 dB over scan angles from $0°$ to $60°$, respectively.

Measured and FEKO moment method simulated center element gain patterns, in the passively terminated array, at 1.2, 1.3, and 1.4 GHz are presented in Figures 2.52 (E-plane) and 2.53 (H-plane). The patterns indicate peak gain (approximately 5 to 6 dBi) occurring in the vicinity of broadside and no more than 4 dB amplitude taper at 60° from broadside. Also, there is no evidence of a blind spot. Over the 15.4% bandwidth shown, the measured and simulated gain patterns are in good agreement.

2.10 Rectangular Waveguide Phased Array Example

Open-ended rectangular or circular waveguide elements are commonly suggested for phased array applications. Single open-ended rectangular and circular waveguide antennas are analyzed in detail in Wolff [123]. The

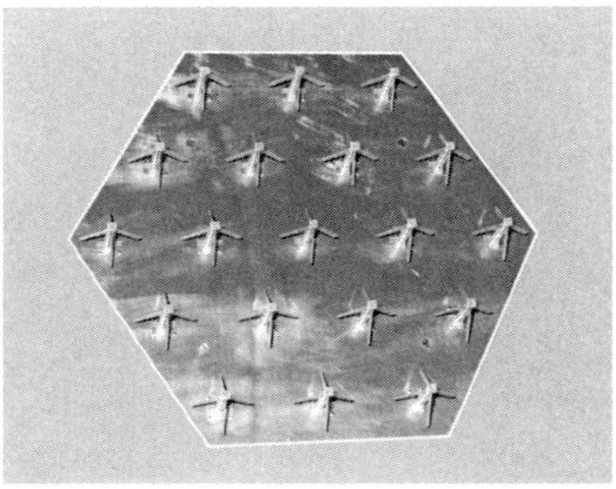

Figure 2.49 Photograph of a 19-element crossed V-dipole array.

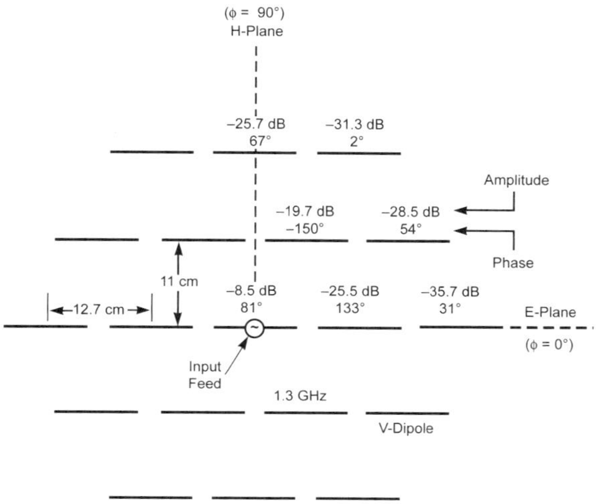

Figure 2.50 Measured center element reflection coefficient and mutual coupling for a 19-element crossed V-dipole array.

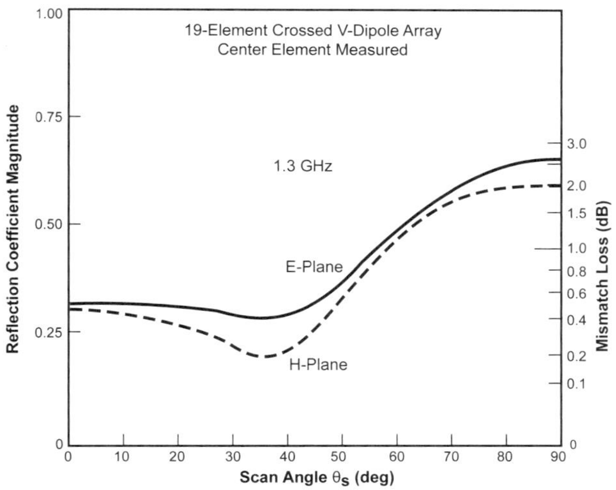

Figure 2.51 Calculated scan reflection coefficient versus scan angle as computed from measured mutual coupling data (Figure 2.50) for the 19-element crossed V-dipole array shown in Figure 2.49.

behavior of planar arrays of open-ended rectangular waveguides has been investigated in numerous studies [6, 30, 31, 39–41, 124–140]. Borgiotti [124]

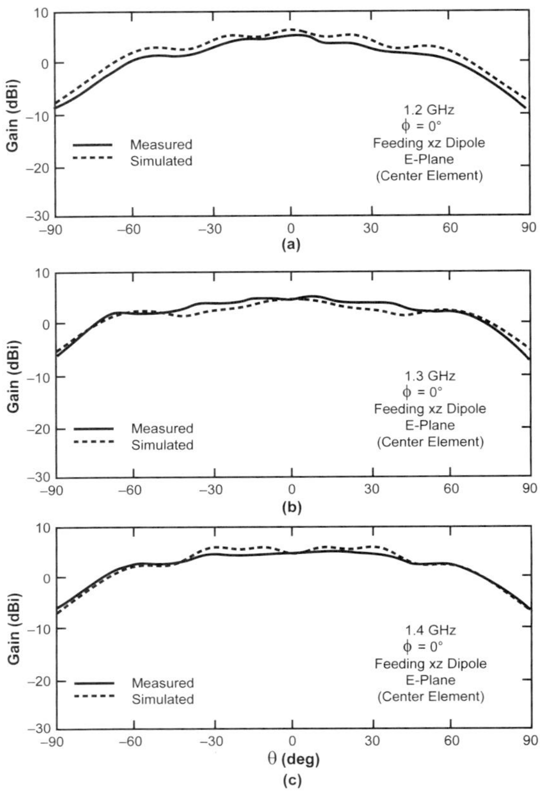

Figure 2.52 Measured and moment method simulated center element gain patterns for a 19-element V-dipole array. Feeding xz dipole, E-plane, $\phi = 0°$ cut. (a) 1.2 GHz, (b) 1.3 GHz, and (c) 1.4 GHz.

obtained a mutual admittance expression between two identical radiating apertures in the form of a Fourier transform of a function related to the power radiation pattern of the element. Mailloux [125] used the method of moments and a single-mode approximation to the aperture field to analyze the coupling between two closely spaced open-ended waveguide slots. Mailloux [126] also analyzed the coupling between collinear open-ended waveguide slots by expanding the aperture field in a Fourier series. Bird [127] has analyzed the coupling between rectangular waveguides of different sizes. An approximation to the behavior of a finite array of rectangular waveguides was obtained by Amitay, Galindo, and Wu [6] by using infinite array techniques with finite array excitation. Wu also analyzed a finite array of parallel plate waveguides [128]. Cha and Hsiao [129] and Hidayet [130] investigated the

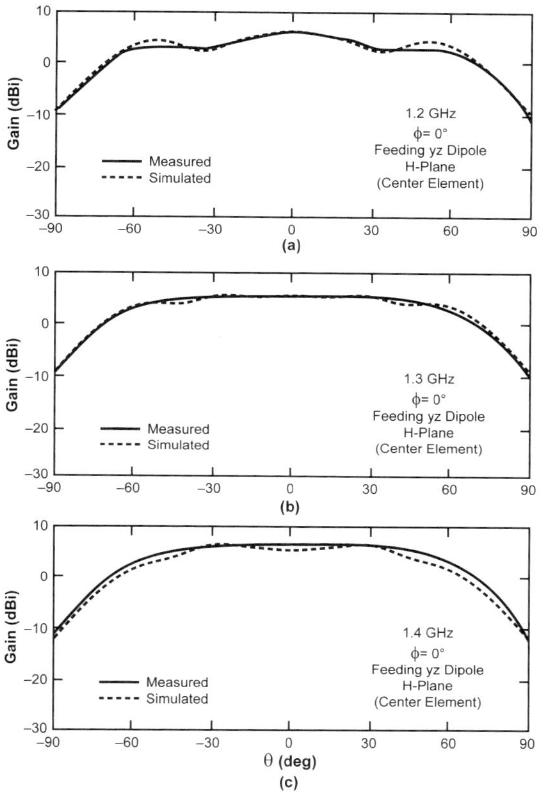

Figure 2.53 Measured and moment method simulated center element gain patterns for a 19-element V-dipole array. Feeding yz dipole, H-plane, $\phi = 0°$ cut. (a) 1.2 GHz, (b) 1.3 GHz, and (c) 1.4 GHz.

reflection coefficients of planar waveguide arrays of size up to 13×13 for a rectangular grid. Hidayet also analyzed a 7×7 planar array arranged in triangular grid. Luzwick and Harrington [131] investigated the coupled power between elements of 7×7 rectangular and triangular grid arrays. Wang [132] investigated the measured E-plane radiation patterns for the center element of a 3×41 array of rectangular waveguides and performed an infinite array analysis of both rectangular and ridged waveguide arrays.

Consider an example 7-element linear array of rectangular waveguide elements, which is analyzed by the FEKO method of moments solver. Figure 2.54(a) shows the geometry of an open-end rectangular waveguide with metal walls and cross sectional dimensions $a \times b$. The waveguide would be driven with a probe to generate an electric field $\boldsymbol{E}$ that radiates through

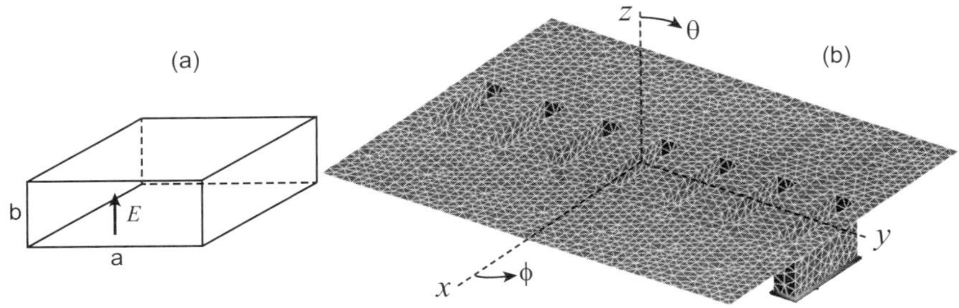

Figure 2.54 (a) Rectangular waveguide geometry, and (b) moment method meshed model of a 7-element rectangular waveguide phased array.

the open end aperture. Figure 2.54(b) shows the seven-element array meshed geometry in which the elements operate at 1.3 GHz and have dimensions a =16.51 cm by b =4.064 cm (6.5 inches by 1.6 inches) that are arranged along the y axis with center-to-center spacing 10.922 cm (4.3 inches). The automatically meshed model in Figure 2.54(b) uses 5,278 metallic triangular patches. The elements radiate through a perfectly conducting rectangular ground plane with dimensions 76.45 cm by 60.96 cm (30.1 inches by 24.0 inches). Each of the seven elements are fed by a rectangular waveguide section 10.16 cm (4.0 cm) long operating in the dominant transverse electric TE10 mode excitation. The TE10 mode cutoff frequency f_c for these air-filled rectangular waveguide elements is determined by the following equation [4]

$$f_c^{\text{TE10}} = c/(2a) \tag{2.129}$$

where c is the speed of light and a is the waveguide broadwall width. Substituting a =16.51 cm in Equation (2.129) yields a waveguide cutoff frequency for the TE10 mode equal to 0.908 GHz. Figure 2.55(a) shows the simulated center element and edge element directivity patterns ($\hat{\theta}$ component) in the yz plane (E-plane) – in this case one element is driven and the surrounding elements are terminated in matched loads. In Figure 2.55(b), the array is uniformly illuminated in amplitude and shows the directivity patterns as the waveguide array is scanned from 0° to 45° from broadside. It is observed that the peak directivity decreases and the beam broadens as the main beam is scanned away from broadside. The peak directivities are equal to 15.26 dBi, 15.06 dBi, 14.49 dBi, and 13.74 dBi at the scan angles $\theta_s = 0°$, 15°, 30°, and 45°, respectively. Compared to broadside scan, at 45° scan it is observed that the peak directivity is reduced by 1.5 dB due to the projected aperture factor $10\log_{10}\cos\theta_s$. The half-power beamwidths are equal to 15.23°, 15.79°, 17.61°, and 21.42° at the scan angles $\theta_s = 0°$, 15°, 30°, and 45°, respectively. The beam broadening factor at 45° scan is equal

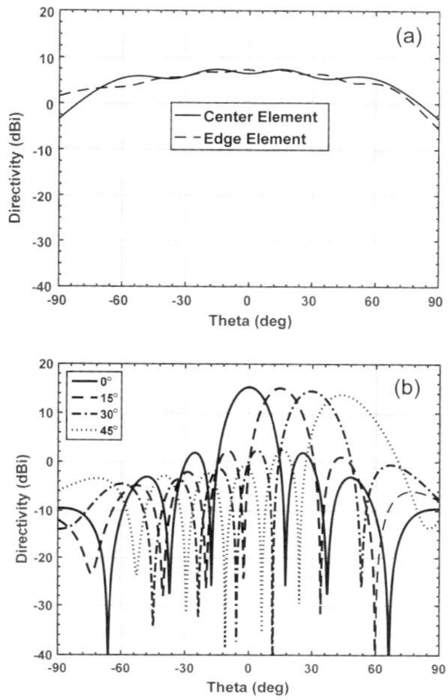

Figure 2.55 Moment method simulations for a 7-element rectangular waveguide phased array. (a) Center element and edge element directivity patterns, and (b) scanned array directivity patterns.

to $\sqrt{2}$, which is due to the projected aperture factor $1/\cos\theta_s$ as the beam is scanned away from broadside.

An example of a 32-element linear array of horizontally polarized open-end rectangular waveguides that has been used in focused near-field adaptive nulling measurements [141] is shown in Figure 2.56. The photograph in Figure 2.56(a) shows a pyramidal horn antenna that was used to illuminate the waveguide array. The waveguide array was mounted on a positioner to collect radiation pattern data before and after adaptive nulling with a single jammer source. Figure 2.56(b) shows a diagram of the array beamformer that consisted of intermediate frequency (IF) modules with phase shifters (indicated as PHS) to electronically steer the main beam, and attenuators (indicated as ATT) to implement an amplitude taper across the array with a 32 to 1 power combiner. Additionally in Figure 2.56(b), after phase shifting the received signal is split into a second path (denoted by the letter B) to selectively route signals from three elements as auxiliary channels, along with the main channel, that are downconverted and digitized within a four-channel nulling receiver. A digital computer then processes the received voltage in the four channels, forming a covariance matrix and then adaptive weights are

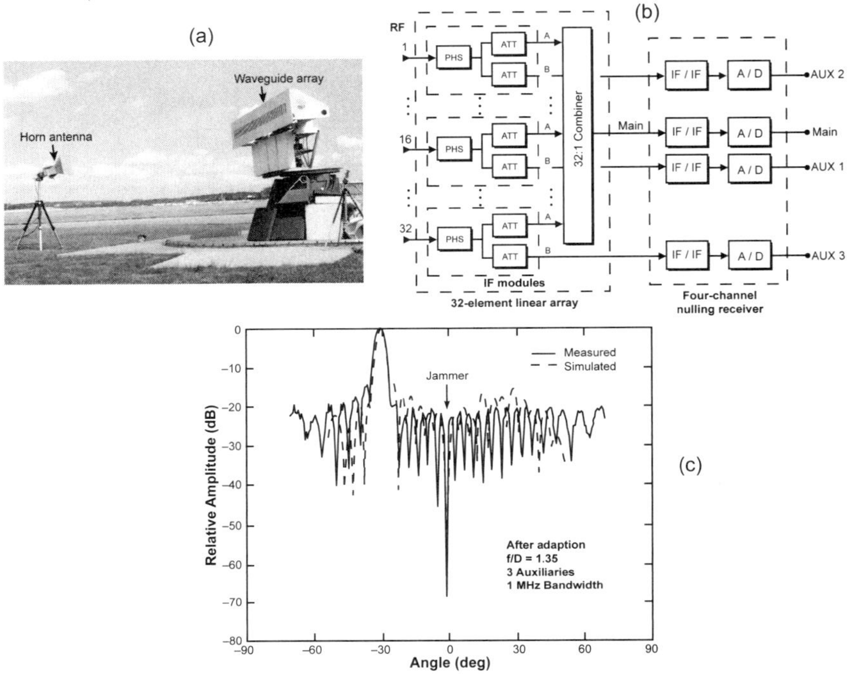

Figure 2.56 Example of an adaptive nulling phased array antenna system. (a) Photograph of a 32-element rectangular waveguide phased array. (b) Block diagram of array beamformer and nulling receiver. (c) Measured and simulated adaptive array radiation patterns.

computed based on Equation (2.12). In Figure 2.56(c), the phased array was phase focused at a center frequency of 1.3 GHz at the near-field source horn position (range distance was 4.6m (1.35 array diameters)) for a $-30°$ scan angle with a -20 dB Chebyshev taper, and then a narrowband noise jamming signal was nulled at broadside (0°) — the measured and simulated adaptive array radiation patterns are in good agreement. Note that the source horn was used as both a CW calibration source and as the jammer. This phased array focused near-field antenna measurements technique [15, pp. 169–174] is also discussed in Chapter 12 (Section 12.4).

2.11 Summary

This chapter has reviewed the theory for application to phased array antenna development. Beamforming architectures, basic array theory, and the method of moments were reviewed. The method of moments allows the analysis of phased array antennas, including the effects of mutual coupling between the antenna elements. Dipole, monopole, and waveguide array elements have been described in this chapter. In the following chapters, the practical design and development of a number of antennas and arrays is described.

2.12 PROBLEM SET

2.1 From Equation (2.24), the magnitude of the array factor is expressed as $|\text{AF}| = \left|\frac{\sin(Nu/2)}{\sin(u/2)}\right|$, where $u = kd(\sin\theta - \sin\theta_s)$, where $k = 2\pi/\lambda$, d is the array spacing, θ is the observation angle from broadside, and θ_s is the scan angle from broadside as depicted in Figure 2.6. Consider an N-element uniformly illuminated broadside array such that $\theta_s = 0°$, so $u = kd\sin\theta$. Determine an expression for the beamwidth between first nulls (BWFN) and the half-power beamwidth (HPBW) for the broadside array. Hint: Set $u = kd\sin\theta_1 = 2\pi/N$ and solve for θ_1 to determine the angle of the first null, and for a large array with narrow beamwidth use the small angle approximation $\sin\theta_1 \approx \theta_1$. Note that for beamwidth calculation, this small angle approximation for the sine function is good within 1% for angles between 0 radians and about 0.244 radians (14°). By symmetry the BWFN is twice θ_1 and the HPBW is approximately equal to one half the BWFN. Answer: $\text{BWFN}_{\text{broadside}} \approx \frac{4\pi}{kdN} \approx \frac{2\lambda}{dN} \approx \frac{2\lambda}{L}$, $\text{HPBW}_{\text{broadside}} \approx \frac{2\pi}{kdN} \approx \frac{\lambda}{dN} \approx \frac{\lambda}{L}$, where L is the length of the broadside array and L is large in terms of wavelengths.

2.2 Consider the moment method simulated beamwidths (HPBW and BWFN) for the $N = 5$ element broadside dipole array with $d = \lambda/2$ element spacing discussed in Section 2.7 (Figures 2.33 to 2.35). Using the equations derived in Problem 2.1, compute the HPBW and BWFN and compare with the moment method simulated results. Answer: The calculated HPBW is 22.9° compared to the moment method HPBW which is 20.7°. The calculated BWFN is 45.8° compared to the moment method BWFN which is 40°.

2.3 Refer to Figure 2.6 and consider the case where an N-element linear array has uniform amplitude illumination and the main beam is steered to endfire. From Equation (2.24), the magnitude of the array factor is expressed as $|\text{AF}| = \left|\frac{\sin(Nu/2)}{\sin(u/2)}\right|$, where now $u = kd(\cos\psi - \cos\psi_s)$, where $k = 2\pi/\lambda$, d is the array spacing, ψ is the observation angle from endfire, and ψ_s is the scan angle from endfire. For the endfire array $\psi_s = 0°$, so $u = kd(\cos\psi - 1)$. Determine an expression for the beamwidth between first nulls (BWFN) and the half-power beamwidth (HPBW) for this endfire array. Hint: Set $u = kd(\cos\psi_1 - 1) = -2\pi/N$ and solve for ψ_1 to determine the angle of the first null, use the small angle approximation for the cosine function $\cos\psi_1 \approx 1 -$

$\psi_1^2/2$. Note that for beamwidth calculation, this small angle approximation for the cosine function is good within 1% for angles between 0 radians and about 0.664 radians (38°). By symmetry the BWFN is twice ψ_1 and the HPBW is approximately equal to one half the BWFN. Answer: $\text{BWFN}_{\text{endfire}} \approx 2\sqrt{\frac{4\pi}{kdN}} \approx 2\sqrt{\frac{2\lambda}{dN}} \approx 2\sqrt{\frac{2\lambda}{L}}$, $\text{HPBW}_{\text{endfire}} \approx \sqrt{\frac{4\pi}{kdN}} \approx \sqrt{\frac{2\lambda}{dN}} \approx \sqrt{\frac{2\lambda}{L}}$, where L is the length of the endfire array and L is large in terms of wavelengths.

2.4 Consider the moment method simulated beamwidths (HPBW and BWFN) for the $N = 5$ element endfire dipole array with $d = \lambda/4$ element spacing discussed in Section 2.7 (Figures 2.33, 2.36, and 2.37). Using the equations derived in Problem 2.3, compute the HPBW and BWFN and compare with the moment method simulated results. Answer: The calculated HPBW is 72.4° compared to the moment method HPBW which is 75.5°. The calculated BWFN is 144.8° compared to the moment method BWFN which is 156°.

References

[1] J.D. Kraus, *Antennas*. New York: McGraw-Hill, 1950.

[2] S.A. Schelkunoff and H.T. Friis, *Antennas: Theory and Practice*. New York: Wiley, 1952.

[3] W.L. Stutzman and G.A. Thiele, *Antenna Theory and Design, 2nd ed.* New York: Wiley, 1998.

[4] J. Volakis, ed., *Antenna Engineering Handbook, 4th ed.* New York: McGraw-Hill, 2007.

[5] C.A. Balanis, *Antenna Theory and Design, 3rd ed.* New York: Wiley, 2005.

[6] N. Amitay, V. Galindo, and C.P. Wu, *Theory and Analysis of Phased Array Antennas.* New York: Wiley, 1972.

[7] A.A. Oliner and G.H. Knittel, eds., *Phased Array Antennas*. Dedham, Mass.: Artech House, 1972.

[8] R.C. Hansen, ed., *Significant Phased Array Papers*. Dedham, Mass.: Artech House, 1973.

[9] R.J. Mailloux, *Phased Array Handbook, 2nd ed.* Norwood, Mass.: Artech House, 2005.

[10] E. Brookner, ed., *Practical Phased-Array Antenna Systems*. Norwood, Mass.: Artech House, 1991.

[11] R.C. Hansen, *Microwave Scanning Antennas, vol. II: Array Theory and Practice.* Los Altos, Calif.: Peninsula Publishing, 1985.

[12] R.C. Hansen, *Phased Array Antennas, 2nd ed.* New York: Wiley, 2009.

[13] A.K. Bhattacharyya, *Phased Array Antennas.* New York: Wiley, 2006.

[14] T. Jeffrey, *Phased-Array Radar Design: Application of Radar Fundamentals.* Raleigh, North Carolina: Scitech Pub Inc., 2009.

[15] A.J. Fenn, *Adaptive Antennas and Phased Arrays for Radar and Communications.* Norwood, Mass.: Artech House, 2008.

[16] D.R. Billetter, *Multifunction Array Radar.* Norwood, Mass.: Artech House, 1989.

[17] S.P. Skobelev, *Phased Array Antennas with Optimized Element Patterns.* Norwood, Mass.: Artech House, 2011.

[18] H.J. Visser, *Array and Phased Array Antenna Basics.* New York: Wiley, 2005.

[19] R.L. Haupt, *Antenna Arrays: A Computational Approach.* New York: Wiley-IEEE Press, 2010.

[20] D. Ehyale, *Novel Approaches to the Design of Phased Array Antennas: Low Cost, Low Complexity Phased Array Antenna Design.* LAP Lambert Academic Publishing, 2012.

[21] E. Brookner, "Phased-Array Radars," *Scientific American,* vol. 252, no. 2, 1985, pp. 94–102.

[22] A.J. Fenn, D.H. Temme, W.P. Delaney, and W.E. Courtney, "The Development of Phased Array Radar Technology," *Lincoln Laboratory Journal*, vol. 12, no. 2, 2000, pp. 321–340.

[23] J.L. Allen, L. Cartledge, W.P. Delaney, and J. Dibartolo, et al., "Phased Array Radar Studies, 1 July 1959 to 1 July 1960," Technical Report 228, MIT Lincoln Laboratory, 12 Aug. 1960, DTIC no. AD-0249470.

[24] J.L. Allen, L. Cartledge, W.P. Delaney, and J. Dibartolo, "Phased Array Radar Studies, 1 July 1960 to 1 July 1961," Technical Report 236, MIT Lincoln Laboratory, 13 Nov. 1961, DTIC no. AD-271724.

[25] J.L. Allen, D.M. Bernella, W.W. Carpenter, and W.P. Delaney, "Phased Array Radar Studies, 1 July 1961 to 1 July 1963," Technical Report 299, MIT Lincoln Laboratory, 20 Feb. 1963, DTIC no. AD-417572.

[26] J.L. Allen, D.M. Bernella, F. Betts, and L. Cartledge, "Phased Array Radar Studies, 1 July 1963 to 1 July 1964," Technical Report 381, MIT Lincoln Laboratory, 31 March 1965, DTIC no. AD-629363.

[27] J.L. Allen, "A Theoretical Limitation on the Formation of Lossless Multiple Beams in Linear Arrays," *IRE Trans. Antennas and Propagation*, vol. 9, no. 4, 1961, pp. 350–352.

[28] W.P. Delaney, "An RF Multiple Beam Forming Technique," Group Report 41G-0012, Lincoln Laboratory, 9 Aug. 1961, DTIC no. AD-262017.

[29] J.L. Allen, "Theory of Array Antennas (with Emphasis on Radar Applications)," Technical Report 323, Lincoln Laboratory, 25 July 1963, DTIC no. AD-422945.

[30] B.L. Diamond, "A Generalized Approach to the Analysis of Infinite Planar Phased Array Antennas," *Proc. IEEE*, vol. 56, no. 11, 1968, pp. 1837–1851.

[31] B.L. Diamond and G.H. Knittel, "A New Procedure for the Design of a Waveguide

Element for a Phased-Array Antenna," in *Phased Array Antennas*, A.A. Oliner and G.H. Knittel, eds. Dedham, Mass.: Artech House, 1972, pp. 149–156.

[32] J.L. Allen, "Gain and Impedance Variation in Scanned Dipole Arrays," *IRE Trans. Antennas and Propagation*, vol. 10, no. 5, 1962, pp. 566–572.

[33] J.L. Allen, "On Surface-Wave Coupling Between Elements of Large Arrays," *IEEE Trans. Antennas and Propagation*, vol. 13, no. 4, 1965, pp. 638–639.

[34] J.L. Allen and B.L. Diamond, "Mutual Coupling in Array Antennas," Technical Report 424, Lincoln Laboratory, 4 Oct. 1966, DTIC no. AD-648153.

[35] B.L. Diamond, "Small Arrays – Their Analysis and Their Use for the Design of Array Elements," in *Phased Array Antennas*, A.A. Oliner and G.H. Knittel, eds. Dedham, Mass.: Artech House, 1972, pp. 127–131.

[36] A.J. Fenn, "Theoretical and Experimental Study of Monopole Phased Array Antennas," *IEEE Trans. Antennas and Propagation*, vol. 34, no. 10, 1985, pp. 1118–1126.

[37] J. Butler, and R. Lowe, "Beam-Forming Matrix Simplifies Design of Electronically Scanned Antennas," *Electron. Des.*, vol. 9, 12 Apr. 1961, pp. 170–173.

[38] R.J. Mailloux, "Phased Array Theory and Technology," *Proc. IEEE*, vol. 70, no. 3, 1982, pp. 246–291.

[39] G.N. Tsandoulas, "Unidimensionally Scanned Phased Arrays," *IEEE Trans. Antennas and Propagation*, vol. 28, no. 1, 1980, pp. 86–99.

[40] G.N. Tsandoulas, "Wideband Limitations of Waveguide Arrays," *Microwave J.*, vol. 15, no. 9, 1972, pp. 49–56.

[41] G.N. Tsandoulas and G.H. Knittel, "The Analysis and Design of Dual-Polarization Square-Waveguide Phased Arrays," *IEEE Trans. Antennas and Propagation*, vol. 21, no. 6, 1973, pp. 796–808.

[42] G.H. Knittel, "Relation of Radar Range Resolution and Signal-to-Noise Ratio to Phased-Array Bandwidth, " *IEEE Trans. Antennas and Propagation*, vol. 22, no. 3, 1974, pp. 418–426.

[43] E. Brookner, ed. *Radar Technology*. Norwood, Mass.: Artech House, 1991.

[44] G.W. Stimson, *Introduction to Airborne Radar*. Raleigh, North Carolina: Scitech Publishing, 1998, pp. 546–562.

[45] E.J. Kelly and G.N. Tsandoulas, "A Displaced Phase Center Antenna Concept for Space Based Radar Applications," *IEEE Eascon*, Sept. 1983, pp. 141–148.

[46] G.N. Tsandoulas, "Space-Based Radar," *Science*, 17 July 1987, pp. 257–262.

[47] M.I. Skolnik, ed., *Radar Handbook, 3rd ed.* New York: McGraw-Hill, 2008.

[48] H.K. Schuman, D.R. Pflug, and L.D. Thompson, "Infinite Planar Arrays of Arbitrarily Bent Thin Wire Radiators," *IEEE Trans. Antennas Propagation*, vol. 32, no. 3, 1984, pp. 364–377.

[49] E.D. Sharp, "A Triangular Arrangement of Planar-Array Elements That Reduces the Number Needed," *IRE Trans. Antennas and Propagat.*, vol. 9, no. 2, 1961, pp. 126–129.

[50] N.R. Brennecke and W.N. Moule, "Uses of Fences to Optimize Operating Impedance of Phased Arrays Using an Improved Measuring Technique," *IEEE Antennas and Propagation Soc. Int. Symp. Dig.*, 1964, pp. 134–142.

[51] B.A. Munk, *Frequency Selective Surfaces: Theory and Design*. New York: Wiley, 2000.

[52] B.A. Munk, *Finite Antenna Arrays and FSS*. New York: Wiley, 2003.

[53] A.J. Fenn and P.T. Hurst, *Ultrawideband Phased Array Antenna Technology for Sensing and Communications Systems,* Cambridge, MA: The MIT Press, 2015.

[54] K.F. Braun, *Electrical Oscillations and Wireless Telegraphy,* Nobel Prize Lecture, 11 Dec. 1909.

[55] R.J. Mailloux, "A History of Phased Array Antennas," Chapter 17 in *History of Wireless*, T.K. Sarkar, R.J. Mailloux, A.A. Oliner, M. Salazar-Palma, and D.L. Sengupta, eds. New Jersey: Wiley, 2006.

[56] W. Delaney, "From Vision to Reality: 50+ Years of Phased Array Development," *2016 IEEE Int. Symp. on Phased Array Systems and Technology.*

[57] E. Brookner, "Developments and Breakthroughs in Radars and Phased-Arrays," *2016 IEEE Radar Conference.*

[58] J.S. Herd and M.D. Conway, "The Evolution to Modern Phased Array Architectures," *Proc. IEEE* vol. 104, no. 3, March 2016, pp. 519–529.

[59] M. Dudek, I. Nasr, G. Bozsik; M. Hamouda, D. Kissinger, G. Fischer, "System Analysis of a Phased-Array Radar Applying Adaptive Beam-Control for Future Automotive Safety Applications," *IEEE Trans. on Vehicular Technology,* vol. 64, no. 1, 2015, pp. 34 – 47.

[60] G.M. Rebeiz, S.-Y. Kim, O. Inac, W. Shin, O. Gurbuz, Y.-C. Ou, F. Golcuk, T. Kanar, B.-H. Ku, "Millimeter-Wave Large-Scale Phased-Arrays for 5G Systems," *2015 IEEE MTT-S Int. Microwave Symp.*, 2015, pp. 1–3.

[61] M.H. Novak, F.A. Miranda, John L. Volakis, "An Ultra-Wideband Millimeter-Wave Phased Array," *2016 10th European Conf. on Antennas and Propagation (EuCAP),* 2016, pp. 1 – 3.

[62] M. Agiwal, A. Roy, N. Saxena, "Next Generation 5G Wireless Networks: A Comprehensive Survey," *IEEE Communications Surveys & Tutorials* vol. 18, no. 3, 2016, pp. 1617 – 1655.

[63] P.W. Howells, "Explorations in Fixed and Adaptive Resolution at GE and SURC," *IEEE Trans. Antennas and Propagation*, vol. 24, no. 5, 1976, pp. 575–584.

[64] S.P. Applebaum, "Adaptive Arrays," *IEEE Trans. Antennas and Propagation*, vol. 24, no. 5, 1976, pp. 585–598.

[65] W.F. Gabriel, "Adaptive Arrays – An Introduction," *Proc. IEEE,* vol. 64, 1976, pp. 239–271.

[66] J.T. Mayhan, "Some Techniques for Evaluating the Bandwidth Characteristics of Adaptive Nulling Systems," *IEEE Trans. Antennas and Propagation,* vol. 27, no. 3, 1979, pp. 363–373.

[67] J.T. Mayhan, A.J. Simmons, and W.C. Cummings, "Wide-Band Adaptive Antenna Nulling Using Tapped Delay Lines," *IEEE Trans. Antennas and Propagation,* vol. 29, no. 6, 1981, pp. 923–936.

[68] A.J. Fenn, "Evaluation of Adaptive Phased Array Far-Field Nulling Performance in the Near-Field Region," *IEEE Trans. Antennas and Propagation*, vol. 38, no. 2, 1990, pp. 173–185.

[69] R.A. Monzingo and T.W. Miller, *Introduction to Adaptive Arrays*. New York: Wiley, 1980.

[70] R.T. Compton, Jr., *Adaptive Antennas, Concepts and Performance*. Upper Saddle River, New Jersey: Prentice-Hall, 1988.

[71] R. Nitzberg, *Adaptive Signal Processing for Radar*. Norwood, Mass.: Artech House, 1992.

[72] A. Farina, *Antenna-Based Signal Processing Techniques for Radar Systems*. Norwood, Mass.: Artech House, 1992.

[73] J.E. Hudson, *Adaptive Array Principles*. New York: Peter Peregrinus, 1981.

[74] A.J. Fenn, *Breast Cancer Treatment by Focused Microwave Thermotherapy*, Sudbury, MA: Jones and Bartlett, 2007.

[75] A.J. Fenn, *Adaptive Phased Array Thermotherapy for Cancer,* Boston, MA: Artech, 2009.

[76] D.H. Schaubert, "Wide-Band Phased Arrays of Vivaldi Notch Antennas," *IEE 10th Int. Conf. on Antennas and Propagation*, 14–17 April 1997, pp. 1.6–1.12.

[77] M. Kragalott, W.R. Pickles, and M.S. Kluskens, "Design of a 5:1 Bandwidth Stripline Notch Array from FDTD Analysis," *IEEE Trans. Antennas and Propagation,* vol. 48, no. 11, 2000, pp. 1733–1741.

[78] D.T. McGrath and C.E. Baum, "Scanning and Impedance Properties of TEM Horn Arrays for Transient Radiation," *IEEE Trans. Antennas and Propagation*, vol. 47, no. 3, 1999, pp. 469–473.

[79] G.A. Brigham, M. Davidovitz, S.M. Duffy, and J.S. Herd, "Ultra-Wideband Step Notch Array Using Stripline Feed," *2007 IEEE Int. Symposium on Antennas and Propagation*, 2007, pp. 3660–3663.

[80] J.J. Lee, R. Koenig, D. Nagata, and L.L. Lai, "Compact Light Weight UHF Arrays Using Long Slot Apertures," *IEEE Trans. Antennas and Propagation,* vol. 54, no. 7, 2006, pp. 2009–2015.

[81] I. Tzanidis, J. P. Doane, K. Sertel and J.L. Volakis, "Wheeler's Current Sheet Concept and Munk's Wideband Arrays," *Proc. IEEE Int. Symp. on Antennas and Propagation,* 2012, pp. 1–2.

[82] M. Jones and J. Rawnick, "A New Approach to Broadband Array Design Using Tightly

Coupled Elements," *IEEE Military Communications Conference, 2007*, pp. 1–7.

[83] I. Tzanidis, S. Kubilay, and J.L. Volakis, "Interwoven Spiral Array (ISPA) with a 10:1 Bandwidth on a Ground Plane," *IEEE Antennas and Wireless Propagation Letters,* vol. 10, 2011, pp. 115–118.

[84] S.S. Holland and M.N. Vouvakis, "The Planar Ultrawideband Modular Antenna (PUMA) Array," *IEEE Trans. Antennas and Propagat.,* vol. 60, no. 1, 2012, pp. 130–140.

[85] J.P. Doane, K. Sertel, and J.L. Volakis, "Bandwidth Limits for Lossless Planar Arrays over Ground Plane," *Electronics Letters,* vol. 48, no. 10, 2012, pp. 540–542.

[86] C.J. Wilson, Electronically Steerable Field Reflector Techniques, Technical Report No. RADC-TR-64-521, Feb. 1965.

[87] W.D. Fitzgerald, "Limited Electronic Scanning with an Offset-Feed Near-Field Gregorian System," MIT Lincoln Laboratory Technical Report 486, 24 Sept. 1971.

[88] J.A. Martinez-Lorenzo, A. Garcia-Pino, B. Gonzalez-Valdes, and C.M. Rappaport, "Zooming and Scanning Gregorian Confocal Dual Reflector Antennas," *IEEE Trans. on Antennas and Propagat.,* vol. 56, no. 9, 2008, pp. 2910-2919.

[89] A.J. Fenn and J.W. Jordan, "Design and Analysis of an Axisymmetric Phased Array Fed Gregorian Reflector System for Limited Scanning," *2016 IEEE Int. Symp. on Phased Array Systems and Technology,* 18 – 21 October 2016, Waltham, MA, pp. 1-3.

[90] M. A. Jensen; J. W. Wallace, "A Review of Antennas and Propagation for MIMO Wireless Communications," *IEEE Trans. on Antennas and Propagation,* vol. 52, no. 11, 2004, pp. 2810 – 2824.

[91] F. C. Robey, S. Coutts, D. Weikle, J. C. McHarg and K. Cuomo, "MIMO radar theory and experimental results," *Conference Record of the Thirty-Eighth Asilomar Conference on Signals, Systems and Computers, 2004*, vol. 1, 2004, pp. 300–304.

[92] Y.I. Abramovich, G.J. Frazer, B.A. Johnson, "Principles of Mode-Selective MIMO OTHR," *IEEE Trans. on Aerospace and Electronic Systems,* vol. 49, no. 3, 2013, pp. 1839–1868.

[93] C. L. Dolph, "A Current Distribution for Broadside Arrays Which Optimizes the Relationship Between Beamwidth and Side Lobe Level," *Proc. IRE,* vol. 31, June 1946, pp. 335–318.

[94] D. Barbiere. "A Method for Calculating the Current Distribution of Tschebyscheff Arrays," *Proc. IRE,* vol. 10, January 1952, pp. 78–82.

[95] A. D. Bresler, "A New Algorithm for Calculating the Current Distributions of Dolph-Chebyshev Arrays," *IEEE Trans. Antennas Propagat.,* vol. AP-28, no. 6, November 1980, pp. 951, 952.

[96] T.T. Taylor, "Design of Circular Apertures for Narrow Beamwidth and Low Side Lobes," *IEEE Trans. on Antennas and Propagat.,* vol AP-8, no. 4, Jan. 1960, pp. 17–22.

[97] A.E.H. Love, "The Integration of the Equations of Propagation of Electric Waves,"

Philosophical Trans., Series A197, 1901, pp. 1-45.

[98] S.A. Schelkunoff, "On Diffraction and Radiation of Electromagnetic Waves," *Phys. Rev.*, vol. 56, 1939, pp. 308–316.

[99] R.F. Harrington, *Time-Harmonic Electromagnetic Fields,* New York: McGraw-Hill, 1961.

[100] J.R. Mautz and R.F. Harrington, "Electromagnetic Scattering from a Homogeneous Material Body of Revolution," *Int. J. Electronics and Communications (AEU),* vol. 33, 1979, pp. 71-80.

[101] E. Arvas, A.Rahhal-Arabi, A. Sadigh, S.M. Rao, "Scattering from Multiple Conducting and Dielectric Bodies of Arbitrary Shape," *IEEE Antennas and Propagat. Magazine*, vol. 33, no. 2, 1991, pp. 29-36.

[102] U. Jakobus, "Comparison of Different Techniques for the Treatment of Lossy Dielectric/Magnetic Bodies within the Method of Moments Formulation," *Int. J. Electronics and Communications (AEU),* vol. 54, no. 3, 2000, pp. 163-173.

[103] C.A. Balanis, *Advanced Engineering Electromagnetics, 2nd ed.* New Jersey: Wiley, 2012.

[104] V.H. Rumsey, "Reaction Concept in Electromagnetic Theory," *Physical Review*, vol. 94, 1954, pp. 1483–1491.

[105] R.F. Harrington, *Field Computation by Moment Methods.* New York: Macmillan, 1968.

[106] S. Rao, D. Wilton, and A. Glisson, "Electromagnetic Scattering by Surfaces of Arbitrary Shape," *IEEE Trans. Antennas and Propagation*, vol. 30, no. 5, 1982, pp. 409–418.

[107] E.H. Newman and K. Kingsley, "An Introduction to the Method of Moments," *Computer Physics Communications,* vol. 68, 1991, pp. 1–18.

[108] J.H. Richmond and N.H. Geary, "Mutual Impedance of Nonplanar-Skew Sinusoidal Dipoles," *IEEE Trans. Antennas and Propagation*, vol. 23, no. 3, 1975, pp. 412–414.

[109] J.H. Richmond, "Radiation and Scattering by Thin-Wire Structures in a Homogeneous Conducting Medium (Computer Program Desc.)," *IEEE Trans. Antennas and Propagation*, vol. 22, no. 2, 1974, p. 365.

[110] K.E. Schmidt, "Mutual Impedance of Nonplanar Skew Dipoles," *IEEE Trans. Antennas and Propagation*, vol. 44, no. 9, 1996, pp. 1298–1299.

[111] N.N. Wang, "Reaction Formulation for Radiation and Scattering from Plates, Corner Reflectors, and Dielectric-Coated Cylinders," The Ohio State University, Project Report 2902-15, Columbus, Ohio, Apr. 1974.

[112] W.C. Gibson, *The Method of Moments in Electromagnetics.* Boca Raton, Florida: Chapman & Hall, 2008.

[113] J. van Tonder; U. Jakobus, "Introduction of Curvilinear Higher-Order Basis Functions for MoM and MLFMM in FEKO," *Computational Electromagnetics Workshop (CEM),* 2013, pp. 13 – 14.

[114] W.C. Chew, J.M. Jin, E. Michielssen, and J. Song, *Fast and Efficient Algorithms in Computational Electromagnetics*. Norwood, Mass.: Artech House, 2001.

[115] Mayer, E.D., and A. Hessel, "Feed Region Modes in Dipole Phased Arrays," *IEEE Trans. Antennas Propagat.*, Vol. 30, No. 1, 1982, pp. 66-75.

[116] Fenn, A.J., "Element Gain Pattern Prediction for Finite Arrays of V-Dipole Antennas over Ground Plane," *IEEE Trans. Antennas Propagat.*, Vol. 36, No. 11, 1988, pp. 1629-1633.

[117] H. Jingjian, Z. Xiaofa, X Shaoyi,W. Weiwei, and Y. Naichang, "Suppression of Cross-Polarization of the Microstrip Integrated Balun-Fed Printed Dipole Antenna," *Int. J. of Antennas and Propagation,* Volume 2014, pp. 1–8.

[118] B. Edward and D. Rees, "A Broadband Printed Dipole with Integrated Balun," *Microwave J.*, vol. 30, no. 5, pp. 339–344, 1987.

[119] R. L. Li, T. Wu, B. Pan, K. Lim, J. Laskar, and M. M. Tentzeris, "Equivalent-Circuit Analysis of a Broadband Printed Dipole with Adjusted Integrated Balun and an Array for Base Station Applications," *IEEE Trans. on Antennas and Propagat.,* vol. 57, no. 7, 2009, pp. 2180–2184.

[120] P. Lindberg, E. Ojefors, Z. Barna, A. Thornell-Pers, and A. Rydberg,"Dual Wideband Printed Dipole Antenna with Integrated Balun," *IET Microwaves, Antennas & Propagation,* vol. 1, no. 3, 2007, pp. 707–711.

[121] M. Scott, "A Printed Dipole for Wide-Scanning Array Application," *Proc. 11th Int. Conf. on Antenna and Propagation,* Manchester, UK, vol. 1, no. 480, 2001, pp. 37–40.

[122] H.-R. Chuang and L.-C. Kuo,"3-D FDTD Design Analysis of a 2.4-GHz Polarization-Diversity Printed Dipole Antenna with Integrated Balun and Polarization-Switching Circuit for WLAN and Wireless Communication Applications," *IEEE Trans. on Microwave Theory and Techniques*, vol. 51, no. 2, 2003, pp. 374–381.

[123] E.A. Wolff, *Antenna Analysis*, New York: John Wiley, pp.192-208.

[124] G.V. Borgiotti, "A Novel Expression for the Mutual Admittance of Planar Radiating Elements," *IEEE Trans. Antennas Propagat.*, Vol. 16, No. 3, 1968, pp. 329-333.

[125] R.J. Mailloux, "First-Order Solutions for Mutual Coupling Between Waveguides Which Propagate Two Orthogonal Modes," *IEEE Trans. Antennas Propagat.*, Vol. 17, No. 6, 1969, pp. 740-746.

[126] R.J. Mailloux, "Radiation and Near-Field Coupling Between Two Collinear Open-Ended Waveguides," *IEEE Trans. Antennas Propagat.*, Vol. 17, No. 1, 1969, pp. 49-55.

[127] T.S. Bird, "Analysis of Mutual Coupling in Finite Arrays of Different-Sized Rectangular Waveguides," *IEEE Trans. Antennas Propagat.*, Vol. 38, No. 2, 1990, pp. 166–172.

[128] C.P. Wu, "Analysis of Finite Parallel-Plate Waveguide Arrays," *IEEE Trans. Antennas Propagat.*, Vol. 18, No. 3, 1970, pp. 328–334.

[129] A.G. Cha and J.K. Hsiao, "A Matrix Formulation for Large Scale Numerical Computation of the Finite Planar Waveguide Array Problem," *IEEE Trans. Antennas*

Propagat., Vol. AP-22, No. 1, January 1974, pp. 106–108.

[130] M.A. Hidayet, "Finite Phased Array Analysis," Ph.D. dissertation, U. Michigan, 1974.

[131] J. Luzwick and R.F. Harrington, "Mutual Coupling Analysis in a Finite Planar Rectangular Waveguide Antenna Array," Syracuse Univ., Tech. Rep., No. 7, June 1978.

[132] S.S. Wang, "Wide-Angle Wide-Band Elements for Phased Arrays," Ph.D. dissertation, Polytechnic Institute of New York, 1975.

[133] J.D. Kraus, *Antennas*, New York: McGraw-Hill, 1950, ch. 4.

[134] R.F. Harrington and J.R. Mautz, "A Generalized Network Formulation for Aperture Problems," A.F. Cambridge Res. Lab., Rep. AFCRL-TR-75-0589, November 1975.

[135] A.J. Fenn, "Moment Method Calculation of Reflection Coefficient for Waveguide Elements in a Finite Planar Phased Antenna Array," Ph.D. dissertation, The Ohio State Univ., 1978.

[136] A. J. Fenn, G.A. Thiele, and B.A. Munk, "Moment Method Analysis of Finite Rectangular Waveguide Phased Arrays," *IEEE Trans. Antennas Propagat.*, Vol. 30, No. 4, 1982, pp. 554–564.

[137] B.L. Diamond, "Resonance Phenomena in Waveguide Arrays," *IEEE G-AP Int. Symp. Dig.*, 1967, pp. 110–115.

[138] H. Yavuz and O.M. Buyukdura, "Mutual Coupling Effects of Finite Rectangular Phased Arrays," *IEEE Antennas and Propagat. Soc. Int. Symp. Digest,* 1994, pp. 418–421.

[139] P. Grassi, R. Mittra, A. Monorchio, and G. Manara, "Characterization of Finite Waveguide Arrays Using a New Generalized Scattering Matrix Approach," *IEEE Antennas and Propagat. Symposium Digest*, 2004, pp. 125–128.

[140] A. Pellegrini, A. Monorchio, G. Manara, R. Mittra, "A Hybrid Mode Matching-Finite Element Method and Spectral Decomposition Approach for the Analysis of Large Finite Phased Arrays of Waveguides", *IEEE Trans. on Antennas and Propagat.,* vol. 62, 2014, pp. 2553–2561.

[141] A.J. Fenn, H.M. Aumann, F.G. Willwerth, and J.R. Johnson, "Focused Near-Field Adaptive Nulling: Experimental Investigation," *1990 IEEE Antennas and Propagation Society International Symposium Digest,* vol. 1, May 7-11, 1990, pp. 186-189.

3

Wideband Impedance Matching of a V-Dipole Antenna

3.1 Introduction

Dipole antennas are commonly used as single elements, as a feed for reflector antennas, or in arrays for communications and radar applications [1-32]. Crossed dipole antennas can be used for example to provide dual linear polarization, as discussed in Chapter 2. With a phase-quadrature feed arrangement that has a 90° phase difference between the crossed dipole ports, circular polarization can be transmitted or received, which provides for nearly omnidirectional coverage as discussed for the turnstile antenna in Chapter 1 (Section 1.13.2). The RF performance of an antenna is often quantified by the reflection coefficient or VSWR at the input port to the antenna. For example, the magnitude of the reflection coefficient of an antenna is a critical design parameter that must be minimized for providing maximum realized antenna gain or efficient power transmission between the antenna and the transmission line connected to the antenna.

Impedance matching is a well established technique for improving the reflection coefficient of antennas over a band of frequencies [33-44]. A shielded transmission line such as a coaxial cable is often used to interface between a dipole antenna and the transmitter or receiver unit. Since coaxial cable is an unbalanced transmission line and the dipole requires a balanced feed, a device known as a balun (abbreviated from balanced to unbalanced) is used to make the connnection between the dipole terminals and the coaxial cable. Numerous types of impedance matching techniques and baluns have been used to provide a balanced (differential) feed for dipole antennas [16, pp. 497-541]. Some common matching techniques for dipoles include T-match, gamma match, and omega match. Some common types of baluns include sleeve balun (Bazooka balun), $\lambda/4$ and $\lambda/2$ coaxial baluns, and ferrite core balun. Dipoles are sometimes fed in a balanced mode by using a folded balun design [8, 16, 44], which was discussed in Chapter 2 (Section 2.9). In this chapter, an impedance matching example for a folded balun-fed swept-back dipole antenna is described in detail. Swept-back dipoles provide a wide illumination pattern that can be used in feeding a parabolic reflector.

Besides the use of impedance matching, to provide an increased amount of inherent bandwidth, the dipole arms are designed to be cylindrical tubes rather than thin wires. An artist's concept of a crossed dipole antenna using tubular arms is depicted in Figure 3.1. Simplified diagrams of a single dipole is shown in Figure 3.2, where the tubular dipole arms have length l and diameter d. In Figure 3.2(a) a uniform coaxial feed line is shown, and in Figure 3.2(b) a stepped coaxial feedline provides a line transformer section to provide impedance matching. The folded balun section consists of two vertical tubes with center-to-center spacing s and height H. The balun tubes are assumed to have outer diameter D and they are short circuited to the ground plane, forming a short-circuit stub in parallel with respect to the dipole feed terminals. In this antenna design, the arms are swept-back at a $45°$ angle to increase the E-plane beamwidth as discussed in Chapter 2. An

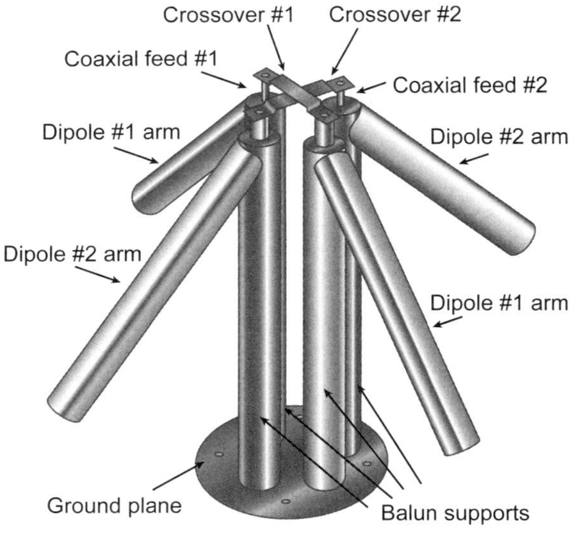

Figure 3.1 Diagram showing a crossed swept-back dipole with coaxial folded balun over a ground plane.

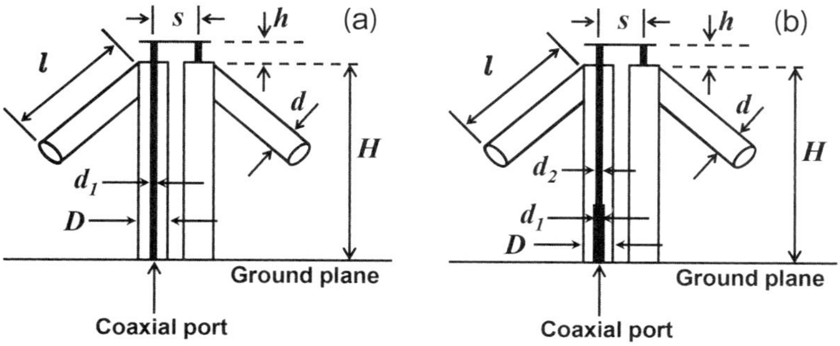

Figure 3.2 Diagrams showing a single swept-back dipole with coaxial folded balun over a ground plane. (a) Uniform coaxial feed line, and (b) stepped coaxial feedline to provide a line transformer section.

electrically conducting feed tab connects the center conductor to the dummy (passive metal tube) portion of the folded balun. The coaxial port connects to an inner conductor of diameter d_1 and a surrounding tube with inner diameter D (an infinitesimal wall thickness is assumed here). In Figure 3.2(a), the values of d_1 and D are adjusted typically to provide a 50Ω characteristic impedance. In practice, the coaxial port is fed typically with a 50-ohm microwave connector such as type N, SMA, or TNC depending on RF power handling requirements and operating frequency. The parallel balun tubes form an open-wire transmission line with characteristic impedance Z_o given by [12]

$$Z_o = \frac{120}{\sqrt{\epsilon_r}} \cosh^{-1} \frac{s}{D} \tag{3.1}$$

where ϵ_r is the dielectric constant. The characteristic impedance of the coaxial feed line is given by [12]

$$Z_o = \frac{138}{\sqrt{\epsilon_r}} \log_{10} \frac{D}{d_1} \tag{3.2}$$

In [33–35], the theory for impedance matching using line transformers, series networks, and parallel networks is provided. This chapter provides a detailed discussion and simulations of a line transformer for providing wideband matching for the input impedance of an example UHF swept-back dipole antenna. In practice, the impedance matching approach discussed here would involve the design of the feeding structure, which in this case would be a stepped coaxial transmission line transformer contained within the balun section, as depicted in Figure 3.2(b). In this case, the coaxial feedline contains an impedance transformer section with an inner conductor with diameter d_2 adjusted for a desired characteristic impedance to match the input impedance of the dipole. Chapter 1 reviewed transmission line theory in detail. The next section (Section 3.2) provides a brief review of transmission line theory as it relates to the derivation of the line transformer matching theory (Section 3.3). In Section 3.4 an application of the line transformer matching theory to designing a wideband swept-back dipole is described along with simulated moment method results. Section 3.5 describes a comparison between simulated and measured reflection coefficient results for a prototype swept-back dipole. Section 3.6 simulates the radiation patterns for the wideband swept-back dipole feeding a parabolic reflector antenna. Section 3.7 provides a summary and Section 3.8 provides a problem set.

3.2 Basic Transmission Line Theory

In this section, a brief review of basic transmission line theory is provided as it relates to the line transformer impedance matching theory [33-35]. Referring to Figure 3.3, consider first a length l of transmission line with a characteristic impedance Z_o. Assume that the transmission line's attenuation constant α is equal to zero, which is a reasonable assumption here since the length of the transmission line matching section is electrically less than one-half wavelength and little attenuation would be expected. The electrical length is given by the quantity βl, where $\beta = 2\pi/\lambda$ is the wavelength-dependent phase propagation constant of the coaxial transmission line. Assume that an antenna has a complex load impedance denoted as Z_L. At a distance $z = -l$ along a lossless transmission line with characteristic impedance Z_o, the input impedance $Z_{\text{in}}(-l)$ as function of position is given by (refer to

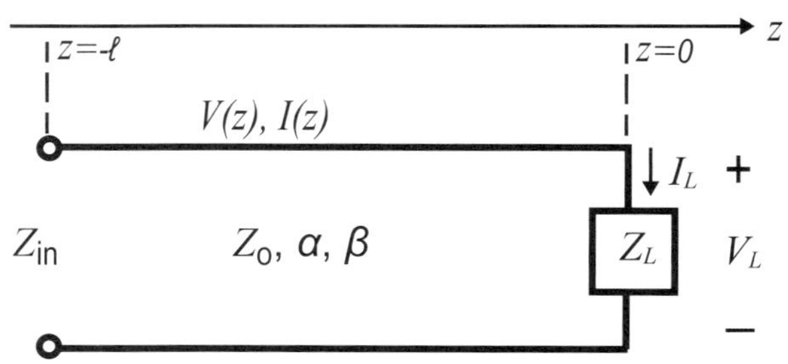

Figure 3.3 Transmission line with characteristic impedance Z_o and length l connected to a load impedance Z_L.

Equation (1.31))

$$Z_{\text{in}}(-l) = Z_o \frac{Z_L + jZ_o \tan \beta l}{Z_o + jZ_L \tan \beta l} \tag{3.3}$$

and in terms of its complex conjugate

$$Z^*_{\text{in}}(-l) = Z_o \frac{Z^*_L - jZ_o \tan \beta l}{Z_o - jZ^*_L \tan \beta l} \tag{3.4}$$

The input admittance Y_{in} is given by

$$Y_{\text{in}} = 1/Z_{\text{in}} \tag{3.5}$$

Solving Equation (3.3) for Z_L yields

$$Z_L(z = 0) = Z_o \frac{Z_{\text{in}} - jZ_o \tan \beta l}{Z_o - jZ_{\text{in}} \tan \beta l} \tag{3.6}$$

In the case where the load is a short circuit ($Z_L = 0$) or an open circuit ($Z_L = \infty$), respectively, Equation (3.3) reduces to

$$Z^{\text{sc}}_{\text{in}}(-l) = jZ_o \tan(\beta l) \tag{3.7}$$

$$Z^{\text{oc}}_{\text{in}}(-l) = -jZ_o \frac{1}{\tan(\beta l)} \tag{3.8}$$

The product of Equations (3.7) and 3.8 leads to an expression for the characteristic impedance in terms of the short-circuit and open-circuit input impedances as

$$Z_o = \sqrt{Z^{\text{sc}}_{\text{in}} \times Z^{\text{oc}}_{\text{in}}} \tag{3.9}$$

The reflection coefficient in terms of the load impedance Z_L and transmission line characteristic impedance Z_o is given by

$$\Gamma = \frac{Z_L - Z_o}{Z_L + Z_o} \tag{3.10}$$

and in normalized form in terms of $z_L = Z_L/Z_o$ as

$$\Gamma = \frac{z_L - 1}{z_L + 1} = |\Gamma| e^{j\phi} \tag{3.11}$$

It follows from Equation (3.11) that the normalized load impedance is expressed as

$$z_L = \frac{1 + \Gamma}{1 - \Gamma} \tag{3.12}$$

Rearranging Equation (3.10), the load impedance in terms of the reflection coefficient is given by

$$Z_L = Z_o \frac{1 + \Gamma}{1 - \Gamma} \tag{3.13}$$

The voltage standing wave ratio (VSWR) is expressed as

$$\text{VSWR} = \frac{1 + |\Gamma|}{1 - |\Gamma|} \tag{3.14}$$

Having reviewed some of the fundamental equations that quantify the input impedance, load impedance, reflection coefficient, and VSWR, the next section addresses line transformer impedance matching theory.

3.3 Line Transformer Impedance Matching Theory

In the line transformer impedance matching procedure that follows, electromagnetic simulations (or measurements) would be used to determine the input impedance at the antenna terminals of the swept-back crossed dipole antenna depicted in Figure 3.1. The geometry of the dipole arms and height of the feed terminals with respect to the ground plane are designed based on theory, which was discussed in Chapter 2 and is also discussed later in this chapter. In the electromagnetic field simulations and line transformer calculations that follow, the dipole antenna is fed initially (before matching) with a uniform $Z_o = 50\Omega$ coaxial feedline (Figure 3.2(a)), which is not expected to provide the desired final reflection coefficient magnitude or VSWR. A voltage source is located at the input port of the 50-ohm coaxial feedline (located at the ground plane), and the input impedance of the antenna simulated at the ground plane reference position is determined. (Note that the same approach can be applied for measurements.) The next step is to transform the input impedance at the ground plane to the complex load Z_L at the antenna terminals using Equation (3.6). This antenna terminal input impedance (load impedance) becomes the starting value for impedance matching. Next, a series stepped coaxial transmission line (Figure 3.2(b)) with derived values of characteristic impedance and length is introduced as a line transformer to improve the reflection coefficient over wide bandwidth. Referring to Figure 3.4, the input impedance matching technique applied to the dipole antenna is generally as follows:

- Plot the antenna load impedance (input impedance at the terminals of the antenna) over wide bandwidth in Cartesian coordinates, where the horizontal axis is resistance and the vertical axis is reactance. In Figure 3.4, load impedance data points are displayed at four example frequencies f_1, f_2, f_3, f_4.

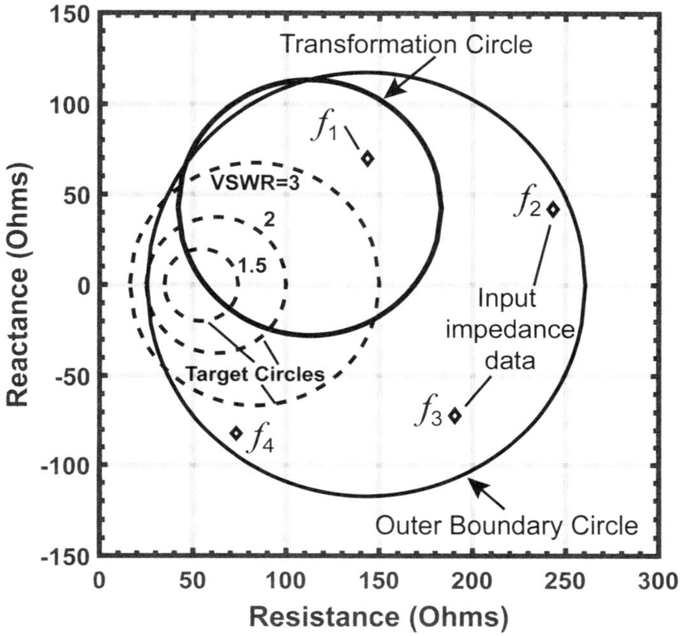

Figure 3.4 Conceptual diagram depicting VSWR (1.5, 2, 3) target circles, an outer boundary circle encompassing four sample input impedance data points, and an example transformation circle.

- Define a target circle as the maximum allowed VSWR (plotted as a circle in the Cartesian coordinate system) at an input port located on a standard 50-ohm transmission line. In Figure 3.4, three example constant VSWR circles (1.5, 2, 3) are shown. A VSWR=2 would be a typical target goal for the impedance matched antenna.
- Draw an outer boundary circle (OBC) in the Cartesian coordinate system encompassing the input impedance data at the antenna terminals and tangent to a target VSWR circle. In Figure 3.4, the outer boundary circle encompasses the four impedance points shown, and this circle is tangent to the minimum value of the VSWR=2 circle.
- Compute the required characteristic impedance of the line transformer Z'_o using the square root (geometric mean) of the product of the resistance minimum (denoted $R_{\text{min}}^{\text{OBC}}$) and resistance maximum ($R_{\text{max}}^{\text{OBC}}$) on the resistance axis intersecting the OBC, that is,

$$Z'_o = \sqrt{R_{\text{min}}^{\text{OBC}} \times R_{\text{max}}^{\text{OBC}}} \tag{3.15}$$

- Compute and plot the family of transformation circles that ensures impedance values versus frequency will transform to within the desired target VSWR (target circle).
- Select a critical frequency in this case where the antenna load impedance intersects the outer boundary circle, and locate the corresponding transformation circle that encompasses the impedance data point at this frequency, From the electrical length βl of the corresponding transformation circle, solve for the transformer critical length $l = l_c$ using $\beta = 2\pi/\lambda$ where λ is the wavelength at the critical frequency. In Figure 3.4, one example transformation circle is shown.
- Tabulate the electrical line length of the coaxial line transformer section over the frequency band of interest, and verify that the input impedance at the terminals of the dipole antenna tracks within a series of transformation circles.
- Transform the antenna load impedance to the input impedance and reflection coefficient looking into the line transformer with characteristic impedance Z_o' and critical length l_c.
- Normalize the reflection coefficient data to the final characteristic impedance (50Ω) of the coaxial feedline attached to the line transformer, and plot as the matched input impedance.

The necessary equations for characterizing the target circle and transformation circles are described in the next two sections.

3.3.1 Target Circle for Maximum Allowed VSWR

Consider first a *target circle*, which is also referred to by some authors as a *definition circle* [33-35]. The target circle defines the maximum allow VSWR, and is expressed mathematically and visually in the Cartesian complex impedance coordinate system. Letting the antenna input impedance be represented as $Z_{\text{in}} = R_{\text{in}} + jX_{\text{in}}$, then the equation for the target circle is given by

$$(R_{\text{in}} - a)^2 + X_{\text{in}}^2 = b^2 \tag{3.16}$$

where R_{in} is the input resistance, X_{in} is the input reactance, a is the offset of the center of the target circle from the value of zero input resistance, and b is the radius of the circle. The a and b parameters are expressed in terms of the target transmission line characteristic impedance Z_o and the VSWR as

$$a = \frac{Z_o}{2}\left(\text{VSWR} + \frac{1}{\text{VSWR}}\right) \tag{3.17}$$

$$b = \frac{Z_o}{2}\left(\text{VSWR} - \frac{1}{\text{VSWR}}\right) \tag{3.18}$$

and note that

$$a^2 - b^2 = Z_o \tag{3.19}$$

$$a - b = \frac{Z_o}{\text{VSWR}} = R_{\text{in}}^{\text{min}} \tag{3.20}$$

$$a + b = Z_o \times \text{ VSWR} = R_{\text{in}}^{\text{max}} \tag{3.21}$$

Equation (3.16) is now expanded as

$$R_{\text{in}}^2 - 2aR_{\text{in}} + X_{\text{in}}^2 = b^2 \tag{3.22}$$

Noting for an arbitrary impedance Z that

$$Z = R + jX \tag{3.23}$$

$$Z^* = R - jX \tag{3.24}$$

and

$$Z + Z^* = 2R \tag{3.25}$$

$$Z - Z^* = j2X \tag{3.26}$$

it follows that

$$ZZ^* = R^2 + X^2 \tag{3.27}$$

Using these impedance relations (Equations (3.23) to (3.27)), Equation (3.22) can be expressed as

$$Z_{\text{in}}Z_{\text{in}}^* - a(Z_{\text{in}} + Z_{\text{in}}^*) = b^2 - a^2 \tag{3.28}$$

Equation (3.28) is the desired expression for the target circle in terms of the input impedance Z_{in} and VSWR matched to the input transmission line. The next section derives a series of transformation circles that will ensure that the terminal impedance of the antenna will be matched to the target VSWR on the input transmission line.

3.3.2 Transformation Circle Derivation

Refer now to Figure 3.5, which shows a line transformer with characteristic impedance Z'_o and an antenna load impedance Z_L. The input impedance Z_{in} at the port defined by nodes 1 and 2 is expressed as,

$$Z_{\text{in}}(-l) = Z'_o \frac{Z'_L + jZ'_o \tan \beta l}{Z'_o + jZ_L \tan \beta l} \tag{3.29}$$

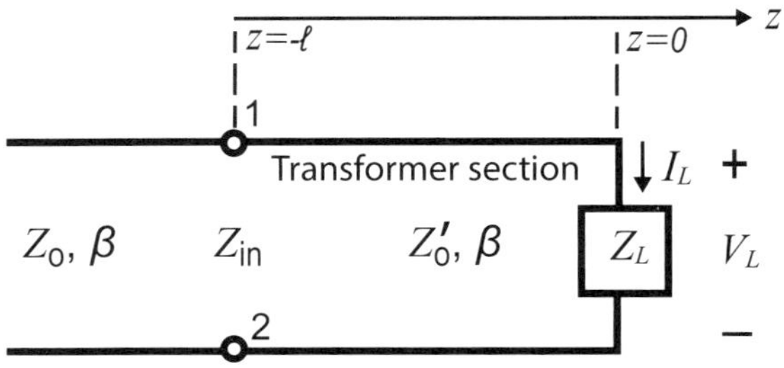

Figure 3.5 Transmission line with characteristic impedance Z_o followed by a transformer section with characteristic impedance Z_o' and length l connected to a load impedance Z_L.

and in terms of its complex conjugate

$$Z_{\text{in}}^*(-l) = Z_o' \frac{Z_L^* - jZ_o' \tan \beta l}{Z_o' - jZ_L^* \tan \beta l} \tag{3.30}$$

At the port defined by nodes 1 and 2, the input impedance can be normalized either by Z_o' or by Z_o, depending on which side the input impedance is being computed.

Substituting Equations (3.29) and (3.30) into Equation (3.28) yields

$$\begin{aligned} &\left(Z_o' \tfrac{Z_L + jZ_o' \tan \beta l}{Z_o' + jZ_L \tan \beta l}\right)\left(Z_o' \tfrac{Z_L^* - jZ_o' \tan \beta l}{Z_o' - jZ_L^* \tan \beta l}\right) \\ &-a\left[\left(Z_o' \tfrac{Z_L + jZ_o' \tan \beta l}{Z_o' + jZ_L \tan \beta l}\right) + \left(Z_o' \tfrac{Z_L^* - jZ_o' \tan \beta l}{Z_o' - jZ_L^* \tan \beta l}\right)\right] = b^2 - a^2 \end{aligned} \tag{3.31}$$

In the book by Reich [33], it is shown that Equation (3.31) can be expressed in the standard form of the equation of a circle center on the resistance axis at R and on the reactance axis at X with radius r as

$$(R_L - R)^2 + (X_L - X)^2 = r^2 \tag{3.32}$$

where

$$R = \frac{a(1 + \tan^2 \beta l)}{1 + \frac{a^2 - b^2}{(Z_o')^2} \tan^2 \beta l} \tag{3.33}$$

$$X = \frac{Z_o \tan \beta l \left(\frac{a^2 - b^2}{(Z_o')^2} - 1\right)}{1 + \frac{a^2 - b^2}{(Z_o')^2} \tan^2 \beta l} \tag{3.34}$$

$$r = \frac{b(1+\tan^2\beta l)}{1+\frac{a^2-b^2}{(Z_o')^2}\tan^2\beta l} \tag{3.35}$$

Equation (3.32) is the desired transformation circle for a line transformer with characteristic impedance Z_o' and electrical length βl. In other words, load impedance values Z_L that are located within the transformation circle will be transformed to impedance values within the desired target circle with characteristic impedance Z_o. To derive Equations (3.33) to (3.35) it is necessary first to multiply all terms in Equation (3.31) by the expression $(Z_o' + jZ_L \tan\beta l)(Z_o' - jZ_L^* \tan\beta l)$. Next, the relations given by Equations (3.25) to (3.27) are used to express the equation in terms of R_L and X_L. Next, bring all terms involving R_L and X_L to the left side and the remaining terms to the right side. By completing the square for the R_L and X_L terms, and a considerable amount of algebraic manipulations, the desired results Equations (3.33) to (3.35) are obtained.

By referring to Equation (3.19) a further simplification of the transformation Equations (3.33) to (3.35) is achieved, that is, replace $a^2 - b^2$ with Z_o, so that

$$R = \frac{a(1+\tan^2\beta l)}{1+\frac{Z_o}{(Z_o')^2}\tan^2\beta l} \tag{3.36}$$

$$X = \frac{Z_o\tan\beta l\left(\frac{Z_o}{(Z_o')^2}-1\right)}{1+\frac{Z_o}{(Z_o')^2}\tan^2\beta l} \tag{3.37}$$

$$r = \frac{b(1+\tan^2\beta l)}{1+\frac{Z_o}{(Z_o')^2}\tan^2\beta l} \tag{3.38}$$

3.4 Wideband V-Dipole Antenna Simulation

3.4.1 Dipole Design and Moment Method Simulation Model

Consider now a computer simulation model of a wideband dual-polarized crossed dipole antenna with swept-back arms. A V-dipole antenna of this type could be mounted on a small ground plane and could be used in this configuration for example as a wideband gain calibration antenna or as a wideband feed for a parabolic reflector antenna. Figure 3.6 shows an example of a linearly polarized swept-back (V-shaped) dipole antenna feeding a reflector antenna.

A dual-polarized dipole antenna over a ground plane in free space is now designed here for wide bandwidth within a portion of the arbitrarily chosen octave frequency range 400 MHz to 800 MHz. At an assumed center

Figure 3.6 Example prototype linearly polarized V-dipole antenna as a feed for a reflector antenna.

frequency of 600 MHz, the free-space wavelength is 50 cm. For this example, consider a dual-polarized swept-back dipole antenna with folded balun over a perfectly conducting finite ground plane that is modeled with the FEKO moment method solver (www.feko.info) as shown in an isometric view in Figure 3.7. A side view of the antenna is shown in Figure 3.8. The dipole arm length was chosen as 13.4 cm, which is 0.268λ at 600 MHz (center frequency). The dipole arms and balun sections are modeled as 1.42 cm diameter (D in Figure 3.2) perfectly conducting tubes. The dipole height is chosen as 17.9 cm (H in Figure 3.2) so that at the center frequency, the

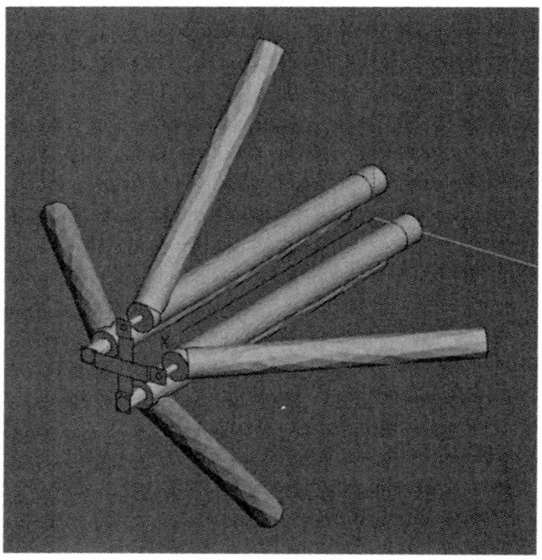

Figure 3.7 Simulation model of a crossed V-dipole antenna. Isometric view.

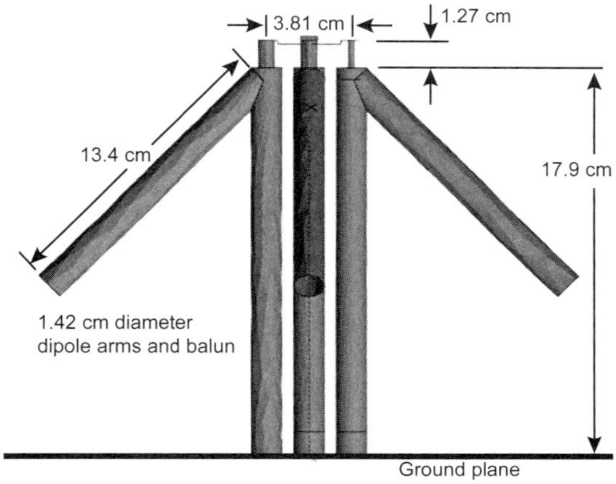

Figure 3.8 Simulation model of a crossed V-dipole antenna. Side view.

height at the midpoint of the dipole arm is approximately $\lambda/4$ over the ground plane. The square ground plane side dimension was 45.72 cm. The folded balun tubes are spaced 3.81 cm center to center (s in Figure 3.2), which by Equation (3.1) provides a 197Ω characteristic impedance transmission line. The folded balun provides mechanical support of the dipole arms and is short circuited at the ground plane. Electrically, this balun behaves as a purely reactive impedance in parallel with the dipole feed terminals. One of the balun tubes contains a uniform diameter perfectly conducting rod that provides a 50Ω characteristic impedance coaxial transmission line. This coaxial line is connected to the dipole terminals using a crossover connection located 1.27 cm (h in Figure 3.2) above the top of the folded balun. A top view of the antenna is shown in Figure 3.9, where the side length of the ground plane in this example is noted to be 0.46m.

3.4.2 Impedance Matching Procedure for the Wideband Dipole

The impedance matching procedure in this example is as follows [33–35]:

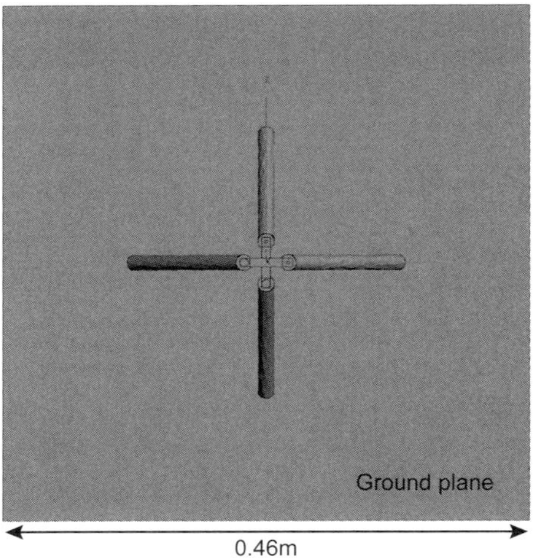

Figure 3.9 Simulation model of a crossed V-dipole antenna. Top view.

- Plot the dipole antenna load impedance (at the dipole terminals) in Cartesian coordinates. where the horizontal axis is resistance and the vertical axis is reactance,
- Define the target circle as the maximum allowed VSWR=2 (plotted as a circle in the Cartesian coordinate system) at an input port located on a standard 50-ohm transmission line.
- Draw an outer boundary circle (OBC) in the Cartesian coordinate system encompassing the antenna load impedance data and tangent to the target VSWR=2 circle at the minimum resistance value.
- Compute the required characteristic impedance of the line transformer Z_o' using the square root (geometric mean) of the product of the resistance minimum (denoted $R_{\text{min}}^{\text{OBC}}$) and resistance maximum ($R_{\text{max}}^{\text{OBC}}$) on the resistance axis intersecting the OBC, that is,

$$Z_o' = \sqrt{R_{\text{min}}^{\text{OBC}} \times R_{\text{max}}^{\text{OBC}}} \tag{3.39}$$

- Compute and plot, as an overlay on the impedance data, the transformation circles, from 15° to 165° in 15° steps. Construct additional transformation circles for finer resolution as needed.
- Select a critical frequency in this case where the antenna load impedance intersects the outer boundary circle, and locate the corresponding transformation circle that encompasses the impedance data point at this frequency. From the electrical length βl of the

corresponding transformation circle, solve for the transformer critical length $l = l_c$ using $\beta = 2\pi/\lambda$ where λ is the wavelength at the critical frequency.

- Tabulate the electrical line length of the transformer section over the frequency band of interest, and verify that the input impedance at the terminals of the dipole antenna tracks within the series of transformation circles.
- Transform the antenna load impedance to the input impedance and reflection coefficient looking into the line transformer with characteristic impedance Z'_o and critical length l_c.
- Normalize the reflection coefficient data to the final characteristic impedance (50Ω) of the feedline attached to the transformer, and plot as the matched input impedance.

3.4.3 Simulated Matching Results for the Wideband Dipole

Before matching, the simulated results versus frequency at one of the terminals of the crossed dipole, with the other port 50Ω resistively loaded, are shown in Figures 3.10 (input impedance in Cartesian coordinates), 3.11 (Smith chart) and 3.12 (reflection coefficient magnitude versus frequency). In this moment method simulation, a voltage source was applied at the input port to the 50Ω coaxial feed line at the ground plane, and then the computed input impedance at the ground plane was transformed to the

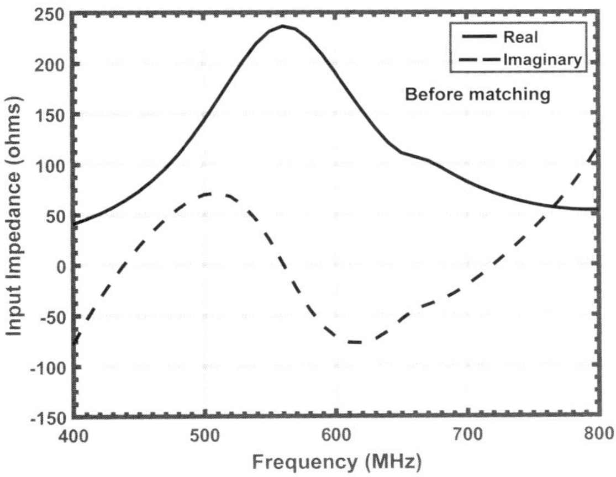

Figure 3.10 Simulated input impedance versus frequency at the terminals of a crossed V-dipole antenna before matching.

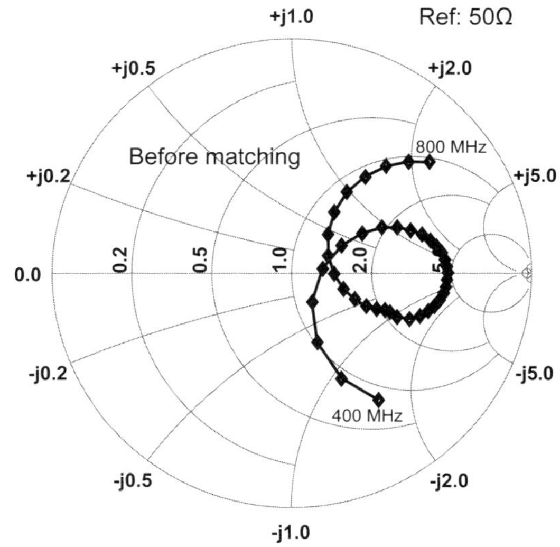

Figure 3.11 Simulated input impedance (Smith chart) at the terminals of a crossed V-dipole antenna before matching.

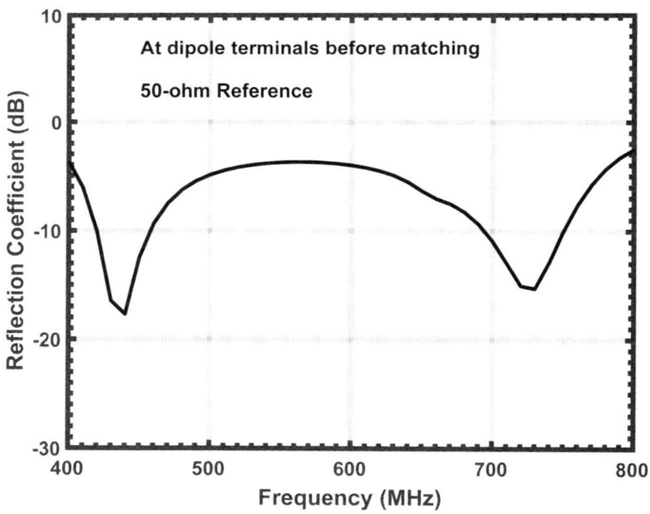

Figure 3.12 Simulated reflection coefficient magnitude in decibels versus frequency at the terminals of a crossed V-dipole antenna before matching.

dipole terminals using Equation (3.6). As observed in Figure 3.10, the input impedance has a first natural resonance at approximately 440 MHz, and is antiresonant at approximately 560 MHz. The antenna has a second natural resonance at approximately 720 MHz. The same simulated input impedance data are plotted in a Smith chart in Figure 3.11. The corresponding reflection coefficient magnitude versus frequency is shown in Figure 3.12. It is observed that the reflection coefficient magnitude in dB has local minimum values occurring near the two natural resonant frequencies. The reflection coefficient magnitude has a peak occurring at the antiresonant frequency as well as large values occurring at the edges of the 400 MHz to 800 MHz band.

Assume now that it is desired to improve the reflection coefficient performance, such that a voltage standing wave ratio (VSWR) ≤ 2 is achieved over a large bandwidth. To start the impedance matching procedure, in Figure 3.13 the complex (real, imaginary) input impedance is plotted as reactance versus resistance. Let the desired upper frequency arbitrarily be chosen as 780 MHz. Then, construct the outer boundary circle (OBC) to

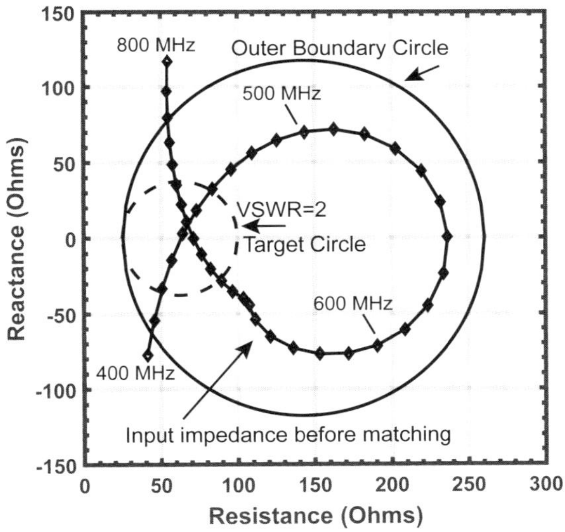

Figure 3.13 Simulated input impedance (solid curve with diamond markers) at the terminals of a UHF crossed V-dipole antenna fed with a 50-ohm coaxial transmission line in a folded balun arrangement. The target circle (dashed curve) corresponds to a VSWR=2 on a 50-ohm characteristic impedance line. An outer boundary circle has been constructed to define the region of the input impedance that will be impedance matched to the target circle region.

encompass the impedance data such that the OBC (solid circle) is tangent to the VSWR=2 circle (dashed circle) where the minimum resistance value is 25Ω, as shown in Figure 3.13. The minimum and maximum resistance values of the outer boundary circle are 25Ω and 260.5Ω, respectively. Then from Equation (3.39) the required line transformer has a characteristic impedance

$$Z'_o = \sqrt{25\Omega \times 260.5\Omega} = 80.7\Omega \tag{3.40}$$

which is rounded here to $Z'_o = 81\Omega$.

Having determined the line transformer characteristic impedance $Z'_o = 81\Omega$, the next step is to construct the corresponding series of transformation circles (refer to Equations (3.32) and (3.36) to (3.38)) from $15°$ to $165°$, which are shown in Figure 3.14. The input impedance data at the dipole terminals are then overlayed with the transformation circles in Figure 3.15.

To determine the required length (l) of the impedance-matching line transformer, it is necessary to select the so-called critical frequency, in this case where the upper frequency (780 MHz) load impedance intersects with one of the transformation circles. The impedance data at 780 MHz is at the edge of the $\beta l = 120°$ circle. Since $\beta = 2\pi/\lambda$, then $l = 12.7$ cm is the length

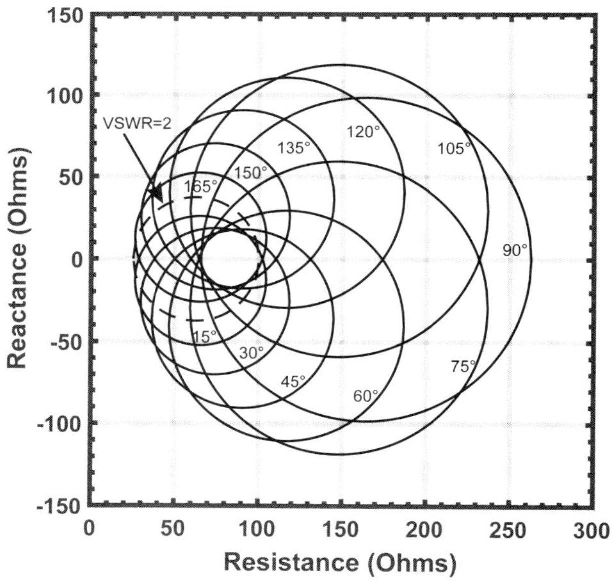

Figure 3.14 The family of transformation circles for a line transformer of 81-ohm characteristic impedance for electrical lengths 15° to 165°. The target circle (dashed curve) corresponds to a VSWR=2.

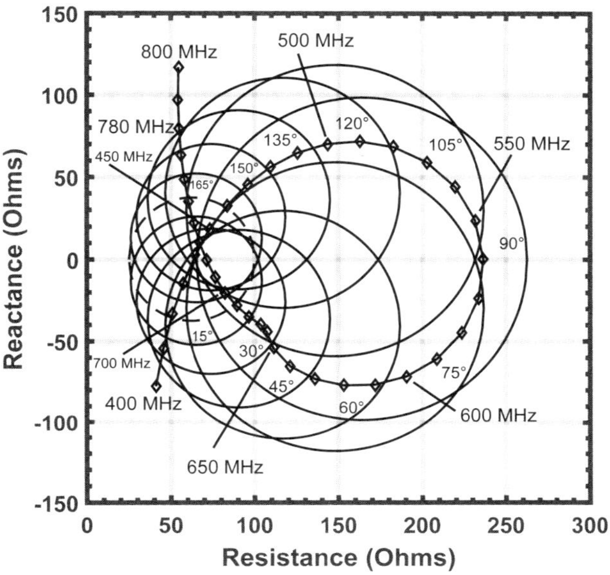

Figure 3.15 Simulated input impedance (solid curve with diamond markers) at the terminals of a UHF crossed V-dipole antenna fed with a 50-ohm coaxial transmission line in a folded balun arrangement. The target constant VSWR=2 circle is shown as a dashed curve. The impedance transformation circles from 15° to 165° are shown.

of the line transformer to be considered. Table 3.1 shows the electrical length of the 81Ω line transformer at several frequencies. Comparing these electrical lengths with the transformation circles and data shown in Figure 3.15, it is observed that the impedance data points fall within, or lay just outside, transformation circles that impedance match to the target VSWR=2 circle (dashed circle). That is, as the complex input impedance varies with frequency as shown in Figure 3.15, the data approximately track within the diameter of one of the transformation circles.

Having determined the desired parameters of the line transformer, that is, characteristic impedance $Z_o' = 80.7\Omega$ and line length $l = 12.7$ cm, Equation (3.3) can be used to compute the dipole antenna input impedance at the end of the line transformer. Figure 3.16 shows the before and after impedance matching results where it is observed that the impedance-matched data generally fall within the target VSWR=2 circle, except (as expected) at the band edges. Figure 3.17 shows the same simulated data in a Smith chart. The VSWR before and after matching is shown in Figure 3.18. The target

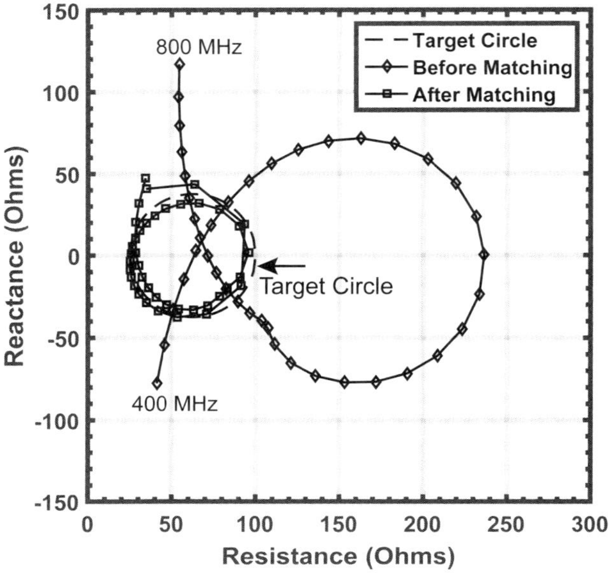

Figure 3.16 Simulated input impedance of a UHF crossed V-dipole antenna before and after impedance matching with an 81-ohm series line transformer.

Table 3.1
Line Transformer Electrical Length βl as a Function of Frequency to Match the Input Impedance to Achieve the Target VSWR<2 for a Crossed V-Dipole Antenna. The Critical Frequency is Taken as 780 MHz Where the Electrical Length of the Line Transformer is 120°.

Frequency	Electrical Length of Line Transfomer (βl)
420 MHz	65°
450 MHz	69°
500 MHz	77°
550 MHz	85°
600 MHz	92°
650 MHz	100°
700 MHz	108°
750 MHz	115°
780 MHz	120°

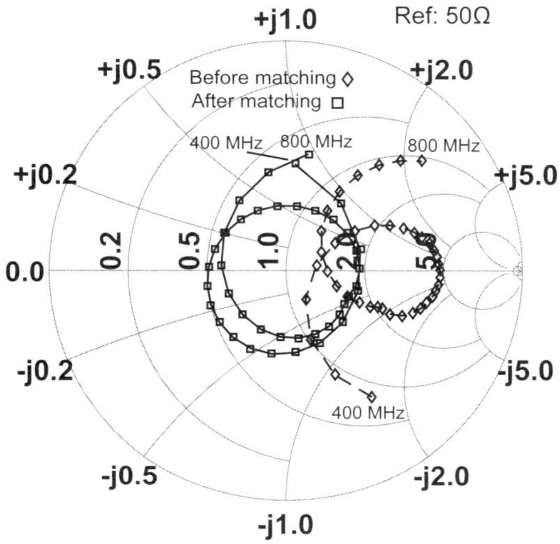

Figure 3.17 Smith chart plot of the simulated input impedance of a UHF crossed V-dipole antenna before and after impedance matching with an 81-ohm series line transformer. Same data as in Figure 3.16.

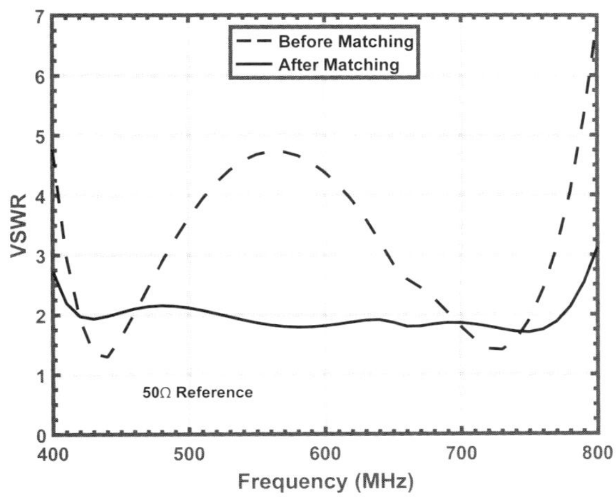

Figure 3.18 Simulated VSWR for a UHF crossed V-dipole antenna before and after impedance matching with an 81-ohm series line transformer. The reference impedance is 50 ohms. The target VSWR was 2.0.

VSWR=2 has generally been achieved over a wide bandwidth as desired. Note that the VSWR increases to just under 2.2 in the vicinity of 480 MHz.

Due to impedance matching, the overall improvement in VSWR is observed to be significant, by about a factor of two over the middle portion of the frequency range. The simulated reflection coefficient magnitude before and after matching is shown in Figure 3.19, and the corresponding transmission mismatch loss is shown in Figure 3.20. The transmission mismatch loss before matching has a maximum value of about 2.4 dB occurring near 560 MHz. After impedance matching the mismatch loss is less than about 0.6 dB from 420 MHz to 780 MHz — due to impedance matching the realized gain would be within 0.6 dB of the ideal antenna directivity with no mismatch loss. For example, with the 81-ohm coaxial line transformer section simulated in FEKO, at the center frequency of 600 MHz the simulated boresight directivity of the swept-back dipole over ground plane was 6.45 dBi and the realized gain was 6.0 dBi — for this case the simulated mismatch loss was 0.45 dB. Prior to matching with the line transformer section, for the constant 50-ohm coaxial feed at 600 MHz the simulated boresight directivity was 6.45 dBi, but the realized gain was only 4.27 dBi due to the mismatch loss being 2.18 dB. Thus, with the line transformer design, at the center frequency the swept-back dipole realized gain increased by 1.75 dB compared to the case where the coaxial feedline had a constant 50-ohm characteristic impedance.

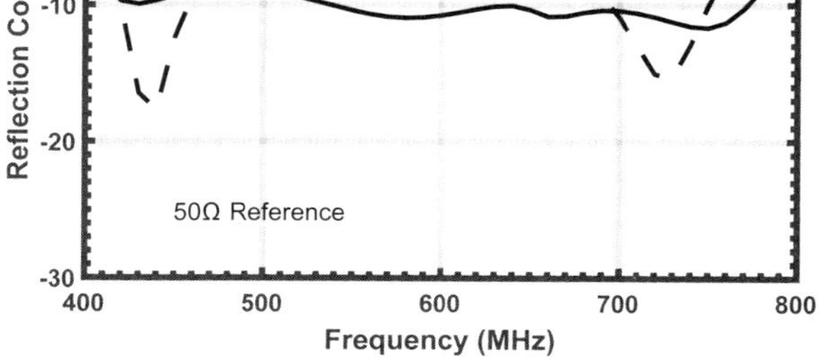

Figure 3.19 Simulated reflection coefficient magnitude for a UHF crossed V-dipole antenna before and after impedance matching with an 81-ohm series line transformer. The reference impedance is 50 ohms.

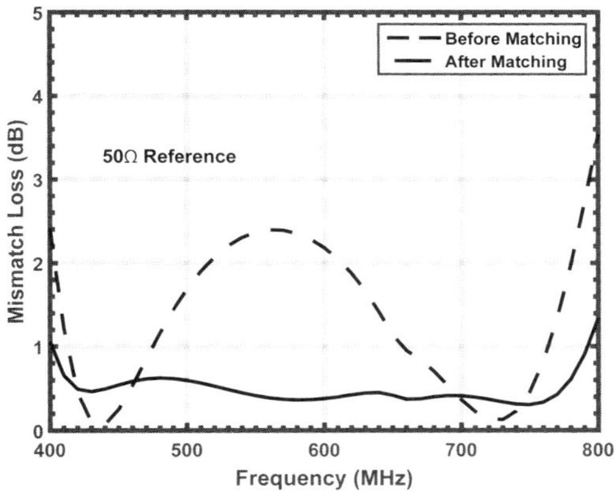

Figure 3.20 Simulated transmission mismatch loss for a UHF crossed V-dipole antenna before and after impedance matching with an 81-ohm series line transformer. The reference impedance is 50 ohms.

3.5 Comparison of Simulations with Measurements

A dual-polarized swept-back dipole prototype was fabricated using metal tubing based on the dimensions shown in Figure 3.8. The prototype dipole used a stepped coaxial center conductor matching section, and Equation (3.2) was used to determine the coaxial conductor dimensions to implement the 81-ohm series line transformer. A comparison of the simulated and measured reflection coefficient of the impedance-matched swept-back dipole over the 400 MHz to 800 MHz band is shown in Figure 3.21 and good agreement is evident.

3.6 Swept-Back Dipole Feeding a Parabolic Reflector

In this section, a simulation of the swept-back dipole (designed in Section 3.4) feeding a 4.57m diameter parabolic reflector with focal distance to diameter ratio (f/D) equal to 0.4 is described. In this simulation the dipole was assumed to be linearly polarized similar to the dipole shown in the photograph in Figure 3.6. Figure 3.22 shows the simulated dipole/reflector model, where the dipole's ground plane is positioned at the focal point of the reflector. The shape of the parabolic reflector is readily derived by equating the length of the on-axis ray path with the length of an off-axis ray path with respect to the focal point (designated as point 3). That is, for a point on the reflector at the

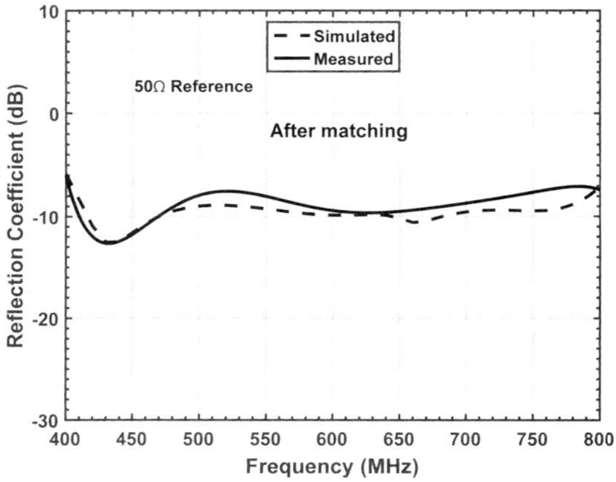

Figure 3.21 Comparison of simulated and measured reflection coefficient for a UHF crossed V-dipole antenna after impedance matching with an 81-ohm series line transformer. The reference impedance is 50 ohms.

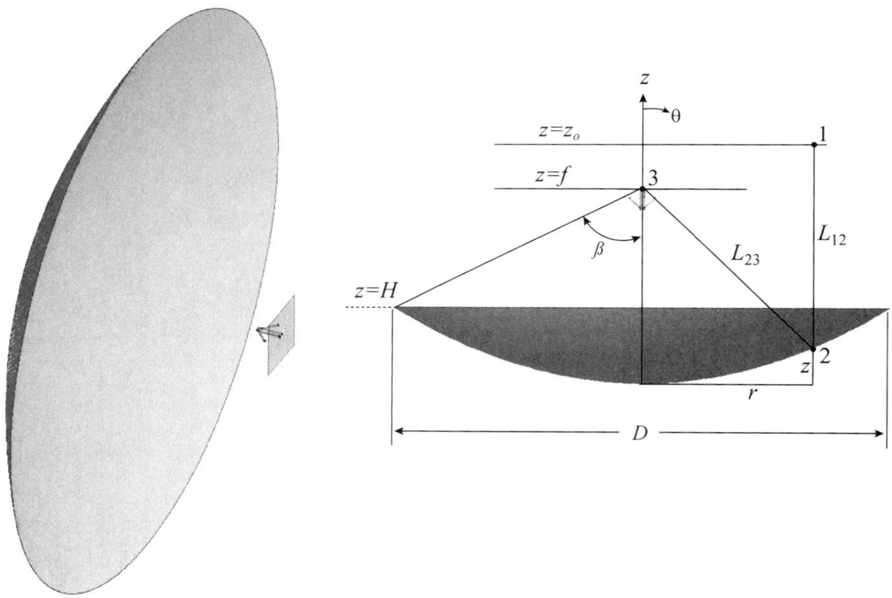

Figure 3.22 Parabolic reflector fed with a V-dipole antenna at the focal point.

radial distance r set

$$(f + z_o) = L_{12} + L_{23} = (z_o - z) + \sqrt{r^2 + (f - z)^2} \tag{3.41}$$

which reduces to

$$f + z = \sqrt{r^2 + (f - z)^2} \tag{3.42}$$

and solving for z yields

$$z = \frac{r^2}{4f} \tag{3.43}$$

where $r = \sqrt{x^2 + y^2}$. The height (or depth) of the reflector is computed by substituting the reflector radius $R = D/2$ in Equation (3.43) with the result

$$H = \frac{R^2}{4f} = \frac{D^2}{16f} \tag{3.44}$$

The angle β from the focal point to the rim of the reflector is expressed as

$$\beta = \tan^{-1}\left(\frac{R}{f - H}\right) \tag{3.45}$$

In this case, the angle from the focal point of the reflector to the rim of the reflector is $\beta = 64°$. The method of moments simulated wideband swept-back dipole far-zone E-plane and H-plane directivity patterns are shown in Figures 3.23 and 3.24, respectively, and wide-angle illumination of the reflector is obtained with this dipole feed antenna. The FEKO physical optics solver was used to compute the far-field E-plane and H-plane radiation patterns of the reflector antenna (Figures 3.25 and 3.26). The simulated reflector peak directivity is 24.7 dBi, 27.4 dBi, and 28.6 dBi at 420 MHz, 600 MHz, 780 MHz, respectively. Wideband performance is observed with this type of dipole feed.

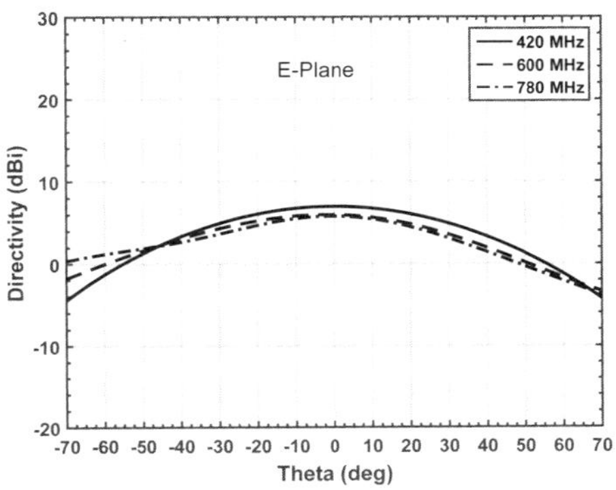

Figure 3.23 Simulated wideband V-dipole E-plane directivity patterns.

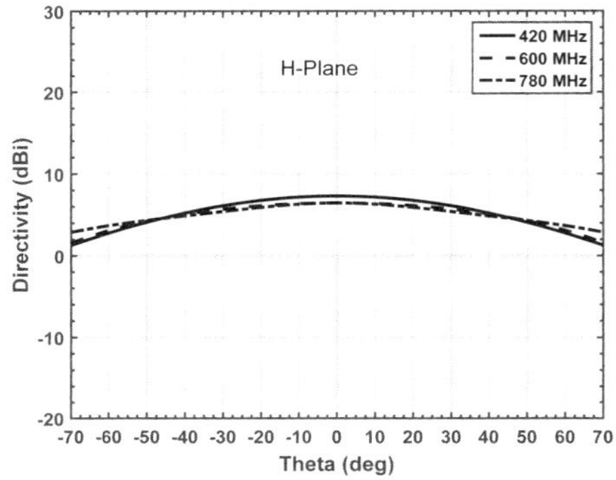

Figure 3.24 Simulated wideband V-dipole H-plane directivity patterns.

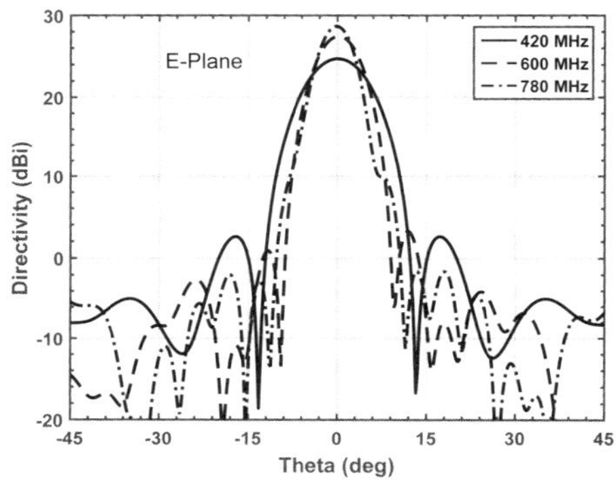

Figure 3.25 Simulated E-plane directivity patterns for wideband V-dipole feeding a parabolic reflector.

3.7 Summary

This chapter has reviewed the line transformer theory and method for impedance matching a complex antenna load to a target VSWR over a wide band of frequencies. The input impedance of an example UHF crossed swept-back dipole antenna was simulated using the method of moments. The line transformer impedance matching method was applied to the simulated antenna and to a prototype antenna and a wide-band match was achieved. This impedance matching technique can be applied to other antennas. This type of swept-back dipole antenna has wide-angle pattern coverage and was shown to provide wideband performance as a feed for a parabolic reflector antenna.

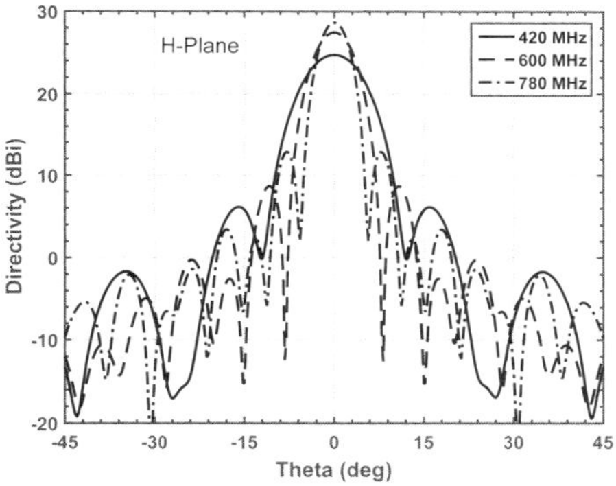

Figure 3.26 Simulated H-plane directivity patterns for wideband V-dipole feeding a parabolic reflector.

3.8 PROBLEM SET

3.1 Consider the UHF crossed V-dipole antenna that was depicted in Figure 3.8. The moment method simulated input impedance versus frequency transformed to the terminals of this antenna before matching was shown in Figure 3.10. The MATLAB curve fitting application was used to generate a sum of sines with 9 terms to model the real and imaginary parts of the input impedance normalized by 50 ohms. Then in MATLAB with the variable

```
x = FreqGHz;
```

the normalized input resistance is expressed as

```
ZL_real =...
a1*sin(b1*x+c1) + a2*sin(b2*x+c2) + a3*sin(b3*x+c3) +...
a4*sin(b4*x+c4) + a5*sin(b5*x+c5) + a6*sin(b6*x+c6) +...
a7*sin(b7*x+c7) + a8*sin(b8*x+c8) + a9*sin(b9*x+c9);
```

where the coefficients are given by

```
a1=3.854;b1=7.452;c1=3.025;
a2=0.6799;b2=29.69;c2=3.538;
a3=0.06904;b3=60.49;c3=3.884;
a4=0.2571;b4=45.48;c4=1.669;
a5=1.784;b5=11.43;c5=2.599;
a6=0.03788;b6=70.61;c6=6.075;
a7=0.02363;b7=91.29;c7=2.745;
a8=0.01695;b8=110.8;c8=1.897;
a9=0.01403;b9=138.9;c9=2.976;
```

and the normalized input reactance is expressed as

```
ZL_imag =...
a1*sin(b1*x+c1) + a2*sin(b2*x+c2) + a3*sin(b3*x+c3) +...
a4*sin(b4*x+c4) + a5*sin(b5*x+c5) + a6*sin(b6*x+c6) +...
a7*sin(b7*x+c7) + a8*sin(b8*x+c8) + a9*sin(b9*x+c9);
```

where the coefficients are given by

```
a1=1.458;b1=14.61;c1=3.512;
a2=2.636;b2=19.48;c2=4.111;
a3=0.6687;b3=27.91;c3=13.84;
a4=0.2666;b4=45.52;c4=8.838;
a5=0.1515;b5=73.81;c5=-1.131;
a6=0.1345;b6=78.73;c6=11.74;
a7=-0.02236;b7=109.1;c7=-3.682;
a8=-0.0002681;b8=116.7;c8=2.222;
a9=-0.006035;b9=149.9;c9=-1.5;
```

Using these two analytic expressions for the simulated input impedance at the terminals of the swept-back dipole, follow the line transformer impedance matching procedure outlined in Section 3.4.2 to design the stepped coaxial transmission line and reproduce the results presented in Figures 3.10 to 3.20.

References

[1] R.C. Hansen, (ed.), *Microwave Scanning Antennas, vol. 2*, Los Altos, CA: Peninsula Publishing, 1985, pp. 162-170, 301-308.

[2] R.C. Hansen, *Phased Array Antennas, 2nd ed.*, New Jersey: Wiley, 2009.

[3] R.C. Hansen, (ed.), *Significant Phased Array Papers,* Dedham, MA: Artech House, 1973.

[4] R.J. Mailloux, "Phased Array Theory and Technology," *Proc. IEEE*, vol. 70, no. 3, 1982, pp. 246-291.

[5] E.D. Mayer, and A. Hessel, "Feed Region Modes in Dipole Phased Arrays," *IEEE Trans. Antennas Propagat.*, vol. 30, no. 1, 1982, pp. 66-75.

[6] L. Stark, "Comparison of Array Element Types," in *Phased Array Antennas (Proc. 1970 Phased Array Antenna Symp.)*, A.A. Oliner and G.H. Knittel, (eds.), Dedham, MA: Artech House, 1972, pp. 51-66.

[7] G.H. Knittel, (E. Brookner, (ed.)), Chapter 21, Phased Array Antennas – An Overview, in *Radar Technology*, Dedham, MA: Artech House, 1977, pp. 289-301.

[8] A.J. Fenn, "Element Gain Pattern Prediction for Finite Arrays of V-Dipole Antennas over Ground Plane," *IEEE Trans. Antennas Propagat.*, vol. 36, no. 11, 1988, pp. 1629-1633.

[9] S.A. Schelkunoff, and H.T. Friis, *Antennas: Theory and Practice*, New York: Wiley, 1952, pp. 499-502.

[10] H.K. Schuman, D. R. Pflug, and L. D. Thompson, "Infinite Planar Arrays of Arbitrarily Bent Thin Wire Radiators," *IEEE Trans. Antennas Propagat.*, vol. 32, no. 4, 1984, pp. 364-377.

[11] E. Brookner, B. Porter, K. Chang, Y.-C. Chang, D. Zwillinger, B. Considine, T. Sikina Demonstration of accurate prediction of PAVE PAWS embedded element gain, *IEEE Int. Symp. on Phased Array Systems and Technology,* 2010, pp. 417 - 422.

[12] H. Jasik, *Antenna Engineering Handbook,* New York: McGraw-Hill, 1961.

[13] J.D. Kraus, *Antennas*. New York: McGraw-Hill, 1950.

[14] W.L. Stutzman and G.A. Thiele, *Antenna Theory and Design, 2nd ed.* New York: Wiley, 1998.

[15] J. Volakis, ed., *Antenna Engineering Handbook, 4th ed.* New York: McGraw-Hill, 2007.

[16] C.A. Balanis, *Antenna Theory and Design, 3rd ed.* New York: Wiley, 2005.

[17] A.J. Fenn, *Adaptive Antennas and Phased Arrays for Radar and Communications.* Norwood, Mass.: Artech House, 2008.

[18] E. Brookner, "Phased-Array Radars," *Scientific American,* vol. 252, no. 2, 1985, pp. 94–102.

[19] A.J. Fenn, D.H. Temme, W.P. Delaney, and W.E. Courtney, "The Development of Phased Array Radar Technology," *Lincoln Laboratory Journal*, vol. 12, no. 2, 2000, pp. 321–340.

[20] J.L. Allen, L. Cartledge, W.P. Delaney, and J. Dibartolo, et al., "Phased Array Radar Studies, 1 July 1959 to 1 July 1960," Technical Report 228, MIT Lincoln Laboratory, 12 Aug. 1960, DTIC no. AD-0249470.

[21] J.L. Allen, L. Cartledge, W.P. Delaney, and J. Dibartolo, "Phased Array Radar Studies, 1 July 1960 to 1 July 1961," Technical Report 236, MIT Lincoln Laboratory, 13 Nov. 1961, DTIC no. AD-271724.

[22] J.L. Allen, D.M. Bernella, W.W. Carpenter, and W.P. Delaney, "Phased Array Radar Studies, 1 July 1961 to 1 July 1963," Technical Report 299, MIT Lincoln Laboratory, 20 Feb. 1963, DTIC no. AD-417572.

[23] J.L. Allen, D.M. Bernella, F. Betts, and L. Cartledge, "Phased Array Radar Studies, 1 July 1963 to 1 July 1964,' Technical Report 381, MIT Lincoln Laboratory, 31 March 1965, DTIC no. AD-629363.

[24] L.V. Blake and M.W. Long, *Antenna, Fundamentals, Design, Measurement, 3rd ed.* Raleigh, NC: SciTech Publishing, 2009.

[25] E.A. Wolff, *Antenna Analysis*, New York: Wiley, 1966.

[26] J.L. Allen, "Gain and Impedance Variation in Scanned Dipole Arrays," *IRE Trans. Antennas and Propagation*, vol. 10, no. 5, 1962, pp. 566–572.

[27] J.L. Allen and W.P. Delaney, "On the Effect of Mutual Coupling on Unequally Spaced

Dipole Arrays," *IRE Trans. Antennas and Propagation,* vol. 10, no. 6, 1962, pp. 784–785.

[28] J.L. Allen, "On Array Element Impedance Variation with Spacing," *IRE Trans. Antennas and Propagation*, vol. 12, no. 3, 1964, pp. 371–372.

[29] J.L. Allen, "On Surface-Wave Coupling Between Elements of Large Arrays," *IEEE Trans. Antennas and Propagation*, vol. 13, no. 4, 1965, pp. 638–639.

[30] H.K. Schuman, D.R. Pflug, and L.D. Thompson, "Infinite Planar Arrays of Arbitrarily Bent Thin Wire Radiators," *IEEE Trans. Antennas Propagation*, vol. 32, no. 3, 1984, pp. 364–377.

[31] A.J. Fenn, P.T. Hurst, J.D. Krieger, J.S. Sandora, and L.I. Parad, "Ultrawideband VHF/UHF Dipole Array Antenna," 2010 IEEE International Symposium on Phased Array Systems and Technology, Waltham-Boston, Mass., 2010, pp. 79–82.

[32] K.E. Schmidt, "Mutual Impedance of Nonplanar Skew Dipoles," *IEEE Trans. Antennas and Propagation*, vol. 44, no. 9, 1996, pp. 1298–1299.

[33] H.J. Reich, ed., Harvard Radio Research Laboratories, *Very High Frequency Techniques*, New York: Mc-Graw Hill, Vol. 1, Chapter 3, 1947.

[34] R.L. Thomas, *A Practical Introduction to Impedance Matching*, Dedham, MA: Artech House, 1976.

[35] W.N. Caron, *Antenna Impedance Matching*, Newington, CT: American Radio Relay League, 1989.

[36] R.M. Fano, "Theoretical Limitations on the Broadband Matching of Arbitrary Impedances," Massachusetts Institute of Technology, Research Laboratory of Electronics, Technical Report 41, Jan. 2, 1948.

[37] H.A. Wheeler, "Wideband Impedance Matching," Wheeler Laboratories Report 418, Great Neck, New York, 1950.

[38] V.H. Rumsey, "The Design of Frequency-Compensating Matching Sections," *Proc. IRE*, October 1950, pp. 1191–1196.

[39] G.L. Matthaei, L. Young, E.M.T. Jones, *Microwave Filters, Impedance-Matching Networks, and Coupling Structures*, New York, McGraw-Hill, 1964.

[40] T.R. Cuthbert, *Circuit Design Using Personal Computers*, New York: John Wiley, 1983.

[41] T.R. Cuthbert, *Broadband Direct-Coupled and Matching RF Networks*, Greenwood, Arkansas: TRCPEP Publications, 1999.

[42] C. Bowick, *RF Circuit Design*, Burlington, MA: Newnes, 2008.

[43] B.A. Munk, *Finite Antenna Arrays and FSS,* Hoboken, New Jersey: John Wiley, 2003.

[44] T.A. Milligan, *Modern Antenna Design, 2nd ed.* Hoboken, NJ: John Wiley, 2005, pp. 252-253.

4

Omnidirectional Antennas

4.1 Introduction

This chapter describes simulations and measurements of antennas that are omnidirectional in azimuth. Omnidirectional coverage is desired in communications and direction finding applications where the desired signal direction is arbitrary. Omnidirectional coverage can be provided for vertical polarization with vertical monopole or vertical dipole antennas which can typically have a cylindrical or conical shape. A monopole antenna with a cone shape is referred to as a monocone or conical monopole. Similarly, a dipole antenna with cone-shaped arms is referred to as a bicone or biconical antenna. Omnidirectional horizontally polarized coverage can be achieved with horizontal loops or slotted cylinder antennas. Section 4.2 describes vertically polarized cylindrical and conical monopole antennas. Section 4.3 describes a vertically polarized discone antenna. Section 4.4 describes an omnidirectional monopole array designed for simultaneous transmit and receive application. Section 4.5 describes omnidirectional horizontally polarized loop and slotted cylinder antennas. The turnstile crossed dipole was described as an omnidirectional antenna in Chapter 1 (Section 1.12.2), where the arms were aligned in a plane. Section 4.6 investigates the turnstile crossed-dipole antenna where the dipole arms are swept back to provide omnidirectional pattern coverage for horizontal and vertical polarization. Each of the antennas described in this chapter can be considered wideband depending on how the bandwidth is defined, either in terms of the radiation pattern or impedance match or both. For transmit applications both the radiation pattern and impedance match will be important. For some receive applications involving communications, only the radiation pattern gain will be of primary importance as long as the communications link closes.

4.2 Monopole Antennas

4.2.1 Background for Cylindrical and Conical Monopoles

In this section, two types of antennas are considered here, cylindrical monopoles and conical monopoles over a ground plane [1-22]. The contrast between a cylindrical monopole and a conical monopole is depicted in the photograph shown in Figure 4.1. In this photograph each monopole has been electrically attached to the center pin of a type-N microwave connector. The conical monopole is shown with a closed-end plate, which is known from the literature to have little effect on the input impedance and radiation pattern. Both cylindrical and conical monopoles are useful as single transmitting and receiving antennas, and in phased arrays for wide-angle scanning [21]. In Figure 4.2(a), a close-up photograph of a cylindrical monopole mounted in an array of like elements on an aluminum ground plane is shown. The optical reflection of the vertical monopole element in the ground plane shows an optical image that appears the same as the electrical image due to the boundary conditions described in Chapter 1 (refer to Section 1.5 and Figure 1.32). The image of the monopole effectively forms an equivalent dipole in free space, which can be analyzed readily with the method of moments. An equivalent dipole array example was shown previously in Chapter 2 (see Figure 2.39). Alternately, the method of moments can be applied to the monopole and ground plane combination. Figure 4.2(b) shows a photograph of an 11×11 passively terminated monopole array, and Figure 4.2(c) shows the corresponding FEKO simulation model. A comparison of the measured and simulated center element gain patterns with the surrounding elements terminated in 50Ω resistive loads are shown in Figure 4.2(d) and good agreement is observed. In the array environment, the embedded monopole array pattern is omnidirectional about the axis of the monopole as was shown in Chapter 2 (see Figure 2.41(c)) and by measurements [21].

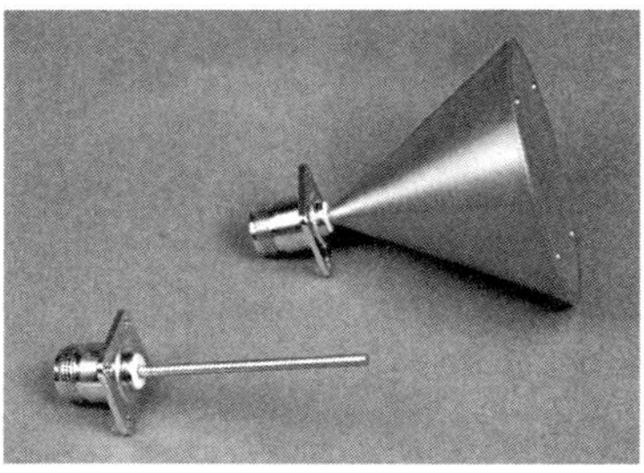

Figure 4.1 Photograph of cylindrical and conical monopole antennas.

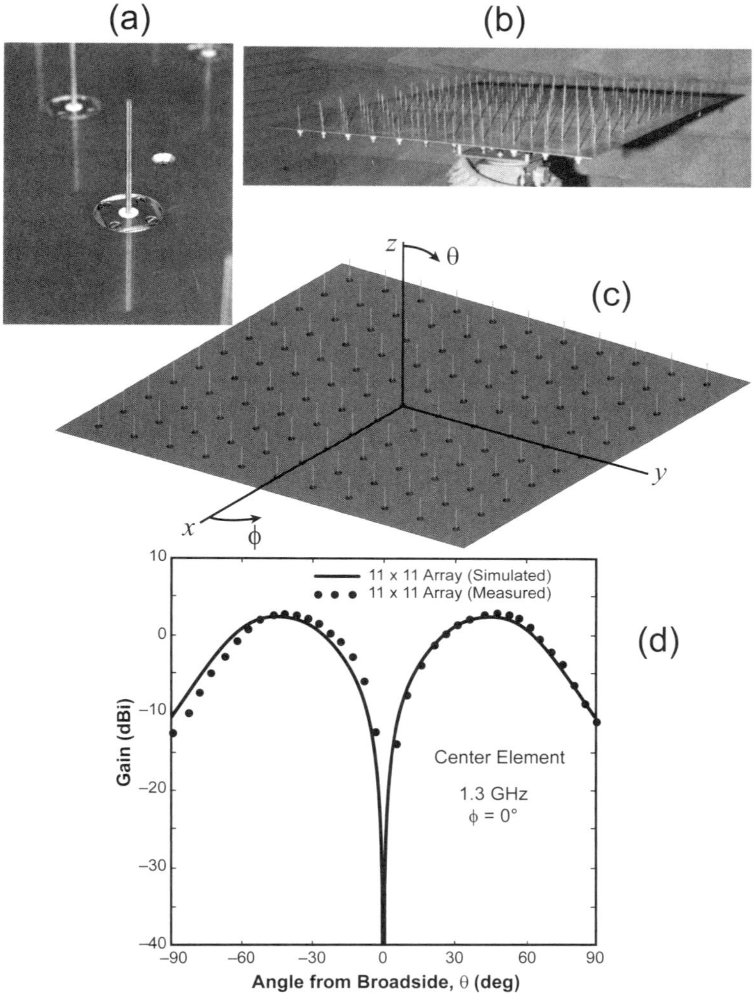

Figure 4.2 (a) Photograph of a cylindrical monopole element in an array, (b) photograph of an 11×11 passively terminated monopole array, (c) simulation model of the monopole array, and (d) comparison of measured and simulated center element gain patterns with the surrounding elements terminated in 50Ω resistive loads. Monopole length is 6.35 cm, wire radius is 1.588 mm, array element spacing in a square lattice is 10.67 cm, and ground plane is 1.22m $\times$ 1.22m.

A sketch of a cylindrical monopole over a conducting ground plane is shown in Figure 4.3. The base of the monopole is connected to a coaxial line that operates in the transverse electromagnetic (TEM) mode. Brown and

Woodward have performed extensive measurements of the input impedance of cylindrical monopoles over a ground plane [22] as shown in Figures 4.4 (input resistance) and 4.5 (input reactance). These data were measured at 60 MHz with numerous physical size cylindrical monopoles with closed-end circular metal plate bases mounted over a circular metallic screen 3.66m in diameter and 6 wires per cm, which was placed on the earth's surface. A large variation in the input impedance of cylindrical monopoles occurs for thin-diameter cylinders where the length to diameter ratio is large. As the diameter of the cylindrical monopole increases for a fixed length, the input impedance tends to have a decreasing variation, which provides more bandwidth.

It should be noted that the input impedance of a monopole over an electrically large perfectly conducting ground plane is one half that of the equivalent dipole. The peak gain of a one-quarter wavelength cylindrical monopole over an infinite perfectly conducting ground plane is 5.1 dBi compared to the peak gain of a one-half wavelength dipole in free space which is 2.1 dBi. Referring to the geometry in Figure 4.3, for an infinite perfectly conducting ground plane the monopole peak gain occurs at $\theta = 90°$. When an infinite ground plane has a finite conductivity such as the earth's surface, the monopole radiated field is attenuated to a null in the far field at $\theta = 90°$. For a monopole mounted on a finite ground plane, the beam peak is lifted up from $\theta = 90°$. The monopole is omnidirectional with respect to the axis of the monopole.

Figure 4.6 shows a sketch of a conical monopole over a ground plane. The conical monopole is fed with a coaxial transmission line operating in the transverse electric mode (TEM). The cone is assumed to have vertical height H, slant length a, upper diameter D, and lower diameter d. The cone

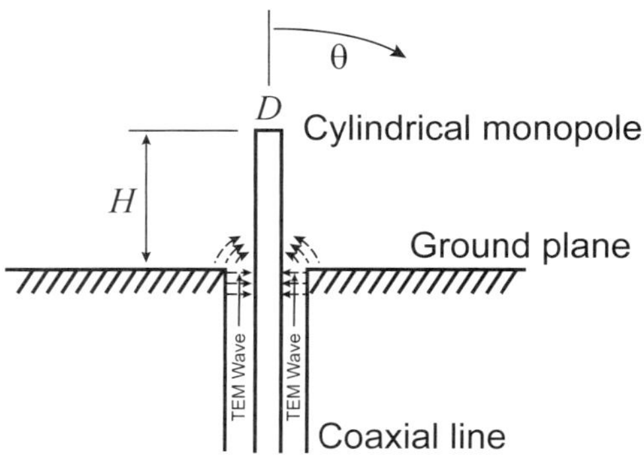

Figure 4.3 Sketch of a cylindrical monopole over a ground plane, fed with a coaxial line.

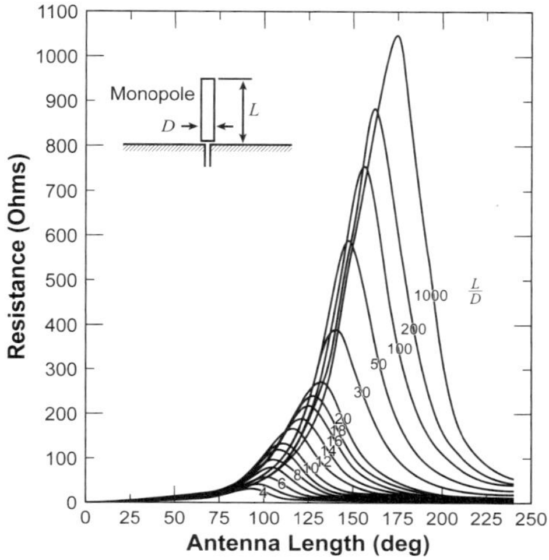

Figure 4.4 Measured input resistance for a cylindrical monopole over a ground plane, fed with a coaxial line [22].

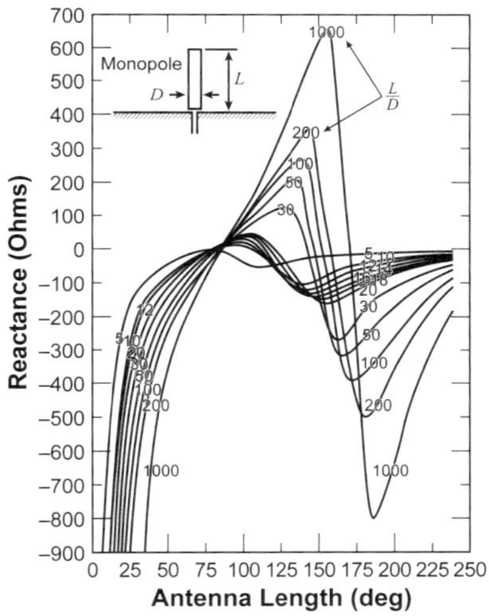

Figure 4.5 Measured input reactance for a cylindrical monopole over a ground plane, fed with a coaxial line [22].

half-angle is given by

$$\theta_o = \tan^{-1}\left[\frac{D-d}{2H}\right] \tag{4.1}$$

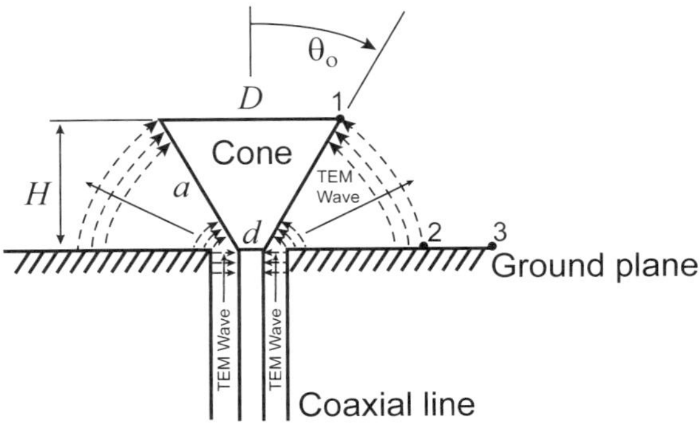

Figure 4.6 Sketch of a conical monopole over a ground plane, fed with a coaxial line.

The cone slant length is expressed in terms of the height and cone half angle as

$$a = \frac{H}{\cos\theta_o} \tag{4.2}$$

Brown and Woodward have performed extensive input impedance (also radiation pattern [15]) measurements of conical monopoles over a ground plane [22] as shown in Figures 4.7 (input resistance) and 4.8 (input reactance). These measured data were collected at 500 MHz with a circular conducting ground plane 2.44m in diameter on the earth's surface. As the cone angle

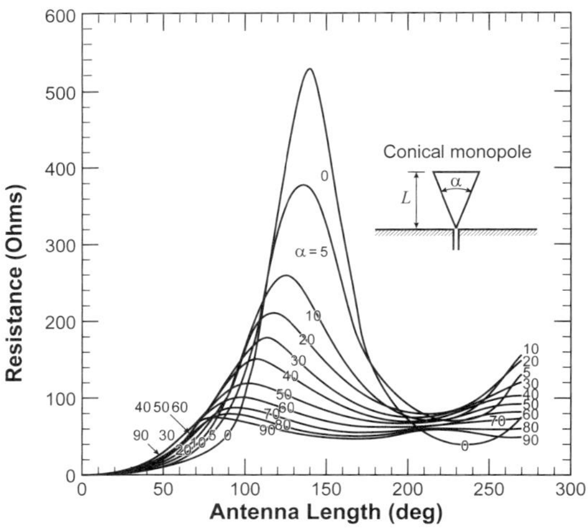

Figure 4.7 Measured input resistance for a conical monopole over a ground plane, fed with a coaxial line [22].

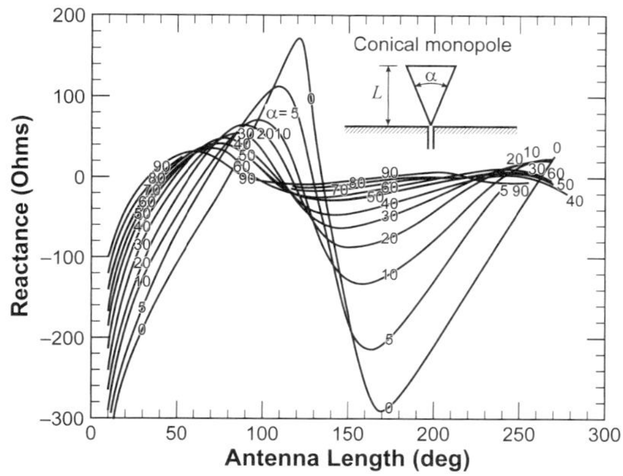

Figure 4.8 Measured input reactance for a conical monopole over a ground plane, fed with a coaxial line [22].

increases there is a decreased variation in the input impedance, which allows an increased bandwidth.

It is noted by Reich [6] that as the slant length (or height) of a conical monopole approaches one wavelength, the input impedance is approximately equal to the characteristic impedance. Referring to Figure 4.6 at the feed gap region, on transmit a TEM wave is launched (from the coaxial feed line) between the cone's base (apex) and ground plane, and the wave propagates outward until it reaches the antenna aperture represented by the curved wavefront between points 1 and 2. At the edge discontinuity of the finite cone (point 1), a portion of the incident field will be scattered (diffracted) back towards the feed region and a portion will be transmitted (diffracted) into free space. When the wave reaches point 3 at the finite rim of the ground plane, a portion of the wave will be scattered (diffracted) back towards the feed and a portion will be transmitted (diffracted by the edge) into free space.

An approximate expression for the reflection coefficient of a finite monocone with slant length a on an infinite ground plane has been derived by Papas and King [10, 11]. The input impedance of a conical monopole can be determined by an eigenfunction expansion of the electric field components E_θ and E_r and magnetic field component H_ϕ. The tangential components are made continuous across the antenna aperture. Only the dominant TEM mode of the coaxial feedline is included at the feed region. The approximate reflection coefficient Γ for cones with a large flare angle (cones with half-angle approximately 30° or larger) can be calculated for the TEM outgoing and TEM reflected wave at the fictitious boundary surface at the cone aperture, as given by

$$\Gamma(ka) = e^{-2ika} \frac{1 + i\frac{60}{Z_o} \sum\limits_{n=1}^{\infty} \frac{2n+1}{n(n+1)} [P_n(\cos\theta_o)]^2 \zeta_n(ka)}{-1 + i\frac{60}{Z_o} \sum\limits_{n=1}^{\infty} \frac{2n+1}{n(n+1)} [P_n(\cos\theta_o)]^2 \zeta_n(ka)} \quad (4.3)$$

where i is the imaginary number, $P_n(x)$ is the Legendre polynomal of order n, $k = 2\pi/\lambda$ is the wavenumber, and

$$\zeta(ka) = \frac{h_n^{(2)}(ka)}{h_{n-1}^{(2)}(ka) - \frac{n}{ka} h_n^{(2)}(ka)} \quad (4.4)$$

In Equation (4.3) the summation is performed over the odd integer values of n. In the zeta function defined by Equation (4.4), $h_n^{(2)}(ka)$ is the spherical Hankel function of the second kind which is given by

$$h_n^{(2)}(ka) = j_n(ka) - iy_n(ka) \quad (4.5)$$

where $j_n(ka)$ and $y_n(ka)$ are the spherical Bessel functions of the first and second kind, respectively. The spherical Bessel functions of the first and second kind are computed from the Bessel functions of the first and second kind as,

$$j_n(ka) = \sqrt{\frac{\pi}{2ka}} J_{n+\frac{1}{2}}(ka) \quad (4.6)$$

$$y_n(ka) = \sqrt{\frac{\pi}{2ka}} Y_{n+\frac{1}{2}}(ka) \quad (4.7)$$

The conical monopole input impedance is then computed as

$$Z_{\text{in}} = Z_o \frac{1 + \Gamma(ka)}{1 - \Gamma(ka)} \quad (4.8)$$

where Z_o is the characteristic impedance of an infinite conical monopole over an infinite ground plane. Papas and King state that Equation (4.3) is valid when the cone flare angle is large and the characteristic impedance of the coaxial feed is equal to the characteristic impedance of the conical monopole.

The characteristic impedance of an infinite conical monopole over an infinite ground plane is given by the following equation [2, p. 105]

$$Z_o = \frac{1}{2} \sqrt{\frac{\mu}{\epsilon}} \frac{1}{\pi} \ln \cot \frac{\theta_o}{2} = \frac{1}{2} \eta \frac{1}{\pi} \ln \cot \frac{\theta_o}{2} \quad (4.9)$$

and in the free space case with $\eta = \eta_o = 120\pi$

$$Z_o = 60 \ln \cot \frac{\theta_o}{2} \tag{4.10}$$

Using Equation (4.10) the characteristic impedance of an infinite monocone versus cone half angle on an infinite ground plane is shown in Figure 4.9. For example, when the infinite monocone has a half angle equal to 45°, the characteristic impedance is 52.9 ohms. A perfect impedance match to 50 ohms will occur for an infinite cone half angle equal to 47°. Figure 4.10 shows the calculated reflection coefficient for the infinite cone assuming a 50-ohm coaxial transmission line is connected to the antenna, and Figure 4.11 shows

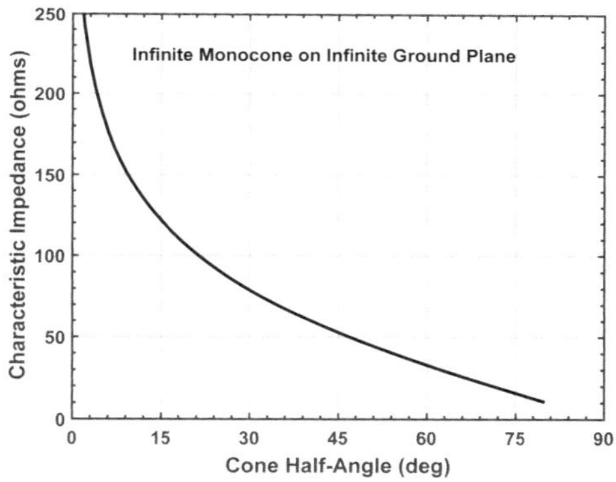

Figure 4.9 Calculated characteristic impedance for an infinite conical monopole over an infinite ground plane.

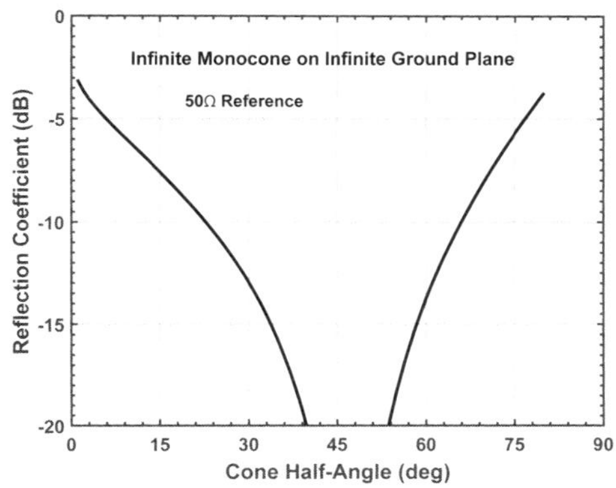

Figure 4.10 Calculated reflection coefficient from the calculated characteristic impedance for an infinite conical monopole over a ground plane.

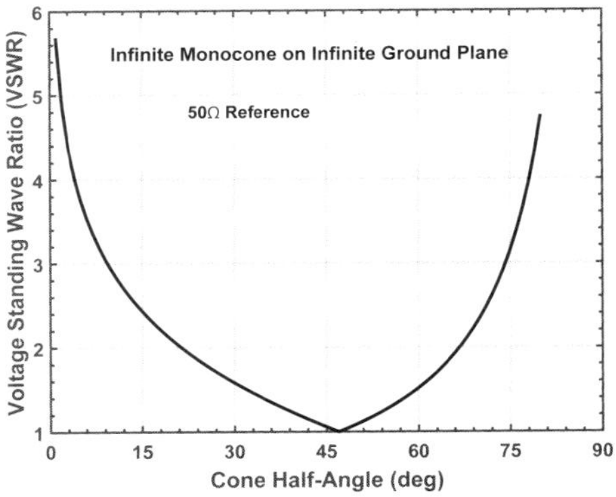

Figure 4.11 Calculated VSWR from the calculated characteristic impedance for an infinite conical monopole over a ground plane.

the corresponding VSWR. According to Figure 4.11, the infinite cone will have a VSWR $\leq$ 2.5 for cone half-angles in the range of about 15° to 70°. For a given coaxial transmission line with a typical characteristic impedance of 50 ohms, a simplified design approach just uses the characteristic impedance of an infinite conical monopole to approximate the cone input impedance for finite cones that have a height one wavelength or longer.

The above discussions provide a theoretical approach to approximating the input impedance of conical monopoles. However, in this chapter, the preferred method for quantifying the input impedance, reflection coefficient, VSWR, mismatch loss, and radiation patterns is by the method of moments that performs a full-wave solution of the dielectric-loaded coaxial feed, finite cone and finite ground plane. The method of moments simulations can be used to adjust the conical monopole parameters to tune the antenna over a desired frequency band.

4.2.2 An Example Ultrawideband Array Application with Monocone Antennas

Ultrawideband (UWB) communications antennas typically must transmit and receive ultrawideband waveforms without introducing significant amplitude and phase distortion. In many applications, it is desirable to have an array of antennas where each of the antennas has an omnidirectional radiation pattern characteristic. An antenna array allows the use of signal processing algorithms for adaptive nulling and direction finding. As is known from the literature, conical monopoles over a ground plane provide a suitable omnidirectional antenna element for very wide bandwidth communications links [23–26].

As depicted in Figure 4.12, a sparse array of conical monopoles has been used in direction finding measurements as reported by Keller [23]. Other array configurations are of general interest such as linear or two-dimensional arrays of ultrawideband conical monopoles, which impacts the design of the individual array elements.

The terms wideband, very wideband, and ultrawideband (as commonly used in the literature) are understood here to be equivalent. A number of ultrawideband applications have received FCC approval for use as described in Federal Communications Commission document FCC 02-48, First Report and Order, Revision of Part 15 of the Commission's Rules Regarding Ultra-Wideband Transmission Systems, Adopted February 14, 2002, Release date April 22, 2002 [27]. The 3.1 GHz to 10.6 GHz band allocated by the FCC for ultrawideband communications is a possible application for the antenna described in the first example given in this chapter. For example, an ultrawideband signal (pulse position modulation (PPM) or orthogonal frequency division multiplex (OFDM)) or multiple RF signals over an ultrawideband would be applied across the antenna feed gap region, inducing a time-varying current that gives rise to electromagnetic radiation for wireless transmission. For wireless ultrawideband reception, an incident time-varying electromagnetic wave signal induces a time-varying electric current on the monocone and a time-varying voltage across the antenna feed gap.

A design of a linearly polarized ultrawideband antenna element capable of operating over the instantaneous 3.1 to 10.6 GHz band for receive and/or transmit wireless communications applications is described in the next subsection. The design uses a conical monopole on a circular ground plane

Figure 4.12 Diagram and photograph showing an ultrawideband field test setup with conical monopole antennas for transmit and receive [23].

with a number of practical features for indoor and outdoor use.

4.2.3 Prototype 1: Monocone Antenna for the 3.1 GHz to 10.6 GHz Band

Referring to the diagram of a conical monopole antenna shown in Figure 4.6, the following parameters allow good performance over the 3.1 GHz to 10.6 GHz instantaneous band. It has been assumed that the coaxial feedline for the monocone is a standard SMA connector, where the center conductor has a diameter equal to 0.127 cm (0.050 inches). The base diameter of the cone then is chosen as 0.254 cm (0.1 inches), which allows a threaded center conductor of the SMA connector to be attached within a corresponding tapped hole in the cone base region. For prototype testing, the conical monopole is made of solid brass with a flat top and a flat base region. The cross section of the monopole is circular for ease of machining and to achieve omnidirectional azimuthal coverage. A standard SMA connector (4-hole flange mount jack receptacle) with a modified center pin that was threaded made the RF antenna connection via a threaded hole in the conical monopole. Thus, solder was not required for this fabricated antenna. The SMA connector introduces a 1.27 cm (0.5 inches) length of Teflon-loaded coaxial line. The moment method simulation includes the same 1.27 cm length of Teflon-loaded coaxial line. For antenna testing, a 15.24 cm (6 inch) diameter circular aluminum plate 0.3175 cm (0.125 inch) thick was used as the ground plane — the simulation model used a perfectly conducting ground plane with the same diameter. A photograph of the fabricated conical monopole antenna (prototype 1) is shown in Figure 4.13.

The height of the prototype 1 cone is 1.5 cm (0.59 inches), and the top truncation diameter is 1.2 cm (0.472 inches). The full cone angle for this design is 35°. At the lowest frequency 3.1 GHz, the wavelength is 9.68 cm

Figure 4.13 Photograph of a conical monopole over a circular ground plane, designed for the 3.1 to 10.6 GHz band (prototype 1).

(3.8 inches) and at the highest frequency 10.6 GHz the wavelength is 2.189 cm (1.11 inches). The top truncation diameter varies from 0.124λ at 3.1 GHz to 0.425λ at 10.6 GHz. The height of the conical monopole varies from 0.155λ at 3.1 GHz to 0.53λ at 10.6 GHz. It should be noted that the conical monopole size in this design does not fit the simple model that could be used for infinite or electrically large cones, that is, the characteristic impedance, reflection coefficient, and VSWR shown in Figures 4.9 to 4.11, respectively, would not apply. Also, the cone half-angle of the conical monopole here is not large, so the finite cone theory of Papas and King [10, 11] would not be expected to apply. Instead, an initial estimate of the reflection coefficient and VSWR performance can be determined from measured input impedance data for finite cones provided by Brown and Woodward [15]. In terms of electrical degrees, the height of the conical monopole varies from 55.8° at 3.1 GHz, to 107.8° at 6 GHz, to 190.8° at 10.6 GHz. From the measured Brown and Woodward data in Figures 4.7 and 4.8, the estimated input impedance for this cone design varies from about $30 - j10$ ohms at 3.1 GHz, to $160 + j10$ ohms, at 6 GHz, to $76 - j25$ ohms at 10.6 GHz. Note that these estimated input impedance values are referenced to the input coaxial feed region of the monocone. The corresponding reflection coefficient and VSWR for this case are computed from Equations (1.15) and (1.27), respectively, as (-11.1 dB, 1.77) at 3.1 GHz, (-5.6 dB, 3.2) at 6 GHz, and (11.0 dB, 1.78) at 10.6 GHz.

An accurate calculation of the conical monopole input impedance versus frequency can be made using the method of moments, as shown by comparisons below with measured data for this antenna. The moment method simulation model for the 3.1 GHz to 10.6 GHz antenna design is shown in Figure 4.14. The simulation model includes a 1.27 cm long Teflon-loaded coaxial feedline to model the SMA connector.

The simulated directivity patterns of prototype 1 at 3.1 GHz, 6 GHz and 10.6 GHz are shown in Figure 4.15. Figure 4.16 shows the simulated

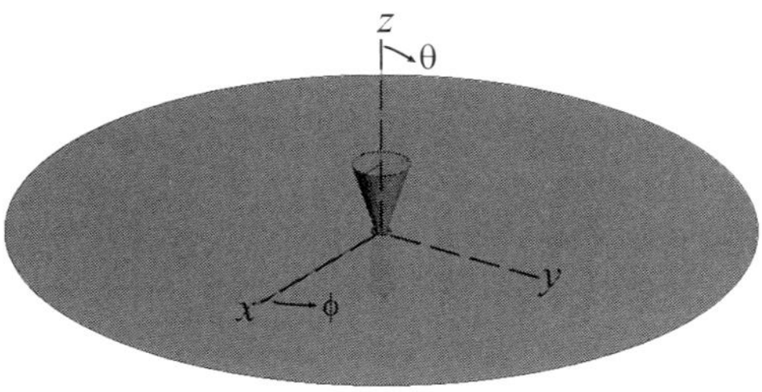

Figure 4.14 Simulation model for a conical monopole over a 15.24 cm diameter circular ground plane fed with a coaxial line, designed for the 3.1 to 10.6 GHz band (prototype 1). Cone height 1.5 cm, cone full angle 35°.

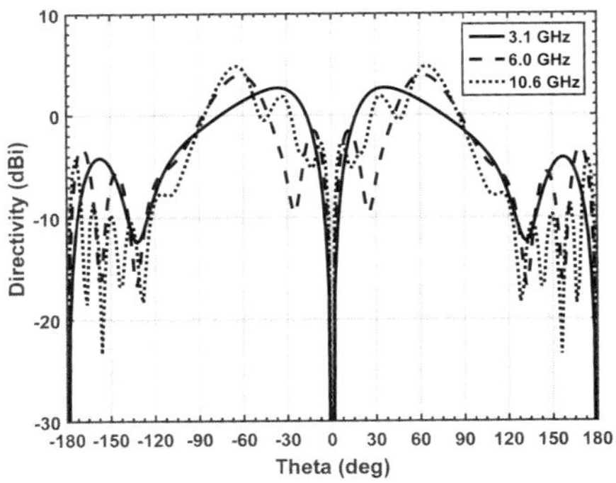

Figure 4.15 Simulated directivity patterns for a conical monopole over a circular ground plane (prototype 1).

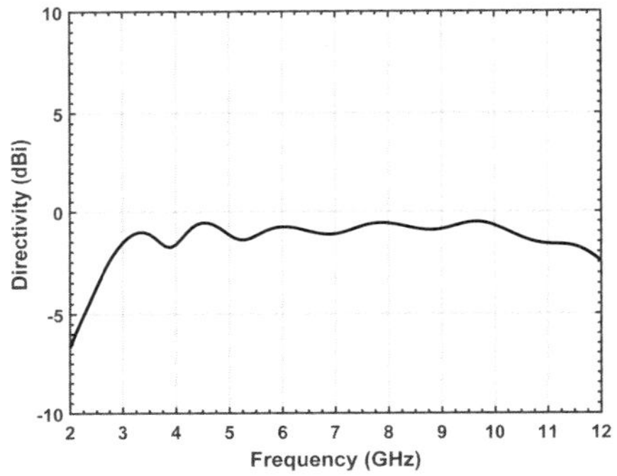

Figure 4.16 Simulated directivity versus frequency at $\theta = 90°$ for a conical monopole over a circular ground plane (prototype 1).

directivity versus frequency at the observation angle $\theta = 90°$, and over the frequency band 3.1 GHz to 10.6 GHz there is less than 2 dB variation in directivity. A comparison of the simulated and measured input resistance and input reactance versus frequency are shown in Figures 4.17 and 4.18, respectively, and generally good agreement is observed. Note that the reference plane here is at the input to the SMA connector. Although not shown here, if the input impedance data (in Figures 4.17 and 4.18) at the input to the SMA connector were transformed to a reference position at the ground plane, they would tend to look more like the measured impedance data shown in Figures 4.7 and 4.8. In the desired frequency range of 3.1 GHz to 10.6 GHz, the measured input resistance has two peaks occurring near 3.5 GHz (79 ohms) and 7.8 GHz (123 ohms). The corresponding simulated and measured reflection coefficient, VSWR, and mismatch loss are shown

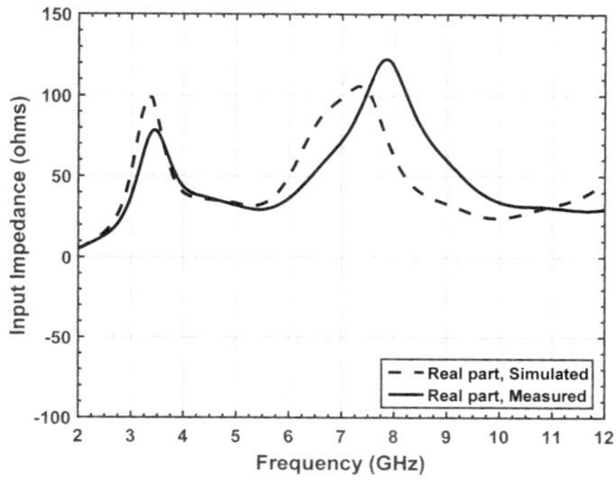

Figure 4.17 Simulated and measured input impedance (real part) for the conical monopole over a circular ground plane, fed with a coaxial line (prototype 1). The reference plane is at the input to the SMA connector.

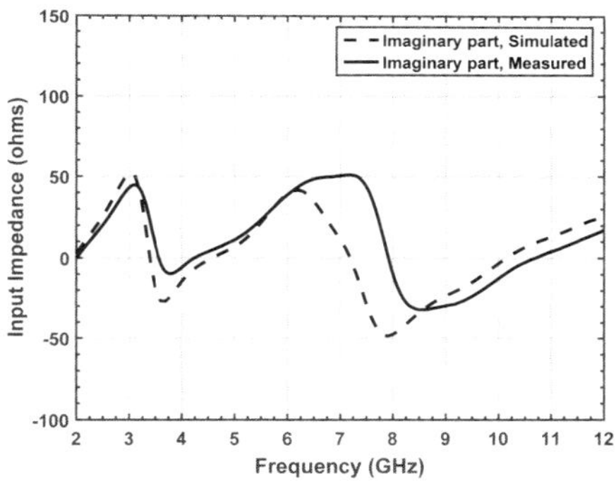

Figure 4.18 Simulated and measured input impedance (imaginary part) for the prototype 1 conical monopole over a circular ground plane, fed with a coaxial line. The reference plane is at the input to the SMA connector.

in Figures 4.19 to 4.21, respectively, and good agreement is observed. The measured and simulated reflection coefficient minimum occurs at 4 GHz and 4.25 GHz, respectively. The measured VSWR is less than about 2.5:1 over the 3.1 GHz to 10.6 GHz band. Figure 4.21 shows that the measured mismatch loss due to impedance mismatch is less than 1 dB over the desired band.

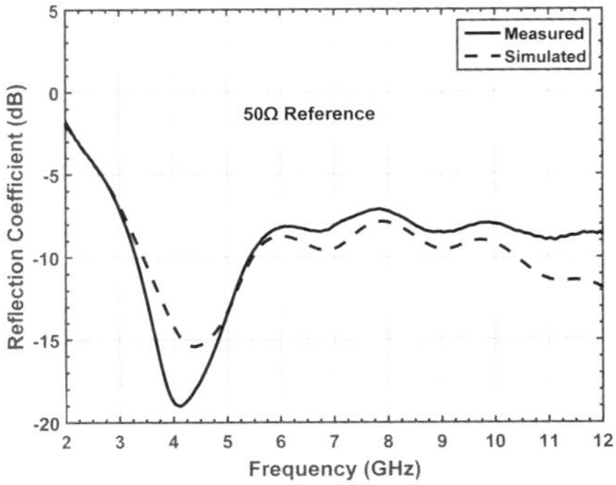

Figure 4.19 Simulated and measured reflection coefficient for a conical monopole over a circular ground plane, fed with a coaxial line (prototype 1).

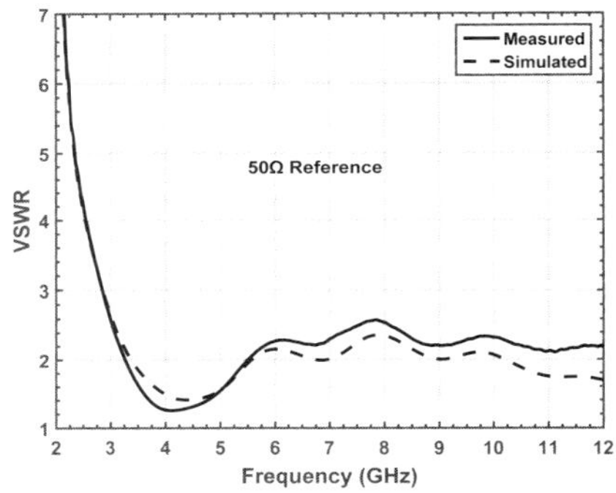

Figure 4.20 Simulated and measured VSWR for a conical monopole over a circular ground plane, fed with a coaxial line (prototype 1).

4.2.4 Prototype 2: Monocone Antenna for X-Band, 8 GHz to 12 GHz

As a second example, consider the design of a conical monopole to cover the X-band frequency range 8 GHz to 12 GHz. A photograph of the prototype 2 X-band conical monopole antenna is shown in Figure 4.22 and the corresponding simulation model is shown in Figure 4.23. The design parameters have been chosen to allow the element to be arranged either as a single element or in an array of conical monopoles over a ground plane. The height of the cone is 0.635 cm (0.25 inches), and the top truncation diameter is 0.508 cm (0.2 inches). The base of the cone is 0.254 cm (0.1 inches). The full cone angle for this design is 22.6°. At the lowest frequency 8 GHz, the wavelength is 3.75 cm (1.48 inches) and at the highest frequency 12 GHz the wavelength

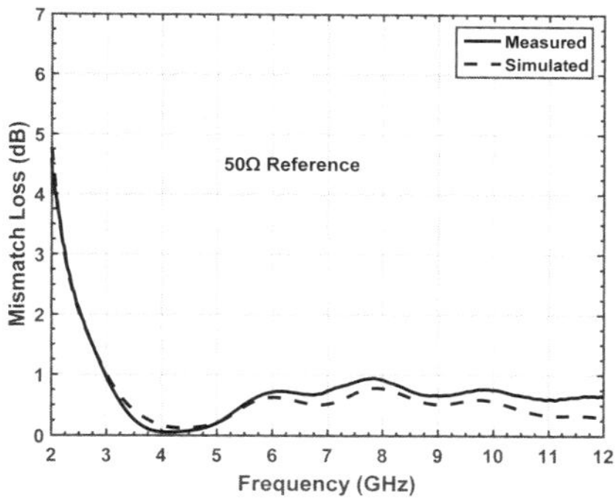

Figure 4.21 Simulated and measured mismatch loss for a conical monopole over a circular ground plane, fed with a coaxial line (prototype 1).

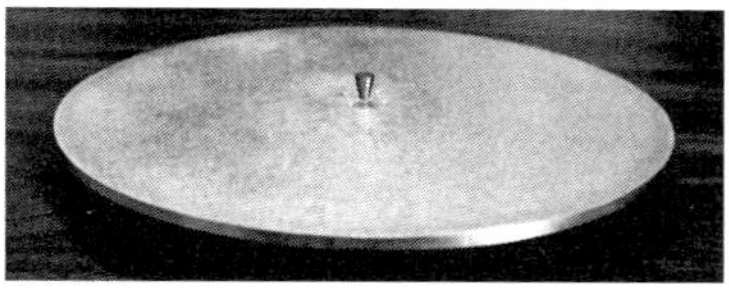

Figure 4.22 Photograph of an X-band conical monopole over a circular ground plane (prototype 2).

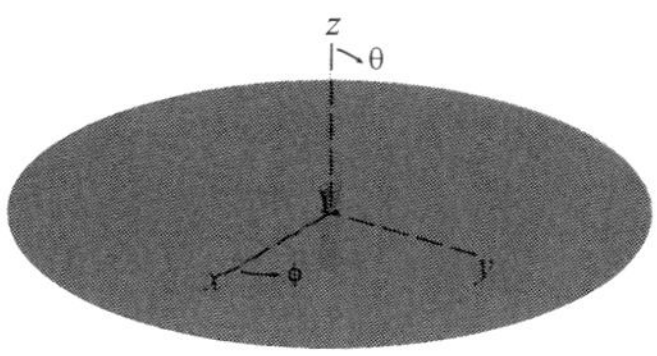

Figure 4.23 Simulation model for an X-band conical monopole over a circular ground plane (prototype 2).

is 2.5 cm (0.984 inches). The top truncation diameter varies from 0.135λ at 8 GHz to 0.2λ at 12 GHz. The height of the conical monopole varies from 0.169λ at 8 GHz to 0.25λ at 12 GHz. It should be noted that the conical monopole size in this design does not fit the simple model that could be used for electrically larger cones, that is, the characteristic impedance, reflection coefficient, and VSWR shown in Figures 4.9 to 4.11, respectively, would not apply. Instead, an initial estimate of performance can be determined from measured input impedance data for finite cones provided by Brown and Woodward [15]. In terms of electrical degrees, the height of the conical monopole varies from 60.8° at 8 GHz, to 76.2° at 10 GHz, to 90° at 12 GHz.

From the measured Brown and Woodward data in Figures 4.7 and 4.8, the estimated input impedance for this cone design varies from about $30 + j10$ ohms at 8 GHz, to $70 + j40$ ohms, at 10 GHz, to $120 + j60$ ohms at 12 GHz. The corresponding estimated reflection coefficient and VSWR for this case (assuming a 50-ohm feedline) are computed from Equations (1.15) and (1.27), respectively, as (-11.1 dB, 1.77) at 8 GHz, (-9 dB, 2.1) at 10 GHz, and (-5.8 dB, 3.1) at 12 GHz.

A comparison of the prototype 2 measured and simulated input impedance over a 15.24 cm (6 inch) diameter ground plane is shown in Figures 4.24 (input resistance) and 4.25 (input reactance), and good agreement is observed. Over the 8 GHz to 12 GHz band, the input resistance is

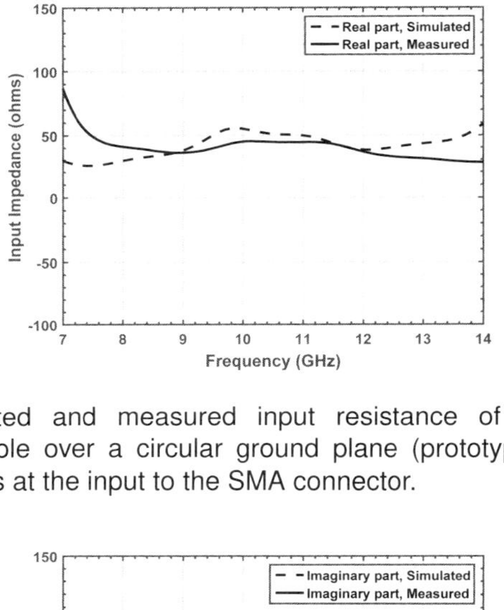

Figure 4.24 Simulated and measured input resistance of an X-band conical monopole over a circular ground plane (prototype 2). The reference plane is at the input to the SMA connector.

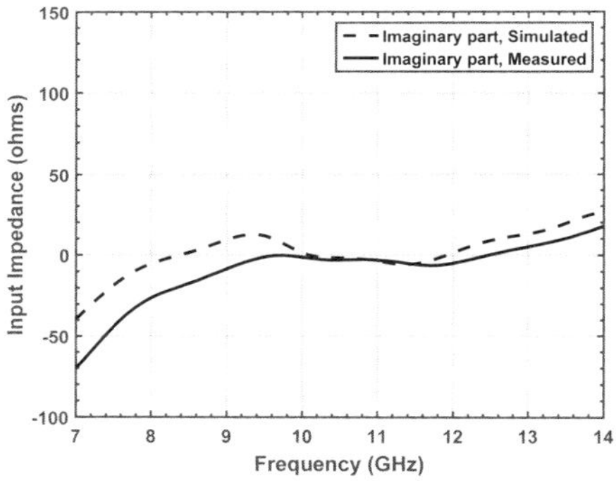

Figure 4.25 Simulated and measured input reactance of an X-band conical monopole over a circular ground plane (prototype 2). The reference plane is at the input to the SMA connector.

in the range of about 35 ohms to 45 ohms and the input reactance is in the range of about −25 ohms to 0 ohms. The measured and simulated reflection coefficients were computed from the input impedance using a 50-ohm reference and good agreement is observed as shown in Figure 4.26. The reflection coefficient magnitude is less than −10 dB over the 8 GHz to 12 GHz band. In Figure 4.27, the measured VSWR is observed to be less than 1.75 over the 8 GHz to 12 GHz band, and in Figure 4.28 the mismatch loss is less than 0.4 dB.

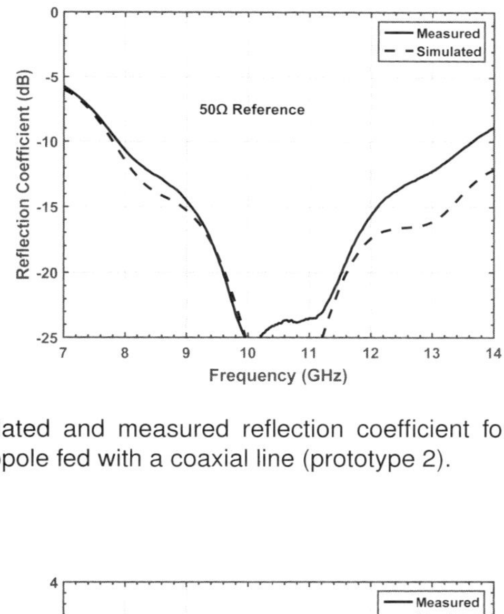

Figure 4.26 Simulated and measured reflection coefficient for an X-band conical monopole fed with a coaxial line (prototype 2).

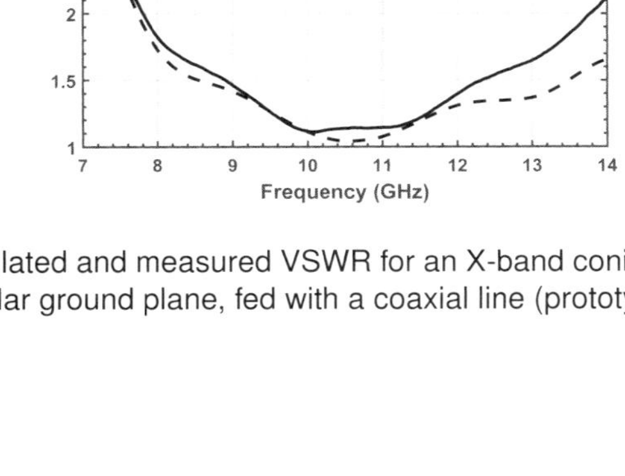

Figure 4.27 Simulated and measured VSWR for an X-band conical monopole over a circular ground plane, fed with a coaxial line (prototype 2).

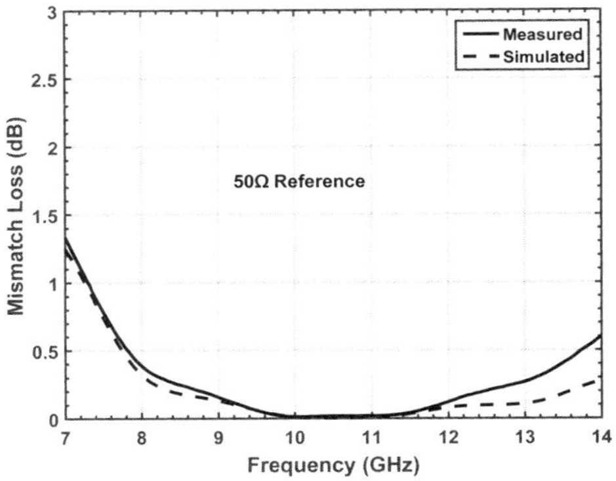

Figure 4.28 Simulated and measured mismatch loss for an X-band conical monopole over a circular ground plane, fed with a coaxial line (prototype 2).

4.2.5 Prototype 3: Monocone Array for X-Band, 8 GHz to 12 GHz

The prototype 2 conical monopole was mounted in a four-element linear array (referred to as prototype 3) with 1.27 cm (0.5 inches) element spacing, centered on a 15.24 cm (6 inch) square aluminum ground plane. At 12 GHz, the array electrical spacing is 0.508λ. A close-up view of the prototype 3 array is shown in Figure 4.29. The measured and simulated input impedance for a central element of the prototype 3 array is shown in Figure 4.30. At 8 GHz, the measured input resistance is approximately 75 ohms and then quickly settles to around 50 ohms at 8.5 GHz and remains close to that value up to 12 GHz. The simulated input resistance is approximately 35 ohms at 8 GHz, and remains close to 50 ohms over the rest of the frequency band. The measured and simulated input reactance is close to 0 ohms over the entire band except for the measurements near 8.25 GHz where the value is approximately -20 ohms. As shown in Figure 4.31, the measured and simulated mismatch loss are in good agreement, and the mismatch loss is less than 0.25 over the 8 GHz to 12 GHz band.

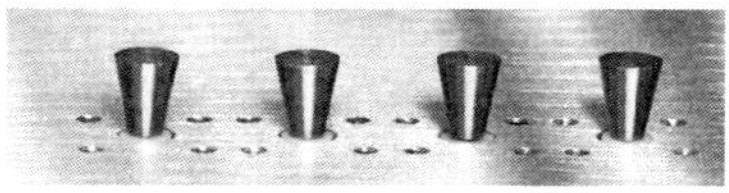

Figure 4.29 Photograph of a 4-element X-band conical monopole linear array (prototype 3).

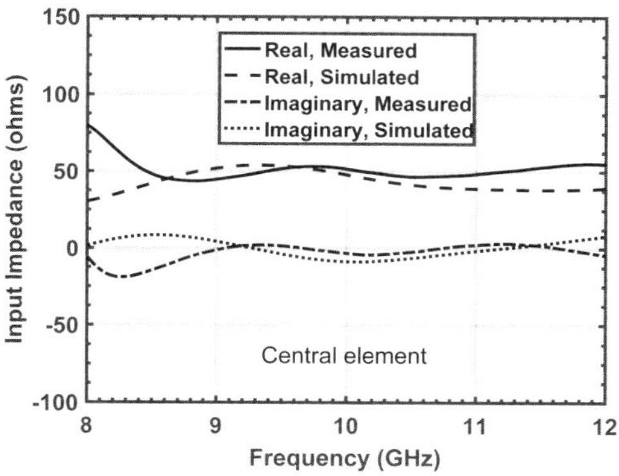

Figure 4.30 Measured and simulated input impedance of a central element in a 4-element X-band conical monopole linear array over a ground plane (prototype 3).

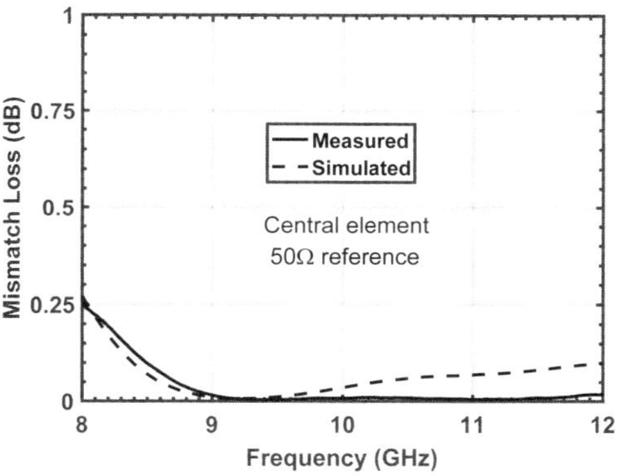

Figure 4.31 Measured and simulated mismatch loss of a central element of the 4-element X-band conical monopole linear array over a ground plane (prototype 3).

4.3 Wire Discone Antenna with Cylindrical Monopole

The previous sections discussed cone antennas and ground planes made of a solid material. For lower frequencies such as VHF, where the cone antenna and ground plane become physically large and heavy, it is desirable to approximate the cone and ground plane geometries with metallic wires that form an outline of the cone and ground plane. An omnidirectional vertically-polarized discone [28–35] antenna design involves a cone fed against a circular disc, where the disc diameter is similar in size to the diameter of the cone. Discone antennas are designed in a number of ways such as a cone with disc, cone with disc and tuning coil, cone and disc with tuning coil and thin monopole (whip) antenna. The tuning coil helps to achieve a lower frequency response and can consist of a helical conductor connecting the disc section and the monopole. In the example shown here, the discone antenna uses a large diameter monopole instead of a thin whip, and a tuning coil is not used. An example of a modified wire discone antenna for VHF operation is depicted in a simulation model shown in Figure 4.32. A photograph of the fabricated wire discone antenna integrated with the cylindrical monopole that was mounted on a vertical mast is shown in Figure 4.33. A close-up view of the antenna feed region is depicted in Figure 4.34, which shows the lower part of the cylindrical monopole, the disc wires, upper hub, insulator, lower hub, and cone wires. The antenna was designed to have a VSWR less than 2.5 over the frequency band 150 MHz to 230 MHz. In this design, there are eight (8) radial wires that approximate a disc and eight (8) radial wires that approximate a cone. The radial disc wires are 27.305 cm long and 0.368 cm diameter, and the radial cone wires are 38.1 cm long and 0.48 cm diameter. The cylindrical monopole is 34.29 cm long and 2.86 cm diameter. These lengths were determined by simulations with the FEKO moment method solver. At the high frequency, 230 MHz, the wavelength is 1.3m so the radial disc wires are 0.21λ, the radial cone wires are 0.29λ, and the monopole is 0.26λ. The simulated and measured voltage standing wave ratio (VSWR) of the wire discone with cylindrical monopole antenna is shown in Figure 4.35 and good agreement is observed.

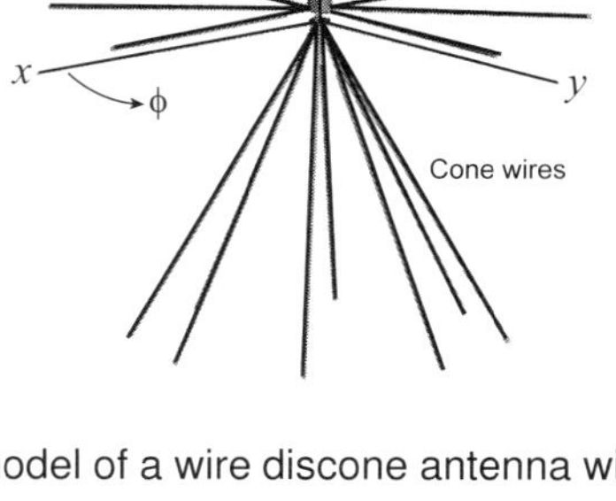

Figure 4.32 Simulation model of a wire discone antenna with a cylindrical monopole.

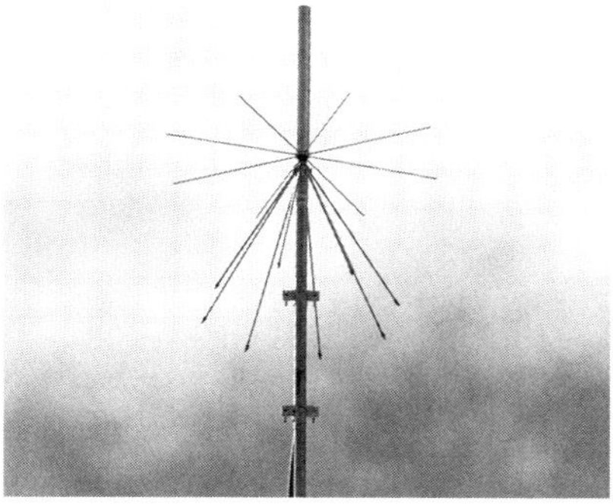

Figure 4.33 Photograph of a wire discone antenna with a cylindrical monopole.

Figure 4.34 Photograph showing a close-up view of the feed region of the wire discone antenna.

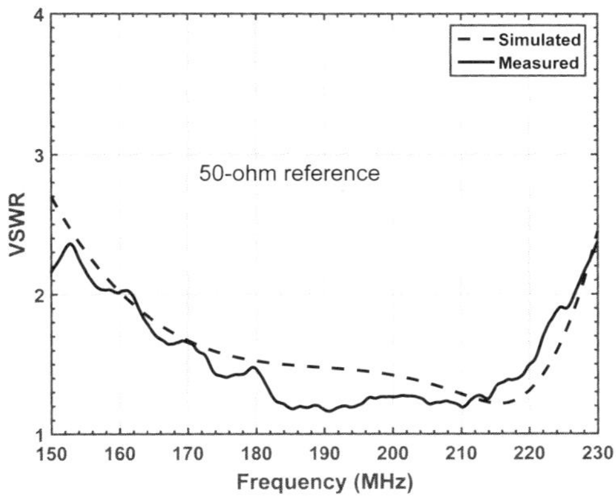

Figure 4.35 Comparison of the simulated and measured VSWR for a wire discone antenna with a cylindrical monopole.

4.4 Monopole Array for Simultaneous Transmit and Receive

This section describes an antenna array composed of an omnidirectional 8-element transmit array and a central receive antenna, for operation in the

2.4 GHz to 2.5 GHz Industrial, Scientific, and Medical (ISM) radio band. A diagram of this array is shown in Figure 4.36. This array has been developed for simultaneous transmit and receive (STAR) applications for wireless communications that require high isolation between transmit and receive functions [36, 37]. The transmit antenna array has eight monopole radiating antenna elements uniformly spaced on a ring. The transmit monopole antenna elements are approximately one-quarter wavelength long at the center frequency and they are oriented perpendicular to an electrically conducting circular ground plane. The transmit monopole array elements are uniformly spaced approximately one-half wavelength apart, and they are each driven at a position approximately one-quarter wavelength in front of a central electrically conducting circular cylinder. Referring to the phasing diagram in Figure 4.37, the transmit monopoles are phased with a linear progressive phase shift between elements to cover 0 to 360° around the ring array, such that opposing transmit elements are 180° out of phase creating a near-field null on the axis of the ring array. With this type of linear progressive phase distribution, the transmit array far-field radiation pattern has an omnidirectional (constant) magnitude in azimuth. It is known from the literature that a ring array antenna with a linear phase progression covering a full 360°s around the array circumference will produce an omnidirectional radiation pattern, as described by Rudge [38]. In the case where there is an even number of elements in the ring array and each opposing pair is fed anti-phase or with a 180° phase shift, there will be a radiation pattern null formed at the midpoint of the array. This ring array midpoint nulling effect is recognized for monopole arrays and other types of antennas with application to simultaneous transmit and receive. A ring array of four progressively

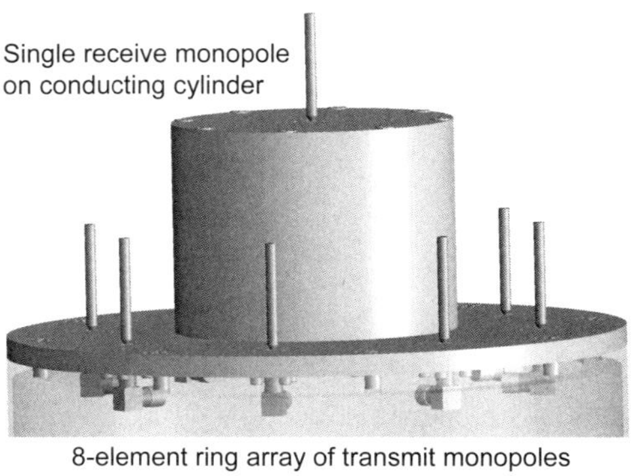

Figure 4.36 Diagram of a monopole transmit array and monopole receive antenna for simultaneous transmit and receive application.

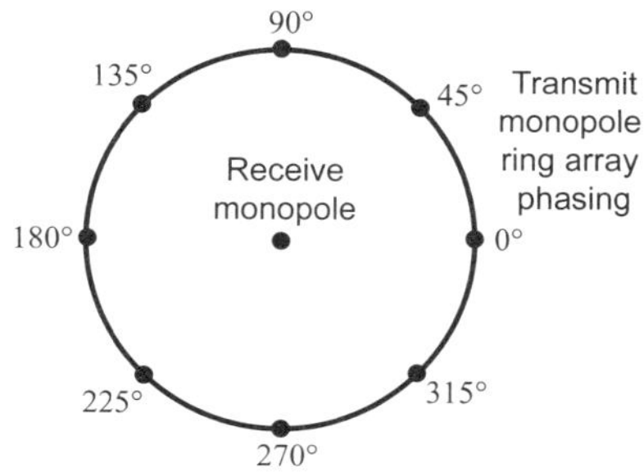

Figure 4.37 Phasing diagram for a transmit monopole ring array and monopole receive antenna for simultaneous transmit and receive application.

phased (0°, 90°, 180°, 270°) dipole antenna elements and a central dipole for improved isolation has been investigated by simulations and measurements by Chiang [39].

In the antenna design described here, to achieve high isolation between transmit and receive antennas, the receive antenna is centrally located on the axis of the transmit ring array and it is elevated, into the shadow zone [40] of the transmit array antennas, on a circular ground plane above the transmit array. In this design, there is no direct ray path from the transmit array monopole elements and the receive monopole element, so the only mutual coupling is due to edge diffraction. The transmit array is designed to operate over a circular ground plane or within a parallel plate radial waveguide. A prototype STAR monopole antenna array has been fabricated as shown in Figure 4.38. In this design, the transmit and receive monopoles each have a

Figure 4.38 Photograph of a monopole transmit array and monopole receive antenna for simultaneous transmit and receive application.

length 3.048 cm (1.2 inches) and diameter 0.3175 cm (0.125 inches). The transmit array has eight monopole arranged in a ring array with diameter 15.68 cm (6.172 inches) that is mounted on an aluminum ground plane with diameter 21.234 cm (8.36 inches). The receive monopole is mounted on the top of a closed-end aluminum cylinder with height 7.315 cm (2.88 inches) and outside diameter 9.672 cm (3.808 inches). The transmit array beamforming was accomplished with analog components as follows. The input signal to the transmit array was divided by a 4-way power divider, and compared to a reference coaxial cable length of 0°, coaxial adapters approximating relative phases of 45°, 90°, and 135° were used. The four phase-delayed signal paths were then further divided using 180° hybrids to generate the desired progressive phase variation of 0°, 45°, 90°, 135°, 180°, 225°, 270°, 315° for the eight-element transmit ring array. The array was tested in an anechoic chamber and provides high isolation on the order of 55 to 60 dB, as described below. The measured results are summarized in Figures 4.39 to 4.42. The measured reflection coefficient at the input port of the 8-way beamformer of the transmit phased array is shown in Figure 4.39. Over the 2.4 GHz to 2.5 GHz ISM band the measured reflection coefficient magnitude is less than −12 dB. As shown in Figure 4.40, the measured mutual coupling between the monopole transmit array and the central monopole receive antenna is less than −56 dB over the 2.4 GHz to 2.5 GHz band. The measured far-field vertically polarized azimuth gain patterns of the transmit phased array are shown in Figure 4.41, which demonstrates an approximate omnidirectional characteristic. In Figure 4.42 as expected, due to the progressive phase steering, the measured far-field azimuth phase patterns for the monopole transmit array demonstrate the 0° to 360° azimuthal phase shift.

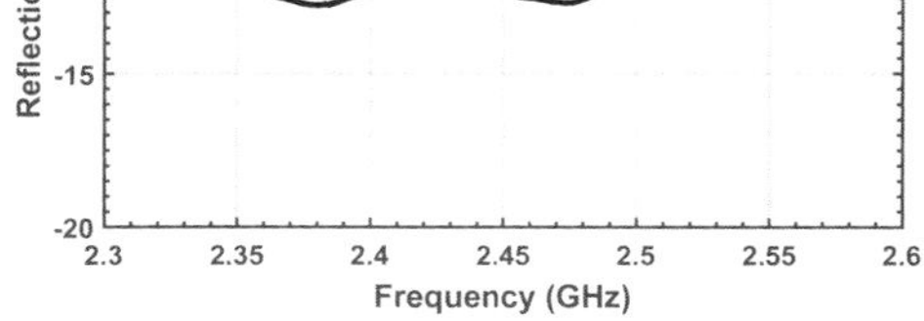

Figure 4.39 Measured reflection coefficient for a monopole transmit array for simultaneous transmit and receive application.

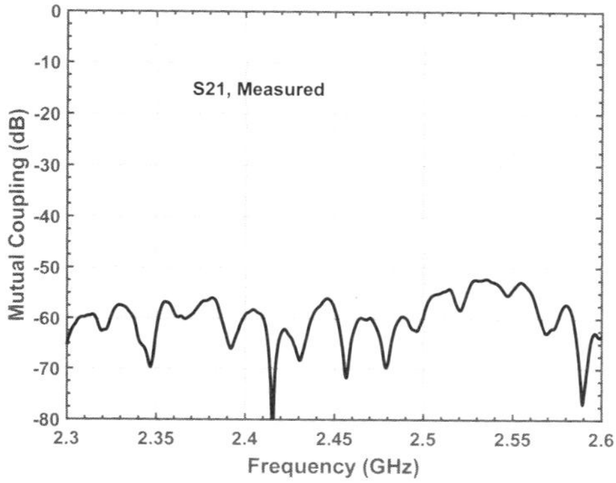

Figure 4.40 Measured mutual coupling (S21) between a monopole transmit array and monopole receive antenna for simultaneous transmit and receive application.

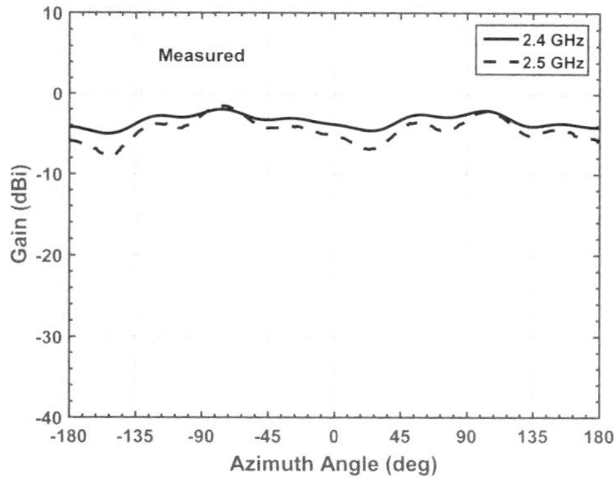

Figure 4.41 Measured far-field azimuth gain patterns for a monopole transmit array for simultaneous transmit and receive application.

4.5 Horizontal Loop and Slotted Cylinder Antennas

Omnidirectional horizontal polarization can be generated in a number of ways. For example a photograph of a horizontal uniform current loop such as an Alford loop antenna [41–47] is shown in Figure 4.43. In this photograph a standard semi-rigid coaxial cable is used to feed a two-way radial power divider implemented on a PC board with etched transmission lines on each side of the board. The spacing between the transmission lines is chosen to achieve a desired characteristic impedance. Each arm of the radial power divider feeds a V-dipole, and the two V-dipoles form an approximate uniform

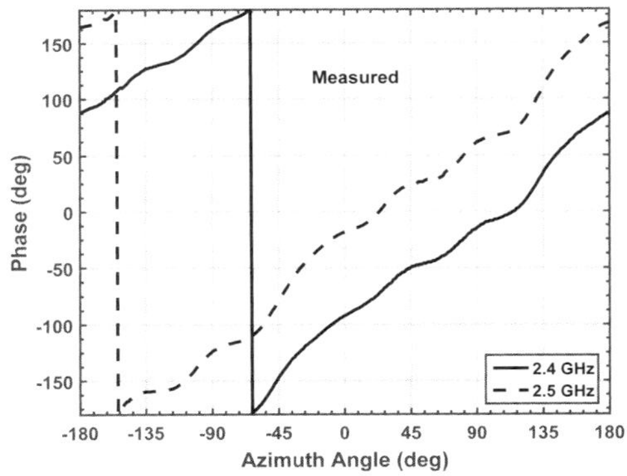

Figure 4.42 Measured far-field azimuth phase patterns for a monopole transmit array for simultaneous transmit and receive application.

Figure 4.43 Photograph of an Alford-type uniform current loop antenna.

current loop. The example shown has a square perimeter, but a circular loop is also possible. With two feeds, the typical loop perimeter length for resonance is on the order of one wavelength. Mutual coupling effects are substantial with these close proximity V-dipoles. To achieve a desired impedance match, capacitively coupled tuning stubs are included at the tips of the V-dipoles. An improved impedance match can be achieved by tapering the ends of the V-dipoles to a sharp tip. As the perimeter of the loop increases, additional feed points are required to maintain an approximate uniform current loop. For example, when the loop perimeter is on the order of two wavelengths four feedpoints are needed. Wideband versions of the horizontally polarized omnidirectional Alford loop antenna have been investigated recently by a number of authors [48–52]. For example, a bandwidth on the order of an octave (−6 dB S11 from 1.5 GHz to 3.0 GHz) has been demonstrated with a loop antenna with four feed points and parasitic elements [49].

An alternate antenna configuration surrounds the uniform current loop with a metallic cylinder with vertical slots as depicted in the artist's concept shown in Figure 4.44. The resonance of the antenna is controlled by the slot width and by the diameter and height of the cylinder as well as dielectric loading from any radome covering the antenna. Typical dimensions for resonance (determined empirically) are cylinder height $H = 0.425\lambda$, cylinder diameter $D = 0.35\lambda$, and slot width $W = 0.04\lambda$. With either the uniform current loop or the loop-fed slot cylinder, the radiation pattern of an embedded element remains omnidirectional in the presence of other like elements in a planar phased array configuration.

In applications, where additional gain is required in the vertical dimension it is possible to stack slotted cylinder antennas in a co-linear array fashion [53, pp. 197–211]. Figure 4.45 depicts a diagram of an example UHF two-element co-linear array of slotted cylinder antennas with three slots per element. The slots can be driven with a uniform current loop or, in this case, directly with coaxial feed cables. To feed the two sets of three slots, two coaxial cables can be routed along the central axis of the slotted cylinders, and then each of the coaxial cables can be connected to three-way power dividers that provide the necessary transmission lines to feed the three slots. Each slot can then be fed by connecting the outer conductor of the coaxial line to one side of the slot and the center pin to the other side of the slot. An alternate

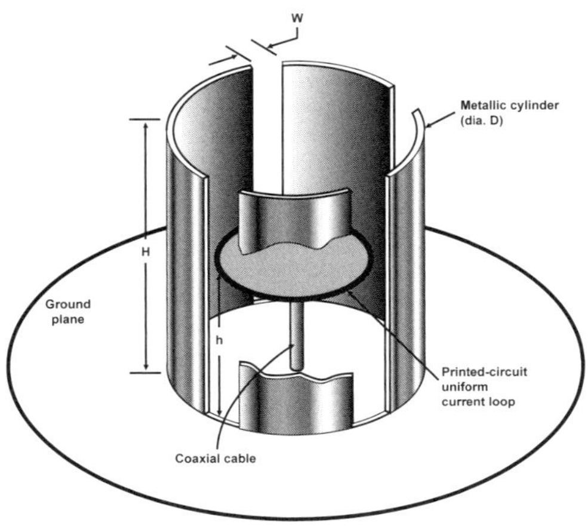

Figure 4.44 Artist's concept of a loop-fed slotted cylinder antenna.

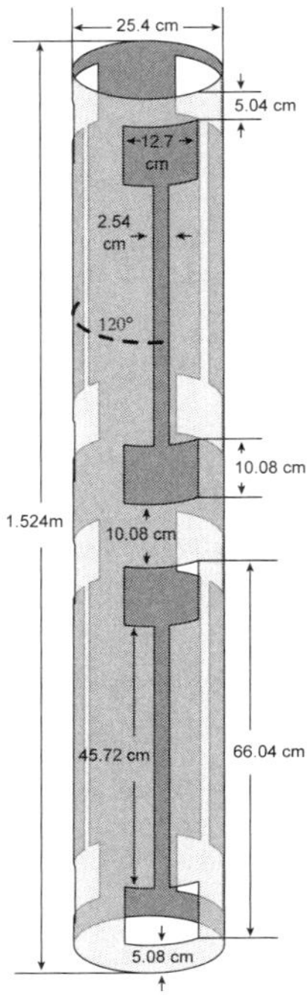

Figure 4.45 Drawing of a UHF two-element colinear array of horizontally polarized slotted cylinder antennas.

approach to feeding the slots is to use a transformer balun. A photograph of the UHF two-element colinear array of horizontally polarized slotted cylinder antennas is shown in Figure 4.46. The measured impedance mismatch loss is shown in Figure 4.47. The measured horizontally polarized omnidirectional gain patterns are shown in Figure 4.48.

4.6 Turnstile Antenna with Crossed V-Dipoles

The turnstile antenna [54] was discussed briefly in Chapter 1 (Section 1.13.2) where crossed Hertzian (short) dipoles with a 90° phase shift between the arms was used to generate an isotropic and omnidirectional radiation pattern. In this section, a practical turnstile dipole antenna is described. A wideband UHF dual-polarized dipole antenna with swept-back arms (V-dipole) and folded balun over a ground plane in free space was described in Chapter 3 (refer to Figure 3.8), and the same design is analyzed here as a turnstile antenna. The simulation model is shown in Figure 4.49. The dipole arm length was chosen as 13.4 cm, which is 0.268λ at 600 MHz (center frequency). The

Figure 4.46 Photograph of a UHF two-element colinear array of horizontally polarized slotted cylinder antennas.

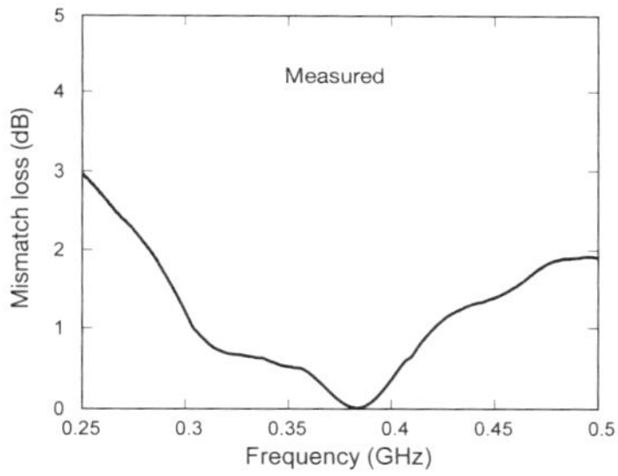

Figure 4.47 Measured impedance mismatch loss of the UHF two-element colinear array of horizontally polarized slotted cylinder antennas.

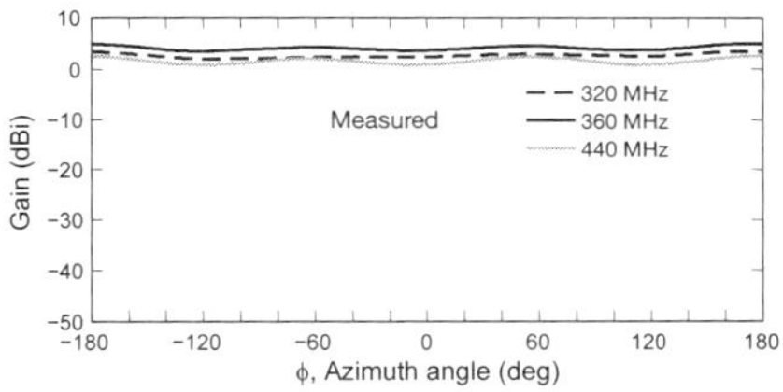

Figure 4.48 Measured horizontally polarized omnidirectional gain pattern of the UHF two-element colinear array of slotted cylinder antennas.

dipole arms and balun sections are modeled as 1.42 cm diameter (d and D in Figure 3.2) perfectly conducting tubes. The dipole height is chosen as 17.9 cm (H in Figure 3.2) so that at the center frequency, the height at the midpoint of the dipole arm is approximately $\lambda/4$ over the ground plane. A circular ground plane with diameter 45.72 cm is used here. The folded balun tubes are spaced 3.81 cm center to center (s in Figure 3.2), which by Equation (3.1)

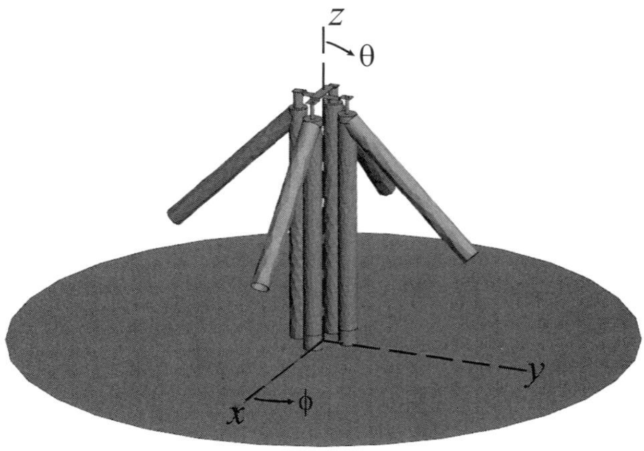

Figure 4.49 Simulation model for a UHF turnstile antenna composed of crossed V-dipoles over a ground plane.

provides a 197Ω characteristic impedance transmission line. The folded balun provides mechanical support of the dipole arms and is short circuited at the ground plane. Electrically, this balun behaves as a purely reactive impedance in parallel with the dipole feed terminals. One of the balun tubes contains a coaxial line that is connected to the dipole terminals using a crossover connection located 1.27 cm (h in Figure 3.2) above the top of the folded balun. The two ports are fed with a relative phase difference of 90°. Figure 4.50 shows the azimuth gain patterns in the xy plane at 450 MHz for vertical and horizontal polarizations. The horizontal component is at the -5 dBi level. If the arms were not swept back, this horizontal component would be closer to the -2 dBi level. The vertical component is on the order of -12.5 dBi, and would otherwise be negligible if the arms were not swept back.

4.7 Summary

This chapter has described simulations and measurements of antennas that are omnidirectional in azimuth. Omnidirectional coverage can be provided for vertical polarization with vertical monopole or dipole antennas which can typically have a cylindrical or conical shape. Similarly, a dipole antenna with cone-shaped arms is referred to as a bicone or biconical antenna. A monopole ring array with progressive phasing can generate omnidirectional far-field radiation patterns for simultaneous transmit and receive. Omnidirectional horizontally polarized coverage can be achieved with horizontal loops or slotted cylinder antennas. The next chapter also discusses loop antennas, but

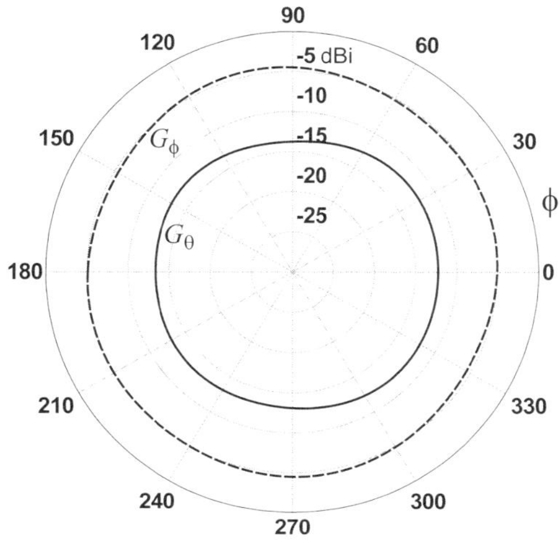

Figure 4.50 Simulated azimuth gain patterns at 450 MHz for a UHF turnstile antenna composed of crossed V-dipoles over a ground plane.

in a configuration of multiple orthogonal loops as a vector sensor antenna for direction finding application.

4.8 PROBLEM SET

4.1 Write computer code to implement Equations (4.3) to (4.8) to determine the input impedance of a finite conical monopole over an infinite ground plane, which is valid for cones with half-angle θ_o approximately 30° or greater. Show that for $\theta_o = 30°$ and ka values ranging from 0 to 8, where $k = 2\pi/\lambda$ and a is the cone slant length, that the finite cone input resistance approaches the infinite cone characteristic impedance 79 ohms (based on Equation (4.10)) and the input reactance approaches 0 ohms as observed by Papas and King [10, Figure 4].

References

[1] J.D. Kraus, *Antennas*. New York: McGraw-Hill, 1950.

[2] S.A. Schelkunoff and H.T. Friis, *Antennas: Theory and Practice*. New York: Wiley, 1952.

[3] W.L. Stutzman and G.A. Thiele, *Antenna Theory and Design, 2nd ed.* New York: Wiley, 1998.

[4] J. Volakis, ed., *Antenna Engineering Handbook, 4th ed.* New York: McGraw-Hill, 2007.

[5] C.A. Balanis, *Antenna Theory, Analysis and Design, 3rd ed.* New York: Wiley, 2005.

[6] H.J. Reich, ed., *Very High Frequency Techniques, vol. 1*, New York: McGraw-Hill, 1947.

[7] S.A. Schelkunoff, *Advanced Antenna Theory,* New York: Wiley, 1952.

[8] S.A. Schelkunoff, "General Theory of Symmetric Biconical Antennas," *J. of Applied Physics,* vol. 22, no. 11, Nov. 1951, pp. 330–332.

[9] C.T. Tai, "On the Theory of Biconical Antennas," *J. Appl. Phys.,* vol. 19, Dec. 1948, pp. 1155–1160.

[10] C. H. Papas and R. King, "Input Impedance of Wide-Angle Conical Antennas Fed By a Coaxial Line," *Proc. I.R.E.* , vol. 39, Nov. 1949, pp. 1269–1271.

[11] C.H. Papas and R. King, "Input Impedance of Wide-Angle Conical Antennas," Harvard University, Cruft Laboratory, Technical Report No. 52, Dec. 1, 1948.

[12] C.H. Papas and R.W.P. King, "Radiation from Wide-Angle Conical Antennas Fed by a Coaxial Line," *Proc. I.R.E.,* vol. 39, Jan 1951, pp. 49–51.

[13] P.D.P. Smith, "The Conical Dipole of Wide Angle," *J. Appl. Physics,* vol. 19, January, 1948, pp. 11–23.

[14] P.D.P. Smith, "Ground Plane Field of the Wide Angle Conical Dipole," J. Appl. Phys., vol. 20, p. 636

[15] G.H. Brown and O.M. Woodward, Jr., "Experimentally Determined Radiation Characteristics of Conical and Triangular Antennas," RCA Review, vol. 13, December, 1952, pp. 425–452.

[16] S. A. Schelkunoff, "Theory of Antennas of Arbitrary Size and Shape," Proc. I.R.E., vol 29, April 1941, pp. 493–521.

[17] C.T. Tai, "Application of a Variational Principle to Biconical Antennas," *J. Appl. Phys.,* vol. 20, 1948, pp. 1076–1084.

[18] D. N. Black and T. A. Brunasso, "An Ultra Wideband Bicone Antenna," *IEEE International Conference on Ultra-Wideband,* 2006, 15 January, 2007.

[19] J.E. Lawrance, "The Fundamental Resonant Frequency and Radiation Characteristics of Wide Angle Conical Antennas," Thesis, Univ. of New Mexico, 2010.

[20] M. Gardill, G. Fischer, R. Weigel, A. Koelpin, "Design of a Robust and Low-Cost Monocone Antenna Element for Use in Vehicle Roof-Mounted Antenna Arrays," *Radio and Wireless Symposium,* 2013, pp. 112–114.

[21] A.J. Fenn, "Theoretical and Experimental Study of Monopole Phased Array Antennas," *IEEE Trans. Antennas and Propagation*, vol. 34, no. 10, 1985, pp. 1118–1126.

[22] G.H. Brown and O.M. Woodward, Jr., "Experimentally Determined Impedance Characteristics of Cylindrical Antennas," *IRE Proc.*, vol. 33, 1945, pp. 257–262.

[23] C.M. Keller, and D.P. Young, "Ultra-Wideband (UWB) Signal Localization Using a

Vehicle-Sized Array," *2005 IEEE International Conference on Ultra-Wideband*, 2005, pp. 290–295.

[24] R.H.T. Bates and G.A. Burrell, "Towards Faithful Radio Transmission of Very Wide Bandwidth Signals," *IEEE Transactions on Antennas and Propagation*, vol. 20, no. 6, November 1972, pp. 684 – 690.

[25] J.G. Maloney, B.L. Shirley, and G.S. Smith, "Physical Description for the Reception of Short Pulses by Antennas," *1994 Antennas and Propagation Society International Symposium, AP-S. Digest*, vol. 3, 1994, pp. 1790-1793.

[26] J.D. Kraus, Antennas, 2nd Edition, New York: McGraw-Hill, 1988, p. 355.

[27] FCC 02-48, First Report and Order, Revision of Part 15 of the Commission's Rules Regarding Ultra-Wideband Transmission Systems, Federal Communications Commission, Adopted February 14, 2002, Release date April 22, 2002.

[28] A.G. Kandoian, "Three New Antenna Types and Their Applications," *Proc. IRE*, vol. 34, Feb. 1946, pp. 70W-75W.

[29] R.A. Burberry, *VHF and UHF Antennas,* London: Peter Perigrinus, Ltd., 1992.

[30] M. Dionigi, M. Mongiardo, C. Tomassoni, Investigation on the Phase Center of Ultra Wideband Discone Antennas, *German Microwave Conference Digest of Papers,* 2010 pp. 59–62.

[31] W. Yan, A. Chen, T. Jiang, "Design of UHF Miniature Discone Antenna," *Proc. 9th Int. Symp. on Antennas, Propagation and EM Theory,* 2010, pp. 356–358.

[32] A. Chen, T. Jiang, Z. Chen, D. Su; Wenxuan Wei; Yanjun Zhang "A Wideband VHF/UHF Discone-Based Antenna," *IEEE Antennas and Wireless Propagation Letters,* vol. 10, 2011, pp. 450–453,

[33] S. Verma, A. Mishra, R. Khan, "Analysis of Variation of Various Parameters on Design of Discone Antenna," *IEEE Applied Electromagnetics Conference (AEMC)*, 2011, pp. 1–6.

[34] R. Goncalves, P. Pinho, N.B. Carvalho, "Design and Implementation of a 3D Printed Discone Antenna for TV Broadcasting System," *IEEE Int. Symp. on Antennas and Propagation & USNC/URSI National Radio Science Meeting,* 2015, pp. 314–315.

[35] Y. Zhao, W. Wang, "Design of a Novel Broadband Skeletal Discone Antenna With a Compact Configuration," *IEEE Antennas and Wireless Propagation Letters,* vol. 13, 2014, pp. 1725–1728.

[36] A.J. Fenn, P.T. Hurst, J.S. Herd, K.E. Kolodziej, L.I. Parad, H. Steyskal, "Simultaneous Transmit and Receive Antenna System", United States Patent No. 8,749,441, June 10, 2014.

[37] K.E. Kolodziej, P.T. Hurst, A.J. Fenn, L.I. Parad, "Ring Array Antenna with Optimized Beamformer for Simultaneous Transmit and Receive," *IEEE Int. Symp. on Antennas and Propagation,* Chicago, Illinois, July 2012, pp. 1–2.

[38] A.W. Rudge, *Handbook of Antenna Design,* London: Peter Peregrinus Ltd, 1983, vol. 2, pp. 305–329.

[39] B.A. Chiang, "A Study of Communication Antenna Isolation," Final Report FAA-RD-73-94, Howard University, Washington, DC, June 1973, AD768848.

[40] R.G. Kouyoumjian and P.H. Pathak, "A Uniform Theory of Diffraction for an Edge in a Perfectly Conducting Surface," *Proc. IEEE,* vol. 62, no. 11, Nov. 1974, pp. 1448–1461.

[41] Alford, A. and A. G. Kandoian, "Ultrahigh-Frequency Loop Antennas," *Trans. AIEE,* vol. 59, 1940, pp. 843–848.

[42] S.R. Anderson, "VHF Omnirange Accuracy Improvements," *IEEE Trans. Aerospace and Navigational Elec.*, vol. 12, no. 1, 1965, pp. 26–35.

[43] D.L. Sengupta, "Theory of V.O.R. Antenna Radiation Patterns," *Electron. Lett.,* vol. 7, no. 15, 1971, pp. 418–420.

[44] Gupta, R.K., "On Radiation Properties of the Alford Loop and a Dipole Antenna," *Proc. Inst. Elec. Eng. India*, 1971, pp. 145–150.

[45] A.J. Fenn, "Arrays of Horizontally Polarized Loop-Fed Slotted Cylinder Antennas," *IEEE Trans. Antennas Propagat.*, vol. 33, no. 4, 1985, pp. 375–382.

[46] T.A. Milligan, *Modern Antenna Design, 2nd ed.* Hoboken, NJ: Wiley-IEEE Press, 2005

[47] A.J. Fenn, *Adaptive Antennas and Phased Arrays for Radar and Communications.* Norwood, Mass.: Artech House, 2008.

[48] Y. Yu, F. Jolani, and Z. Chen, "A Wideband Omnidirectional Horizontally Polarized Antenna for 4G LTE Applications," *IEEE Antennas and Wireless Propagation Letters,* vol. 12, 2013, pp. 686–689.

[49] Q. Zhu, S. Yang, and Z. Chen, "A Wideband Horizontally Polarized Omnidirectional Antenna for LTE Indoor Base Stations" *Microwave and Optical Technology Letters*, vol. 54, 2015, pp. 2112–2116.

[50] X.L. Quan, R.L. Li, J.Y. Wang, and Y.H. Cui, "Development of a Broadband Horizontally Polarized Omnidirectional Planar Antenna and its Array for Base Stations," *Progress In Electromagnetics Research*, vol. 128, 2012, pp. 441–456.

[51] X. Zhang, X. Zhong, Z. Liu, and B. Li, "A Horizontally Polarized Omnidirectional Antenna with a Reflector for Ceiling-Mounted Indoor Applications," *Progress In Electromagnetics Research Letters,* vol. 54, 2015, pp. 27–32.

[52] S. Zhang and M. Ye, "A Wideband Horizontally Polarized Omnidirectional Antenna with Coupling Lines," *Progress In Electromagnetics Research Letters,* vol. 62, 2016, pp. 111–116.

[53] A.J. Fenn and P.T. Hurst, *Ultrawideband Phased Array Antenna Technology for Sensing and Communications Systems*, Cambridge, MA: The MIT Press, 2015.

[54] G. Brown, "The Turnstile Antenna," *Electronics*, 1936, pp. 14–17.

5

Vector Sensor Array Antennas

5.1 Introduction

Vector sensor array antenna systems can be used to measure multiple polarization components of an incident electromagnetic field for direction finding [1–7]. Polarization diversity and spatial diversity offered by sparsely spaced multipolarized array elements can be useful in providing frequency reuse capability and in mitigating signal dropouts due to multipath or polarization mismatch for providing reliable data communications. Vector sensor array antenna systems provide the capability for receiving and processing linearly polarized, circularly polarized, elliptically polarized, partially polarized, and unpolarized fields. A vector sensor antenna that uses multipolarized diverse elements with diverse antenna gain patterns can be calibrated for determining the angle of arrival for radiofrequency sources. A tripolarized antenna with three mutually perpendicular dipoles has been investigated for adaptive nulling applications [8]. This chapter describes the electromagnetic design of a vector sensor HF antenna array [9] that could be integrated on a small satellite referred to in the literature as a nanosatellite or CubeSat. A potential future application for this technology is for performing radio astronomy mapping missions [9–17]. The antenna array described here could be applied to other platforms such as airborne vehicles, towers, ground vehicles, and balloon-based applications.

This chapter is organized as follows. In Section 5.2 some examples of vector sensor array prototype antennas are given. In Section 5.3, the operating modes of vector sensor antennas are described. In Section 5.4, vector sensor antenna design and simulated radiation patterns are given for potential application to celestial RF source mapping at HF from a nanosatellite. Section 5.5 describes the polarization response for a full six-mode vector sensor array antenna. Section 5.6 reviews the technique of source mapping from vector sensor time-series data. Section 5.7 provides a summary.

5.2 Background

There are various implementations of a full six-mode vector sensor antenna that include three orthogonal loop modes and three orthogonal dipole modes. In some applications with structural limitations, only a partial vector sensor antenna with less than six modes is practical – in this case multiple partial vector sensor antennas can be used to implement all six modes. Examples of full and partial vector sensor antenna prototypes are described below.

5.2.1 Prototype 1: Vector Sensor with Three Orthogonal Rings

An example of a full vector sensor antenna for radio astronomy [11] is depicted in the photograph shown in Figure 5.1. This electromagnetic vector sensor was designed for ground-based measurements of galactic sources [11–15] within the HF and low VHF bands, and it is composed of three orthogonal rings approximately 1m in diameter. At 30 MHz, the wavelength is 10 meters so the rings are considered electrically small (0.1λ diameter) and they will have a frequency independent radiation pattern shape. In this case, the rings operate below resonance in a dual-mode fashion as a Hertzian electric loop and a Hertzian electric dipole. The rings are supported above the ground by three adjustable fiberglass legs. Each of the rings has two ports that are connected via a transformer balun and coaxial transmission lines within the

Figure 5.1 Photograph of a full vector sensor antenna composed of three orthogonal dual-mode loops. Each loop has two ports that allow dipole and loop currents to be generated.

ring's copper tubing to a sum and difference hybrid. The result is that each physical loop serves as both a dipole and a loop as will be described in Section 5.3. The full vector sensor array antenna produces six output channels allowing full characterization of the incident electromagnetic field. A similar ground-based HF vector sensor antenna system that uses three orthogonal two-port loops operating in loop and dipole modes has been developed by Meloling [18].

5.2.2 Prototype 2: Array of Partial Vector Sensor Antennas

In some cases, only partial vector sensors may be practical. An example of a sparse array of partial vector sensor antennas installed on an unmanned air vehicle (UAV) [7] with electrically conducting wings is depicted in the photograph shown in Figure 5.2. In this photograph, there are crossed loops mounted on the wing tips and they generate a dipole mode and two crossed loop antenna modes. Note that since the wings in this case are electrically conducting, a horizontal loop or dipole would be shorted out and would not be an effective receive antenna. Single loops are mounted on the tail booms and generate dipole and loop antenna modes – one of the loops is mounted horizontally and the other vertically.

5.2.3 Prototype 3: Full Vector Sensor with Vertical Crossed Loops, Uniform Current Loop, and a Monopole for a Nanosatellite

Examples of nanosatellite structures in the form of commercial CubeSats are depicted in the photograph shown in Figure 5.3. In this photograph two sizes

Figure 5.2 Photograph of a sparse array of partial vector sensor antennas installed on an unmanned air vehicle.

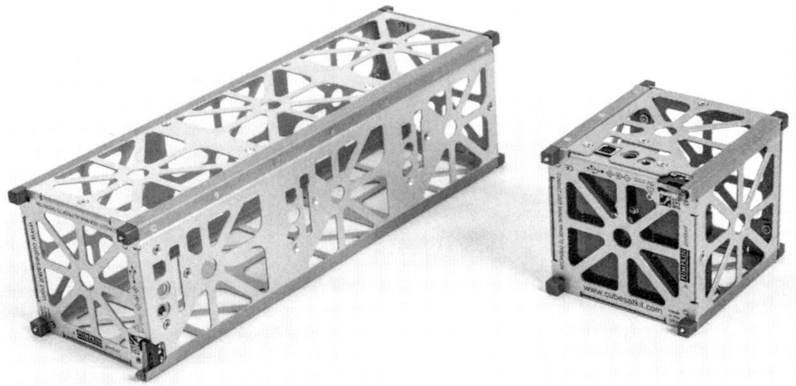

Figure 5.3 Photograph of commercial CubeSat mechanical structures. Left: 10 cm x 10 cm x 30 cm version (3U). Right: 10 cm x 10 cm x 10 cm (1U).

of CubeSats are shown. On the left is a 10 cm x 10 cm x 30 cm version (referred to as a 3U structure), and on the right is a 10 cm x 10 cm x 10 cm version (1U structure). These structures are shown in a skeletonized form where other components can be mounted. A photograph of a deployable vector sensor mechanical mockup with 3m long electrically conducting metal tape measure loops (tip-to-tip) on a 3U CubeSat structure during preliminary ground-based mechanical testing [11, 14] is shown in Figure 5.4. In this photograph, gravity-offload structural members are included to provide support to the metal tape loops as they deploy. The electromagnetic design and simulations of this prototype CubeSat vector sensor antenna are given in Section 5.4. The next section describes the vector sensor antenna modes.

Figure 5.4 Photograph of a vector sensor antenna deployed from a CubeSat 3U mockup during preliminary mechanical testing for a CubeSat application.

5.3 Vector Sensor Antenna Modes

To describe the operating modes of vector sensor antennas, consider Figure 5.5(a) that shows an example simplified partial vector sensor antenna. This antenna has the appearance of crossed electrically conducting loops operating in the presence of an electrically conducting ground plane. In general, a ground plane is not required, but in the case of the UAV in Figure 5.2 the wing acts as a ground plane and support structure. Additionally, the ground plane here will also be considered a nanosatellite structure as shown in the next section. The antenna is assumed to be electrically small in terms of wavelength and will have a frequency independent radiation pattern for each of the operating modes.

Each physical loop has four RF ports and they are driven with a modeforming (beamforming) RF circuit that generates independent dipole and loop modes. By symmetry, the crossed loops can be electrically connected at the orthogonal intersection points in this diagram, without disturbing the two modes. That is, the two modes will not couple significantly with one another. By driving the ports in a common phase direction as shown Figure 5.5(b), a dipole mode is generated from the vertically flowing currents. In this dipole mode, the currents that flow on the horizontal sections of the loop and on the ground plane are out of phase and would not contribute to the far-zone radiation pattern. In Figure 5.5(c), the ports are now driven in a phase

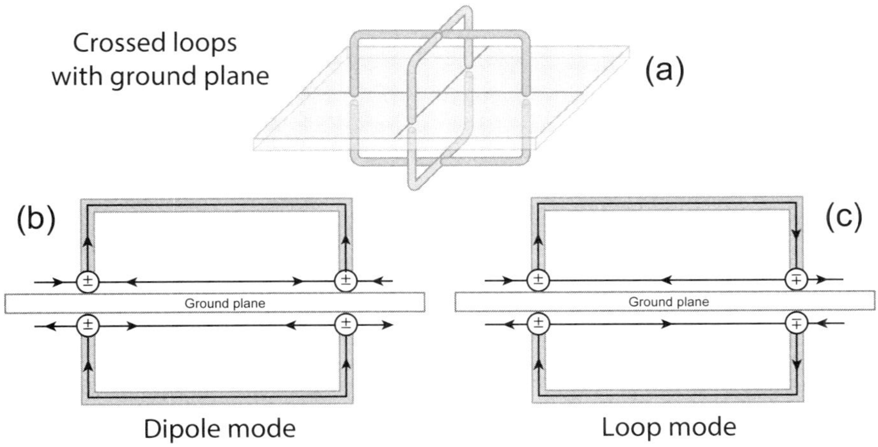

Figure 5.5 (a) Conceptual diagram showing a crossed-loop vector sensor antenna on a conducting ground plane. (b) Operating in dipole mode. (c) Operating in loop mode.

difference arrangement causing the electric currents to flow in a clockwise or loop fashion. Again, in this loop mode, currents flowing on the ground plane are out of phase and do not contribute to the far-zone radiation pattern.

In the case where the two crossed loops are driven in a common phase condition, the currents on the four vertical segments of the loops are in phase as depicted in Figure 5.6, which generates an effective dipole mode. When the loops are driven with a 180° phase difference, the loop mode is generated as shown in Figure 5.7.

5.4 Vector Sensor Antenna Design and Simulated Radiation Patterns for a Nanosatellite Application

It is known from the literature that mapping of radio frequency galactic noise-like sources can be determined by ground based large diameter antenna array measurements for frequencies above about 25 MHz [20–23]. Due to the total electron content of the ionosphere [21], radio frequency sources radiating below about 25 MHz are partially or almost completely reflected by the ionosphere. Therefore, the electromagnetic waves from galactic radiofrequency sources are partially or almost completely blocked by the earth's ionosphere up to an altitude of about 300 km. Terrestrial radio frequency emissions are a significant source of interference for ground-based low-frequency mapping sensors. The natural radio frequency

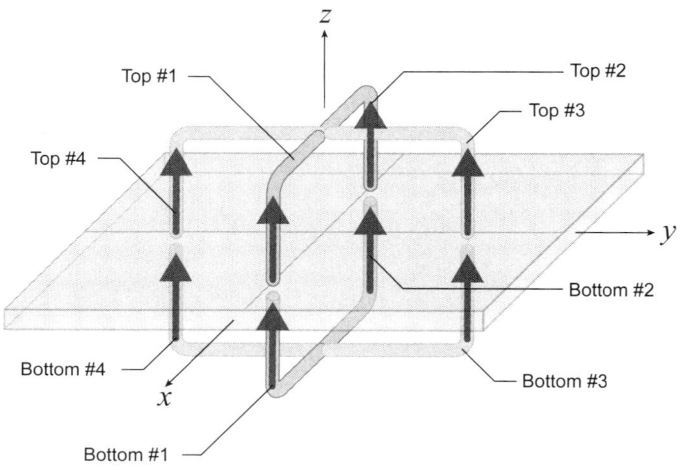

Figure 5.6 Conceptual diagram showing a crossed-loop partial vector sensor antenna operating in the dipole mode.

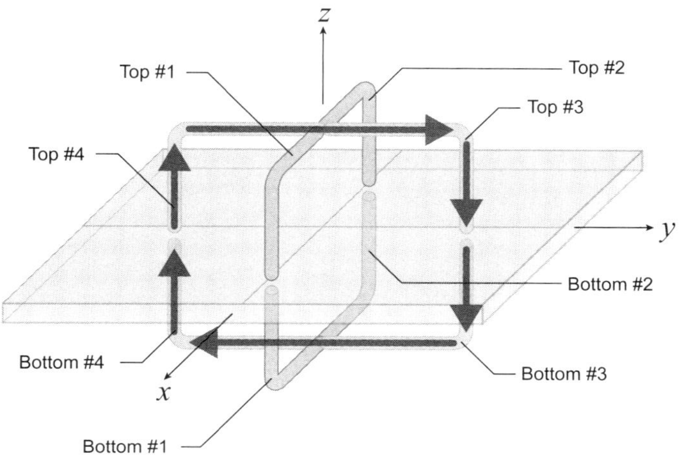

Figure 5.7 Conceptual diagram showing a crossed-loop partial vector sensor antenna operating in the loop mode.

shielding provided by the ionosphere reduces terrestrial interference that would be received by a low-frequency satellite sensor orbiting above the ionosphere. Thus, mapping of galactic RF sources below 25 MHz can best be accomplished from a spacecraft above 300 km altitude. Multipolarized vector sensor antenna systems have been explored for a variety of direction finding applications [1–7, 9–18] and these sensors when deployed on a nanosatellite in orbit above the ionosphere are an alternate approach to mapping galactic sources. Ideally, measurements from the vector sensor antenna elements should be dominated by external sources rather than by internal thermal noise due to electronic components such as low noise amplifiers (LNAs). HF systems operating as a beacon have been designed for CubeSats [19], and could potentially be used in calibrating an orbiting vector sensor array.

A full electromagnetic vector sensor antenna [1, 2] characterizes the electromagnetic field at a single location in space and with a common phase center. A full vector sensor is composed of three orthogonal dipole modes and three orthogonal loop modes. These six modes allow for a complete measurement of the electromagnetic field amplitude and phase of incoming radiation. The full six-mode vector sensor antenna design here uses the following approach for a CubeSat application as shown in Figure 5.8. The axis of the CubeSat is aligned with the z axis with the radiation patterns characterized in spherical coordinates by $\hat{\theta}$ and $\hat{\phi}$ polarizations. King has shown that a loop antenna can also operate as a dipole [24]. Loop and dipole

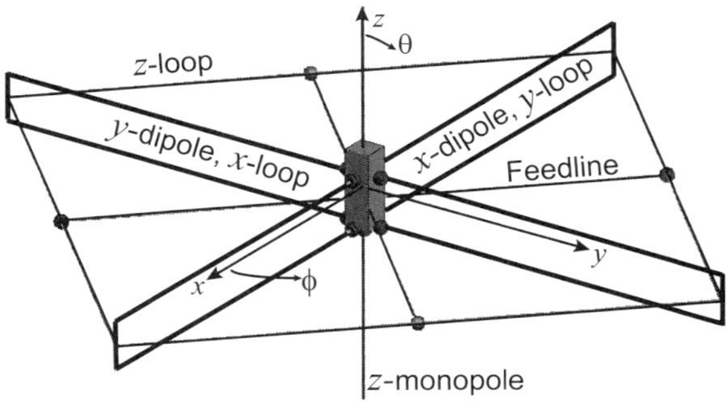

Figure 5.8 Conceptual diagram for a full vector sensor deployed from a CubeSat structure.

antenna modes are generated from crossed vertical dual mode loop elements. As shown in this diagram there are two dipole modes labeled as x-dipole (Mode 1) and y-dipole (Mode 3) that are aligned with the x and y axes, respectively. There are two loop modes labeled as x-loop (Mode 2) and y-loop (Mode 4), with the axes of the loops aligned with the x and y axes, respectively. Note that a wide bandwidth for signal to noise ratio (SNR) can be achieved for the vector sensor dipole and loop modes with active impedance matching circuitry as discussed by Robey et al. [13]. Similar active matching circuits for antennas have been described by Grubb [25], Nordholt [26], and Li [27]. The z-monopole antenna (Mode 5) is fed against the CubeSat structure and is omnidirectional with respect to the z axis. The horizontal z-loop antenna (Mode 6) is fed at four ports (by the feedlines shown) and is omnidirectional with respect to the z axis. To achieve a uniform electric current, the four ports of the horizontal loop are fed with the same positive and negative polarity with respect to rotation about the z axis. To illuminate the four ports of the horizontal loop with equal amplitude, the four feedlines are driven by a four-way power combiner.

The wire length of the vector sensor antenna arms is chosen in the range of 1.5m to 2m and keeps the antennas electrically small, below resonance, so that constant radiation pattern shapes are maintained over the desired frequency range. The four modes associated with the two crossed wire loop antennas and two crossed horizontal dipole antennas are generated by sum and difference modeformers. Figure 5.9 shows a diagram of a simplified circuit model to provide sum and difference outputs for a four-port loop attached to a CubeSat structure. The dipole currents are generated by the sum port of the hybrid component. Similarly, Figure 5.10 depicts the loop currents that are generated by the difference port of the hybrid component.

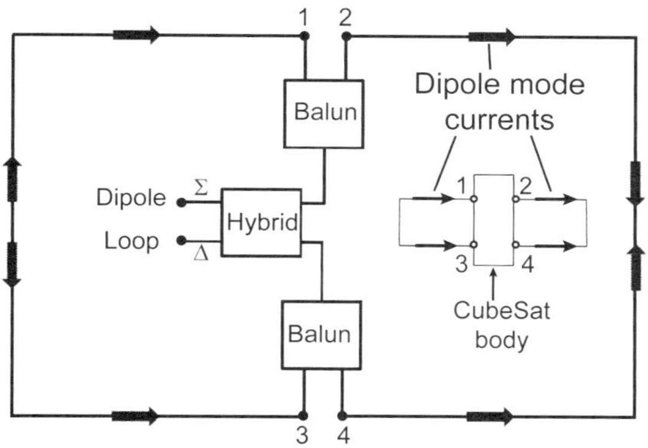

Figure 5.9 Diagram showing a simplified circuit model to provide sum and difference outputs for a four-port loop. The dipole currents are generated by the sum port of the hybrid component.

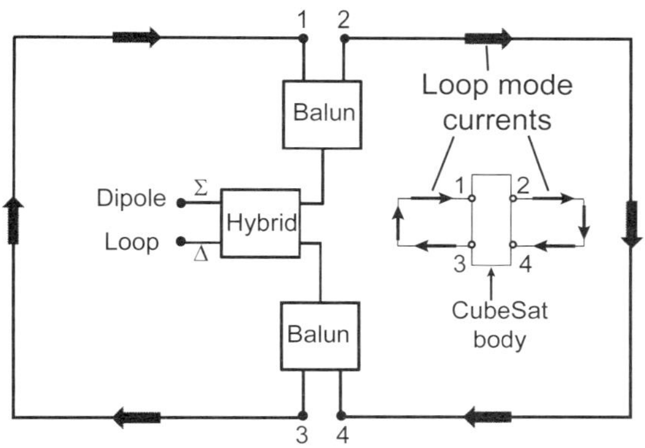

Figure 5.10 Diagram showing a simplified circuit model to provide sum and difference outputs for a four-port loop. The loop currents are generated by the difference port of the hybrid component.

Radiation patterns of the vector sensor array shown in Figure 5.8 were computed using the FEKO method of moments solver. The 10 cm by 10 cm by 30 cm CubeSat body was modeled as a perfect electrica conductor with triangular patch basis functions. The loop and monopole antennas were modeled using perfectly conducting wires with triangular linear basis functions. In this simulation, the arms of the vertical loops are 1.5m (4.9 ft) long and the parallel spacing between the wires forming the loops is 17.8 cm (7 inches). The vertical monopole length is 1.5m along the z axis. The horizontal loop axis is in the z direction and it has a perimeter length of

8.4m. An example of the 10 MHz normalized 3D radiation patterns of the six modes are shown in Figure 5.11. It is observed that the radiation patterns of Modes 1 and 4 have equivalent shapes, but they are orthogonally polarized – the Mode 1 (x dipole) relative gain pattern is $\hat{\phi}$ polarized and the Mode 4 (y loop) is $\hat{\theta}$ polarized. Similarly, Modes 2 and 3 have equivalent radiation pattern shapes and are orthogonally polarized. Finally, Modes 5 and 6 have the same pattern shapes and are orthogonally polarized. To demonstrate the wideband pattern characteristics, azimuth radiation patterns of all six modes are shown at 5 MHz and 40 MHz in Figure 5.12. The radiation pattern of each mode is observed to be relatively invariant to frequency.

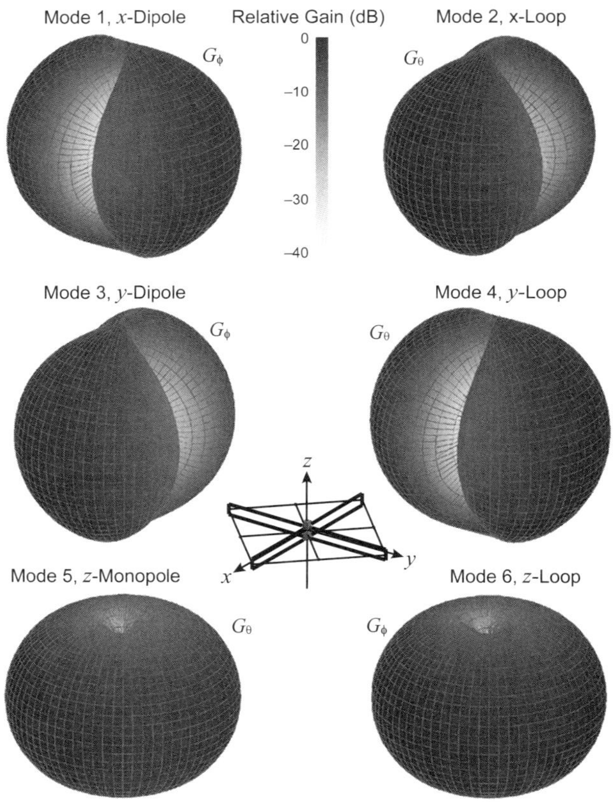

Figure 5.11 Normalized 3D radiation patterns at 10 MHz for all six vector sensor modes.

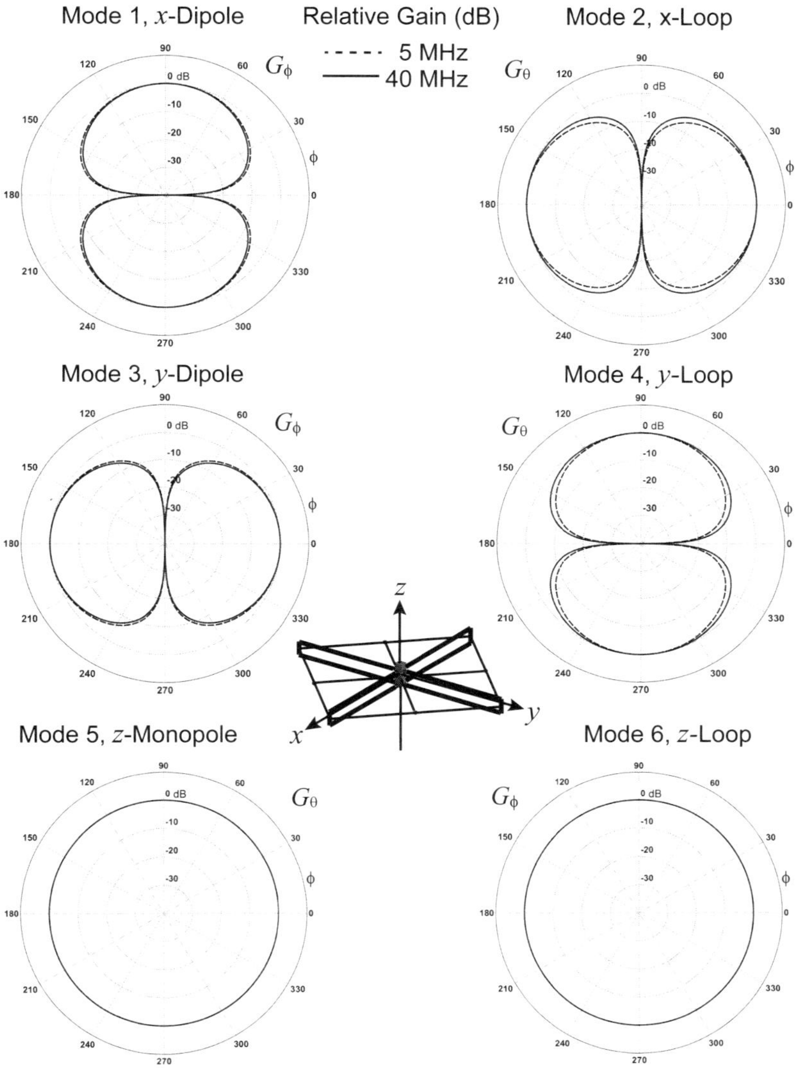

Figure 5.12 Normalized azimuth radiation patterns at 5 MHz and 40 MHz for all six vector sensor modes.

5.5 Vector Sensor Antenna Polarization Response

Assume that a vector sensor antenna has six receive modes consisting of $x, y,$ and z oriented dipoles and $x, y,$ and z oriented loops. The dipole and loop antennas have a common phase center that is assumed to be located at the origin of the standard spherical coordinate system. The angle θ is measured with respect to the z axis and ϕ is the angle with respect to the xy plane. The transmitted electromagnetic fields of ideal electrically short dipole and small loop antennas can be expressed in closed form as given below. By reciprocity, the receive antenna radiation patterns are equivalent to the transmit antenna radiation patterns.

When the electric dipole moment has only an x orientation, that is, $\boldsymbol{I}_e l = \hat{\boldsymbol{x}} I_e l$, the far-zone fields were derived previously in Chapter 1 in Equations (1.501) and (1.502) as

$$\boldsymbol{E}^{FF}_{\text{x-dipole}}(\boldsymbol{r}) = \left[-\hat{\boldsymbol{\theta}}\cos\theta\cos\phi + \hat{\boldsymbol{\phi}}\sin\phi\right] jkI_e l\eta\frac{e^{-jkr}}{4\pi r} \tag{5.1}$$

$$\boldsymbol{H}^{FF}_{\text{x-dipole}}(\boldsymbol{r}) = -\left[\hat{\boldsymbol{\theta}}\sin\phi + \hat{\boldsymbol{\phi}}\cos\theta\cos\phi\right] jkI_e l\frac{e^{-jkr}}{4\pi r}. \tag{5.2}$$

The $\hat{\boldsymbol{x}}$-loop fields are given by (refer to Equations (1.503) and (1.504))

$$\boldsymbol{E}^{FF}_{\text{x-loop}}(\boldsymbol{r}) = -\left[\hat{\boldsymbol{\theta}}\sin\phi + \hat{\boldsymbol{\phi}}\cos\theta\cos\phi\right] \eta k^2 IA\frac{e^{-jkr}}{4\pi r} \tag{5.3}$$

$$\boldsymbol{H}^{FF}_{\text{x-loop}}(\boldsymbol{r}) = \left[\hat{\boldsymbol{\theta}}\cos\theta\cos\phi - \hat{\boldsymbol{\phi}}\sin\phi\right] k^2 IA\frac{e^{-jkr}}{4\pi r}. \tag{5.4}$$

When the electric dipole moment has only a y orientation, that is, $\boldsymbol{I}_e l = \hat{\boldsymbol{y}} I_e l$, the far-zone fields are given by (refer to Equations (1.505) and (1.506))

$$\boldsymbol{E}^{FF}_{\text{y-dipole}}(\boldsymbol{r}) = -\left[\hat{\boldsymbol{\theta}}\cos\theta\sin\phi + \hat{\boldsymbol{\phi}}\cos\phi\right] jk\eta I_e l\eta\frac{e^{-jkr}}{4\pi r} \tag{5.5}$$

$$\boldsymbol{H}^{FF}_{\text{y-dipole}}(\boldsymbol{r}) = \left[\hat{\boldsymbol{\theta}}\cos\phi - \hat{\boldsymbol{\phi}}\cos\theta\sin\phi\right] jkI_e l\frac{e^{-jkr}}{4\pi r}. \tag{5.6}$$

The $\hat{\boldsymbol{y}}$-loop fields are given by (refer to Equations (1.507) and (1.508)

$$\boldsymbol{E}^{FF}_{\text{y-loop}}(\boldsymbol{r}) = \left[\hat{\boldsymbol{\theta}}\cos\phi - \hat{\boldsymbol{\phi}}\cos\theta\sin\phi\right] \eta k^2 IA\frac{e^{-jkr}}{4\pi r} \tag{5.7}$$

$$\boldsymbol{H}^{FF}_{\text{y-loop}}(\boldsymbol{r}) = \left[\hat{\boldsymbol{\theta}}\cos\theta\sin\phi + \hat{\boldsymbol{\phi}}\cos\phi\right] k^2 IA\frac{e^{-jkr}}{4\pi r}. \tag{5.8}$$

When the electric dipole moment has only a z orientation, that is, $\boldsymbol{I}_e l = \hat{\boldsymbol{z}} I_e l$, the far-field vector components are given by (refer to Equations (1.493) and (1.494))

$$\boldsymbol{E}^{FF}_{\text{z-dipole}}(\boldsymbol{r}) = \left[\hat{\boldsymbol{\theta}}\sin\theta\right] j\omega\mu I_e l\frac{e^{-jkr}}{4\pi r} \tag{5.9}$$

$$\boldsymbol{H}^{FF}_{\text{z-dipole}}(\boldsymbol{r}) = \left[\hat{\boldsymbol{\phi}} \sin\theta\right] I_e l j k \frac{e^{-jkr}}{4\pi r}. \tag{5.10}$$

The far-zone fields for the Hertzian magnetic dipole (electric current loop) oriented with the axis in the z direction, that is, $\boldsymbol{m} = m\hat{\boldsymbol{z}}$, where $m = IA$, are given by (refer to Equations (1.495) and (1.496))

$$\boldsymbol{E}^{FF}_{\text{z-loop}}(\boldsymbol{r}) = \left[\hat{\boldsymbol{\phi}} \sin\theta\right] IA\omega\mu k \frac{e^{-jkr}}{4\pi r}. \tag{5.11}$$

$$\boldsymbol{H}^{FF}_{\text{z-loop}}(\boldsymbol{r}) = \left[-\hat{\boldsymbol{\theta}} \sin\theta\right] k^2 IA \frac{e^{-jkr}}{4\pi r} \tag{5.12}$$

In Equations (5.1) to (5.12) the terms in the bracketed expressions are the vector polarization response of the antenna under transmitting conditions. Note that for the dipole modes, there is a j factor multiplying the bracketed expression, but there is no j factor for the loop modes. On receive, the vector polarization response is the conjugate of the transmit response.

Now consider the received voltage for a plane wave impinging on a vector sensor antenna with ideal short dipole and small loop elements. Utilizing the generalized antenna vector effective height ($\boldsymbol{h}$), which has units of meters, the induced open-circuit voltage across the terminals of an antenna is given by

$$V = \boldsymbol{E}_{\text{inc}} \cdot \boldsymbol{h}^* = \boldsymbol{h}^* \cdot \boldsymbol{E}_{\text{inc}} \tag{5.13}$$

where

$$\boldsymbol{h} = h\hat{\boldsymbol{p}}_{\text{a}} = \boldsymbol{p}_{\text{a}} \tag{5.14}$$

and where h is the magnitude of the vector effective height and $\hat{\boldsymbol{p}}_{\text{a}}$ is the unit vector polarization response of the antenna.

From Equations (5.1), (5.5), and (5.9) for the dipole modes and (5.3), (5.7), and (5.11) for the loop modes, the polarization responses of the vector sensor modes are expressed as

$$\boldsymbol{p}_{\text{x-dipole}} = \left[-\hat{\boldsymbol{\theta}} \cos\theta \cos\phi + \hat{\boldsymbol{\phi}} \sin\phi\right] \tag{5.15}$$

$$\boldsymbol{p}_{\text{y-dipole}} = -\left[\hat{\boldsymbol{\theta}} \cos\theta \sin\phi + \hat{\boldsymbol{\phi}} \cos\phi\right] \tag{5.16}$$

$$\boldsymbol{p}_{\text{z-dipole}} = \left[\hat{\boldsymbol{\theta}} \sin\theta\right] \tag{5.17}$$

$$\boldsymbol{p}_{\text{x-loop}} = -\left[\hat{\boldsymbol{\theta}} \sin\phi + \hat{\boldsymbol{\phi}} \cos\theta \cos\phi\right] \tag{5.18}$$

$$\boldsymbol{p}_{\text{y-loop}} = \left[\hat{\boldsymbol{\theta}} \cos\phi - \hat{\boldsymbol{\phi}} \cos\theta \sin\phi\right] \tag{5.19}$$

$$\boldsymbol{p}_{\text{z-loop}} = \left[\hat{\boldsymbol{\phi}} \sin\theta\right] \tag{5.20}$$

Equivalently, from Equations (5.1), (5.5), and (5.9) for the dipole modes and (5.3), (5.7), and (5.11) for the loop modes, the polarization response matrix of the vector sensor antenna six-mode array is given in terms of the $\hat{\theta}$ (first column) and $\hat{\phi}$ (second column) components by

$$\boldsymbol{p}_a = \begin{bmatrix} -\cos\theta\cos\phi & \sin\phi \\ -\cos\theta\sin\phi & -\cos\phi \\ \sin\theta & 0 \\ -\sin\phi & -\cos\theta\cos\phi \\ \cos\phi & -\cos\theta\sin\phi \\ 0 & \sin\theta \end{bmatrix} \tag{5.21}$$

Note that the first three rows of the $\boldsymbol{p}_a$ matrix in Equation (5.21) correspond to the $\hat{x}$, $\hat{y}$, and $\hat{z}$ dipole modes and the last three rows correspond to the $\hat{x}$, $\hat{y}$, and $\hat{z}$ loop modes.

Now from the derivation given in Chapter 1 (Equation 1.522), recall that that the incident wave electric-field unit vector $\hat{\boldsymbol{e}}^{\text{inc}}$ or incident polarization unit vector $\hat{\boldsymbol{p}}^{\text{inc}}$ in spherical coordinates is given by

$$\hat{\boldsymbol{e}}^{\text{inc}} = \hat{\boldsymbol{p}}^{\text{inc}} = \frac{E_\theta^{\text{inc}}}{|\boldsymbol{E}|}\hat{\boldsymbol{\theta}} + \frac{E_\phi^{\text{inc}}}{|\boldsymbol{E}|}\hat{\boldsymbol{\phi}} = \sin\gamma_p e^{j\delta_p}\hat{\boldsymbol{\theta}} + \cos\gamma_p\hat{\boldsymbol{\phi}}. \tag{5.22}$$

and observe that

$$\sin\gamma_p = \frac{E_\theta^{\text{inc}}}{|\boldsymbol{E}|} \tag{5.23}$$

$$\cos\gamma_p = \frac{E_\phi^{\text{inc}}}{|\boldsymbol{E}|} \tag{5.24}$$

where

$$|\boldsymbol{E}| = \sqrt{E_\theta^2 + E_\phi^2} \tag{5.25}$$

In the matrix form of Equation (5.22), the incident polarization unit vector for the kth source is given by

$$\hat{\boldsymbol{p}}_k^{\text{inc}} = \begin{bmatrix} \sin\gamma_k e^{j\delta_k} \\ \cos\gamma_k \end{bmatrix} \tag{5.26}$$

The full vector sensor received signal voltage 6 x 1 matrix $\boldsymbol{V}$ is formed by the product of the 6 x 2 matrix for the antenna polarization response $\boldsymbol{p}_a$

given by Equation (5.21) and the 2 x 1 matrix for the incident field polarization $\hat{\boldsymbol{p}}_k^{\rm inc}$ as

$$\boldsymbol{V} = \hat{\boldsymbol{p}}_{\rm a}^{*} \cdot \hat{\boldsymbol{p}}_k^{\rm inc} = \begin{bmatrix} -\cos\theta_k \cos\phi_k & \sin\phi_k \\ -\cos\theta_k \sin\phi_k & -\cos\phi_k \\ \sin\theta_k & 0 \\ -\sin\phi_k & -\cos\theta_k\cos\phi_k \\ \cos\phi_k & -\cos\theta_k \sin\phi_k \\ 0 & \sin\theta_k \end{bmatrix} \begin{bmatrix} \sin\gamma_k e^{j\delta_k} \\ \cos\gamma_k \end{bmatrix}. \tag{5.27}$$

The voltage matrix given by Equation (5.27) is expressed in a normalized form. Taking account of the incident field strength, the absolute form of the voltage matrix can be used in computing a 6 x 6 covariance matrix. The cross correlation R_{mn}^k of the received voltages for the mth and nth vector sensor modes, due to the kth source, is given by

$$R_{mn}^k = \mathcal{E}(V_{mk} V_{nk}^*) \tag{5.28}$$

where * means complex conjugate and $\mathcal{E}(\cdot)$ means mathematical expectation.

Let the channel or source covariance matrix be denoted $\boldsymbol{R}$. If there are K uncorrelated narrowband or broadband sources, then the K-source covariance matrix is the sum of the covariance matrices for the individual sources, that is,

$$\boldsymbol{R} = \sum_{k=1}^{K} \boldsymbol{R}^k(\theta_k, \phi_k, \gamma_k, \delta_k) + \boldsymbol{I} \tag{5.29}$$

where $\boldsymbol{R}^k$ is the covariance matrix of the kth source, $\boldsymbol{I}$ is the identity matrix which is used to represent the thermal noise level of the receiver. The covariance matrix contains the necessary signal properties to determine the angles θ_k and ϕ_k, and polarization parameters γ_k and δ_k.

Next, the covariance matrix defined by the elements in Equation (5.28) is Hermitian (that is, $\boldsymbol{R} = \boldsymbol{R}^{\dagger}$) which, by the spectral theorem, can be decomposed in eigenspace as [28]

$$\boldsymbol{R} = \sum_{n=1}^{N} \lambda_n \boldsymbol{e}_n \boldsymbol{e}_n^{\dagger} \tag{5.30}$$

where $\lambda_n, n = 1, 2, \cdots, N$ are eigenvalues of $\boldsymbol{R}$, and $\boldsymbol{e}_n, n = 1, 2, \cdots, N$ are the associated eigenvectors of $\boldsymbol{R}$. The multiple-source covariance matrix eigenvalues $(\lambda_1, \lambda_2, \cdots, \lambda_N)$ are a convenient quantitative measure of the utilization of the adaptive array degrees of freedom. Taking one of the N elements as a reference, then there are $N - 1$ degrees of freedom available in the adaptive process for adaptively nulling or mapping signals. Expressed in decibels, the minimum amplitude that an eigenvalue can have is 0 dB (the receiver noise level) – this is because the identity matrix has been added to the covariance matrix. Counting the number of eigenvalues above the receiver noise level is a direct indication of how many degrees of freedom are engaged by incident signals [29, 30].

5.6 Source Mapping From Vector Sensor Time-Series Data

A number of nonlinear iterative imaging (mapping) algorithms have been developed for the vector sensor array antenna, and these algorithms have been used in mapping simulated celestial sources as described by Knapp et al. [11]. For example, a maximum likelihood approach known as expectation-maximization (EM) [31, 32] has been applied to estimate and map sources [11], and the technique is reviewed below.

In modeling vector sensor array data, it is assumed that there are K temporal samples that are either measured or simulated from N vector sensor elements. In this case, $N = 6$ for the full vector sensor. The field of view here is the full sky and it is represented by I pixels (cells). Each source cell is characterized by estimates of the two spatial angles and polarization state (linear and circular polarizations for example). The ith source cell radiation is characterized here by a white Gaussian process that is complex normal with zero mean and standard deviation σ_i, CN(0,σ_i).

The received data vector for the kth temporal sample, denoted $\boldsymbol{z}(k)$, is expressed as a summation over all of the I cells as

$$\boldsymbol{z}(k) = \sum_{i=1}^{I} c_i(k)\boldsymbol{d}_i + \boldsymbol{n}(k) \tag{5.31}$$

where $\boldsymbol{d}_i$ are the directional steering vectors for the ith cell, $c_i(k)$ are the temporal samples of the source process, and $\boldsymbol{n}(k)$ represents the receiver noise as a white complex process that is zero mean. In matrix-vector form, the received data matrix $\boldsymbol{Z}$ is expressed as

$$\boldsymbol{Z} = \boldsymbol{DC} + \boldsymbol{N} \tag{5.32}$$

where $\boldsymbol{Z}$ is a matrix of the column vectors $\boldsymbol{z}(k)$, $\boldsymbol{D}$ is a matrix of the directional steering vector columns $\boldsymbol{d}_i$, $\boldsymbol{C}$ is composed of the column vectors of $c_i(k)$, and $\boldsymbol{N}$ is composed of the columns of vector noise samples $\boldsymbol{n}(k)$.

For K samples, the sample covariance matrix is expressed as the following summation

$$\boldsymbol{S} = \frac{1}{K}\sum_{k=1}^{K} z_k z_k^{\dagger} \tag{5.33}$$

or in matrix form

$$\boldsymbol{S} = \frac{1}{K}\boldsymbol{Z}\boldsymbol{Z}^{\dagger} \tag{5.34}$$

It is assumed here that the receive system is narrowband and does not depend on frequency. All of the K samples are assumed to belong to the same covariance matrix.

Now, let θ and ϕ be parameters for the source angles that are to be estimated from vector sensor data. The maximum likelihood estimates for the incident source angles θ and ϕ are denoted as $\tilde{\theta}$, and $\tilde{\phi}$, respectively.

The log-likelihood function for the vector sensor data is expressed as,

$$\log \mathcal{L}(\theta, \phi|\boldsymbol{Z}) = KN\log\pi - K\log|\boldsymbol{R}| - K\mathrm{tr}\left(\boldsymbol{R}^{-1}\right)\boldsymbol{S} \tag{5.35}$$

where the trace (tr) of an $N \times N$ square matrix $\boldsymbol{A}$, with elements a_{mn}, is equal to the sum of the main diagonal elements, that is,

$$\mathrm{tr}(\boldsymbol{A}) = \sum_{i=1}^{N} a_{ii} \tag{5.36}$$

and $|\cdot|$ denotes matrix determinant.

The steps of the expectation-maximum algorithm are as follows:

- In the first step, initialize the complete data spectral estimates as $\hat{\boldsymbol{\Sigma}}^1$ and covariance matrix as

$$\hat{\boldsymbol{R}}^1 = \boldsymbol{D}\hat{\boldsymbol{\Sigma}}^1\boldsymbol{D}^{\dagger} + \boldsymbol{R}_n \tag{5.37}$$

- Iterate in the second and subsequent steps for the complete spectral estimates for P iterations, that is, $p = 1, 2, \cdots, P$ as,

$$\hat{\boldsymbol{\Sigma}}^{p+1} = \mathrm{diag}\left[\hat{\boldsymbol{\Sigma}}^p + \hat{\boldsymbol{\Sigma}}^p\boldsymbol{D}^{\dagger}\left(\hat{\boldsymbol{R}^p}^{-1}\boldsymbol{S}^p\hat{\boldsymbol{R}^p}^{-1} - \hat{\boldsymbol{R}^p}^{-1}\right)\boldsymbol{D}\hat{\boldsymbol{\Sigma}}^p\right] \tag{5.38}$$

$$\hat{\boldsymbol{R}}^{p+1} = \boldsymbol{D}\hat{\boldsymbol{\Sigma}}^{p+1}\boldsymbol{D}^{\dagger} + \boldsymbol{R}_n \tag{5.39}$$

The diagonal terms of

$$\boldsymbol{D}^{\dagger}\left(\hat{\boldsymbol{R}^p}^{-1}\boldsymbol{S}^p\hat{\boldsymbol{R}^p}^{-1} - \hat{\boldsymbol{R}^p}^{-1}\right)\boldsymbol{D} \tag{5.40}$$

are the gradients of the log-likelihood expression with respect to the elements of $\hat{\boldsymbol{\Sigma}}$. Note that Pre- and post-multiplying this gradient expression by $\hat{\boldsymbol{\Sigma}}$ avoids infeasible solutions that can occur in other algorithms such as the projected gradient ascent algorithm [11].

To increase the number of sources that can be mapped, second-order statistics are used as new inputs to estimate the spectrum and the covariance matrix by using fourth-order moments (also referred to as kurtosis). In this case, a new matrix $\boldsymbol{Z}'$ is generated with columns that are formed from the first-order array data samples as the matrix elements of a vectorization (denoted vec$(\cdot)$) of $\boldsymbol{z}(k)\boldsymbol{z}^{\dagger}(k)$. As there are 20 unique such matrix elements in the upper right triangle of the 6×6 covariance matrix, the new matrix $\boldsymbol{Z}'$ is a vector of length $N' = 20$. Thus, using second-order statistics a new covariance matrix $\hat{\boldsymbol{R}}'$ is formed, which allows mapping of an increased number of sources. The new 20×20 covariance matrix has 20 eigenvalues, and so up to $N' - 1 = 19$ sources can be mapped.

Two example simulations [11] are now briefly described, one where 5 celestial sources are white Gaussian processes (for example, a sun) and the other where 19 sources are impulsive (such as a pulsar). Applying the EM algorithm to these two cases, the simulated results are displayed as the Stokes total power density parameter ($I = S_{\text{ave}}$) (Chapter 1, Equation (1.545)) in Figure 5.13 in a Mollweide equal-area projection map [32] that represents a complete Poincaré sphere as an ellipse. In Figure 5.13(a), the EM algorithm correctly maps all five white Gaussian sources. In Figure 5.13(b), 18 of 19 impulsive sources are correctly mapped with the EM algorithm.

5.7 Summary

Vector sensor array antenna systems are used to measure multiple polarization components of an incident electromagnetic field. A few examples of vector sensor antenna prototypes have been described in this chapter. The design of a vector sensor HF antenna array that could be integrated on a small satellite has been described. Moment method simulated radiation patterns for the various vector sensor modes have been shown. A derivation of the receive voltage matrix for the vector sensor array has been given. For mapping sources, the expectation maximum algorithm has been briefly reviewed and simulated results from the literature were shown. A potential future application is for performing radio astronomy missions. The antenna technology described here could be applied to other platforms such as airborne vehicles, towers, ground vehicles, and balloon-based applications.

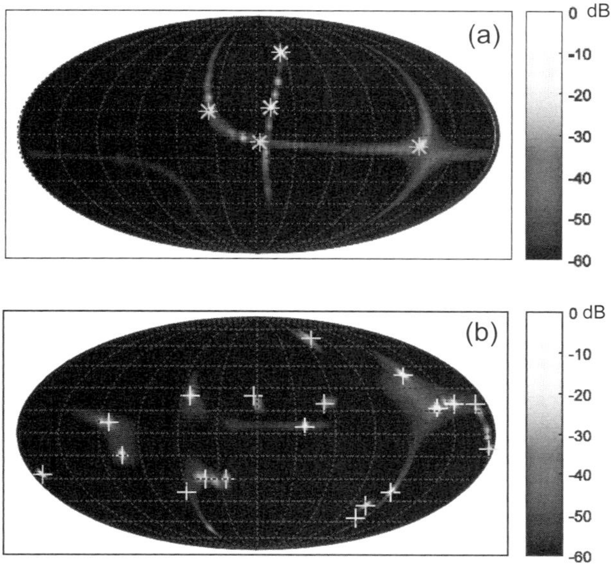

Figure 5.13 Simulated vector sensor expectation maximum mapping results [11] displayed as total power density in a Mollweide elliptical projection, with the true source positions indicated by markers. (a) 5 white Gaussian sources, and (b) 19 impulsive sources.

References

[1] G.F. Hatke, "Conditions for Unambiguous Source Location Using Polarization Diverse Arrays," *Proc. 27th Asilomar Conference on Signals, Systems and Computers,* 1993, pp. 1365–1369.

[2] A. Nehorai and E. Paldi, "Vector-Sensor Array Processing for Electromagnetic Source Localization," *IEEE Trans. Signal Proc.,* vol. 42, no. 2, 1994, pp. 376–398.

[3] S. Appadwedula and C. M. Keller, "Direction-Finding Results for a Vector Sensor Antenna on a Small UAV," *Proc. IEEE Workshop on Sensor Array and Multichannel Signal Proc.*, July 2006, pp. 74–78.

[4] H.S. Mir, "Transfer Function Based Approaches to Array Calibration," *J. of Communications*, vol. 2, no. 3, 2007, pp. 58–63.

[5] H. Mir and J. Sahr, "Calibration of a Polarization Diverse Array," *Int. Conf. on Acous. Speech and Sig. Proc.*, May 2006, pp. 1065–1068.

[6] H. Mir and J. Sahr, "Passive Direction Finding Using Airborne Vector Sensors in the Presence of Manifold Perturbations," *IEEE Trans. Sig. Proc.*, vol. 55, no. 1, 2007, pp. 156–164.

[7] A.J. Fenn and P.T. Hurst, *Ultrawideband Phased Array Antenna Technology for Sensing and Communications Systems,* MIT Lincoln Laboratory Book Series, Cambridge, MA: The MIT Press, 2015, pp. 255–273.

[8] R.T. Compton, "The Tripole Antenna: An Adaptive Array with Full Polarization Flexibility," *IEEE Trans. Antennas and Propagation*, vol. 29, no. 11, 1981, pp. 944–952.

[9] A.J. Fenn, F.C. Robey, P.T. Hurst, M.J. Silver, "Multipolarized Vector Sensor Array Antenna System for Radio Astronomy Applications," Provisional Application number US62243343, Filing date: Oct. 19, 2015.

[10] A.J. Fenn, B. Zhang, P.T. Hurst, K.E. Kolodziej, L.L. Retherford, C.D. Austin, "Multipolarized Vector Sensor Array Antenna System for Search and Rescue Applications," Application number PCT/US2016/027269, Publication date, Oct 20, 2016.

[11] M. Knapp, F. Robey, R. Volz, F. Lind, A. Fenn, A. Morris, M. Silver, S. Klein, and S. Seager, "Vector Antenna and Maximum Likelihood Imaging for Radio Astronomy," *2016 IEEE Aerospace Conference,* Big Sky, Montana, 5-12 March 2016, pp. 1–17.

[12] M. Knapp, PhD Thesis, Massachusetts Institute of Technology, 2018.

[13] F. Robey C., M. Knapp, A.J. Fenn, M. Silver, K. Johnson, F.J. Lind, R. Volz, S. Seager, and F. Neylon-Azad."High Frequency (HF) Radio Astronomy from a Small Satellite," *Small Sat Conference,* 2016.

[14] M. Knapp, R. Volz, Frank D. F. Lind C. Robey, A. Fenn, K. Johnson, M. Silver, A. Morris, and S. Klein. "HF Vector Sensor for Radio Astronomy: Ground Testing Results," *AIAA SPACE 2016,* p. 5498.

[15] R. Volz, M. Knapp, F. Lind, F. Robey. "Covariance Estimation in Terms of Stokes Parameters with Application to Vector Sensor Imaging," *Proc. 51st Asilomar Conference on Signals and Systems,* Nov. 2016.

[16] M. Knapp, D. Gary, M. Hecht, F. Lind, C. Lonsdale, F. Robey, "HERO: A Spacebased Low Frequency Interferometric Observatory for Heliophysics Enabled by Novel Vector Sensor Technology" *8th International Workshop on Planetary, Solar, and Heliospheric Radio Emissions (PRE8),* November 2016

[17] M. Knapp, F. Robey, R. Volz, Frank Lind, A. Fenn, A. Morris, M. Silver, S. Klein, and S. Seager. "HeRO: Heliophysics Radio Observer." *Science at Low Frequencies,* Dec. 2016.

[18] J.H. Meloling, J.W. Rockway, M.P. Daly, et. al., "A Vector-Sensing Antenna System: A high-frequency, vector-sensing array based on the two-port loop antenna element," *IEEE Antennas and Propagation Magazine,* vol. 58, no. 6, 2016, pp. 57–63.

[19] R. Lehmensiek, R. R. van Zyl and D. F. Visser, "The Design of an HF Antenna on a 1U CubeSat," Africon 9–12 Sept. 2013, pp. 1–4.

[20] A.S. Cohen, W.M. Lane, W.D. Cotton, N.E. Kassim, T.J. W. Lazio, R.A. Perley, J.J. Condon, and W.C. Erickson, "The VLA Low-Frequency Sky Survey," *The Astronomical J.,* vol. 134, Sept. 2007, pp. 1245–1262.

[21] N.E. Kassim, T. Joseph W. Lazio, W.C. Erickson, R.A. Perley, W.D. Cotton, E.W.

Greisen, A.S. Cohen, B. Hicks, H.R. Schmitt, and D. Katz, "The 74 MHz System on the Very Large Array," *The Astrophysical J. Supplement Series,* vol. 172, Oct., 2007, pp. 686–719.

[22] W.M. Lane, W.D. Cotton, S. van Velzen, T.E. Clarke, N.E. Kassim, J.F. Helmboldt, T.J.W. Lanzio, and A.S. Cohen, "The Very Large Array Low-frequency Sky Survey Redux (VLSSr)," *Monthly Notices of the Royal Astronomical Soc.,* 4 April 2014, Accessed online: http://arxiv.org/pdf/1404.0694v1.pdf

[23] S. Jester and H. Falcke, "Science with a Lunar Low-Frequency Array: From the Dark Ages of the Universe to Nearby Exoplanets," *New Astronomy Reviews,* vol. 53, 2009, pp. 1–26.

[24] R. King,"The Rectangular Loop Antenna as a Dipole," *IRE Trans. on Antennas and Propagation,* vol. 7 , no. 1, Jan. 1959, pp. 53–61.

[25] R.N. Grubb, R. Livingston, T.W. Bullett, "A New General Purpose High Performance HF Radar," *Proc. 2008 URSI General Assy Comm. G,* http://www.ursi.org/proceedings/procGA08/papers/GHp4.pdf, [Accessed: 2 March 2017], pp. 1–4.

[26] E.H. Nordholt and F. Van Willigen. "A New Approach to Active Antenna Design," *IEEE Trans. on Antennas and Propagation,* vol. AP-28, no. 6, Nov. 1980, pp. 904–909.

[27] W. Li, R. Chen, N. Zhai, S. Li, R. Mittra, "Wideband Matching of an Electrically Small Antenna Using a Negative Impedance Converter Technique," *Proc. 2012 IEEE Int. Symp. on Antennas and Propagation*, 2012, pp. 1–2.

[28] G. Strang, *Linear Algebra and Its Applications,* New York: Academic, 1976, p. 213.

[29] J.T. Mayhan, "Some Techniques for Evaluating the Bandwidth Characteristics of Adaptive Nulling Systems," *IEEE Trans. Antennas Propagat.,* vol. AP-27, no. 3, May 1979, pp. 363-373.

[30] A.J. Fenn, "Maximizing Jammer Effectiveness for Evaluating the Performance of Adaptive Nulling Array Antennas," *IEEE Trans. Antennas Propagat.,* vol. 33, no. 10, October 1985, pp. 1131-1142.

[31] F.C. Robey, "A Covariance Modeling Approach to Adaptive Beanmforming and Detection" MIT Lincoln Laboratory Technical Report 918, 1991.

[32] T.K. Moon, "The Expectation-Maximization Algorithm," *IEEE Signal Processing Magazine,* vol. 13, no. 6, 1996, pp. 47–60.

[33] H. Mott, *Antennas for Radar and Communications: A Polarimetric Approach,* New York: John Wiley, 1992.

6

Lumped-Element Impedance Matched Antennas

6.1 Introduction

Small, handheld wireless and portable devices are used in many communications applications. Achieving desired antenna input impedance characteristics in these devices can be accomplished by using impedance matching techniques including antenna geometry, transmission lines, transformers, and lumped elements. In small devices, there are limited options for the antenna geometry. The line transformer technique was described in Chapter 3. This chapter concentrates on the lumped-element impedance matching approach. Practical lumped elements are reactive components that are low-loss inductors (L) and/or low-loss capacitors (C) arranged in various series and/or parallel networks depending on the antenna design and the impedance matching requirements. A practical approach investigated by Wheeler, Fano, and other groups is to design parallel L and C, series L and C, parallel L series C, or series L parallel C networks that when connected to the antenna provides a maximum allowed reflection coefficient magnitude or VSWR over the desired band of frequencies [1–20]. In antenna matching problems where multiple networks of LC lumped-elements are required, optimization software becomes desirable as described by Rahola [12]. This chapter concentrates on the lumped element parallel LC analytic solution described by Andujar et al. for a handset monopole antenna [16].

Lumped element impedance matching techniques for improving small handheld device antenna bandwidth characteristics for single and multiple bands are being explored by a number of groups [13–19]. This chapter reviews an analytic method for using lumped element circuits to improve the impedance match of antennas over a band of frequencies. The example described in this chapter would apply over communications standards such as GSM1900 (1,850 MHz to 1,990 MHz), UMTS and LTE2100 (1,920 MHz to 2,170 MHz), LTE2300 (2,300 MHz to 2,400 MHz), and ISM, WiFi (2,400 MHz to 2,500 MHz). Section 6.2 describes the design of a monopole antenna for mobile communications. Section 6.3 derives an analytic impedance matching solution with a parallel LC network, and results are presented in Section 6.4. Section 6.5 provides a summary and a problem set is provided in Section 6.6.

6.2 A Monopole Antenna Design for Mobile Communications

Andujar et al. [16] have provided an analytic approach to designing an LC broadband matching circuit applicable to a monopole or dipole type antenna, and their work is reviewed here.

Consider Figure 6.1 that shows the handset antenna geometry investigated by Andujar [16]. This handset design has an antenna that is modeled as a strip monopole with dimensions 2 mm × 23 mm fed against the corner of a planar perfectly conducting ground plane with dimensions 40 mm

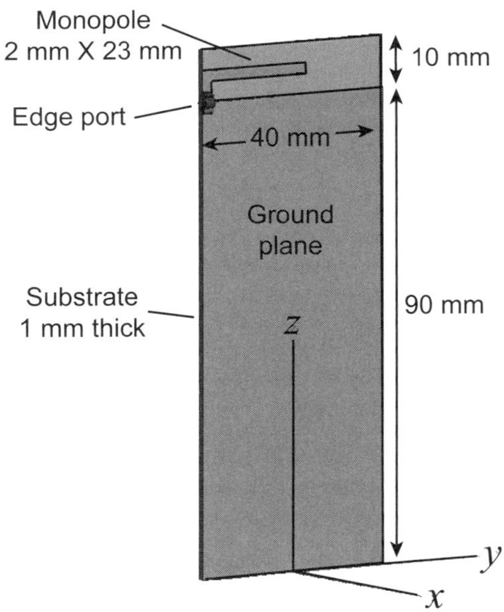

Figure 6.1 Monopole antenna geometry for personal communications application.

× 90 mm. A 1 mm thick substrate with assumed dielectric parameters dielectric constant $\epsilon_r = 4.15$, and loss tangent $\tan\delta = 0.013$. Here, the FEKO method of moments solver is used to compute the radiation pattern and input impedance of the monopole handset model. Since the monopole is oriented in the y direction, the principal polarization is expected to be E_ϕ, with a peak in the x direction and a null in the y direction. as shown in Figure 6.2. The simulated input impedance before matching is shown in Figure 6.3, where it is observed that the first natural resonance occurs at 2,100 MHz with the real part equal to approximately 16 ohms. It should be noted that the effect of the dielectric substrate is to load the monopole antenna, which lowers the resonant frequency that would otherwise occur in free space, from about 2.4 GHz to 2.1 GHz. Below resonance, the monopole antenna has a capacitive reactance (negative imaginary component) and above resonance the reactance is inductive (positive imaginary component). For these impedance data, the computed reflection coefficient relative to 50 ohms is shown in Figure 6.4, where the minimum value occurs at 2,200 MHz. A reflection coefficient magnitude of −6 dB corresponds to a VSWR of 3:1. A maximum VSWR of 3:1 occurs in the frequency range of 2.14 GHz to 2.3 GHz, which corresponds to a percent bandwidth of 7.2%. This monopole antenna input impedance variation with frequency can be represented approximately as a series RLC circuit and the ideal impedance matching circuit that compensates for the input reactance is a parallel LC network as shown in Figure 6.5.

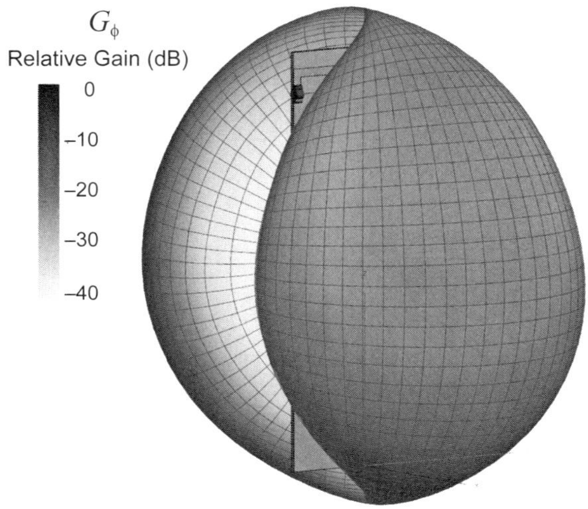

Figure 6.2 Moment method simulated 3D radiation pattern for the E_ϕ component of the monopole antenna.

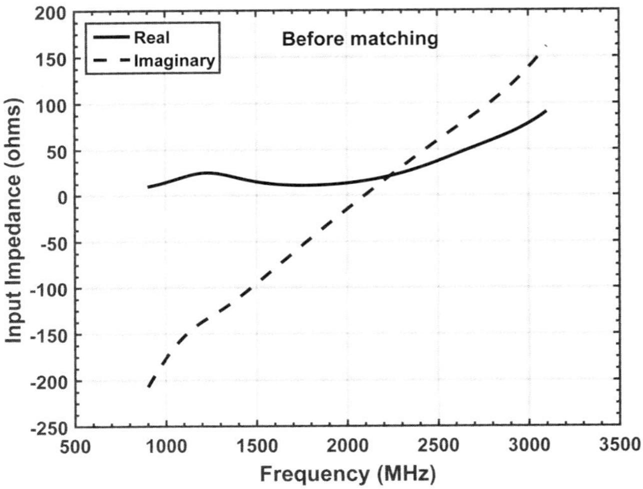

Figure 6.3 Moment method simulated input impedance of a monopole antenna before matching.

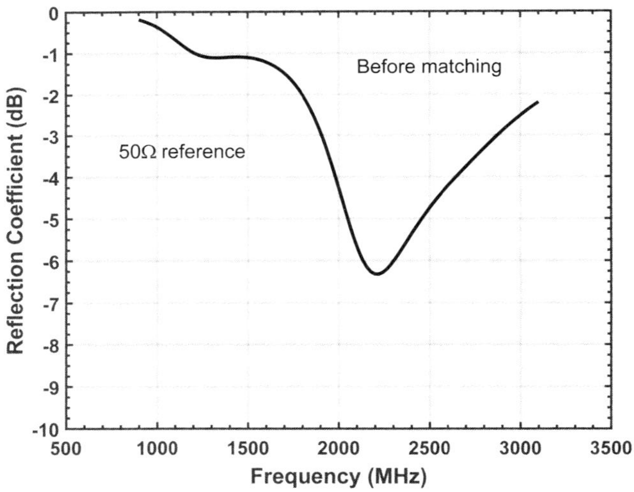

Figure 6.4 Moment method simulated reflection coefficient of a monopole antenna before impedance matching.

6.3 Analytic Impedance Matching Solution with a Parallel LC Network

In the analytic approach deveoped by Andujar [16], the input impedance looking into a parallel LC network connected to the monopole antenna is

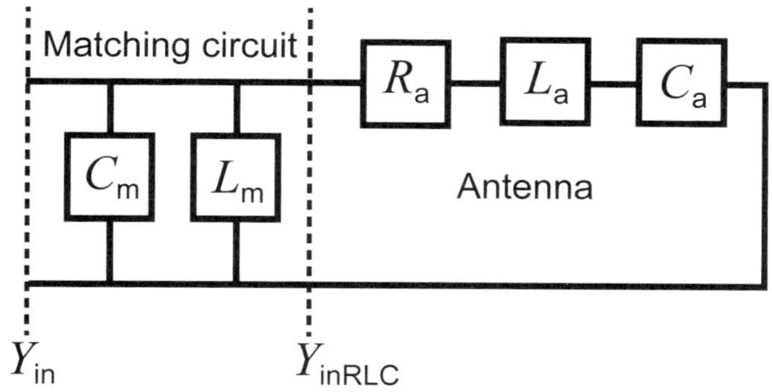

Figure 6.5 Diagram for an antenna represented by a series RLC circuit that is impedance matched with a parallel LC network.

forced to have three resonances where f_o is the natural resonant frequency and f_1 and f_2 are the low and high frequencies, respectively. Figure 6.6 shows this approach as an artist's concept where the matched input impedance in the frequency range f_1 to f_2 wraps around the center of the Smith chart, constrained by a maximum VSWR of 3.

Referring to the diagram in Figure 6.5, near the first resonance of either a monopole or dipole, the input impedance as a function of radian frequency ω can be expressed approximately in terms of an RLC series circuit as

$$Z_{\text{inRLC}}(\omega) = R_{\text{a}} + jX_{\text{a}} = R_{\text{a}} + j\omega L + \frac{1}{j\omega C} = R_{\text{a}} + j\left(\omega L - \frac{1}{\omega C}\right) \tag{6.1}$$

At resonance where $\omega = \omega_o$,

$$Z_{\text{inRLC}}(\omega_o) = R_a \tag{6.2}$$

since the imaginary term is zero, that is,

$$\omega_o L - \frac{1}{\omega_o C} = 0 \tag{6.3}$$

Thus, at resonance the radian frequency is

$$\omega_o = \frac{1}{\sqrt{LC}} \tag{6.4}$$

and the corresponding resonant frequency in cycles per second is given by

$$f_o = \frac{1}{2\pi\sqrt{LC}} \tag{6.5}$$

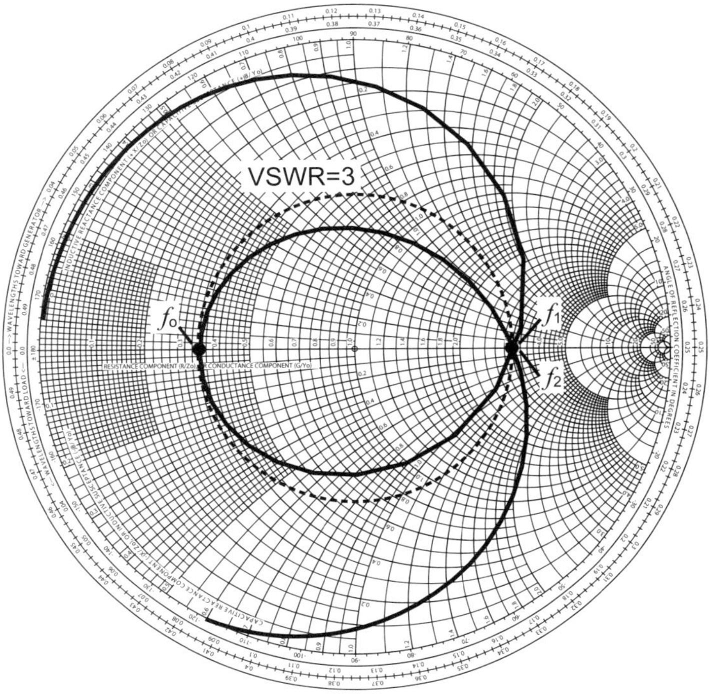

Figure 6.6 Artist concept for forcing the input impedance of an antenna to be resonant at three frequencies and constrained by a maximum VSWR=3, where f_o is the natural resonant frequency and f_1 and f_2 are the low and high frequencies, respectively.

For a given radian frequency, the imaginary part X_a of the antenna impedance can be expressed as

$$X_{\rm a} = \left(\omega L - \frac{1}{\omega C}\right) = \omega L \frac{\sqrt{LC}}{\sqrt{LC}} - \frac{1}{\omega C}\frac{\sqrt{LC}}{\sqrt{LC}} = \sqrt{\frac{L}{C}}\left(\frac{\omega}{\omega_o} - \frac{\omega_o}{\omega}\right) \tag{6.6}$$

The characteristic impedance of a transmission line is given by the general expression [20]

$$Z_o = \sqrt{\frac{R + j\omega L}{G + j\omega C}} \tag{6.7}$$

where the transmission line has the following parameters R is the series resistance per unit length (ohms/m) due to finite conductivity of the transmission line, L is the series inductance per unit length (henries/m), G is the shunt conductance per unit length (siemens/m), and C is the shunt capacitance per unit length (farads/m) due to dielectric losses. In the case where the transmission line is lossless, both R and G are zero, so referring to Equation (6.5) the characteristic impedance given by Equation (6.7) reduces to

$$Z_o = \sqrt{\frac{L}{C}} = \omega_o L = \frac{1}{\omega_o C} \tag{6.8}$$

Thus, the antenna reactance term in Equation (6.1) can now be expressed as

$$X_{\mathrm{a}} = Z_o \left(\frac{\omega}{\omega_o} - \frac{\omega_o}{\omega} \right) \tag{6.9}$$

and thus

$$Z_{\mathrm{inRLC}} = R_{\mathrm{a}} + jX_{\mathrm{a}} = R_{\mathrm{a}} + jZ_o \left(\frac{\omega}{\omega_o} - \frac{\omega_o}{\omega} \right) \tag{6.10}$$

Now the power dissipated P_{diss} by the antenna input resistance R_{a} is given by

$$P_{\mathrm{diss}} = \frac{1}{2}|I|^2 R_{\mathrm{a}} \tag{6.11}$$

The average magnetic energy W_{m} stored in the inductance L is given by

$$W_{\mathrm{m}} = \frac{1}{4}|I|^2 L \tag{6.12}$$

The average electric energy W_{e} stored in the capacitance C is given by

$$W_{\mathrm{e}} = \frac{1}{4}|V_{\mathrm{c}}|^2 C = \frac{1}{4}|I|^2 \frac{1}{\omega^2 C} \tag{6.13}$$

The quality factor here is defined as [20, p. 274]

$$Q = \omega \frac{\text{Average energy stored}}{\text{Energy dissipated per second}} = \omega \frac{W_{\mathrm{m}} + W_{\mathrm{e}}}{P_{\mathrm{diss}}} \tag{6.14}$$

At resonance, the average magnetic and electric stored energies are equal, that is,

$$W_{\mathrm{m}} = W_{\mathrm{e}} \tag{6.15}$$

Thus, the unloaded Q of the antenna at resonance is given by

$$Q_a = \omega_o \frac{2W_{\mathrm{m}}}{P_{\mathrm{diss}}} = \frac{\omega_o L}{R_a} = \frac{1}{\omega R_a C} = \frac{Z_o}{R_a} \tag{6.16}$$

Substituting Equation (6.16) into (6.10) yields

$$Z_{\mathrm{inRLC}} = R_{\mathrm{a}} + jR_{\mathrm{a}} Q_a \left(\frac{\omega}{\omega_o} - \frac{\omega_o}{\omega} \right) \tag{6.17}$$

and then simplifying and expressing in terms of frequency yields

$$Z_{\text{inRLC}} = R_{\text{a}}\left[1 + jQ_a\left(\frac{f}{f_o} - \frac{f_o}{f}\right)\right] \tag{6.18}$$

Now let

$$\nu = \frac{f}{f_o} - \frac{f_o}{f} \tag{6.19}$$

so that

$$Z_{\text{inRLC}} = R_{\text{a}}(1 + jQ_a\nu) \tag{6.20}$$

Now the input admittance of the RLC circuit can be expressed as

$$Y_{\text{inRLC}} = \frac{1}{R_{\text{a}}(1 + jQ_a\nu)} = \frac{R_a - jR_aQ_a\nu}{R_a^2 + (R_aQ_a\nu)^2} \tag{6.21}$$

Now the input admittance looking into the matching section connected to the antenna is given by

$$Y_{\text{in}}(f) = Y_{\text{inRLC}} + j\omega C_m + \frac{1}{j\omega L_m} \tag{6.22}$$

Substituting Equation (6.21) into Equation (6.22) yields for the imaginary part

$$\text{Im}(Y_{\text{in}}(f)) = \frac{-R_aQ_a\nu}{R_a^2 + (R_aQ_a\nu)^2} + \omega C_m - \frac{1}{\omega L_m} \tag{6.23}$$

and for the real part

$$\text{Re}(Y_{\text{in}}(f)) = \frac{G_a}{1 + (Q_a\nu)^2} \tag{6.24}$$

where $G_a = 1/R_a$ is the conductance of the antenna at resonance.

The approach here is to force the imaginary part of the input admittance to zero at three frequencies, f_o the natural resonant frequency, and f_1 and f_2 the possible band-edge frequencies.

$$\text{Im}(Y_{\text{in}}(f_o)) = \text{Im}(Y_{\text{in}}(f_1)) = \text{Im}(Y_{\text{in}}(f_2)) = 0 \tag{6.25}$$

Now, let the VSWR=S, then when the forced resonances occur, the real part of the admitances are set as

$$\text{Re}(Y_{\text{in}}(f_1)) = \text{Re}(Y_{\text{in}}(f_2)) = \frac{1}{Z_oS} \tag{6.26}$$

and

$$\text{Re}(Y_{\text{in}}(f_o)) = \frac{S}{Z_o} \tag{6.27}$$

$$\mathrm{Re}(Y_{\mathrm{in}}(f_1)) = \frac{G_a}{1 + (Q_a \nu_1)^2} = \frac{1}{Z_o S} \tag{6.28}$$

Now

$$\nu_1 = \frac{f_1}{f_o} - \frac{f_o}{f_1} = \frac{f_1^2 - f_o^2}{f_o f_1} \tag{6.29}$$

and then it follows that

$$\nu_1^2 = \frac{G_a Z_o S - 1}{Q_a^2} = \left(\frac{f_1^2 - f_o^2}{f_o f_1} \right)^2 \tag{6.30}$$

which simplifies as

$$f_1^2 - f_o^2 = -f_o f_1 \sqrt{\frac{G_a Z_o S - 1}{Q_a^2}} \tag{6.31}$$

or in the form

$$f_1^2 - f_1 \left(\sqrt{\frac{G_a Z_o S - 1}{Q_a^2}} \right) - f_o^2 = 0 \tag{6.32}$$

which has the solution

$$f_1 = -\frac{f_o \sqrt{G_a Z_o S - 1} - f_o \sqrt{G_a Z_o S - 1 + 4Q_a^2}}{2Q_a} \tag{6.33}$$

and similarly it follows that

$$f_2 = \frac{f_o \sqrt{G_a Z_o S - 1} - f_o \sqrt{G_a Z_o S - 1 + 4Q_a^2}}{2Q_a} \tag{6.34}$$

The bandwidth (BW) is then given by

$$\mathrm{BW} = \frac{f_2 - f_1}{f_o} = \frac{\sqrt{G_a Z_o S - 1}}{Q_a} \tag{6.35}$$

but

$$\mathrm{Re}(Y_{\mathrm{in}}(f_o)) = \frac{S}{Z_o} = G_a \tag{6.36}$$

so it follows that

$$G_a Z_o S = \frac{S}{Z_o} Z_o S = S^2 \tag{6.37}$$

Substituting Equation (6.37) in (6.35) yields

$$\mathrm{BW} = \frac{\sqrt{S^2 - 1}}{Q_a} \tag{6.38}$$

To solve for the value of C_m in terms of L_m, from Equation (6.23) for resonance it follows that

$$\text{Im}(Y_{\text{in}}(f_0)) = 0 = \frac{-R_a Q_a \nu_o}{R_a^2 + (R_a Q_a \nu_o)^2} + \omega C_m - \frac{1}{\omega L_m} \tag{6.39}$$

and noting that $\nu_o = \nu(f = f_o) = 0$ the matching capacitor value is determined as

$$C_m = \frac{1}{\omega_o^2 L_m} \tag{6.40}$$

Using this relation for C_m in terms of L_m it follows that the last two terms in Equation (6.23) can be expressed as

$$\omega C_m - \frac{1}{\omega L_m} = \frac{\omega}{\omega_o^2 L_m} - \frac{1}{\omega L_m} = \frac{\omega^2 - \omega_o^2}{\omega_o^2 \omega L_m} \tag{6.41}$$

Using the result of Equation (6.41), Equation (6.23) can now be expressed as

$$\text{Im}(Y_{\text{in}}(f)) = \frac{-R_a Q_a \nu}{R_a^2 + (R_a Q_a \nu)^2} + \frac{\omega^2 - \omega_o^2}{\omega_o^2 \omega L_m} \tag{6.42}$$

and then at the forced resonance frequency f_1

$$\text{Im}(Y_{\text{in}}(f_1)) = \frac{-R_a Q_a \nu_1}{R_a^2 + (R_a Q_a \nu_1)^2} + \frac{\omega_1^2 - \omega_o^2}{\omega_o^2 \omega_1 L_m} = 0 \tag{6.43}$$

Solving Equation (6.43) for L_m yields

$$L_m = \frac{R_a \left[1 + (Q_a \nu_1)^2\right] (\omega_1^2 - \omega_o^2)}{Q_a \nu_1 (\omega_o^2 \omega_1)} \tag{6.44}$$

Noting that

$$R_a = R_a(f_o) = \frac{Z_o}{S} \tag{6.45}$$

Substituting R_a in Equation (6.44) yields the desired result for the matching inductor as

$$L_m = \frac{Z_o \left[1 + (Q_a \nu_1)^2\right] (\omega_1^2 - \omega_o^2)}{S Q_a \nu_1 (\omega_o^2 \omega_1)} \tag{6.46}$$

As pointed out by Andujar, the above equations apply for an ideal RLC load impedance. In the case where the antenna only approximates an ideal RLC impedance curve, the derived matching circuit will not fully compensate for the frequency variations, and the targeted maximum allowed VSWR will not necessarily be achieved over the bandwidth.

As discussed in Chapter 1 (refer to Equation (1.647)) the antenna quality factor Q_a can be expressed as

$$Q_a(\omega) = Q_z(\omega) = \frac{\omega}{2R(\omega)}\sqrt{\left[\frac{dR(\omega)}{d\omega}\right]^2 + \left[\frac{dX(\omega)}{d\omega} + \left|\frac{X(\omega)}{\omega}\right|\right]^2} \quad (6.47)$$

From Equations (6.33) and (6.34), the fractional bandwidth is determined as

$$\text{FBW} = \frac{f_2 - f_1}{f_o} = \frac{S-1}{Q_a\sqrt{S}} \quad (6.48)$$

which agrees with Equation (1.649) in Chapter 1.

6.4 Calculated Impedance Matching Circuit Parameters and Matched Antenna Results

The equations presented in the previous section were programmed in MATLAB and the following results were obtained for the monopole input impedance data before matching presented in Figure 6.3, which is replotted in Smith chart form in Figure 6.7. As shown in Figure 6.8, the quality factor Q_z was calculated from moment method simulated input impedance of the monopole antenna before matching using Equation (6.47).

The maximum allowed VSWR was assumed to be 3:1, that is, $S = 3$, and the fractional bandwidth was computed from Equation (6.48), which is plotted versus frequency in Figure 6.9. From Equation (6.33) and (6.34), the low and high frequencies are computed as $f_1 = 1.93$ GHz and $f_2 = 2.5$ GHz, respectively. For these values of lower and upper frequencies, the center frequency f_c is computed as

$$f_c = \frac{f_1 + f_2}{2} = 2.22 \text{ GHz} \quad (6.49)$$

Thus, the center frequency is $f_c = 2.22$ GHz whereas the first natural resonant frequency of the handset monopole is $f_o = 2.1$ GHz. By trial and error it was observed that an improved impedance matching result is obtained by substituting the center frequency f_c for f_o and $\omega_c = 2\pi f_c$ for ω_o in the equations presented in the previous section.

The values of the matching inductor and capacitor were computed from Equations (6.46) and (6.40), respectively, with the results

$$L_m = 0.966 \text{ nH} \quad (6.50)$$

$$C_m = 5.32 \text{ pF} \quad (6.51)$$

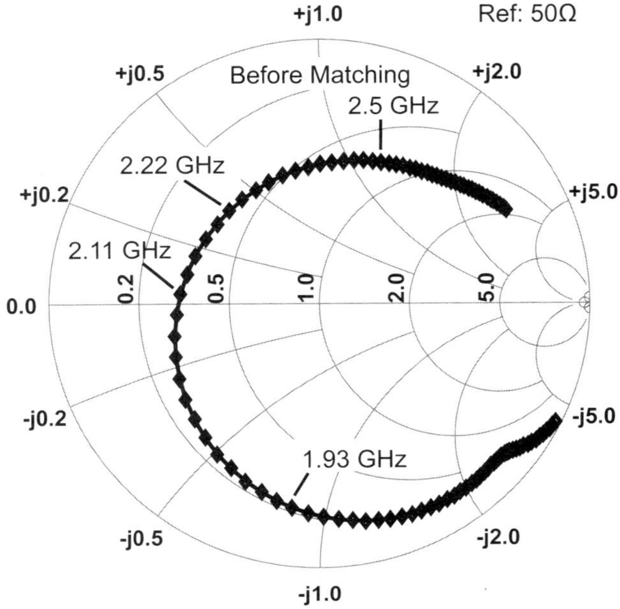

Figure 6.7 Moment method simulated input impedance (Smith chart) of a monopole antenna before matching.

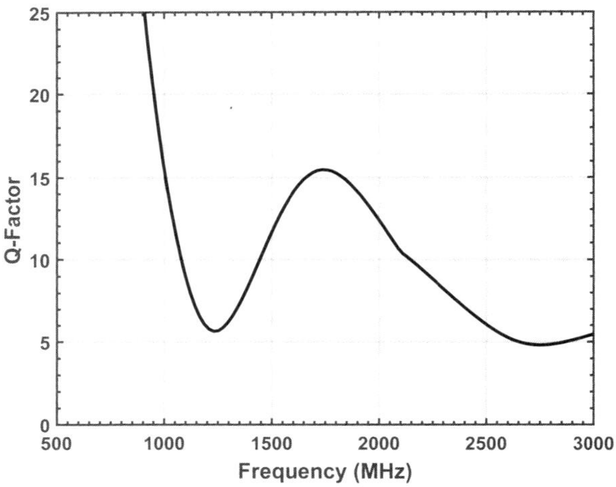

Figure 6.8 Quality factor Q_z calculated from moment method simulated input impedance of a monopole antenna before matching.

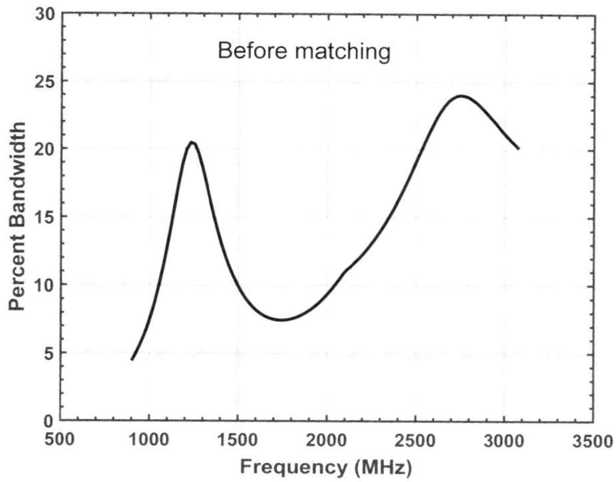

Figure 6.9 Percent bandwidth for VSWR=3 calculated from moment method simulated input impedance of a monopole antenna before matching.

Using these values for L_m and C_m, the matched input impedance is computed from Equation (6.22) as

$$Z_{\text{in}}(f) = \frac{1}{Y_{\text{in}}(f)} \tag{6.52}$$

which is plotted in Figure 6.10. The corresponding Smith chart plot of the input impedance relative to 50 ohms is shown in Figure 6.11. The corresponding reflection coefficient, mismatch loss, and VSWR before and after matching are summarized in Figures 6.12, 6.13, and 6.14, respectively. The VSWR plot in Figure 6.14 shows a widening of the bandwidth due to the parallel LC matching circuit; however, the VSWR does increase to 3.25 at 2.11 GHz due to the approximation made about the RLC behavior of the monopole. At the matched band edges where the VSWR is 3.0, the low and high frequencies are 1.85 GHz and 2.5 GHz, which represents a bandwidth of 30% centered at 2.22 GHz. Without impedance matching, as was discussed for the simulated data in Figure 6.4 the percent bandwidth for a 50-ohm reference impedance was 7.2%, so a significant increase in bandwidth is achieved here with analytically determined lumped element impedance matching.

Finally, it is useful to compare in Figure 6.15 the analytic 2-component solution with the solutions obtained with an impedance matching optimizer for complex loads such as Optenni Lab [12]. The monopole impedance data were imported into Optenni Lab and using the same criteria of VSWR 3.0 and low and high frequencies 1.85 GHz and 2.5 GHz, solutions were

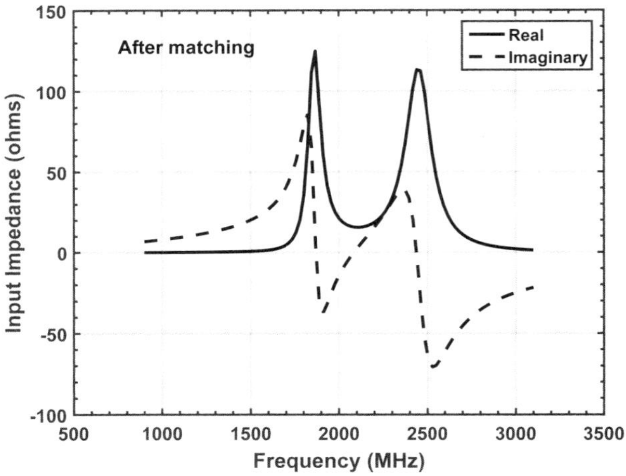

Figure 6.10 Moment method simulated input impedance of a monopole antenna after matching with a parallel LC network.

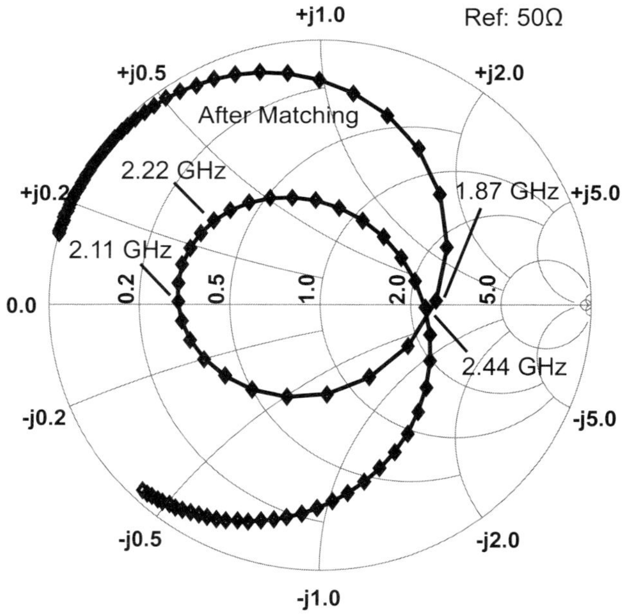

Figure 6.11 Moment method simulated input impedance (Smith chart) of a monopole antenna after matching.

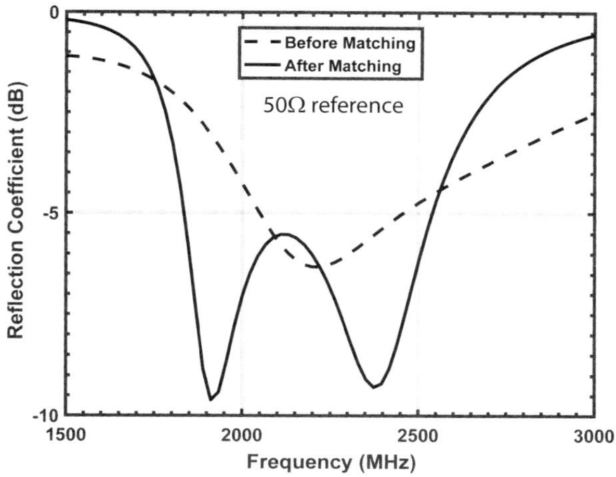

Figure 6.12 Moment method simulated reflection coefficient of a monopole antenna before and after matching.

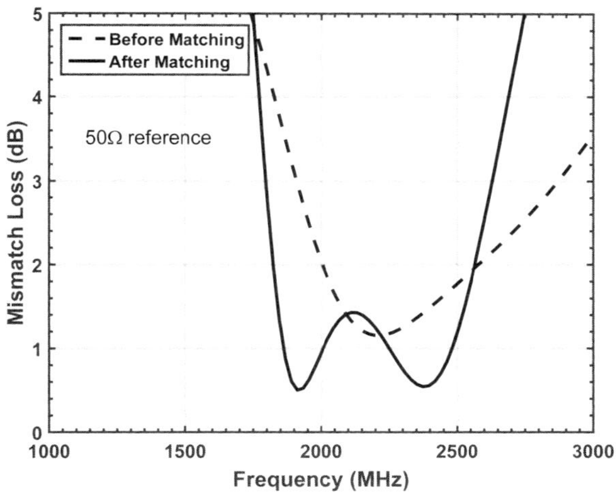

Figure 6.13 Moment method simulated mismatch loss of a monopole antenna before and after matching.

obtained for 2, 4, and 6 LC components. For this example, the analytic 2-component analytic solution compared favorably with the multicomponent Optenni solutions.

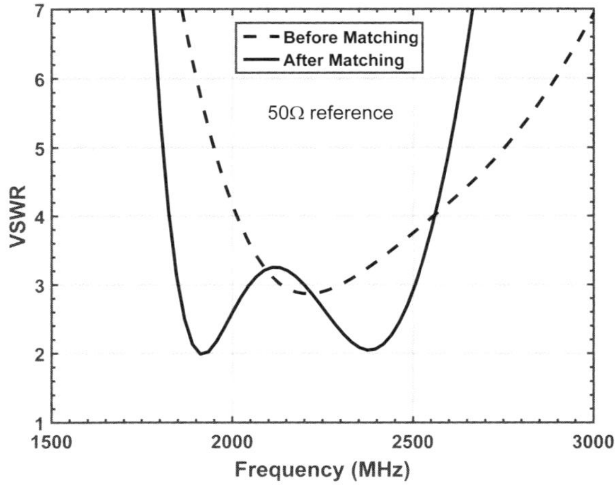

Figure 6.14 Moment method simulated VSWR of a monopole antenna before and after matching.

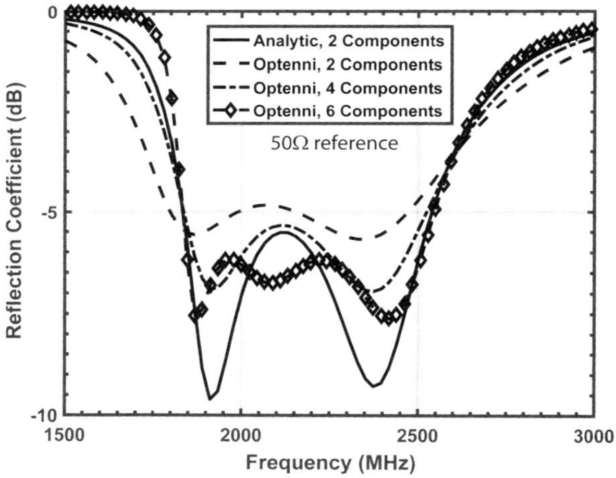

Figure 6.15 Comparison of the impedance-matched reflection coefficient for the 2-component analytic solution and Optenni solutions for 2, 4, and 6 LC components.

6.5 Summary

This chapter has discussed an analytic lumped-element impedance matching technique. An example from the literature has been given for the case where the lumped elements are an inductor and capacitor in parallel with a monopole antenna. The analytic solution assumes that the monopole antenna can be modeled approximately as an ideal series RLC circuit. The method of moments was used to simulate the radiation pattern and input impedance of a handset monopole antenna, and the analytically derived impedance matching circuit significantly improved the bandwidth characteristics of this antenna. Thus for the handset monopole with this impedance matching approach, multiple communications bands from 1,850 MHz to 2,500 MHz encompassing the GSM1900, UMTS, LTE2100, LTE2300, ISM, and WiFi bands can be addressed. For the example shown, the two-component analytic solution provides similar performance compared to optimizer-based multicomponent solutions.

6.6 PROBLEM SET

6.1 Write a computer program to implement the analytic impedance matching solution given in Section 6.3 using an analytic approximation of the moment method input impedance data shown in Figure 6.3, where $\mathrm{Re}(Z_{\mathrm{in}}(f)) = 12.5 + 40(f - 1.75)$ and $\mathrm{Im}(Z_{\mathrm{in}}(f)) = -100 + 160(f - 1.5)^2$, where f is in GHz. The impedance matching results will be very similar to the results presented in Section 6.4. Note that the Smith chart plot in Figure 6.11 can be produced in MATLAB for the variable named "ReflectionCoefficient" using the following statements: l = smithchart(ReflectionCoefficient); l(1).LineWidth=1.0; l(1).Color=[0 0 0]; l(1).Marker='d';

References

[1] H. A. Wheeler, "Wideband Impedance Matching',' Wheeler Laboratories Report 418, Great Neck, NY, May 1950.

[2] R. M. Fano, "Theoretical Limitations on the Broadband Matching of Arbitrary Impedances," J. Franklin Institute, 249, January 1950, pp. 57-83.

[3] R. M. Fano, "Theoretical Limitations on the Broadband Matching of Arbitrary Impedances," J. Franklin Institute, 249, February 1950, pp. 139-154.

[4] V.H. Rumsey, "The Design of Frequency-Compensating Matching Sections," *Proc. IRE*, October 1950, pp. 1191–1196.

[5] R.L. Thomas, *A Practical Introduction to Impedance Matching*, Dedham, MA: Artech House, 1976.

[6] W.N. Caron, *Antenna Impedance Matching*, Newington, CT: American Radio Relay League, 1989.]

[7] G.L. Matthaei, L. Young, E.M.T. Jones, *Microwave Filters, Impedance-Matching Networks, and Coupling Structures*, New York, McGraw-Hill, 1964.

[8] T.R. Cuthbert, *Circuit Design Using Personal Computers*, New York: John Wiley, 1983.

[9] T.R. Cuthbert, *Broadband Direct-Coupled and Matching RF Networks*, Greenwood, Arkansas: TRCPEP Publications, 1999.

[10] A. R. Lopez, "Wheeler and Fano Impedance Matching," *IEEE Antennas and Propagation Magazine*, vol. 49, no. 4, August 2007, pp. 116–119.

[11] A.R. Lopez, "Double-Tuned Impedance Matching," IEEE Antennas and Propagation Magazine, vol. 54, no. 2, April 2012 pp. 109–116.

[12] J. Rahola, "Advances in Impedance Matching Tools for Antenna Applications," *Proc. 2012 IEEE Int. Symp. on Antennas and Propagation,* 2012, pp. 1–2.

[13] J. Anguera, A. Andujar, and C. Garcia, "Multiband and Small Coplanar Antenna System for Wireless Handheld Devices," *IEEE Trans. Antennas and Propagat.*, vol. 61, no. 7, 2013, pp. 3782–3789.

[14] J. Anguera, C. Puente, C. Borja, G. Font, J. Soler, "A Systematic Method to Design Single-Patch Broadband Microstrip Patch Antennas," *Microwave and Optical Technology Letters*, vol. 31, no. 3, 2001, pp. 185–188.

[15] R. Valkonen and A. Lehtovuori, "Determining Bandwidth Estimators and Matching Circuits for Evaluation of Chassis Antennas," *IEEE Trans. Antennas and Propagat.*, vol. 63, no. 9, 2015, pp. 4111–4120.

[16] A. Andujar, J. Anguera, C. Puente, "A Systematic Method to Design Broadband Matching Networks,," *Proc. Fourth European Conf. on Antennas and Propagat.*, 2010, pp. 1–5.

[17] H. F. Pues, A. R. Van de Capelle, "An Impedance-Matching Technique for Increasing the Bandwidth of Microstrip Antennas," *IEEE Trans. on Antennas and Propagat.*, vol. AP-37, no. 11, 1989, pp. 1345–1354.

[18] F.G.S. Silva, R.N. de Lima, R.C.S. Freire and C. Plett, "A Switchless Multiband Impedance Matching Technique Based on Multiresonant Circuits," *IEEE Trans. Circuits and Systems II: Express Briefs,* vol. 60, no. 7, July 2013, pp. 417–421.

[19] A. Lehtovuori, R. Valkonen, M. Valtonen, "Dual-Band Matching of Arbitrary Loads," *Microwave and Optical Technology Letters,* vol. 56, pp. 2958–2966.

[20] D.M. Pozar, *Microwave Engineering, 4th ed.*, Hoboken, NJ: Wiley, 2012.

7

Dipole Array Analysis for the Chain Home Radar

7.1 Introduction

As described in a review article by Neale [1], the Chain Home radar system was developed in England during the 1930s and 1940s, and it was the first early-warning defense system based on radar technology [2]. The many deployed Chain Home radar systems operated near the coast of England in the HF frequency band from 20 MHz to 30 MHz. Artist depictions of the Chain Home transmit and receive antennas are shown in Figures 7.1 to 7.3.

Figure 7.1 Artist depiction of the Chain Home radar transmit and receive antennas.

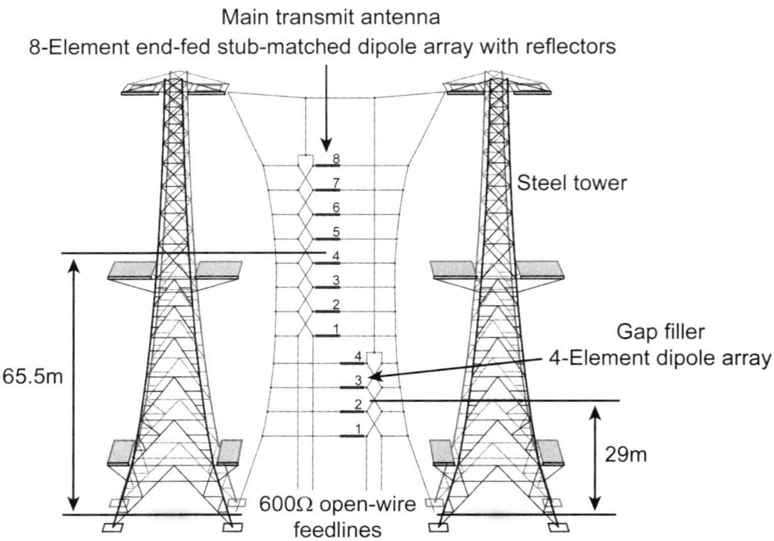

Figure 7.2 Artist depiction of the Chain Home radar transmit dipole arrays.

In Figure 7.1, the design of the Chain Home transmit antenna system is in the form of horizontally polarized curtain arrays [3–11] that had been commonly used in low-frequency communications applications. Three transmit towers are shown with two transmit antenna systems, where one antenna system would be in operation with the other in a standby mode. As shown in Figure 7.2, the transmit antenna system was composed of an upper main 8-element end-fed dipole array and a lower gap filler 4-element end-fed dipole array, and both arrays had a fixed beam that would be directed in free space at broadside. Due to reflection for the earth's surface, the direct free-space beam and the reflected beam cause an interference pattern in elevation creating multiple peaks and nulls at angles that depend on frequency. The receive antenna system utilized pairs of center-fed dipoles with switchable reflectors, at heights corresponding to that of the transmit array, as shown in Figure 7.3. Known as an Adcock array antenna [12], the receive antenna system had four such array towers with crossed dipoles arranged for direction finding and height finding of radar targets.

This chapter is organized as follows. Section 7.2 describes the main transmit array that was used in the Air Ministry Experimental Station (AMES) Type 1, Chain Home (CH) - Early Warning Radar System. To take account of reflection from the earth's surface, Section 7.3 reviews the Fresnel reflection coefficients for plane wave reflection from a lossy half space. In Section 7.4,

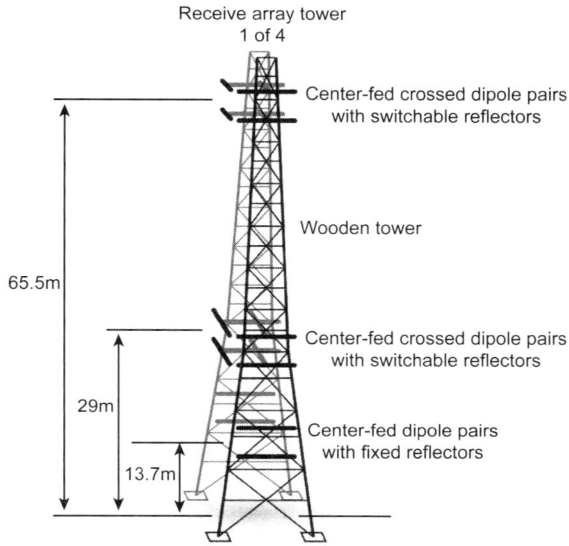

Figure 7.3 Artist depiction of the Chain Home radar receive dipole arrays.

simulated radiation patterns for the Chain Home main transmit array in free-space and over a lossy half space are presented. The transmit array simulations are perfomed in a configuration that would be similar to the way in which the Chain Home radar main transmit antenna would have been designed [1]. Section 7.5 provides a summary and a problem set is provided in Section 7.6.

7.2 Chain Home Main Transmit Dipole Array

The Chain Home main transmit array consisted of eight horizontally polarized end-fed dipole antennas with reflectors (Figure 7.2). The phase center of the main array was located 65.5m over the ground. Due to reflection from the earth's surface, the elevation radiation pattern had a first peak at 2.6° above the horizon and a first null at 5.2° that represented a gap in radar coverage. As shown in Figure 7.2, a *gap filler* four-element dipole array was located 29m over the ground, which had a first peak at 5.2°. In the antenna array analysis that follows, a center frequency of 25 MHz ($\lambda = 12$m) is assumed.

Center-fed dipoles and arrays of dipoles [13–22] typically are fed with a 50-ohm transmission line. In the case of the Chain Home transmit antenna elements, an end-fed dipole was preferred in order to handle high peak powers with a balanced high-voltage transmission line feed as discussed below. Up to about 1 MW peak power, at low duty cycle, from the transmitter was fed to the main transmit array antenna by a 600 ohm, open-wire, balanced transmission line, which was stub matched to the array of *end-fed* dipole antennas. In the conceptual form shown in Figure 7.4, two types of end-fed dipole antennas, are compared to a center-fed dipole (Figure 7.4(a)). Figure 7.4(b) depicts an offset-fed dipole [18] (referred to here as an end-fed dipole), where the feed point location from the dipole tip is set to 15% of the dipole length to achieve an approximate 600 ohm input resistance. In Figure 7.4(c), a J-pole antenna (or J-shaped antenna) [23–25] concept is shown with a radiating dipole section and a non-radiating stub matching section where the feed point location would be adjusted to 600 ohm resistive. In the Chain Home transmit array design, the dipoles have a nominal length of one-half wavelength and are spaced in a linear array nominally one-half wavelength apart. In free space at an azimuth direction broadside to the array, the Chain Home horizontally polarized dipole elements would have omnidirectional gain patterns in elevation with peak gain equal to 1.64 (2.15 dBi). However, with a reflector element (0.55λ wire length and spaced 0.18λ from a 0.5λ dipole), the peak gain in the forward direction ($\theta = 90°$) increases to approximately 3.15 (5 dBi) and in the backward direction decreases to 0.55 (-2.6 dBi) based on a FEKO method of moments simulation as shown in Figure 7.5. The wire radius for the dipole and reflector were assumed to be electrically thin and equal to 0.001λ. The simulated front-to-back ratio is approximately 7.5 dB, and the elevation half power beamwidth is $178°$. In Figure 7.6, the simulated azimuth half-power beamwidth is $72°$. Although not shown, for a dipole length 0.5λ the center-fed dipole with reflector and end-fed dipole with reflector have equivalent radiation patterns. An analytical approximation to the simulated elevation gain pattern in Figure 7.5 as a function of observation angle θ in the xz plane for the Chain Home dipole with reflector is given by

$$g_e(\theta) = 0.55 + 1.3(1 + \sin\theta) \tag{7.1}$$

Since $\theta = \pi/2 - \psi$, the gain function of the Chain Home dipole with reflector can be expressed as

$$g_e(\psi) = 0.55 + 1.3(1 + \cos\psi) \tag{7.2}$$

This element gain equation will be used in Section 7.3 in developing an equation for the Chain Home main transmit array elevation pattern in free space and over a lossy half space.

Consider now the high-voltage feeding system used for the main transmit array. In an open-wire air transmission line, the phase shift is $180°$

every one-half wavelength distance along the line. The dipoles in the 8-element array were illuminated with equal phase for broadside transmission. Based on a photograph (not shown here) of the Chain Home transmit array kindly provided by the UK Air Defence Radar Museum, the two transmission line electrical conductors followed a helical pattern in the vertical direction to alternate which of the two open-wire conductors are attached to the dipole terminals to excite adjacent dipoles in phase. An example calculation of how the Chain Home open-wire feedlines would have been designed is given below.

An open-wire transmission line with wire diameter D and parallel wire spacing s has a characteristic impedance Z_o given by [12]

$$Z_o = \frac{120}{\sqrt{\epsilon_r}} \cosh^{-1} \frac{s}{D} \tag{7.3}$$

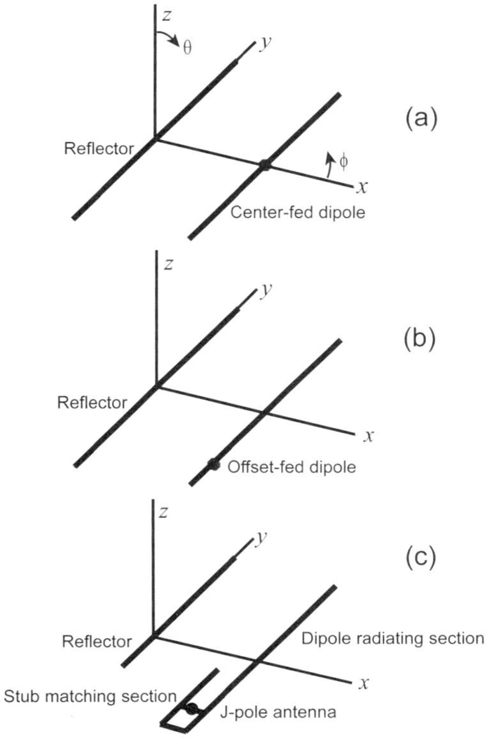

Figure 7.4 Sketch of center-fed dipole, offset-fed dipole, and J-pole antennas with reflectors.

where ϵ_r is the dielectric constant. For a desired characteristic impedance and given wire diameter, the required spacing is determined by solving Equation (7.3) for s with the result

$$s = D \cosh\left(\frac{Z_o\sqrt{\epsilon_r}}{120}\right) \tag{7.4}$$

The power handling of an open-wire transmission line is given by [12]

$$P = \frac{E_{\mathrm{a}}^2 D^2 \sqrt{\epsilon_r}}{240} \cosh^{-1}\frac{s}{D} \tag{7.5}$$

where E_{a} is the maximum allowed electric field in the dielectric, where a derated value of

$$E_{\mathrm{a}} = 2 \times 10^6 \text{ V/m} \tag{7.6}$$

is typically used at normal atmospheric pressure and temperature.

For a 600-ohm open-wire transmission line, the required wire spacing to wire diameter ratio is $s/D = 75$. Using $s/D = 75$, it follows that

$$\cosh^{-1}\frac{s}{D} = \cosh^{-1} 75 = 5.0106 \approx 5 \tag{7.7}$$

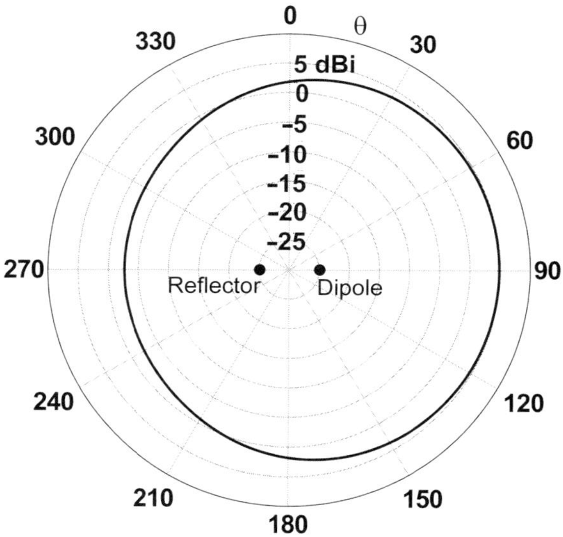

Figure 7.5 Moment method simulated H-plane gain pattern as a function of the angle θ for an offset-fed dipole antenna with reflector in free space.

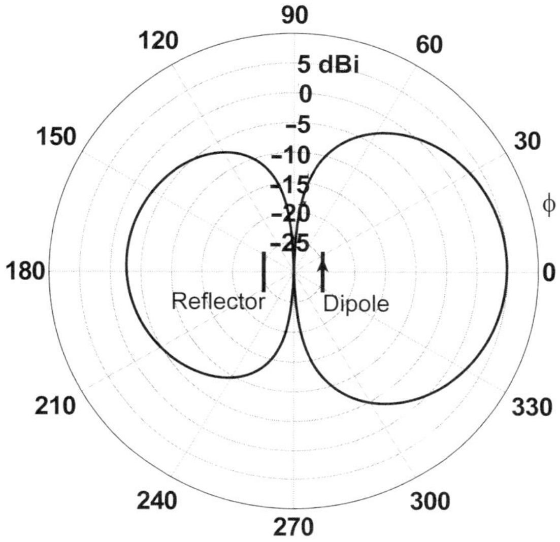

Figure 7.6 Moment method simulated E-plane gain pattern as a function of azimuth angle ϕ for an offset-fed dipole antenna with reflector in free space.

Thus, for a 600-ohm open-wire air transmission line, Equation (7.5) reduces to

$$P_{600\Omega}^{\text{open-wire air}} = \frac{E_{\text{a}}^2 D^2}{48} \tag{7.8}$$

and solving for the wire diameter D yields

$$D_{600\Omega}^{\text{open-wire air}} = \sqrt{\frac{48P}{E_{\text{a}}^2}} \tag{7.9}$$

With the Chain Home radar operating at up to 1 MW peak power, by Equation (7.9), the 600Ω open-wire transmission line could use a wire diameter on the order of 3.5 mm which is rounded up to AWG number 6 (4.11 mm). Now, from the above discussion, using $s = 75D$, the 600-ohm required open-wire transmission line spacing is $s = 30.8$ cm.

7.3 Reflection From a Planar Lossy Half Space

The elevation pattern shape of the Chain Home radar antenna is due to two components, a direct free-space term and a contribution from the reflected field from the half-space lossy dielectric medium in front of the antenna. The free-space region has permeability μ_o and permittivity ϵ_o, and the half-space region has permeability $\mu_c = \mu_o$ and permittivity ϵ_c. Consider Figure 7.7, where a plane wave with wavenumber k_o in free space is incident with grazing angle ψ upon a half space that has a wavenumber k_1. The plane

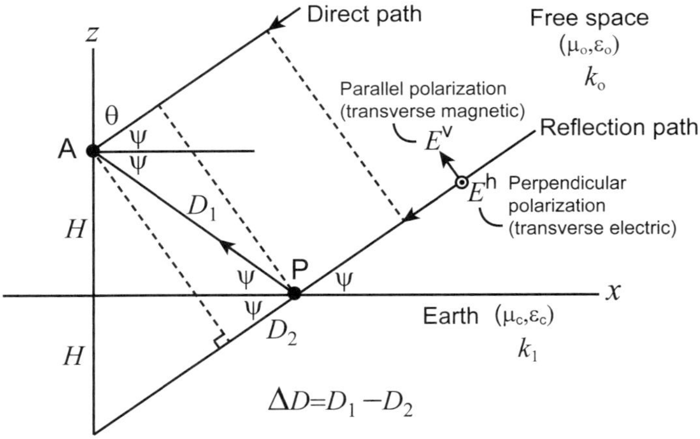

Figure 7.7 Diagram showing plane-wave incidence with a direct path and a reflected path.

of incidence is taken here to be the xz plane. The polarization sense of the electric field is defined, with respect to a flat earth, such that the incident field is polarized horizontally as E^h or vertically polarized as E^v. For plane waves, the electric and magnetic fields are orthogonal to each other. In Figure 7.7, the horizontally polarized electric field E^h is perpendicular to the plane of incidence and is referred to in the literature as perpendicular polarization or transverse electric (TE) polarization. Similarly, the vertically polarized electric field E^v is parallel to the plane of incidence and is referred to in the literature as parallel polarization or transverse magnetic (TM) polarization. In Figure 7.7, the antenna is located at position A with height H above the half space and the point of reflection at the interface between free space and the half space is at position P. In the antenna's coordinate system, the angle with respect to zenith is denoted θ, and it follows that the grazing angle ψ (also referred to here as the elevation angle) is related as $\psi = \frac{\pi}{2} - \theta$. The distance D_1 is quantified according to

$$D_1 = \frac{H}{\sin \psi} \tag{7.10}$$

and the distance D_2 is expressed in terms of D_1 by

$$D_2 = D_1 \cos 2\psi \tag{7.11}$$

and substituting Equation (7.10) in (7.11) yields

$$D_2 = H\frac{\cos 2\psi}{\sin\psi} \tag{7.12}$$

The path length distance ΔD between D_1 and D_2 is computed as

$$\Delta D = D_1 - D_2 = \frac{H}{\sin\psi}(1 - \cos 2\psi) = 2H\sin\psi \tag{7.13}$$

The total field incident at the antenna is equal to the sum of the direct field component and the reflected field component that depends on the polarization. Let the amplitude of the plane wave vertically and horizontally polarized electric fields be denoted E_o^v and E_o^h, respectively, The total vertically and horizontally polarized incident fields at the antenna are given by

$$E_A^v = E_o^v(1 + \Gamma_v e^{-j2k_o H\sin\psi}) \tag{7.14}$$

$$E_A^h = E_o^h(1 + \Gamma_h e^{-j2k_o H\sin\psi}) \tag{7.15}$$

From Kerr [26, p. 99], the Fresnel reflection coefficients Γ_v for vertical polarization and Γ_h for horizontal polarization are expressed as

$$\Gamma_v = \frac{k_1^2\sin\psi - k_o\sqrt{k_1^2 - k_o^2\cos^2\psi}}{k_1^2\sin\psi - k_o\sqrt{k_1^2 + k_o^2\cos^2\psi}} \tag{7.16}$$

$$\Gamma_h = \frac{k_o\sin\psi - \sqrt{k_1^2 - k_o^2\cos^2\psi}}{k_o\sin\psi - \sqrt{k_1^2 + k_o^2\cos^2\psi}} \tag{7.17}$$

which can be rewritten as

$$\Gamma_v = \frac{\left(\frac{k_1}{k_o}\right)^2\sin\psi - \sqrt{\left(\frac{k_1}{k_o}\right)^2 - \cos^2\psi}}{\left(\frac{k_1}{k_o}\right)^2\sin\psi + \sqrt{\left(\frac{k_1}{k_o}\right)^2 - \cos^2\psi}} \tag{7.18}$$

$$\Gamma_h = \frac{\sin\psi - \sqrt{\left(\frac{k_1}{k_o}\right)^2 - \cos^2\psi}}{\sin\psi + \sqrt{\left(\frac{k_1}{k_o}\right)^2 - \cos^2\psi}} \tag{7.19}$$

Now, from Equation (1.356) in Chapter 1, in general the complex wavenumber k_1 squared is equal to the radian frequency ω squared multiplied by the complex permeability μ_c and the complex permittivity ϵ_c as

$$k_1^2 = \omega^2\mu_c\epsilon_c \tag{7.20}$$

but it will be assumed that the complex permeability of the reflecting medium is that of free space, that is,

$$\mu_c = \mu_o \tag{7.21}$$

Thus, it follows that

$$k_1^2 = \omega^2 \mu_o \epsilon_c \tag{7.22}$$

Now since

$$\epsilon' = \epsilon_o \epsilon_r' \tag{7.23}$$

and

$$\epsilon'' = \frac{\sigma}{\omega} \tag{7.24}$$

the complex permittivity can be expressed as

$$\epsilon_c = \epsilon' - j\epsilon'' = \epsilon_o \epsilon_r' - j\frac{\sigma}{\omega} \tag{7.25}$$

The wavenumber k_o squared in free space is given by

$$k_o^2 = \omega^2 \mu_o \epsilon_o \tag{7.26}$$

Taking the ratio of Equations (7.22) and (7.26) yields

$$\frac{k_1^2}{k_o^2} = \frac{\epsilon_c}{\epsilon_o} = \epsilon_r' - j\frac{\sigma}{\omega\epsilon_o} = \epsilon_{cn} \tag{7.27}$$

where ϵ_{cn} is referred to as the normalized complex permittivity. Thus, the vertical and horizontal reflection coefficients can be expressed in terms of the angle of incidence and the normalized complex permittivity of the half space as

$$\Gamma_v = \frac{\epsilon_{cn}\sin\psi - \sqrt{\epsilon_{cn} - \cos^2\psi}}{\epsilon_{cn}\sin\psi + \sqrt{\epsilon_{cn} - \cos^2\psi}} \tag{7.28}$$

$$\Gamma_h = \frac{\sin\psi - \sqrt{\epsilon_{cn} - \cos^2\psi}}{\sin\psi + \sqrt{\epsilon_{cn} - \cos^2\psi}} \tag{7.29}$$

The Chain Home radar antennas operated near the coast, so reflection from seawater can be calculated using Equations (7.28) and (7.29). For seawater with salinity 35 g/kg at 10°C the calculated dielectric constant is $\epsilon_r = 72$ and electrical conductivity $\sigma = 4$ S/m [27]. For these dielectric parameter values, the normalized complex permittivity of the seawater is computed as $\epsilon_{cn} = 72 - j2870$. The calculated reflection coefficients for vertical and horizontal polarizations at 25 MHz for seawater are shown in Figures 7.8 to 7.11.

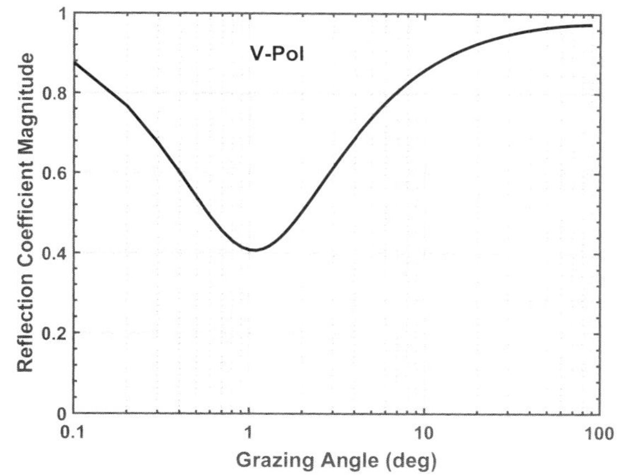

Figure 7.8 Calculated reflection coefficient magnitude for vertical polarization versus grazing angle for seawater at 25 MHz.

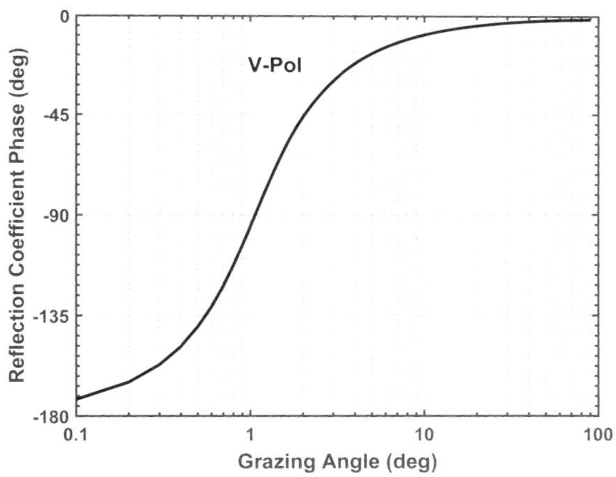

Figure 7.9 Calculated reflection coefficient phase for vertical polarization versus grazing angle for seawater at 25 MHz.

The magnitude of the reflection coefficient for vertical polarization shown in Figure 7.8 varies from about 0.4 to 0.9 for grazing angles in the range of 0.1° to 10°. Over the same range of grazing angles, in Figure 7.9 the vertically polarized reflection coefficient phase varies by about 160°. In contrast, for horizontal polarization the reflection coefficient is nearly constant with magnitude unity (Figure 7.10) and phase (Figure 7.11) approximately 180°. That is for horizontal polarization it follows that $\Gamma_h \approx -1$. The Chain

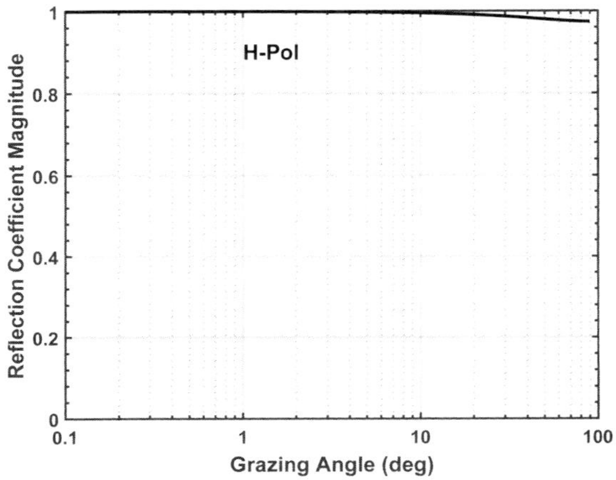

Figure 7.10 Calculated reflection coefficient magnitude for horizontal polarization versus grazing angle for seawater at 25 MHz.

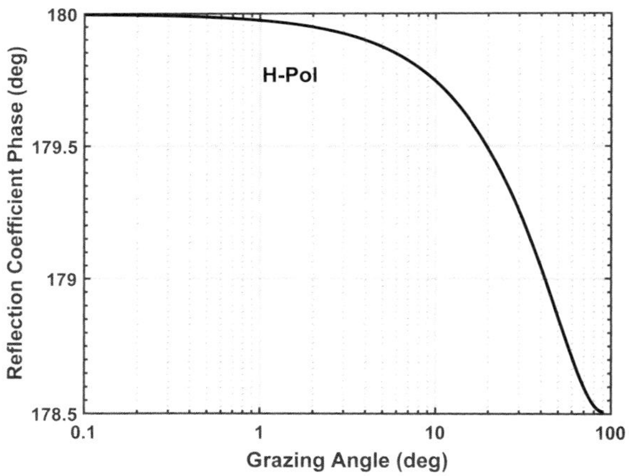

Figure 7.11 Calculated reflection coefficient phase for horizontal polarization versus grazing angle for seawater at 25 MHz.

Home radar antennas were designed for horizontal polarization, in part, because Γ_h is a stable nearly constant value. Thus, for horizontal polarization the incident field given by Equation (7.15) is approximated with $\Gamma_h = -1$ as

$$E_A^h \approx E_o^h(1 - e^{-j2k_oH\sin\psi}) = E_o^h e^{jk_oH\sin\psi} 2\sin(k_oH\sin\psi) \qquad (7.30)$$

Now, consider the case of a horizontally polarized transmit antenna like the Chain Home eight-element dipole array. It is assumed that the transmit array is uniformly illuminated. In this application, the elevation angle with respect to the horizon is the same as the grazing angle. The Chain Home transmit array antenna far-field gain pattern, as a function of the elevation angle ψ including reflection from the seawater, follows from Equation (7.30) as

$$G_h^{\text{Total}}(\psi) \approx 4\sin^2(k_o H \sin\psi) G_h^{\text{Free Space}}(\psi) \tag{7.31}$$

where $G_h^{\text{Free Space}}(\psi)$ is the gain radiation pattern of the Chain Home transmit array in free space. Now, recall from Equation (2.24) in Chapter 2 that the array factor is given by

$$\text{AF} = e^{j(N-1)u/2} \frac{\sin(Nu/2)}{\sin(u/2)} \tag{7.32}$$

In this case,

$$u = k_o d \sin\psi \tag{7.33}$$

where d is the array element spacing. Taking the magnitude of Equation (7.32) and then normalizing to unity yields the normalized form of the array factor magnitude as

$$|\text{AF}|_{\text{Normalized}} = \left| \frac{\sin(Nu/2)}{N\sin(u/2)} \right| \tag{7.34}$$

Since the free-space antenna gain is equal to the element gain times the number of elements times the normalized array factor magnitude squared, for the azimuth broadside case the free-space elevation gain pattern can now be expressed as

$$G_h^{\text{Free Space}}(\psi) = g_e(\psi) N \left| \frac{\sin(Nu/2)}{N\sin(u/2)} \right|^2 \tag{7.35}$$

For the Chain Home transmit array, the number of dipoles is $N = 8$ and the spacing between the dipoles is $d = \lambda/2$. Thus, the factor u reduces to

$$u = k_o d \sin\psi = \frac{2\pi}{\lambda}\frac{\lambda}{2}\sin\psi = \pi\sin\psi \tag{7.36}$$

Now the phase center position, or height, of the Chain Home transmit array is $H = 215$ ft (65.532m). At 25 MHz, the wavelength is 12 meters, so the height of the transmit array phase center is 5.461λ. Thus, the factor $k_o H$ reduces to

$$k_o H = \frac{2\pi}{\lambda} 5.461\lambda = 10.92\pi \tag{7.37}$$

Substituting Equation (7.35) into Equation (7.31), yields

$$G_h^{\text{Total}}(\psi) \approx 4\sin^2(k_o H \sin\psi) g_e(\psi) N \left| \frac{\sin(Nu/2)}{N\sin(u/2)} \right|^2 \tag{7.38}$$

which with Equations (7.36) and (7.37) and with $N = 8$ reduces to

$$G_h^{\text{Total}}(\psi) \approx 32 g_e(\psi) \sin^2(10.92\pi \sin\psi) \left| \frac{\sin(4\pi \sin\psi)}{8 \sin(\frac{\pi}{2} \sin\psi)} \right|^2 \tag{7.39}$$

where the element gain $g_e(\psi)$ for the dipole/reflector combination is given by Equation (7.2). Equation (7.39) is an approximate analytic expression for the H-plane gain elevation pattern of the Chain Home main transmit dipole array with reflectors including the multipath contribution from a lossy half space. This equation does not include mutual coupling effects. In the next section, the array is analyzed by the method of moments, which includes array mutual coupling.

7.4 Simulated Radiation Patterns for the Main Transmit Array

A moment method simulation model for the Chain Home main transmit eight-element array over a lossy dielectric half space is shown in Figure 7.12. In this simulation, the elements are assumed to be center fed (depicted by the dot on the element) and the far-field directivity is assumed to be representative of the end-fed dipoles used in the Chain Home main antenna. The height of the array is assumed to be 65.5m as in Figure 7.2. The center frequency is assumed here to be 25 MHz and the wires that are modeled are assumed electrically

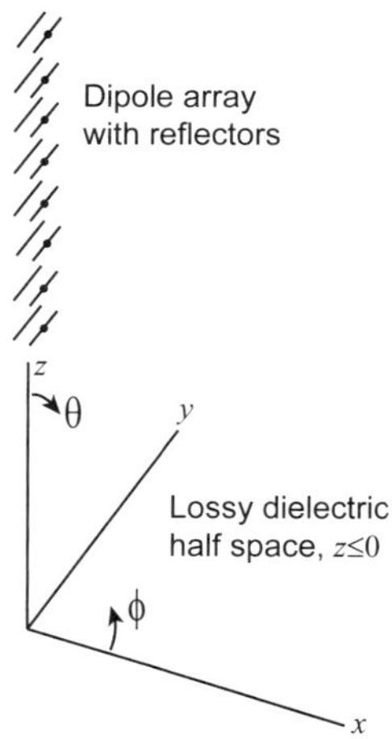

Figure 7.12 Moment method simulation model for the Chain Home main transmit eight-element array over a lossy dielectric half space.

thin (0.001λ wire radius). The thin-wire dipoles are $\lambda/2$ at 25 MHz which corresponds to a physical length of 6m, and is the same as the array element spacing. The thin-wire reflectors are 0.55λ at 25 MHz, which corresponds to a physical length of 6.6m. The spacing between the dipoles and reflectors is 0.18λ or 2.16m.

The simulated horizontally polarized ($\hat{\phi}$ component) gain radiation patterns for the Chain Home main transmit eight-element array in free space are summarized in Figures 7.13 to 7.15. The simulated three-dimensional

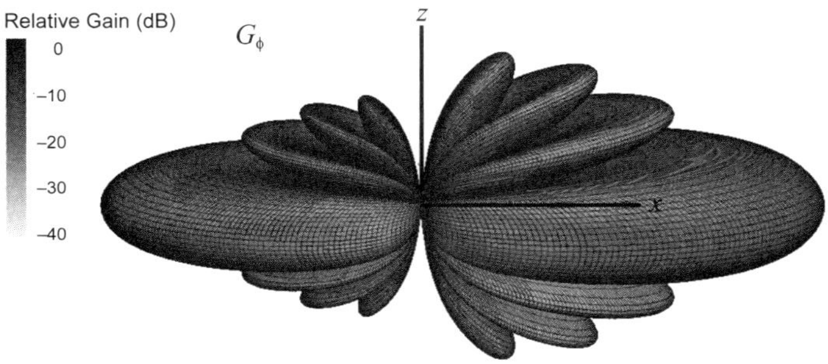

Figure 7.13 Simulated three-dimensional radiation pattern (horizontal polarization) for the Chain Home main transmit eight-element array in free space.

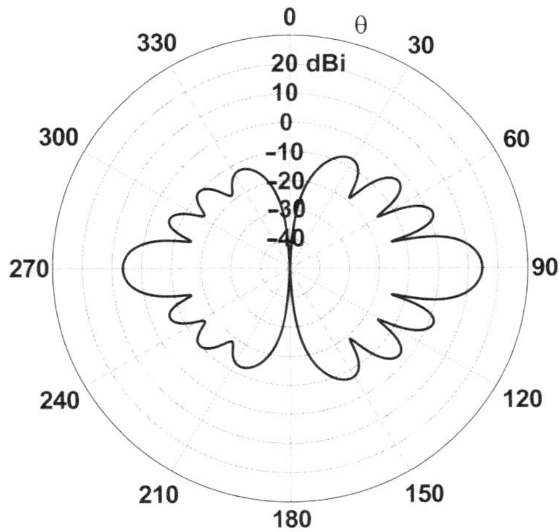

Figure 7.14 Simulated H-plane polar gain radiation pattern (horizontal polarization) in elevation at 25 MHz for the Chain Home main transmit eight-element array in free space.

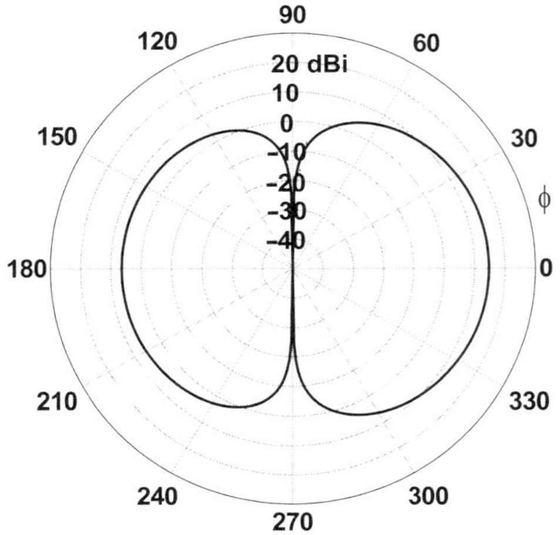

Figure 7.15 Simulated E-plane polar gain radiation pattern (horizontal polarization) in azimuth at 25 MHz for the Chain Home main transmit eight-element array in free space.

radiation pattern for the Chain Home main transmit eight-element array in free space is shown in Figure 7.13. The corresponding free-space H-plane polar radiation pattern as a function of the angle θ is shown in Figure 7.14. The simulated array peak directivity is 14.4 dBi and the elevation half-power beamwidth is 10.9° at 25 MHz. Similarly, the free-space E-plane azimuthal polar radiation pattern as a function of the angle ϕ is shown in Figure 7.15. The simulated free-space array azimuth beamwidth is 75°.

Next, with the array phase center located 65.5m over the earth's surface and the electrically conducting half space modeled as seawater, the moment method simulated far-field gain patterns are summarized in Figures 7.16 to 7.18. The simulated 3D radiation pattern at 25 MHz is shown in Figure 7.16. The corresponding far-field elevation (H-plane) gain pattern is shown in Figure 7.17. With the reflection from the lossy half space, the far-field gain has a null at $\theta = 90°$ and additional lobes are formed in elevation. The simulated first peak above the horizon occurs at $\theta = 87.5°$ ($\psi = 2.5°$), the peak gain is 20.3 dBi, and the half-power beamwidth is 2.6°. The first null (gap) occurs at $\theta = 84.8°$ ($\psi = 5.2°$). Both the first peak and first null in elevation are in agreement with the peak and null angles stated in the paper by Neale [1]. Finally, the azimuth gain pattern cut taken through the main beam peak at 2.5° elevation is shown in Figure 7.18, and the half-power beamwidth is 75°. The simulated backlobe gain at $\phi = 180°$ is 12 dBi, so the front-to-back ratio is 8.3 dB. Finally, Figure 7.19 provides a comparison between the Chain Home

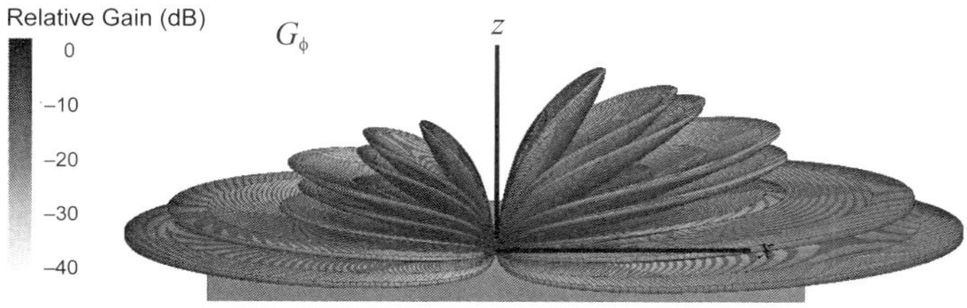

Figure 7.16 Simulated 3D radiation pattern at 25 MHz for the Chain Home main transmit eight-element array over a lossy dielectric half space.

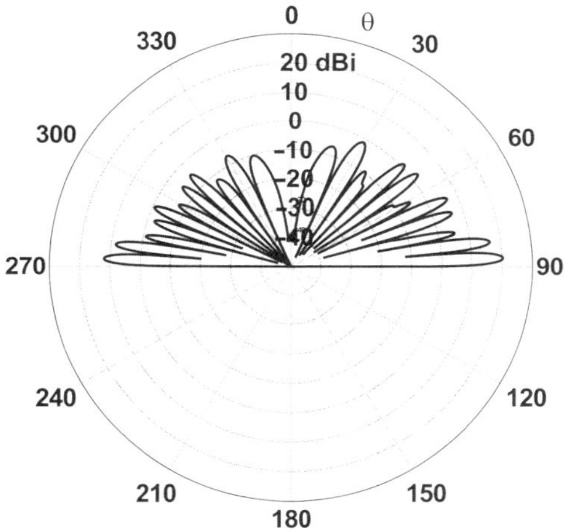

Figure 7.17 Simulated H-plane polar radiation pattern in elevation at 25 MHz for the Chain Home main transmit eight-element array over a lossy dielectric half space.

analytic elevation gain pattern computed from Equations (7.39) and (7.2) and the moment method simulated gain pattern. Good agreement between the analytic method and the moment method simulation is observed in terms of absolute peak gain and location of the sidelobe peaks and nulls in elevation.

7.5 Summary

This chapter has analyzed an HF dipole array with reflectors in a configuration that would be similar to the way in which the Chain Home radar main transmit antenna would have been designed. The equations for the Fresnel reflection coefficients for horizontal and vertical polarization were reviewed.

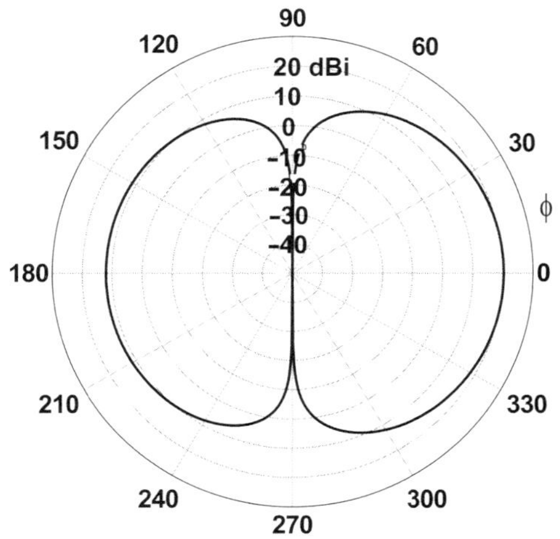

Figure 7.18 Simulated azimuth polar radiation pattern at an elevation angle of 2.5° at 25 MHz for the Chain Home main transmit eight-element array over a lossy dielectric half space.

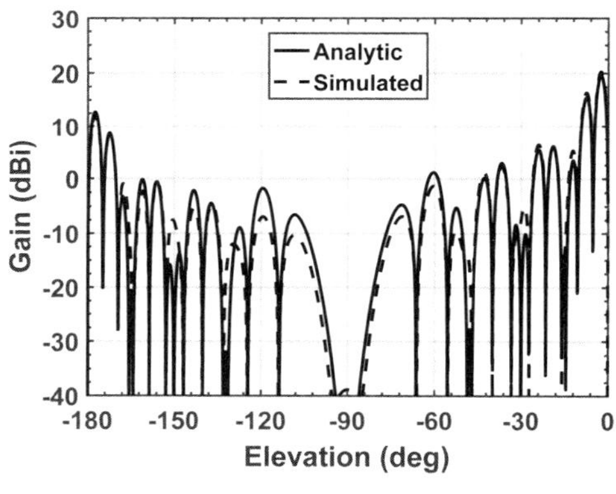

Figure 7.19 Comparison of the analytic computed and moment method simulated H-plane gain radiation pattern at 25 MHz as a function of elevation angle (ψ) for the Chain Home main transmit eight-element array over a lossy dielectric half space.

The amplitude and phase of the reflection coefficient for a lossy half-space medium were computed at the Chain Home center frequency, 25 MHz. The main transmit array performance was simulated using the method of moments in free space and over a lossy half space, and the gain pattern characteristics were quantified. The analytic derived elevation gain pattern was shown to be in good agreement with the moment method simulation.

7.6 PROBLEM SET

7.1 Write a computer program to compute the vertical and horizontal Fresnel reflection coefficients given by Equations (7.28) and (7.29), respectively. Plot the magnitude and phase of the reflection coefficients at 25 MHz for two cases: (a) Seawater with dielectric constant $\epsilon_r' = 72$ and electrical conductivity $\sigma =$ 4 S/m. The results should be identical to Figures 7.8 to 7.11. (b) Dry soil with dielectric constant $\epsilon_r' = 4$ and electrical conductivity $\sigma = 0.1$ S/m.

7.2 Write a computer program to compute and plot the Chain Home main transmit array elevation gain pattern $G_h^{\text{Total}}(\psi)$ given by Equation (7.38), with $N = 8$, $H = 65.5$m and frequency 25 MHz, and using the element elevation gain pattern $g_e(\psi)$ given by Equation (7.2). Compare the elevation gain pattern results with the results shown in Figure 7.19. Then repeat the calculations at 20 MHz and 30 MHz and verify that the elevation angle of the main beam peak is frequency dependent.

7.3 Repeat Problem 7.2, but now for the the Chain Home gap-filler transmit array elevation gain pattern $G_h^{\text{Total}}(\psi)$ given by Equation (7.38), with $N = 4$, $H = 29$m and frequency 25 MHz and show that the peak gain occurs at 5.2° to fill in the null of the main transmit array.

References

[1] B.T. Neale, "CH–The First Operational Radar," *The GEC Journal of Research (Incorporating the Marconi Review),* 1985, vol. 3, no. 2, pp. 73-83.

[2] R.J. Mailloux, "A History of Phased Array Antennas," Chapter 17 in *History of Wireless,* T.K. Sarkar, R.J. Mailloux, A.A. Oliner, M. Salazar-Palma, and D.L. Sengupta, eds. New Jersey: Wiley, 2006.

[3] J. Wood, *History of International Broadcasting, Vol. 2,* London, UK: The Institution of Electrical Engineers, 2000.

[4] R.W.P. King and S.S. Sandler, "Antennas with Transmission-Line Interconnections," *IEEE Trans. Antennas and Propagat.*, 1970, pp. 616–621.

[5] A.G.P. Boswell, "Analysis of Curtain Arrays for HF Broadcasting," *IEE Proceedings,* vol. 136, part H, no. 2, April 1989.

[6] S.W. Kershner, "Curtain Antennas for High-Power HF Broadcasting Applications," *IEEE Trans. on Broadcasting,* vol. BC-14, no. 2, June 1968, pp. 85–94.

[7] J.M. Martin and F. Guilbault, "A New Design for High Performance HF Curtain Antennas," *IEEE Trans. on Broadcasting,* vol. 34, no. 2, June 1988, pp. 180–185.

[8] S.T. Baily, K.L. Beeke and R.G. Manton, "The Performance of HF Curtain Antennas With Irregular Foregrounds," *1988 Fourth Int. Conf. on HF Radio Systems and Techniques,* London, UK, 1988, pp. 132–136.

[9] E.J. Sterba, "Short-Wave Transmitting Antennas," *Bell Laboratories Record,* vol. 7, no. 12, August 1929, pp. 502–509.

[10] P. Knight, R.E. Davies, R.G. Manton, "The Vertical Radiation Patterns of High-Frequency Curtain Arrays on Plateau Sites," The British Broadcasting Corporation, Research Report No. E-084, 1963.

[11] R. Wilensky, "High-Power, Broad-Bandwidth HF Dipole Curtain Array with Extensive Vertical and Azimuthal Beam Control," *IEEE Trans. on Broadcasting,* vol. 34, no. 2, June 1988, pp. 201–209.

[12] J. Volakis, ed., *Antenna Engineering Handbook, 4th ed.* New York: McGraw-Hill, 2007.

[13] Allen, J., "Gain and Impedance Variation in Scanned Dipole Arrays," *IEEE Trans. Antennas and Propagat.*, vol. 10, no. 5, 1962, pp. 566-572.

[14] Hansen, R.C., (ed.), *Microwave Scanning Antennas, Vol. 2*, Los Altos, CA: Peninsula Publishing, 1985.

[15] Kraus, J.D., *Antennas,* 2nd ed., New York: McGraw-Hill, 1988.

[16] Mailloux, R.J., "Phased Array Theory and Technology," *Proc. IEEE*, Vol. 70, No. 3, 1982, pp. 246-291.

[17] S.A. Schelkunoff and H.T. Friis, *Antennas: Theory and Practice*. New York: Wiley, 1952.

[18] W.L. Stutzman and G.A. Thiele, *Antenna Theory and Design, 2nd ed.* New York: Wiley, 1998.

[19] C.A. Balanis, *Antenna Theory and Design, 3rd ed.* New York: Wiley, 2005.

[20] A.A. Oliner and G.H. Knittel, eds., *Phased Array Antennas*. Dedham, Mass.: Artech House, 1972.

[21] R.J. Mailloux, *Phased Array Handbook, 2nd ed.* Norwood, Mass.: Artech House, 2005.

[22] A.J. Fenn, *Adaptive Antennas and Phased Arrays for Radar and Communications.* Norwood, Mass.: Artech House, 2008.

[23] D. Richardson, "The J-Pole Revisited," *CQ Amateur Radio*, March 1998, pp. 34–41.

[24] S.H. Choudhury, M.I. Momtaz, and M.A. Matin, "Analytical Deduction of the Salient Properties of a Half Wavelength J-Pole Antenna," *2010 International Conference on Computational Intelligence and Communication Networks,* pp. 32–35.

[25] "Analysis of J-Pole Antenna Configurations for Underwater Communications," *OCEANS 2016 MTS/IEEE Monterey*, pp. 1–5.

[26] D.E. Kerr (ed.), *Propagation of Short Radio Waves,* IEE Electromagnetic Wave Series 24, London, UK: Peter Perigrinus Ltd., 1990.

[27] A. Stogryn, "Equations for Calculating the Dielectric Constant of Saline Water," *IEEE Trans. on Microwave Theory and Techniques,* August, 1971, pp. 733-736.

8

Circular Waveguide Antenna for a Portable Computer-Based Radar

8.1 Introduction

A custom radar kit that integrates with a portable computer (laptop) for assembly and operation by students and researchers has recently been developed at MIT Lincoln Laboratory. This small radar system [1–4] shown in Figure 8.1 operates in three different modes: Doppler, range, and synthetic aperture radar (SAR) modes. The radar operates in the Industrial, Scientific, and Medical (ISM) band from 2.4 GHz to 2.5 GHz. A block diagram of the radar is shown in Figure 8.1(a). On transmit, a linear voltage ramp is generated by the digital-to-analog (DAC) device in the microcontroller and is amplified to a desired driving level for the voltage controlled oscillator (VCO). The resulting RF signal from the VCO is a linear frequency modulated (LFM) signal (referred to as a chirp). An attenuator then sets the desired signal level as the input to a power amplifier. The output signal from the power amplifier is divided into a path to the transmitting antenna and to a mixer as a local oscillator reference signal. The radar signal transmitted by the transmit antenna will propagate to a target and reflect back toward the receive antenna. The radar target signal output from the receive antenna is passed through a low noise amplifier (LNA) and is then mixed with the local oscillator waveform to downconvert the signal to baseband. The baseband signal is amplified and passed through a low pass filter and is then analog-to-digital (ADC) converted to data that can be processed in the portable computer. As shown in Figure 8.1(b), the assembled radar kit uses two identical cylindrical metal cans that serve as the antennas, one for transmitting and one for receiving radar signals. Separate transmit and receive antennas are utilized

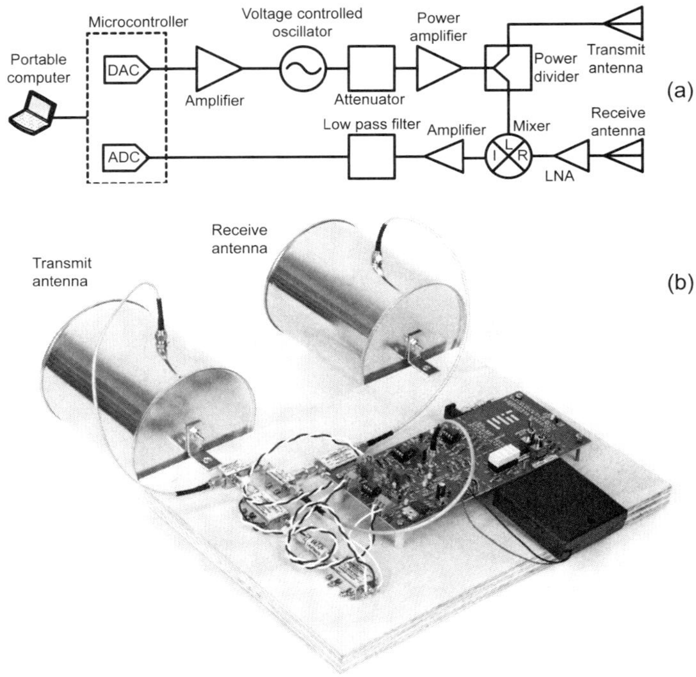

Figure 8.1 Portable computer-based radar kit: (a) Block diagram and (b) photograph showing radar system hardware and two linearly-polarized circular waveguide antennas for transmit and receive.

here to reduce the transmit-to-receive mutual coupling, which helps avoid self interference. These antennas operate as probe-fed linearly polarized open-end circular waveguides. Open-end circular waveguides can be dual-polarized and are commonly used in feeding reflector antennas (see for example the feeds used for the compact range reflector discussed in Chapter 12) or in large phased arrays [5, 6]. For the laptop radar, this antenna design is desirable due to its low cost and good performance in terms of reflection coefficient, peak gain, and beamwidth. This chapter describes the theory, simulations, and measurements of these open-end circular waveguide antennas used in the laptop radar system.

An open-end circular waveguide antenna is depicted in Figure 8.2. The waveguide is assumed to have a diameter D and length L. In the cylindrical coordinate system, which is appropriate for analysis of this waveguide structure, ρ is the radial distance from the z axis of the circular cylinder and the angle ϕ is taken with respect to the x axis. A thin-wire monopole probe with length l is used to feed the circular waveguide at a distance s from the

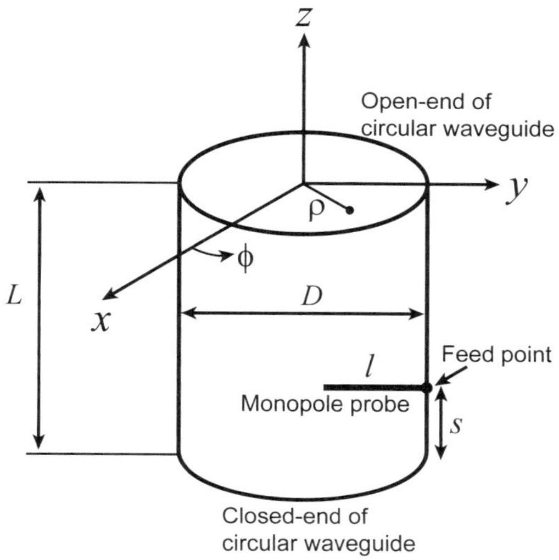

Figure 8.2 Diagram of a probe-fed open-end circular waveguide antenna.

closed-end of the circular waveguide.

In this chapter, the method of moments will be used to simulate the circular waveguide fields, as well as the reflection coefficient, far-field gain radiation patterns, and mutual coupling. The circular waveguide is assumed to be perfectly conducting in the analysis. The theory for fields propagating in a circular waveguide is described in the next section. The design of the open-end circular waveguide antenna for the portable radar is described in Section 8.3, and simulations and measurements are presented in Section 8.4. Section 8.5 provides a summary and Section 8.6 provides a problem set.

8.2 Theory

8.2.1 Background

As described in detail by Pozar [7, pp. 121-130], Maxwell's frequency-domain curl equations in a source-free region can be used to derive the vector electric $\boldsymbol{E}$ and magnetic $\boldsymbol{H}$ fields that propagate in the circular waveguide, that is,

$$\nabla \times \boldsymbol{E} = -j\omega\mu\boldsymbol{H} \tag{8.1}$$

$$\nabla \times \boldsymbol{H} = j\omega\epsilon\boldsymbol{E} \tag{8.2}$$

where the medium filling the waveguide is characterized by the permeability μ and permittivity ϵ, and ω is the radian frequency.

In cylindrical coordinates (ρ, ϕ, z), as depicted in Figure 8.2, the vector electric and magnetic fields can be expressed in terms of electric $\boldsymbol{e}$ and magnetic $\boldsymbol{h}$ transverse $(\hat{\rho}, \hat{\phi})$ and longitudinal $(\hat{z})$ components as

$$\boldsymbol{E}(\rho, \phi, z) = [\boldsymbol{e}(\rho, \phi) + \hat{\boldsymbol{z}} e_z(\rho, \phi)]e^{-\gamma z} \tag{8.3}$$

$$\boldsymbol{H}(\rho, \phi, z) = [\boldsymbol{h}(\rho, \phi) + \hat{\boldsymbol{z}} h_z(\rho, \phi)]e^{-\gamma z} \tag{8.4}$$

where in general the propagation constant γ is a function of the attenuation constant α and the phase propagation constant β as

$$\gamma = \alpha + j\beta \tag{8.5}$$

In the present case, the waveguide attenuation (loss) is assumed to be equal to zero, that is, $\alpha = 0$, so $\gamma = j\beta$.

Depending on the diameter of the circular waveguide, the medium filling the waveguide, the operating frequency, and method of excitation, a number of different modes [7–12] can propagate in the waveguide These modes are designated here by the indices nm and are referred to as transverse electric TE_{nm} and transverse magnetic TM_{nm}. Below a certain frequency, each mode will be cutoff and not propagate in the waveguide. Assume now a circular waveguide with diameter D. The waveguide is driven by a linearly polarized wire probe with length l and diameter d that is inserted into the waveguide at a distance s from the closed-end backwall of the waveguide. The lowest order, dominant, mode that propagates in a circular waveguide is the TE_{11} mode, which is depicted in Figure 8.3. The next mode that propagates is the TM_{01} mode, which is depicted in Figure 8.4. The next three modes that can propagate in a circular waveguide are TE_{21}, TE_{01}, and TM_{11}.

The wavenumber in a medium with general values of permeability μ and permittivity ϵ is given in terms of the radian frequency ω by

$$k = \omega\sqrt{\mu\epsilon} = 2\pi f\sqrt{\mu\epsilon} \tag{8.6}$$

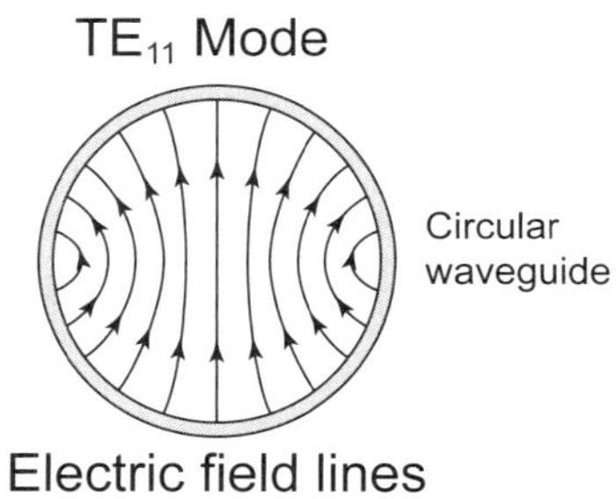

Figure 8.3 Diagram of the TE_{11} mode electric field lines in a circular waveguide. Redrawn from [9].

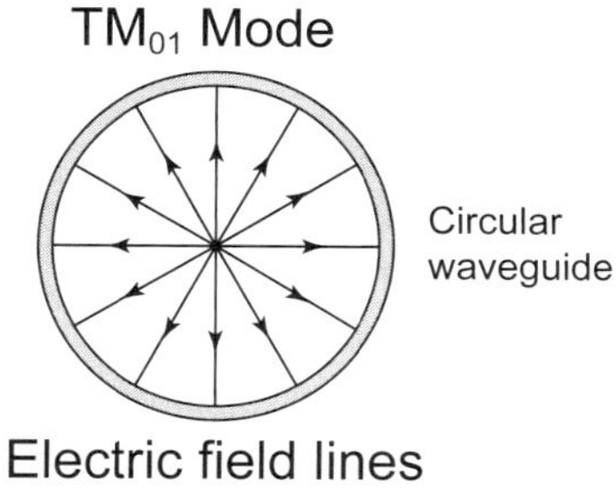

Figure 8.4 Diagram of the TM_{01} mode electric field lines in a circular waveguide. Redrawn from [9].

where $\omega = 2\pi f$ is the radian frequency and f is the frequency. The free-space permittivity is given by ϵ_o and the free-space permeability is given by μ_o. In the case where the medium has a relative permittivity ϵ_r and a relative permeability μ_r, the wavenumber can be expressed as

$$k = \omega\sqrt{\mu_o\mu_r\epsilon_o\epsilon_r} = \omega\sqrt{\mu_o\epsilon_o}\sqrt{\mu_r\epsilon_r} = \frac{\omega}{c}\sqrt{\mu_r\epsilon_r} = \frac{2\pi f}{c}\sqrt{\mu_r\epsilon_r} \tag{8.7}$$

where c is the speed of light. Solving Equation (8.7) for the frequency in terms of the wavenumber yields

$$f = \frac{ck}{2\pi\sqrt{\mu_r\epsilon_r}} \tag{8.8}$$

With an air-filled waveguide as used in the laptop radar antenna the wavenumber is expressed as

$$k = k_o = \omega\sqrt{\mu_o\epsilon_o} = \frac{2\pi}{\lambda_o} \tag{8.9}$$

where the free-space wavelength is given by

$$\lambda_o = \frac{c}{f} \tag{8.10}$$

The general form of the transverse electric and magnetic fields in cylindrical coordinates are derived from the longitudinal components as

$$E_\rho = -\frac{j}{k_c^2}\left(\beta\frac{\partial E_z}{\partial\rho} + \frac{\omega\mu}{\rho}\frac{\partial H_z}{\partial\phi}\right) \tag{8.11}$$

$$E_\phi = -\frac{j}{k_c^2}\left(\frac{\beta}{\rho}\frac{\partial E_z}{\partial\phi} - \omega\mu\frac{\partial H_z}{\partial\rho}\right) \tag{8.12}$$

$$H_\rho = \frac{j}{k_c^2}\left(\frac{\omega\epsilon}{\rho}\frac{\partial E_z}{\partial \phi} - \beta\frac{\partial H_z}{\partial \rho}\right) \tag{8.13}$$

$$H_\phi = -\frac{j}{k_c^2}\left(\omega\epsilon\frac{\partial E_z}{\partial \rho} + \frac{\beta}{\rho}\frac{\partial H_z}{\partial \phi}\right) \tag{8.14}$$

where k_c is referred to as the cutoff wavenumber and is related to the wavenumber k and the phase propagation constant β as

$$k_c = \sqrt{k^2 - \beta^2} \tag{8.15}$$

Solving Equation (8.15) for β yields

$$\beta = \sqrt{k^2 - k_c^2} \tag{8.16}$$

The cutoff wavelength λ_c in terms of the cutoff wavenumber is given by

$$\lambda_c = \frac{2\pi}{k_c} \tag{8.17}$$

The guide wavelength λ_g is the distance along the length of the waveguide, in which the phase of the field cycles through a full 360°, and is expressed as

$$\lambda_g = \frac{2\pi}{\beta} = \frac{2\pi}{\sqrt{k^2 - k_c^2}} = \frac{\lambda}{\sqrt{1 - \left(\frac{\lambda}{\lambda_c}\right)^2}} \tag{8.18}$$

The guide wavelength will be used to set the desired distance, one quarter of a guide wavelength, from the feed probe to the backwall of the circular waveguide.

8.2.2 Transverse Electric Field Modes

The transverse electric field modes in a circular waveguide are such that

$$E_z = 0 \tag{8.19}$$

and the magnetic field component H_z is a solution to the wave equation

$$\nabla^2 H_z + k^2 H_z = 0 \tag{8.20}$$

Letting

$$H_z(\rho, \phi, z) = h_z(\rho, \phi)e^{-j\beta z} \tag{8.21}$$

and applying separation of variables to Equation (8.20) in cylindrical coordinates. The longitudinal magnetic field is then solved for as

$$h_z(\rho, \phi) = (A \sin n\phi + B \cos n\phi) J_n(k_c \rho) \tag{8.22}$$

The fields then in a circular waveguide in cylindrical coordinates are given by [7, pp. 122–125]

$$E_\rho^{\mathrm{TE}} = -\frac{j\omega\mu n}{k_c^2 \rho}(A \cos n\phi - B \sin n\phi) J_n(k_c \rho) e^{-j\beta z} \tag{8.23}$$

$$E_\phi^{\mathrm{TE}} = \frac{j\omega\mu}{k_c}(A \sin n\phi + B \cos n\phi) J_n'(k_c \rho) e^{-j\beta z} \tag{8.24}$$

$$H_\rho^{\mathrm{TE}} = -\frac{j\beta}{k_c}(A \sin n\phi + B \cos n\phi) J_n'(k_c \rho) e^{-j\beta z} \tag{8.25}$$

$$H_\phi^{\mathrm{TE}} = -\frac{j\beta n}{k_c^2 \rho}(A \cos n\phi - B \sin n\phi) J_n(k_c \rho) e^{-j\beta z} \tag{8.26}$$

where $J_n(k_c\rho)$ is the Bessel function of the first kind of order n, and $J_n'(k_c\rho)$ is the derivative of J_n. The amplitude constants, A and B depend on the excitation applied to the waveguide, such as a probe inserted into the waveguide.

Now the boundary condition that the tangential component of the electric field, in this case E_ϕ according to Equation (8.24), must be zero at the cylindrical waveguide wall at the radial distance $\rho = r = D/2$. This boundary condition is enforced by setting

$$J_n'(k_c D/2) = 0 \tag{8.27}$$

Next, denote a parameter q_{nm} as the mth root, or value, of J_n' that enforces the boundary condition, that is, the zeros of the derivative of the nth order Bessel function $J_n(x)$. It follows then that

$$J_n'(k_c D/2) = J_n'(q_{nm}) = 0 \tag{8.28}$$

so that the root parameter for the nmth TE mode is expressed as

$$q_{nm} = k_c \frac{D}{2} \tag{8.29}$$

and then the cutoff wavenumber for the TE_{nm} mode can be expressed as

$$k_c^{\mathrm{TE}_{nm}} = \frac{2q_{nm}}{D} \tag{8.30}$$

The q_{nm} roots are found in mathematical tables. The roots for the first three TE modes are as follows: TE_{11} ($q_{11} = 1.841$), TE_{21} ($q_{21} = 3.054$), TE_{01} ($q_{01} = 3.832$). From Equations (8.8) and (8.30), the cutoff frequency for the TE_{nm} mode is expressed as,

$$f_c^{\text{TE}_{nm}} = \frac{ck_c^{\text{TE}_{nm}}}{2\pi\sqrt{\mu_r\epsilon_r}} = \frac{cq_{nm}}{\pi D\sqrt{\mu_r\epsilon_r}} \tag{8.31}$$

8.2.2.1 TE_{11} Mode

The TE_{11} cutoff wavenumber is expressed in terms of the circular waveguide radius r or diameter D from Equation (8.30) as

$$k_c = \frac{q_{11}}{r} = \frac{2q_{11}}{D} \tag{8.32}$$

where

$$q_{11} = 1.841 \tag{8.33}$$

Substituting Equation (8.33) in (8.32), the TE_{11} cutoff wavenumber is then expressed in terms of the circular waveguide diameter D as

$$k_c^{\text{TE}_{11}} = \frac{3.682}{D} \tag{8.34}$$

The corresponding cutoff frequency of the TE_{11} mode is given by

$$f_c^{\text{TE}_{11}} = \frac{q_{11}c}{\pi D\sqrt{\epsilon_r}} = \frac{1.841c}{\pi D\sqrt{\epsilon_r}} \tag{8.35}$$

Substituting Equation (8.34) in (8.17) yields the TE_{11} mode cutoff wavelength in terms of the waveguide diameter as

$$\lambda_c^{\text{TE}_{11}} = \frac{2\pi}{k_c^{\text{TE}_{11}}} = \frac{2\pi}{(3.682/D)} = 1.7065D \tag{8.36}$$

For propagation along the length of the circular waveguide the guide wavelength for the TE_{11} mode is expressed using Equation (8.18) as

$$\lambda_g^{\text{TE}_{11}} = \frac{\lambda}{\sqrt{1-\left(\frac{\lambda}{\lambda_c}\right)^2}} = \frac{\lambda}{\sqrt{1-\left(\frac{\lambda}{1.7065D}\right)^2}} \tag{8.37}$$

8.2.3 Transverse Magnetic Field Modes

The transverse electric and magnetic field components for the TM modes in a circular waveguide in cylindrical coordinates are given by [7, pp. 124, 125]

$$E_\rho^{\mathrm{TM}} = -\frac{j\beta}{k_c}(A\sin n\phi + B\cos n\phi)J_n'(k_c\rho)e^{-j\beta z} \tag{8.38}$$

$$E_\phi^{\mathrm{TM}} = -\frac{j\beta n}{k_c^2\rho}(A\cos n\phi - B\sin n\phi)J_n(k_c\rho)e^{-j\beta z} \tag{8.39}$$

$$H_\rho^{\mathrm{TM}} = \frac{j\omega\epsilon n}{k_c^2\rho}(A\cos n\phi - B\sin n\phi)J_n(k_c\rho)e^{-j\beta z} \tag{8.40}$$

$$H_\phi^{\mathrm{TM}} = -\frac{j\omega\epsilon}{k_c}(A\sin n\phi + B\cos n\phi)J_n'(k_c\rho)e^{-j\beta z} \tag{8.41}$$

As for the TE case, the boundary condition that the tangential component of the electric field, in this case E_ϕ according to Equation (8.39), must be zero at the cylindrical waveguide wall at the radial distance $\rho = r = D/2$. This boundary condition is enforced by setting

$$J_n(k_c D/2) = 0 \tag{8.42}$$

Next, denote a parameter p_{nm} as the mth root, or value, of J_n that enforces the boundary condition, that is, the zeros of the nth order Bessel function $J_n(x)$. It follows then that

$$J_n(k_c D/2) = J_n(p_{nm}) = 0 \tag{8.43}$$

so that the root parameter for the nmth TM mode is expressed as

$$p_{nm} = k_c\frac{D}{2} \tag{8.44}$$

and then the cutoff wavenumber for the TM_{nm} mode can be expressed as

$$k_c^{\mathrm{TM}_{nm}} = \frac{2p_{nm}}{D} \tag{8.45}$$

The p_{nm} roots are found in mathematical tables. The roots for the first two TM modes are as follows: TM_{01} ($p_{01} = 2.405$), TM_{11} ($p_{11} = 3.832$). From Equations (8.8) and (8.45), the cutoff frequency for the TM_{nm} mode is expressed as,

$$f_c^{\mathrm{TM}_{nm}} = \frac{ck_c^{\mathrm{TM}_{nm}}}{2\pi\sqrt{\mu_r\epsilon_r}} = \frac{cp_{nm}}{\pi D\sqrt{\mu_r\epsilon_r}} \tag{8.46}$$

8.2.3.1 TM_{01} Mode

The TM_{01} cutoff wavenumber is expressed in terms of the circular waveguide radius r or diameter D from Equation (8.45) as

$$k_c = \frac{p_{01}}{r} = \frac{2p_{01}}{D} \tag{8.47}$$

where

$$p_{01} = 2.405 \tag{8.48}$$

From Equation (8.8), the corresponding cutoff frequency of the TM_{01} mode is given by

$$f_c^{TM_{01}} = \frac{p_{01}c}{\pi D\sqrt{\epsilon_r}} = \frac{2.405c}{\pi D\sqrt{\epsilon_r}} \tag{8.49}$$

The cutoff wavenumber in terms of the circular waveguide diameter D is expressed as

$$k_c^{TM_{01}} = \frac{4.81}{D} \tag{8.50}$$

Substituting Equation (8.50) in (8.17) yields the TM_{01} mode cutoff wavelength in terms of the waveguide diameter as

$$\lambda_c^{TM_{01}} = \frac{2\pi}{k_c^{TM_{01}}} = \frac{2\pi}{(4.81/D)} = 1.306D \tag{8.51}$$

For propagation along the length of the circular waveguide the guide wavelength for the TM_{01} mode is expressed using Equation (8.18) as

$$\lambda_g^{TM_{01}} = \frac{\lambda}{\sqrt{1-\left(\frac{\lambda}{\lambda_c}\right)^2}} = \frac{\lambda}{\sqrt{1-\left(\frac{\lambda}{1.306D}\right)^2}} \tag{8.52}$$

8.3 Design

The laptop radar circular waveguide antennas were chosen to have an inside diameter of approximately D =10 cm to be above the cutoff frequency of the dominant TE_{11} mode. The circular waveguide length L =13.3 cm was chosen to be greater than one half of the guide wavelength of the TE_{11} mode. These dimensions correspond approximately to a metal coffee can that was originally used in the laptop radar design [1], but other cans such as a 0.95 liter (one-quart) paint can as shown in Figure 8.1 have similar dimensions. An SMA bulkhead receptacle jack is connected to the outer wall of the circular waveguide, and the connector's center pin, acting as a probe feed, is extended with a straight wire inside the waveguide. The probe feed uses a 0.644 mm diameter wire (#22 AWG), and the probe overall length as measured from the tip of the probe to the inside metal surface of the waveguide is $l = 3$ cm, which is approximately one-quarter of the free space wavelength at 2.4 GHz.

For a 10 cm diameter circular waveguide, the cutoff frequency for the TE_{11} mode is computed from Equation (8.35) as

$$f_c^{TE_{11}} = \frac{1.841c}{\pi D\sqrt{\epsilon_r}} = \frac{1.841(3 \times 10^{10})}{10\pi\sqrt{1}} \approx 1.8 \text{ GHz} \tag{8.53}$$

Similarly, from Equation (8.49) the cutoff frequency of the TM_{01} mode is computed as

$$f_c^{TM_{01}} = \frac{2.405c}{\pi D\sqrt{\epsilon_r}} = \frac{2.405(3 \times 10^{10})}{10\pi\sqrt{1}} = 2.3 \text{ GHz} \tag{8.54}$$

Thus, with this diameter circular waveguide both the TE_{11} and TM_{01} modes can exist in the 2.4 GHz to 2.5 GHz laptop radar operating frequency band.

The probe distance to the closed-end (backwall) of the circular waveguide is determined in a simplified analysis using Equation (8.37) to first compute the guide wavelength λ_g of the dominant TE_{11} mode at 2.4 GHz ($\lambda = 12.5$ cm) as

$$\lambda_g^{TE_{11}} = \frac{\lambda}{\sqrt{1 - \left(\frac{\lambda}{1.7065D}\right)^2}} = \frac{12.5}{\sqrt{1 - \left(\frac{12.5}{1.7065(10)}\right)^2}} = 18.4 \text{ cm} \tag{8.55}$$

The monopole feed probe will radiate with equal field amplitude toward the open and closed ends of the circular waveguide. The TE_{11} electric field that is polarized parallel to and incident at the closed end will reflect with a 180° phase shift, due to the boundary condition. When the wave travels one-quarter of a guide wavelength, the phase of the wave will shift by 90°. To achieve maximum field amplitude toward the open end, it is required that the reflected field from the closed end be in phase with the direct field from the probe. Thus, the desired spacing of the monopole probe from the backwall is one-quarter guide wavelength for the dominant TE_{11} mode or s =18.4 cm/4=4.6 cm. In this case, the reflected wave from the closed end has a total of 360° of phase shift as referenced to the position of the monopole feed probe (90° phase shift from the probe to closed end, 180° due to the reflection coefficient at the closed end, and 90° from the closed end to the reference position of the monopole feed probe).

Now the guide wavelength of the next mode, the TM_{01} mode, is computed using Equation (8.52), again at 2.4 GHz ($\lambda = 12.5$ cm) as

$$\lambda_g^{TM_{01}} = \frac{\lambda}{\sqrt{1-\left(\frac{\lambda}{1.306D}\right)^2}} = \frac{12.5}{\sqrt{1-\left(\frac{12.5}{1.306(10)}\right)^2}} = 43.2 \text{ cm} \tag{8.56}$$

Since the probe to closed-end spacing has been set to $s = 4.6$ cm, the TM_{01} mode will have only a 38.3° phase shift from the probe to the closed end. Thus, the reflected field from the TM_{01} mode that arrives at the reference probe position will have a total phase shift of 257°, and will not be in phase with the field directly radiated toward the open end. Based on this simplified analysis, qualitatively the total electric field incident at the open end will be a combination of the dominant TE_{11} mode amplitude and a lesser amplitude contribution from the TM_{01} mode.

8.4 Simulations and Measurements

The FEKO moment method meshed model of the laptop radar circular waveguide antenna is shown in Figure 8.5. The wire monopole probe is oriented radially inward and is parallel to the y axis. Simulations included field patterns in the waveguide, reflection coefficient and mutual coupling versus frequency, and gain radiation patterns. Measurements of the reflection coefficient, mutual coupling, and gain patterns were conducted in an anechoic chamber.

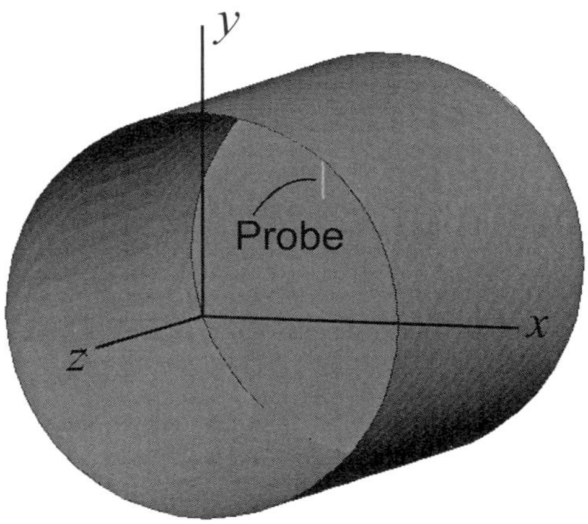

Figure 8.5 Moment method meshed model for the probe-fed open-end circular waveguide antenna.

The simulated total transverse electric field patterns inside the circular waveguide at a distance equal to 2.54 cm from the open-end are shown in Figures 8.6 and 8.7, at 2 GHz and 2.45 GHz, respectively. Based on the

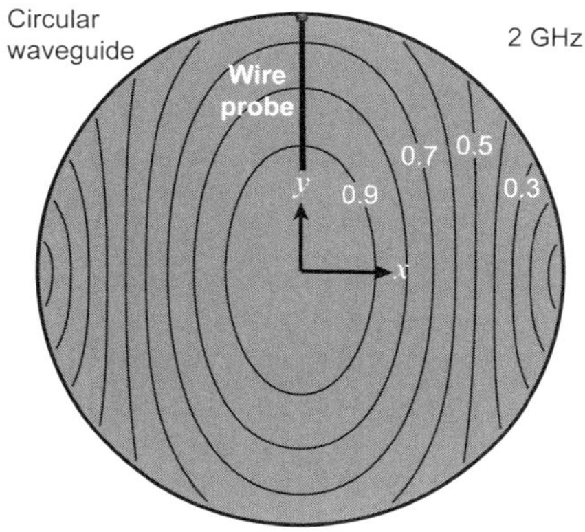

Figure 8.6 Simulated total transverse electric field magnitude in a probe-fed circular waveguide at 2 GHz. Only the TE_{11} mode is generated.

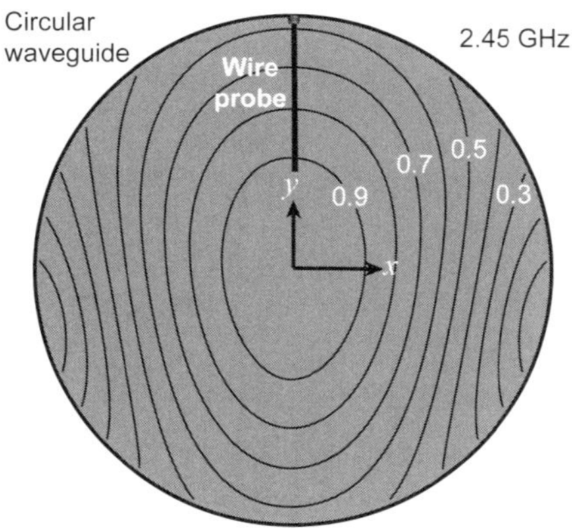

Figure 8.7 Simulated total transverse electric field magnitude in a probe-fed circular waveguide at 2.45 GHz. The field pattern is a combination of the TE_{11} and TM_{01} modes.

previous cutoff wavelength calculations, only the dominant TE_{11} mode will be propagating in the 10 cm diameter circular waveguide at 2 GHz, as observed in Figure 8.6. In Figure 8.6, the field pattern exhibits symmetry with respect to the x and y axes as expected for the TE_{11} mode. At the center frequency

(2.45 GHz) of the laptop radar, both the TE_{11} and TM_{01} modes will propagate in the waveguide, which is observed as a multimode field pattern in Figure 8.7.

Measured and moment method simulated far-field gain radiation patterns at 2.4 GHz are shown in Figures 8.8 (E-plane) and 8.9 (H-plane) and good agreement is observed. The measured peak gain is 7.2 dBi and the half-power beamwidth in both the E-plane and H-plane is 72°. The simulated peak realized gain is 8.1 dBi and the half-power beamwidth is 70° in both the E-plane and H-plane. The moment method simulated and measured reflection coefficient and mismatch loss versus frequency for the open-end circular waveguide are shown in Figures 8.10 and 8.11, respectively. Over the 2.4 GHz to 2.5 GHz Industrial, Scientific, Medical (ISM) band, the simulated

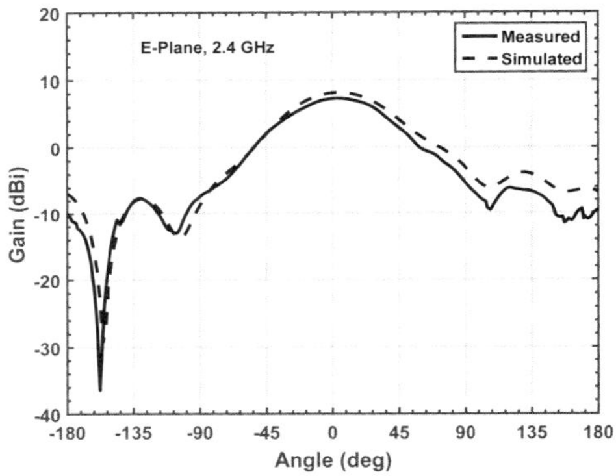

Figure 8.8 Comparison of measured and moment method simulated E-plane gain patterns for a probe-fed circular waveguide at 2.4 GHz.

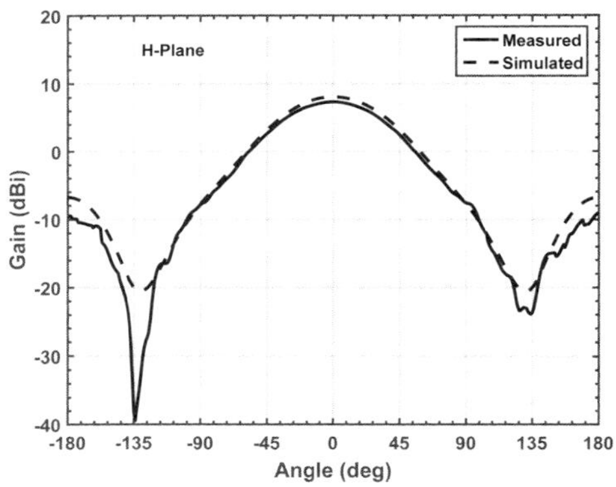

Figure 8.9 Comparison of measured and moment method simulated H-plane gain patterns for a probe-fed circular waveguide at 2.4 GHz.

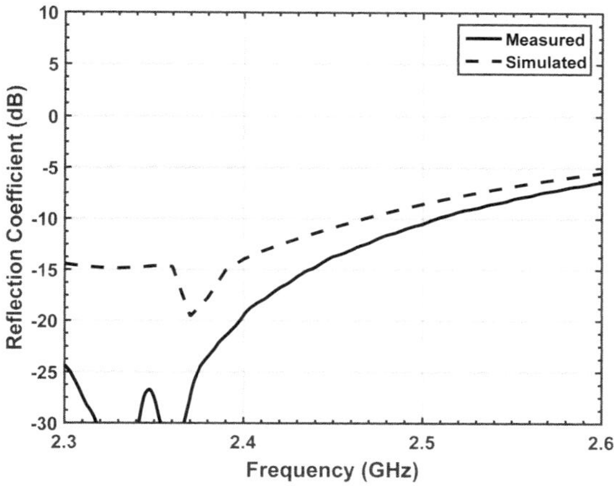

Figure 8.10 Comparison of measured and moment method simulated reflection coefficient magnitude for a probe-fed circular waveguide.

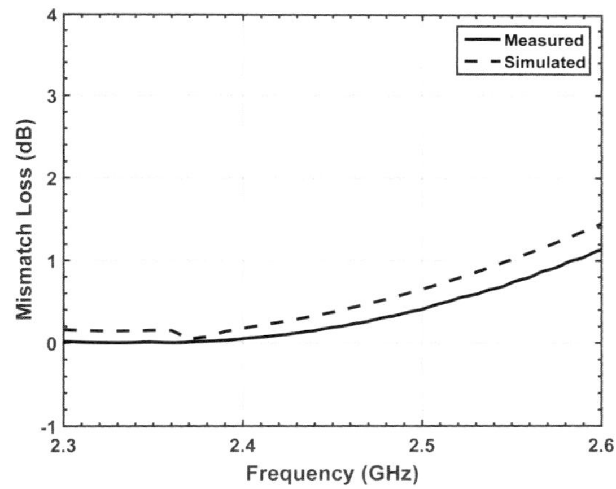

Figure 8.11 Comparison of measured and moment method simulated mismatch loss for a probe-fed circular waveguide.

and measured reflection coefficient responses are in good agreement. The measured reflection coefficient is less than -10 dB over the full ISM band. The corresponding simulated and measured mismatch loss is shown in Figure 8.11 and good agreement is observed. The measured mismatch loss is less than 0.4 dB over the ISM band. Lee and Yung [13] have used a dyadic Green's function approach to determine the input impedance of a coaxial probe-fed circular waveguide as a function of the waveguide diameter,

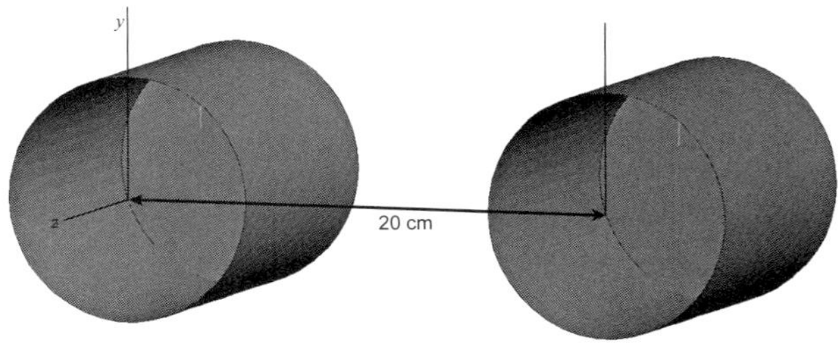

Figure 8.12 Moment method simulation model for two probe-fed circular waveguides separated by 20 cm.

monopole probe length, and monopole probe spacing to the backwall. They showed an example at X-band (9 GHz) where to provide an impedance match the optimum probe length was 0.225λ with the probe spacing to the backwall equal to $0.2\lambda_g$. Further improvements in the reflection coefficients shown in Figure 8.10 could be achieved by adjusting the monopole probe length and spacing to the backwall. The mutual coupling between the two laptop open-end circular waveguides spaced 20 cm apart (refer to Figure 8.1) was simulated with the method of moments and measured with a network analyzer as the S-parameter S21. The simulation geometry is shown in Figure 8.12. The measured and simulated mutual coupling magnitudes are at the -40 dB level and they agree to within about 1 dB over the ISM band as shown in Figure 8.13.

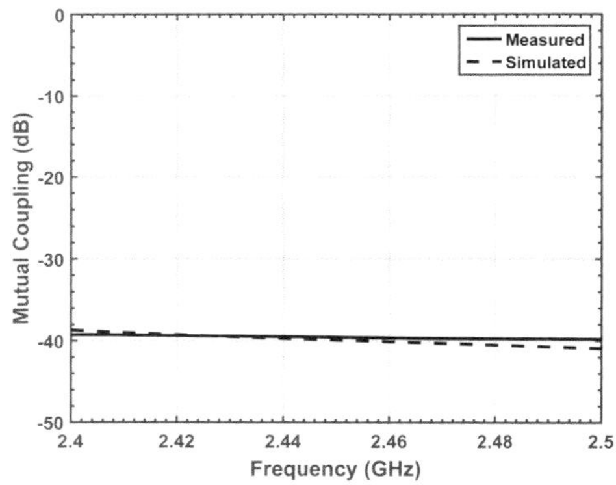

Figure 8.13 Comparison of measured and moment method simulated mutual coupling between two probe-fed circular waveguides separated by 20 cm.

8.5 Summary

This chapter has described the theory, design, simulations and measurements of open-end circular waveguide antennas used in a custom portable radar kit. The linearly polarized antennas operate in the ISM band from 2.4 GHz to 2.5 GHz. Equations for the dominant TE_{11} mode and the next propagating mode TM_{01} have been discussed. Moment method simulations and measurements of the reflection coefficient, radiation patterns, and mutual coupling of the laptop radar antennas have been shown to be in good agreement.

8.6 PROBLEM SET

8.1 With a 10 cm diameter air-filled circular waveguide, it was shown by Equations (8.53) and (8.54) that both the TE_{11} and TM_{01} modes will propagate in the 2.4 GHz to 2.5 GHz ISM band. Determine the diameter of an air-filled circular waveguide so that the cutoff frequency of the TM_{01} mode is 2.6 GHz, such that only the TE_{11} mode will propagate in the waveguide over the 2.4 GHz to 2.5 GHz ISM band. Answer: $D = 8.83$ cm. For this diameter, the cutoff frequency is 1.99 GHz for the TE_{11} mode.

8.2 Based on the measured H-plane gain pattern data at 2.4 GHz shown in Figure 8.11, the open-end circular waveguide antennas have a gain of -8 dBi at an angle of $\pm 90°$. Figures 8.1 and 8.12 show two such antennas located 20 cm apart. Using the Friis power transmission equation $\frac{P_r}{P_t} = \frac{G_t G_r \lambda^2}{(4\pi r)^2}$, as derived in Chapter 1 (Equation (1.701)), calculate an estimate of the mutual coupling in dB between the two antennas at (a) $r = 20$ cm spacing and (b) $r = 40$ cm spacing. Answers: (a) -42.1 dB which is within about 3 dB of the measured value of mutual coupling shown in Figure 8.13. (b) -48.1 dB — doubling the range distance reduces the power coupled between the two antennas by 6 dB.

8.3 The portable radar shown in Figure 8.1 determines the radial velocity of a radar target by means of the Doppler frequency shift. The Doppler radian frequency shift ω_D of an electromagnetic field is given by the radar target signal phase change $\Delta\Phi$ during an interval of time ΔT as $\omega_D = \Delta\Phi/\Delta T$. Assume that a radar target is moving at a radial velocity V with respect to the radar and moves a range distance ΔR in time interval ΔT. Let f_D denote the Doppler frequency shift in cycles per second, and note that $\omega_D = 2\pi f_D$. Determine an expression for the Doppler frequency shift in terms of target radial velocity V and wavelength λ. Hint: The phase change $\Delta\Phi$ is equal to $k\Delta d$, where $k = 2\pi/\lambda$ and Δd is the two-way path length difference to the target that occurs from reference time $t = 0$ to time $t = \Delta T$. In the same time interval the target range distance varies by $\Delta R = V\Delta T$. Solution: The

Doppler radian frequency shift can be expressed as $\omega_D = 2\pi f_D = \Delta\Phi/\Delta T$. The two-way path difference for the target is given by $\Delta d = 2\Delta R = 2V\Delta T$. The Doppler frequency is then given by $f_D = 2V/\lambda$.

8.4 The portable radar shown in Figure 8.1 operates at 2.4 GHz. Using the result from Problem 8.3, compute the Doppler frequency shift for a target moving at a radial velocity of 14 m/s. Answer: The wavelength is $\lambda = 0.125$m, so $f_D = 2 \times 14/0.125 = 112$ Hz.

References

[1] G.L. Charvat, A.J. Fenn, B.T. Perry, "The MIT IAP Radar Course: Build a Small Radar System Capable of Sensing Range, Doppler, and Synthetic Aperture (SAR) Imaging," *2012 IEEE Radar Conference*, 2012 pp. 138–144.

[2] D.R. Granchelli, A.J. Fenn, C. Agbasi-Porter "Lincoln Laboratory Radar Introduction for Student Engineers," *2013 IEEE Integrated STEM Education Conference (ISEC)*, 2013, pp. 1–4.

[3] G.L. Charvat, *Small and Short-Range Radar Systems,* Boca Raton, Florida: CRC Press, 2014, pp. 46–52, 161–168.

[4] J.S. Mullen, L. Milechin, M. Houle, P. Bell, A. Fenn, K.E. Kolodziej, J. Meklenburg, J. Nguyen, B. Perry and A. Reuther, "Bringing Physical Construction and Real-World Data Collection into a Massively Open Online Course (MOOC)," *2017 IEEE Frontiers in Education Conf.*, Indianapolis, Indiana, 18–21 October 2017, pp. 1–5.

[5] R.J. Mailloux, "Phased Array Theory and Technology," *Proc. IEEE*, vol. 70, no. 3, 1982, pp. 246–291.

[6] N. Amitay, V. Galindo, and C.P. Wu, *Theory and Analysis of Phased Array Antennas.* New York: Wiley, 1972.

[7] D.M. Pozar, *Microwave Engineering, 4th ed.*, Hoboken, NJ: Wiley, 2012.

[8] N. Marcuvitz (ed.), *Waveguide Handbook*, New York: McGraw-Hill, 1951, pp. 66–72.

[9] C.S. Lee, S.W. Lee, and S.L. Chaung, "Plot of Modal Field Distribution in Rectangular and Circular Waveguides," *IEEE Trans. Microwave Theory and Techniques,* vol. MTT-22, no. 3, March 1985, pp. 271–274.

[10] S. Ramo, J.R. Whinnery, and T. Van Duzer, *Field and Waves in Communications Electronics,* New York: John Wiley, 1965, p. 432 (foldout page).

[11] J. Volakis, ed., *Antenna Engineering Handbook, 4th ed.* New York: McGraw-Hill, 2007, pp. 51-37–51-38.

[12] A.D. Olver, P.J.B. Clarricoats, A.A. Kishk, and L. Shafai, *Microwave Horns and Feeds,* New York: IEEE Press, 1994, pp. 230–240.

[13] W.W.S. Lee and E.K.N. Yung, "The Input Impedance of a Coaxial Line Fed Probe in a Cylindrical Waveguide," *IEEE Trans. Microwave Theory and Techniques,* vol. 42, no. 8, August 1994, pp. 1468–1473.

9

3D-Printed Nonplanar Circular Patch Array Antenna

9.1 Introduction

Patch antennas, or microstrip antennas, have been explored extensively for a large variety of antenna designs for single elements and in phased arrays [1]. These antennas are particularly useful in applications requiring a thin profile or antennas that must conform to the surface of a curved structure. Common patch antenna shapes are rectangular, square, and circular [1–8]. Patch antennas are used for linear, dual linear or circular polarization applications. Patch antennas can be fed with an electrically conducting pin or a microstrip line directly connected to the patch or proximity coupled to the patch. In this chapter, the focus is on the design, analysis, and measurements of a conformal array of circular patch antennas. A nonplanar, conformal array antenna is described that uses a combination of 3D printer and copper plating techniques, referred to as an additive manufacturing process. Circular patch antenna elements are copper plated onto a 3D printed dielectric substrate made of ABS-M30 material. Measured and simulated element reflection coefficient, element gain patterns, and array scanned beam radiation patterns are shown.

Nonplanar phased arrays are of interest for applications requiring antennas that are conformal to the skin surface of a structure. Using 3D printing techniques [9–22], also referred to as additive manufacturing, structures that can be considered can have an arbitrary shape such as a cylinder, cone, sphere, or other more general irregular shapes. 3D printer technology for fabrication of electronic circuitry and antennas is a subject of great interest for rapid prototyping and has been explored by a number of researchers in the last few years. A conformal array can offer 1) lower aerodynamic drag, 2) reduced radar cross section (RCS), 3) enhanced angular coverage, and 4) potentially lower fabrication and installation costs. Low-profile microstrip patch array antenna elements on a dielectric substrate are a candidate for these applications [23]. Braatan et al. [24] have developed a flexible antenna array concept that has been fabricated and measured with a built-in sensing circuit to adapt to different conformal shapes. The present work in this chapter describes a nonplanar antenna concept in which the conformal array is a rigid structure with a 3D printed substrate and electrically conducting plated patches.

Nonplanar microstrip patch antenna arrays using 3D printer and copper plating techniques have recently been explored [20]. Element reflection coefficient, element gain and phase patterns have been measured and simulated, and array radiation patterns were generated by superposition for an example array.

The chapter is organized as follows. Section 9.2 describes the theory for circular patch antennas. An example moment method simulation for a single circular patch antenna is given in Section 9.3. Section 9.4 describes the design and fabrication of a nonplanar array with a circular arc shape and describes the array simulation model. Section 9.5 compares measured and simulated results. Section 9.6 provides a summary and Section 9.7 provides a problem set.

9.2 Circular Patch Antenna Theory

Figure 9.1 shows a diagram of a circular patch antenna over a ground plane. The circular electrically conducting patch has radius a and is located on a

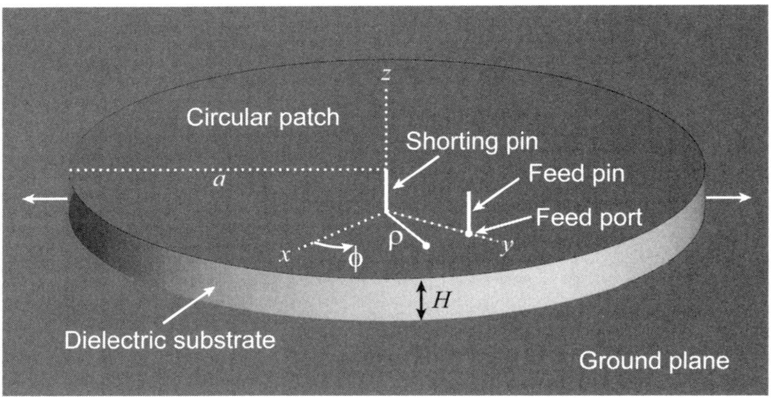

Figure 9.1 Circular patch antenna diagram.

dielectric substrate with height H and dielectric constant ϵ_r. This conducting patch geometry has been analyzed as a circular disk resonator where a number of transverse magnetic modes TM_{nmo} $[TM_{\phi,\rho,z}]$ can exist. An electrically conducting feed pin located at a radial distance s from the center of the patch is used to excite the patch. An optional electrically conducting shorting pin is connected between the center of the patch and the ground plane. The dominant mode, with or without the shorting pin, is the transverse magnetic TM_{110} mode that has a null at the z axis. In the literature, the TM_{110} mode is often abbreviated by dropping the third index, that is, it is written as TM_{11}

The electric and magnetic fields inside the dielectric substrate below the circular patch are given by Derneryd [6]

$$E_z(\rho,\phi) = E_o J_n(k\rho)\cos n\phi \tag{9.1}$$

$$H_\rho(\rho,\phi) = -\frac{j\omega\epsilon n}{k^2\rho} E_o J_n(k\rho)\sin n\phi \tag{9.2}$$

$$H_\phi(\rho,\phi) = -\frac{j\omega\epsilon}{k} E_o J_n'(k\rho)\cos n\phi \tag{9.3}$$

where $J_n(X)$ is the nth order Bessel function and $J_n'(X)$ is the derivative of the nth order Bessel function. The Bessel function and its derivative are given by the following recurrence equations [25]

$$J_n(X) = \frac{X}{2n}\left(J_{n-1}(X) + J_{n+1}(X)\right) \tag{9.4}$$

$$J_n'(X) = \frac{1}{2}\left(J_{n-1}(X) - J_{n+1}(X)\right) \tag{9.5}$$

It is noted that $J_1(\rho = 0) = 0$ confirming that the E_z component has a null when the radial distance is zero.

The far-zone electric field components are given by Derneryd [6]

$$E_\theta(\theta,\phi) = -j^n k_o \frac{e^{-jk_o r}}{r}\frac{V_o a}{2} B_M(k_o a\sin\theta)\cos n\phi \tag{9.6}$$

$$E_\phi(\theta,\phi) = j^n k_o \frac{e^{-jk_o r}}{r}\frac{V_o a}{2} B_P(k_o a\sin\theta)\cos\theta\sin n\phi \tag{9.7}$$

where

$$B_P(X) = J_{n-1}(X) + J_{n+1}(X) = \frac{2n}{X}J_n(X) \tag{9.8}$$

$$B_M(X) = J_{n-1}(X) - J_{n+1}(X) = 2J_n'(X) \tag{9.9}$$

The resonant frequencies of a circular patch antenna are computed according to a cavity model as described by Derneryd [6]

$$f_{nm} = \frac{q_{nm}c}{2\pi a_{\text{eff}}\sqrt{\epsilon_r}} \tag{9.10}$$

where c is the speed of light and a_{eff} is an effective radius of the circular patch that is larger than the actual patch radius. In Equation (9.10), the parameter q_{nm} is the mth root, or value, of J'_n that enforces the boundary condition, that is, the zeros of the derivative of the nth order Bessel function $J_n(x)$. For the dominant TM_{11} mode, $q_{11} = 1.841$. Then from Equation (9.10) it follows that the effective radius is determined as

$$a_{\text{eff}} = \frac{q_{nm}c}{2\pi f_{nm}\sqrt{\epsilon_r}} \tag{9.11}$$

Substituting the values for q_{11} and c in Equation (9.11) and simplifying yields

$$a_{\text{eff}} = \frac{8.79}{f_{\text{GHz}}\sqrt{\epsilon_r}} \tag{9.12}$$

The effective radius a_{eff} takes account of the fringing fields at the edge of the patch and is given in terms of the actual patch radius a by [4]

$$a_{\text{eff}} = a\sqrt{1 + \frac{2H}{\pi a \epsilon_r}\left(\ln\frac{\pi a}{2H} + 1.7726\right)} \tag{9.13}$$

Solving Equation (9.13) for a yields the desired value for the actual patch radius, that is,

$$a = \frac{a_{\text{eff}}}{\sqrt{1 + \frac{2H}{\pi a \epsilon_r}\left(\ln\frac{\pi a}{2H} + 1.7726\right)}} \tag{9.14}$$

An iterative approximation for the actual patch radius is to substitute Equation (9.12) for a in the two quantities in the denominator of Equation (9.14), which then allows the actual patch radius to be computed. Assuming the patch radius is much greater than the substrate height, that is, $a/H >> 1$, Equation (9.14) gives the actual patch radius to within an error of about 2.5%. An approximate feed pin spacing from the center shorting pin is chosen to be equal to about 33% of the radius of the patch [26]. A quantitative method to determine the radial location (ρ) of the feed pin is as follows.

The circular patch antenna input impedance at resonance for the TM_{11} mode (with $n = 1$) is purely real and is computed in terms of the radial position ρ and patch radius a as [6]

$$Z_{\text{in}}(\rho) = \frac{1}{G_{\text{total}}}\frac{J_1^2(k\rho)}{J_1^2(ka)} \tag{9.15}$$

where G_{total} is the antenna total input conductance expressed as

$$G_{\text{total}} = G_{\text{rad}} + G_{\text{dielectric}} + G_{\text{finite conductivity}} \tag{9.16}$$

where G_{rad} is the radiation conductance given by

$$G_{\text{rad}} = \frac{(k_o a)^2}{480} \int_0^{\frac{\pi}{2}} \left[B_M^2(k_o a \sin\theta) + \cos^2\theta B_P^2(k_o a \sin\theta) \right] \sin\theta d\theta \tag{9.17}$$

$G_{\text{dielectric}}$ is the dielectric conductance given by

$$G_{\text{dielectric}} = \frac{\tan\delta}{4\mu_o H f_{11}} \left[(ka)^2 - 1 \right] \tag{9.18}$$

where $\tan\delta$ is the dielectric loss tangent. $G_{\text{finite conductivity}}$ is the equivalent conductance due to the finite conductivity σ of the metal patch antenna given by

$$G_{\text{finite conductivity}} = \frac{\pi(\pi f_{11}\mu_o)^{-3/2}}{4H^2\sqrt{\sigma}} \left[(ka)^2 - 1 \right] \tag{9.19}$$

For a desired patch antenna design, the conductances in Equations (9.17) to (9.19) are evaluated to determine the total conductance G_{total}, which is then substituted into Equation (9.15). Then, the value of ρ is varied until a desired value of Z_{in} is achieved.

9.3 Single Planar Patch Element Simulation Example

Referring to Figure 9.1, assume a circular patch antenna with diameter 8.46 cm ($a = 4.23$ cm) and substrate thickness $H = 0.3175$ cm with a dielectric constant $\epsilon_r = 2.56$ and loss tangent $\tan\delta = 0.008$. Based on Equation (9.13) the effective patch diameter is computed as 8.83 cm, which is about 4.4% greater than the assumed patch diameter. Then, based on Equation (9.10), the resonant frequency of the TM_{11} mode is computed as 1.244 GHz. Assume a copper conducting patch with electrical conductivity $\sigma = 5.9 \times 10^7$ S/m at room temperature. By setting the input impedance equal to 50 ohms (real) at resonance in Equation (9.15), the radial distance for the feed point is computed to be approximately $\rho = 1.15$ cm.

Consider Figure 9.2 that shows a simulation model of the circular patch antenna, which was analyzed with the FEKO moment method solver. In this simulation, the actual feed pin radial position was $\rho = 1.232$ cm. To characterize the dominant TM_{11} mode generated in the patch antenna substrate, Figure 9.3 shows the simulated z-component of the electric field amplitude in the mid-plane ($z = -H/2$) cross section of the antenna. The

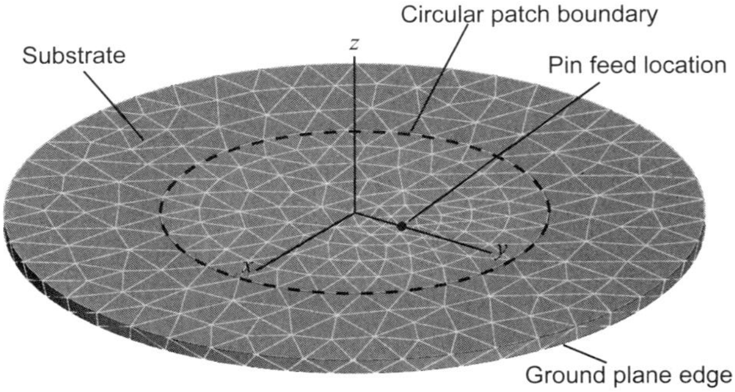

Figure 9.2 Circular patch antenna meshed model for moment method analysis.

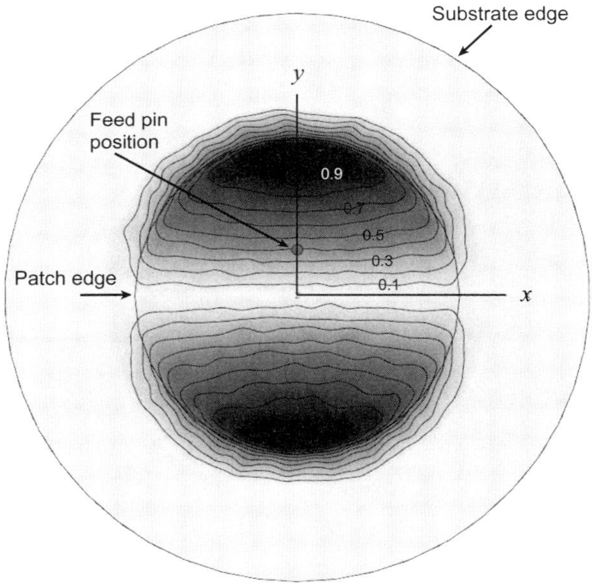

Figure 9.3 Moment method simulated electric field (z component) in the dielectric substrate of the circular patch antenna.

electric field minimum occurs along the x axis. The maximum field occurs at symmetric positions along the y axis near the edge of the patch. Fringing field contours are observed to extend beyond the patch edge, which qualitatively shows the characteristic of the effective radius used in Equation (9.13). The simulated input impedance versus frequency is shown in Figure 9.4.

Resonance occurs at 1.24 GHz which is in agreement with the predicted value

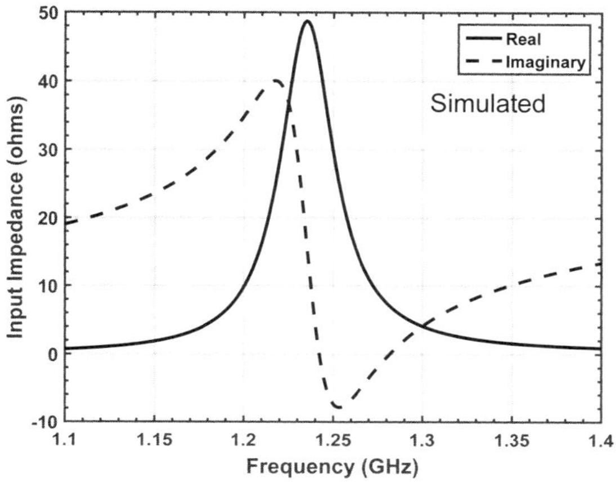

Figure 9.4 Moment method simulated input impedance for the circular patch antenna.

of resonance by Equation (9.10). Assuming a 50-ohm reference impedance, the corresponding simulated reflection coefficient magnitude versus frequency is shown in Figure 9.5, where the minimum value is observed to occur at 1.24 GHz. The next section describes the same type of patch antenna design, but now as an element in a nonplanar array configuration.

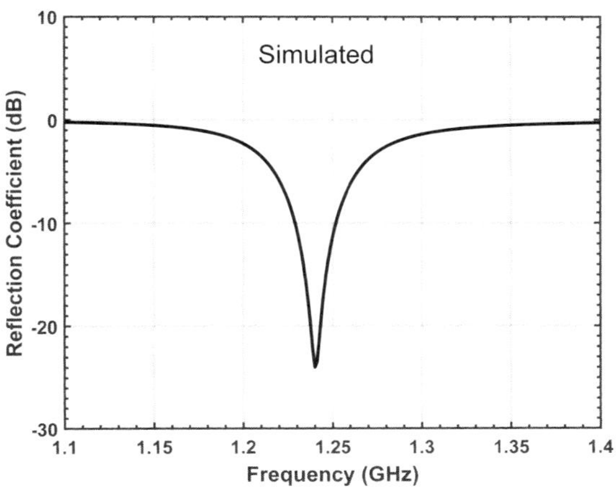

Figure 9.5 Moment method simulated reflection coefficient magnitude for the circular patch antenna.

9.4 Nonplanar Array Antenna Design, Simulation Model, and Fabrication

The additive manufactured nonplanar patch antenna array design is as follows [20]. The substrate material is assumed to be 0.3175 cm thick ABS-M30 with a measured dielectric constant $\epsilon_r = 2.56$ and loss tangent $\tan\delta = 0.008$ in the range of 1.2 to 1.4 GHz. The chemical name for ABS is acrylonitrile butadiene styrene. The desired resonant frequency was selected as 1.24 GHz. The fabricated patch diameter was 8.46 cm corresponding to the example simulated patch antenna diameter described in Section 9.3. The nonplanar substrate has a radius of curvature 1.128m, arc length 1.219m, and width 18.034 cm. The array is composed of 10 electrically conducting patch antennas, numbered 1 to 10. The feed-pin arc offset distance s from the center of each patch antenna was 1.232 cm. The array elements had center-to-center arc spacing 12.002 cm, and they were polarized in the E-plane array direction. Figure 9.6 shows a FEKO moment method model of the nonplanar array of pin-fed circular patches. The array substrate was designed to be fabricated in two sections that would be bonded together as depicted in Figure 9.7.

An overview of the additive-manufactured array fabrication process is summarized by three photographs in Figure 9.8. The ABS-M30 substrate was 3D printed by RedEye (Eden Prairie, Minnesota, USA) using fused deposition modeling (FDM). The front-surface copper-plated circular antenna elements (8.46 cm diameter) and rear-surface plated-copper ground plane were fabricated by RePliForm (Baltimore, Maryland, USA). The 3D printing process used a T12 extrusion tip with a resolution of 0.1778 mm (0.007 inches). A photograph of the fabricated ABS-M30 substrate prior to plating is shown in Figure 9.8(a). Due to the fused deposition modeling (FDM) process, tiny air pockets can exist in the 3D-printed material. These air

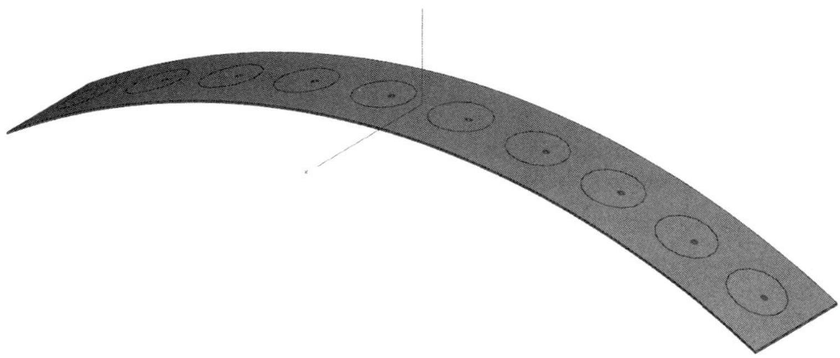

Figure 9.6 Simulation model for a nonplanar array of circular patches.

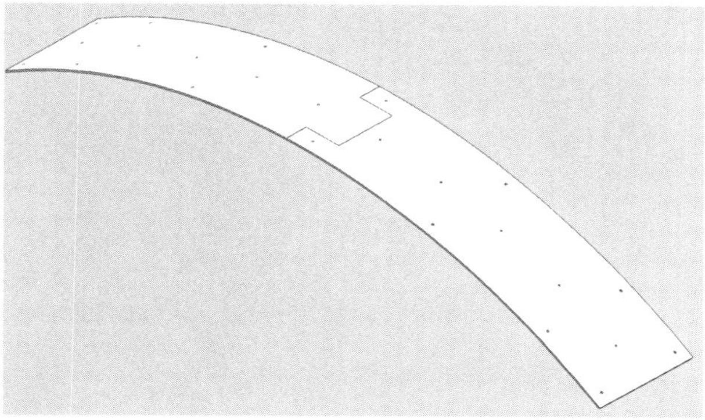

Figure 9.7 Drawing of a nonplanar substrate composed of two joined sections.

Figure 9.8 Photographs of nonplanar array during fabrication. (a) 3D-printed substrate. (b) Substrate with mask for array elements prior to plating. (c) Copper plated array elements on 3D-printed substrate.

pockets have the potential to allow some of the plating tank material to enter into the 3D-printed substrate and alter the dielectric properties. An acetone vapor treatment was used to seal the 3D-printed ABS substrate, and the substrate surface was lightly sanded to prepare for electroplating. To prepare for copper plating of the patch elements, the substrate was masked as shown in Figure 9.8(b). The final copper plated patch array is shown in Figure 9.8(c). A 3D-printed ABS-M30 support structure was attached to the array backside to maintain the array shape during handling and anechoic chamber measurements. The patch antennas were pin fed using SMA coaxial connectors that were electrically connected to the patch antennas and ground plane using silver epoxy. A photograph of the 3D-printed nonplanar array under test is shown in Figure 9.9. Coherent element gain radiation pattern measurements were performed with the antenna mounted on a foam column

Figure 9.9 3D-printed nonplanar array under test.

in a tapered anechoic chamber at a range distance of 8.5m [27].

The central eight elements were considered the active array, with one 50-ohm terminated element at each edge (elements 1 and 10). Embedded element performance was simulated and measured with the surrounding elements terminated in 50-ohm resistive loads. Array radiation patterns were computed from the central eight element gain and phase patterns by superposition (summation) using phase conjugation at each scan angle.

9.5 Measured and Simulated Results

Measurements and simulations of the nonplanar array were performed with one element driven and the surrounding elements terminated in 50-ohm resistive loads [20]. A comparison of the measured and simulated reflection coefficients for central element number 5 is shown in Figure 9.10 and good agreement is observed. The measured and simulated minimum reflection coefficient (resonance) occurs at 1.23 GHz and 1.24 GHz, respectively, which agree within 1%. The -10 dB reflection coefficient bandwidth is approximately 20 MHz or 1.6%. The corresponding measured and simulated mismatch loss for the central element number 5 is shown in Figure 9.11. Element gain patterns for elements 2 to 9 were measured with the surrounding elements terminated in 50-ohm resistive loads. A comparison of the measured and simulated E-plane gain radiation patterns for central element number 5 in the array is shown in Figure 9.12. The central element number 5 peak gain is 4.9 dBi simulated compared to 4.7 dBi measured. A comparison of the measured and simulated E-plane gain radiation patterns for near-edge element number 2 is shown in Figure 9.13. Both the central element and edge element exhibit a broad radiation pattern that allows wide-angle scanning. Referring to Equation (2.68), the coherent embedded element gain patterns were combined by superposition with uniform amplitude illumination

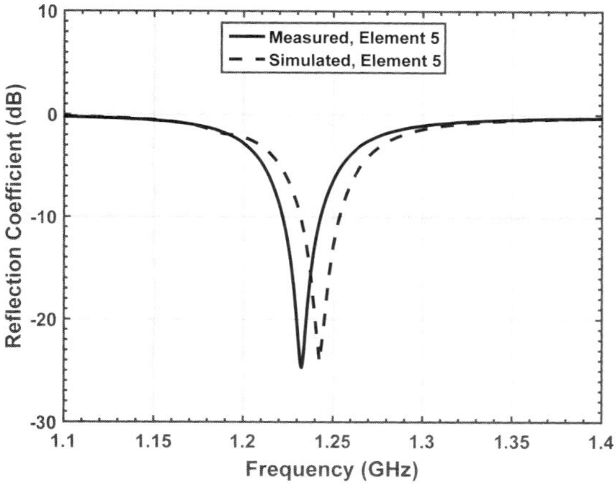

Figure 9.10 Comparison of measured and simulated reflection coefficient magnitude for central element number 5 in the 3D-printed nonplanar array.

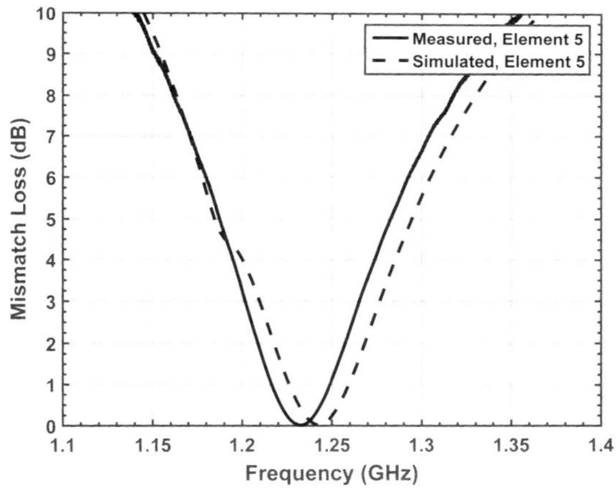

Figure 9.11 Comparison of measured and simulated mismatch loss for central element number 5 in the 3D-printed nonplanar array.

to form scanned radiation patterns. Figure 9.14 shows a comparison of measured and simulated E-plane normalized radiation patterns for broadside scan. Figure 9.15 shows a comparison of measured and simulated E-plane normalized radiation patterns for a 45° scan angle. Good agreement between the measurements and simulations is observed in the scanned main beam and sidelobe regions.

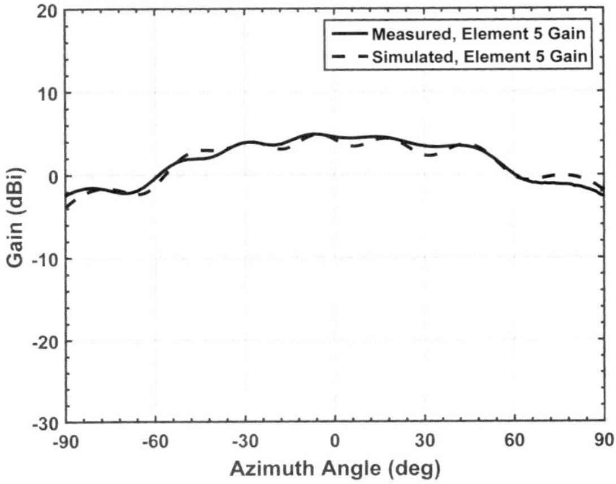

Figure 9.12 Comparison of measured and simulated E-plane gain radiation patterns for central element number 5 in the 3D-printed nonplanar array.

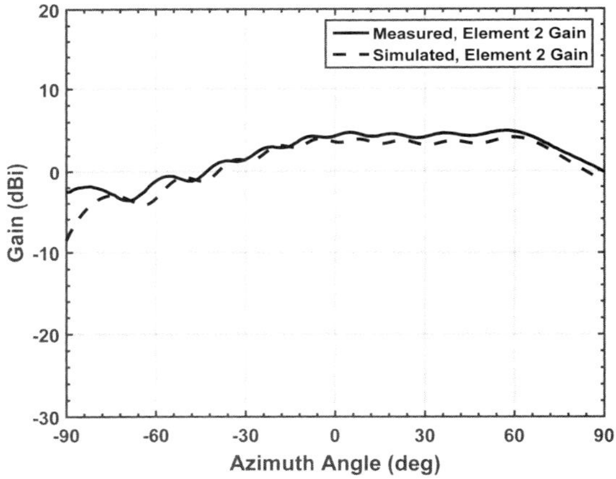

Figure 9.13 Comparison of measured and simulated E-plane gain radiation patterns for near-edge element number 2 in the 3D-printed nonplanar array.

9.6 Summary

The theory for circular patch antennas has been reviewed in this chapter. A nonplanar circular patch array antenna has been described that uses additive manufacturing with a combination of 3D printer and copper-plating techniques. Measured and moment method simulated array element reflection coefficient, element gain patterns, and scanned beam radiation patterns are in good agreement.

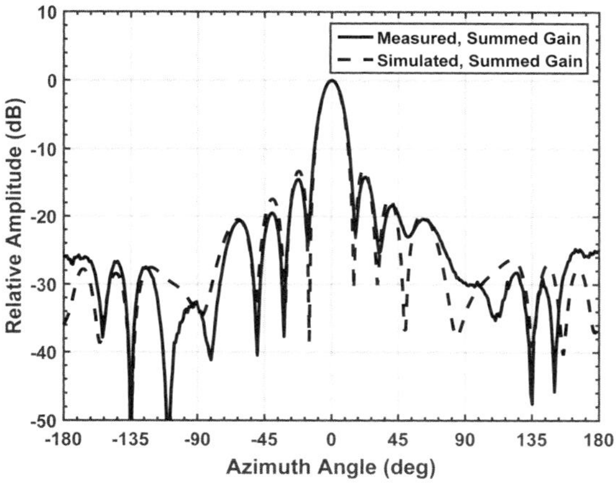

Figure 9.14 Comparison of measured and simulated E-plane normalized radiation patterns of the 3D-printed nonplanar array for broadside scan.

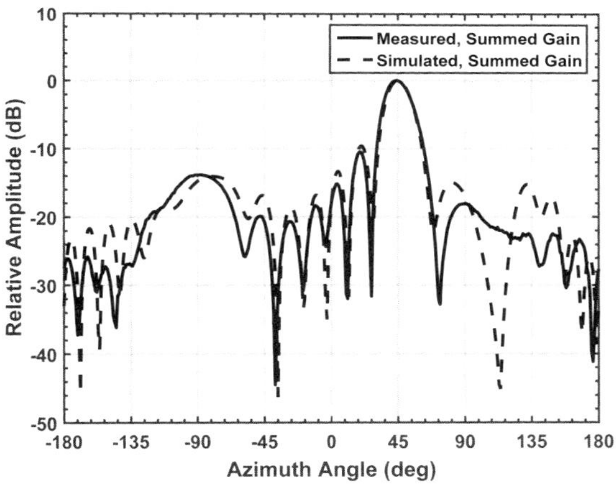

Figure 9.15 Comparison of measured and simulated E-plane normalized radiation patterns of the 3D-printed nonplanar array for 45° scan from broadside.

9.7 PROBLEM SET

9.1 The Global Positioning System (GPS) utilizes frequencies at 1575.42 MHz (L1) and 1227.6 MHz (L2). (a) Assume a pin-fed circular copper patch antenna for L1 operation is desired to be fabricated on a substrate with high dielectric constant 11.0, loss tangent 0.002, and thickness 0.127 cm (0.050 inches). Determine the diameter and feed pin position for the circular patch antenna, and (b) repeat for the GPS L2 frequency. Hint: Use Equations (9.12) and (9.14) to compute the actual patch radius, and then compute the feed pin position from the center using Equations (9.15) to (9.19) assuming a 50-ohm reference. Answers: (a) L1: 3.33 cm diameter, feed pin 0.3 cm from the center. (b) L2: 4.28 cm diameter, feed pin 0.42 cm from the center.

9.2 Another common type of patch antenna is the rectangular patch that can be analyzed approximately with a transmission-line model [8, 26]. The

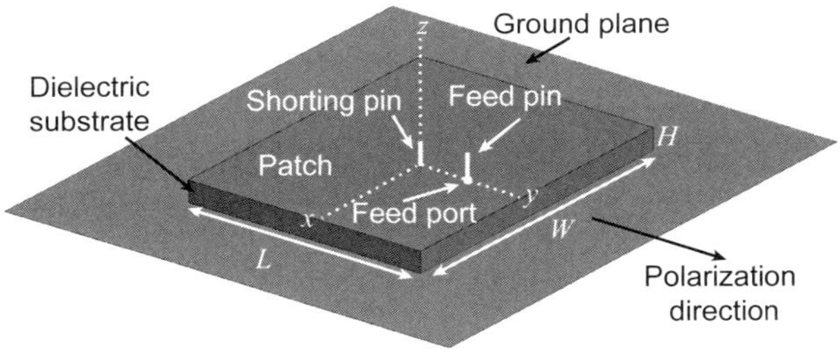

conducting patch has dimensions L (length) and W (width), and radiates above a dielectric substrate with height H and dielectric constant ϵ_r. The effective dielectric constant [26, p. 299] and width [8, p. 819] are given by

$$\epsilon_{\text{eff}} = \frac{\epsilon_r + 1}{2} + \frac{\epsilon_r - 1}{2}\left(1 + \frac{10H}{W}\right)^{-1/2}$$

$$W = \frac{c}{2f_r}\sqrt{\frac{2}{\epsilon_r + 1}}$$

where c is the speed of light and f_r is the resonant frequency. At each edge in the y direction, the fringing field is extended beyond the metal patch by the distance ΔL, and is computed from [26, p. 299]

$$\Delta L = 0.412H\left[\frac{\epsilon_{\text{eff}} + 0.3}{\epsilon_{\text{eff}} - 0.258}\right]\left[\frac{W/H + 0.262}{W/H + 0.813}\right]$$

and the physical length of the patch is then given by [26, p. 301]

$$L = \frac{c}{2f_r\sqrt{\epsilon_{\text{eff}}}} - 2\Delta L$$

As in Problem 9.1, given a dielectric constant of 11.0 and substrate height 0.127 cm (0.050 inches), compute W and L for GPS frequencies (a) L1 and (b) L2. Answers: (a) L1: $W = 3.88$ cm, $L = 2.85$ cm. (b) L2: $W = 5.0$ cm, $L = 3.67$ cm. Note that the rectangular patch is larger than the circular patch.

References

[1] D.M. Pozar and D.H. Schaubert, *Microstrip Antennas: The Analysis and Design of Microstrip Antennas and Arrays,* New York: IEEE Press, 1995.

[2] A. K. Bhattacharjee, S. R. Bhadra Chaudhuri, D. R. Poddar, S. K. Chowdhury, "Equivalence of Impedance and Radiation Properties of Square and Circular Microstrip Patch Antennas," *IEE Proceedings H - Microwaves, Antennas and Propagation,* vol. 136, no. 4, 1989, pp. 338 – 342.

[3] I. Wolff and N. Knoppik, "Rectangular and Circular Microstrip Disk Capacitors," *IEEE Trans. on Microwave Theory and Techniques,* vol. MTT-22, no. 10, October 1974, pp. 857–864.

[4] L.C. Shen, S.A. Long, M.R. Allerding, and M.D. Walton, "Resonant Frequency of a Circular Disc, Printed-Circuit Antenna," *IEEE Trans. Antennas and Propagat.*, vol. 25, 1977, pp. 595–596.

[5] L.C. Shen, "Analysis of a Circular-Disc Printed-Circuit Antenna," *Proc. IEE,* vol. 126, no. 12, 1979, pp. 1220–1222.

[6] A.G. Derneryd, "Analysis of the Microstrip Disk Antenna Element," *IEEE Trans. Antennas and Propagat.*, vol. 27, 1979, pp. 660–664.

[7] J. Watkins, "Circular Resonant Structures in Microstrip," *Electron Letters,* vol. 5, October 1969, pp. 524–525.

[8] C.A. Balanis, *Antenna Theory and Design, 3rd ed.* New York: Wiley, 2005.

[9] J.J. Adams, E.B. Duoss, T.F. Malkowski, M.J. Motala, B.Y. Ahn, R.G. Nuzzo, J.T. Bernhard, and J.A. Lewis, "Conformal Printing of Electrically Small Antennas on Three-Dimensional Surfaces," *Advanced Materials,* vol. 23, no. 11, 2011, pp. 1335–1340.

[10] E. MacDonald, R. Salas, D. Espalin, M. Perez, E. Aguilera, D. Muse, R.B. Wicker, "3D Printing for the Rapid Prototyping of Structural Electronics," *Access, IEEE,* vol.2, Dec. 2014, pp.234–242.

[11] M. Ahmadloo, "Design and Fabrication of Geometrically Complicated Multiband Microwave Devices Using a Novel Integrated 3D Printing Technique," *IEEE Electrical Performance of Electronic Packaging and Systems (EPEPS) Conference,* San Jose, CA, 2013, pp. 29–32.

[12] G. Shaker, H-S. Lee, S. Safavi-Naeini, M. Tentzeris, "Printed Electronics for Next Generation Wireless Devices," *IEEE Loughborough Antennas and Propagation Conference (LAPC),* November 2011, pp. 1–5.

[13] J. Mei, M. Lovell, and M. Mickle, "Formulation and Processing of Novel Conductive Solution inks in Continuous Inkjet Printing of 3-D Electric Circuits," *IEEE Transactions on Electronics Packaging Manufacturing,* vol. 28, no. 3, July 2005, pp. 265–273.

[14] G. Shaker, S. Safavi-Naeini, N. Sangary, and M. Tentzeris, "Inkjet Printing of Ultrawideband (UWB) Antennas on Paper-based Substrates," *IEEE Antennas and Wireless Propagation Letters,* vol. 10, 2011, pp. 111–114.

[15] J. Siden, M. Fein, A. Koptyug, and H. Nilsson, "Printed Antennas with Variable Conductive Ink Layer thickness," *IET Microwaves, Antennas Propagat.,* vol. 1, no. 2, April 2007, pp. 401–407.

[16] J. Mei, M.R. Lovell, and M.H. Mickle, "Formulation and Processing of Novel Conductive Solution Inks in Continuous Inkjet Printing of 3-D Electric Circuits," *IEEE Trans. on Electronics Packaging,* vol. 28, no. 3, July 2005, pp. 265–273.

[17] M. Ahmadloo, and P. Mousavi, "A Novel Integrated Dielectric-and-Conductive Ink 3D Printing Technique for Fabrication of Microwave Devices," *IEEE MTT International Microwave Symposium (IMS) 2013,* Seattle, June 2–7, 2013, pp. 1–3.

[18] M. Ahmadloo, and P. Mousavi, "Application of Novel Integrated Dielectric and Conductive Ink 3D Printing Technique for Fabrication of Conical Spiral Antennas," *IEEE APS/USNC-URSI 2013,* Orlando, Florida, July 7–13, 2013, pp. 780–781.

[19] K.A. Nate, J. Hester, M. Isakov, R. Bahr, M.M. Tentzeris, "A Fully Printed Multilayer Aperture-Coupled Patch Antenna Using Hybrid 3D / Inkjet Additive Manufacturing Technique," *Proc. of the 45th European Microwave Conference,* 7–10 Sept 2015, Paris, France, pp. 610–613.

[20] A.J. Fenn, D.J. Pippin, D. Hogan, C.M. Lamb, F.G. Willwerth, H.M. Aumann, J.P. Doane, "3D Printed Conformal Array Antenna: Simulations and Measurements," *2016 IEEE Int. Symp. on Phased Array Systems and Technology,* Waltham, MA: October 2016, pp. 1–4.

[21] M.W. Elsallal, J. Hood and I. McMichael, T. Busbee, "3D Printed Material Characterization for Complex Phased Arrays and Metamaterials," *Microwave Journal,* October 14, 2016, pp. 20–34.

[22] M. Liang and H. Xin, "Three-Dimensionally Printed/Additive Manufactured Antennas," in *Handbook of Antenna Technologies*, Z.N. Chen, (ed.), Singapore: Springer Science, 2016, pp. 661–697.

[23] S.D. Gupta, "Effect of Mutual Coupling in E and H Plane on Microstrip Antenna Array Conformal on Cylindrical Surface," *2013 Int. Conf. on Signal Processing and Communication,* 12–14 Dec. 2013, pp. 149–154.

[24] B.D. Braaten, S. Roy, S. Nariyal, M.A. Aziz, N.F. Chamberlain, I. Irfanullah, M.T. Reich, and D.E. Anagnostou, "A Self-Adapting Flexible (SELFLEX) Antenna Array for Changing Conformal Surface Applications," *IEEE Trans. on Antennas and Propagation*, vol. 61, no. 2, Feb. 2013, pp. 655 – 665.

[25] G.N. Watson, *A Treatise on the Theory of Bessel Functions*, London: Cambridge University Press, 1966, p. 17.

[26] T.A. Milligan, *Modern Antenna Design, 2nd ed.*, Hoboken, NJ: Wiley-IEEE Press, 2005.

[27] A.J. Fenn, M.W. Shields, and G.A. Somers, "Introduction to the New MIT Lincoln Laboratory Suite of Ranges," *26th Annual Antenna Measurement Techniques Association Meeting & Symposium, AMTA 2004,* Stone Mountain Park, Georgia, 17–22 October 2004.

10

Design and Analysis of a Planar Array-Fed Axisymmetric Gregorian Reflector System

10.1 Introduction

In this chapter, an axisymmetric array-fed confocal parabolic Gregorian reflector antenna system for potential deployment in space from a small satellite is explored. The antenna utilizes a planar array located near the vertex of the primary parabolic reflector. The planar array illuminates a parabolic subreflector that has the same focal point as the primary reflector. Electromagnetic simulations are used to analyze and optimize the antenna parameters for fixed on-axis peak directivity performance. Simulations of the radiation pattern performance of a high magnification dual-reflector system with an electrically large primary reflector and small subreflector are presented. The chapter is organized as follows. Section 10.2 describes the antenna design. Section 10.3 shows simulated results. Section 10.4 provides a summary.

10.2 Antenna Design

Reflector antennas are of interest for space communications and sensing applications requiring high gain and low sidelobes. In the case of a space-deployable antenna, a reduction in mass is an important goal, which might be achieved using inflatable structures. Inflatable antennas for space have been investigated by a number of researchers [1–7]. Thin-film materials with and without electrically conducting coatings can be considered for designing an inflatable space-deployable antenna and supporting structures [3–6].

In the present chapter, an axisymmetric dual-reflector antenna design with confocal paraboloids and a planar array feed is described, which is depicted as an artist's concept in Figure 10.1. It is assumed here that a satellite has accurate attitude control pointing so that errors in the antenna main beam pointing direction can be neglected. A thin-profile planar array would be located near the vertex of the main reflector and attached to the satellite body. An example offset Gregorian reflector antenna system with a planar phased array feed was shown previously in Chapter 2, Figure 2.13. However, here an axisymmetric Gregorian antenna system design with confocal paraboloids and planar array feed was desired, which could allow easier fabrication and inflatable deployment compared to an offset Gregorian or offset Cassegrain reflector design [8–11]. Limited electronic scanning with an axisymmetric Gregorian dual reflector with confocal paraboloids and a planar phased array feed has been demonstrated by simulations [12]. An ideal planar array source feeding the Gregorian subreflector is assumed in the simulations that follow. In the case where the large primary reflector surface could have distortion that creates phase errors, a phased array feed could provide phase compensation. When a large focal magnification is used in the Gregorian design, the blockage of the main beam by the subreflector can be relatively small. The primary reflector diameter is assumed here to be 2.4m with an ideal (undistorted) parabolic surface operating at Ku-band at 16 GHz ($\lambda = 1.875$ cm). With this 2.4m diameter, the antenna aperture is electrically large (128λ at 16 GHz). For an electrically large aperture, sparse moment method simulations with the multilevel fast multipole method (MLFMM) can be used to perform the analysis and optimization of the antenna design parameters [13, 14].

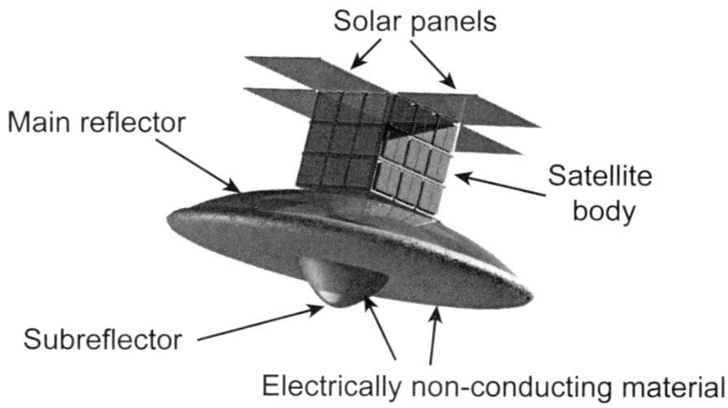

Figure 10.1 Artist depiction of a concept for an inflatable axisymmetric dual-reflector antenna deployed from a satellite.

Figure 10.2 shows a side view of a general axisymmetric Gregorian antenna with confocal parabolic reflectors. The common focal point is located close to the subreflector along the axis of both reflectors. The primary reflector has a diameter D and a focal distance f_p. The subreflector has a diameter d and a focal distance f_s. In Figure 10.3, ray paths are shown for the case where a field is incident along the axial direction of the reflector system. In general, a ray incident on the reflector surface will have the reflected angle equal to the incident angle. Since the subreflector and primary reflector are confocal, the ray path from the subreflector to the array is also in the axial direction. Thus, a uniform plane wave with constant amplitude and phase distribution received across the aperture of the primary reflector will be received at the planar array with constant amplitude and phase distribution.

As described by Fitzgerald [9], the magnification factor m for the Gregorian dual-reflector antenna system with confocal paraboloids is given

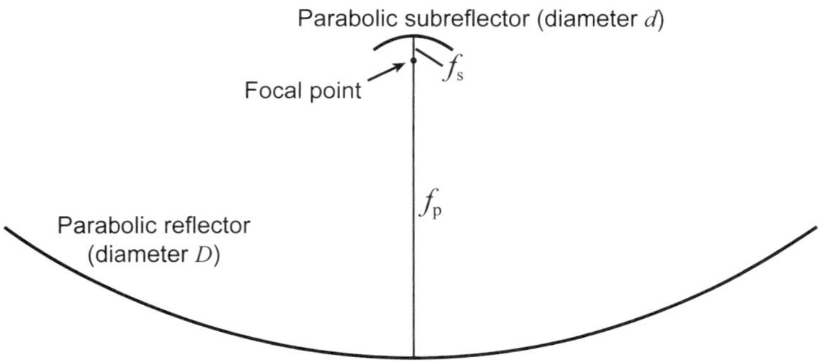

Figure 10.2 Side view of an axisymmetric Gregorian antenna with confocal parabolic reflectors.

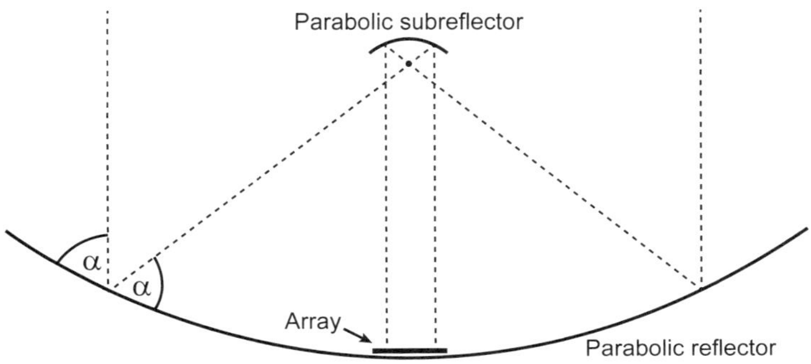

Figure 10.3 Side view of an array-fed axisymmetric Gregorian antenna with confocal parabolic reflectors and ray paths shown as dashed lines.

by the ratio of the primary to subreflector focal distances or

$$m = f_p / f_s \tag{10.1}$$

In the case where limited main beam scanning from the Gregorian antenna system boresight would be desired, the array would need to be phased such that the array scan angle $\theta_{s,\text{Array}}$ is, to first order, equal to the Gregorian main beam scan angle $\theta_{s,\text{Gregorian}}$ times the magnification factor m [9, 12]. That is,

$$\theta_{s,\text{Array}} = m\theta_{s,\text{Gregorian}} \tag{10.2}$$

In the limited scanning case, as Fitzgerald has described, the phased array location relative to the location of the subreflector is a significant design parameter. Additionally, as the array scan angle increases the subreflector must be oversized to avoid significant spillover, otherwise the Gregorian main beam gain would decrease and the sidelobe levels would increase. However, in this chapter only the fixed-boresight case is analyzed.

As shown in Figure 10.4, an example fixed-boresight Ku-band Gregorian confocal reflector system with a primary parabolic reflector diameter D=2.4m and focal distance $f_p = 0.9\text{m}$ ($f_p/D = 0.375$), parabolic subreflector diameter $d = 0.25\text{m}$ and focal distance $f_s = 0.08\text{m}$ ($f_s/d = 0.32$), and planar array diameter 0.2m, has been analyzed and was optimized using numerical simulations conducted with the FEKO software (www.feko.info) MLFMM solver. The feed array geometry is depicted in Figure 10.5. The 20 cm diameter array feed was modeled in FEKO as a near-field source that

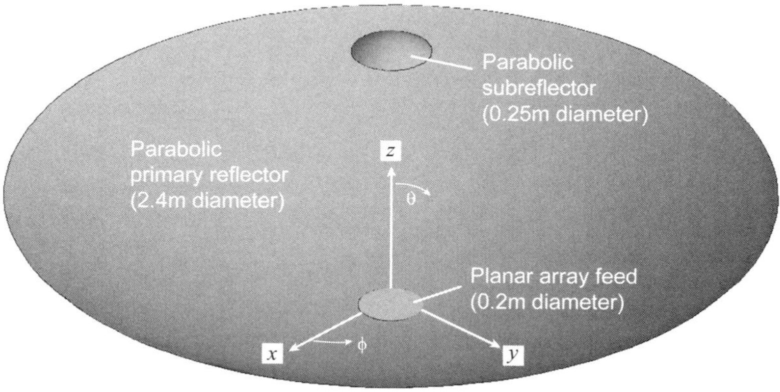

Figure 10.4 Simulation model for the axisymmetric Gregorian antenna with confocal parabolic reflectors and planar array feed.

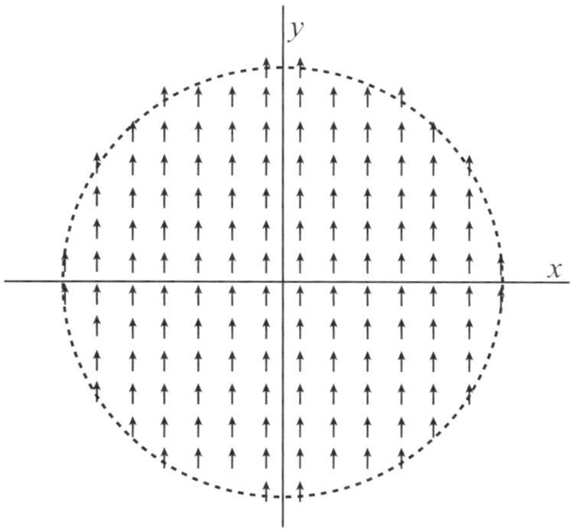

Figure 10.5 Aperture excitation model for the 20 cm diameter planar array feed.

approximates a uniformly illuminated planar array of 140 ideal Hertzian (short) electric dipoles polarized in the y direction. For this feed, the array elements are assumed to be spaced on a truncated square lattice with element spacing 1.5 cm which corresponds to 0.8λ at 16 GHz. Note that for the uniformly illuminated circular aperture feed array geometry shown in Figure 10.5, the E-plane and H-plane half-power beamwidths are given by [15, pp. 688–689]

$$\text{HPBW}_{\text{uniform circular aperture}} = \frac{58.4}{(D/\lambda)} \tag{10.3}$$

Since the aperture diameter is 20 cm, at 16 GHz the aperture diameter is 10.67λ and half-power beamwidth calculated by Equation (10.3) is $5.5°$. From Equation (1.670), the peak directivity of a uniformly illuminated circular aperture with diameter 10.67λ is computed as 30.5 dBi. A uniformly illuminated circular aperture will have -17.6 dBi first sidelobe levels.

The primary and secondary reflectors were analyzed as perfect electric conductors with ideal parabolic shape. The primary reflector had a both fixed diameter and a fixed focal distance. The array feed was assumed to have a fixed diameter of 20 cm. Optimization was performed as a grid search with the subreflector diameter d and subreflector focal distance f_s taken as the search parameters. The search optimization goal was assumed here to be peak directivity at boresight. After optimization was performed, the parabolic subreflector diameter was found to be $d = 0.25$m and the focal distance was $f_s = 0.08$m. The magnification factor m for this dual-reflector antenna system

(Figure 10.4) is given by the ratio of the primary to subreflector focal distances from Equation (10.4) as

$$m = f_p/f_s = 11.25 \tag{10.4}$$

The angle from the center of the feed array to the edge of the subreflector is approximately 8°.

10.3 Electromagnetic Simulation Results

The dual-reflector Gregorian reflector antenna system was simulated with 3,810,646 moment method basis functions using the FEKO fast multipole method solver on a computer system utilizing a total of 34 parallel processors on two CPUs at 2.3 GHz. The peak memory usage was 5.2 GBytes and the execution time was 392 seconds. The simulated far-field (infinite range) and near-field (range 0.98m at the subreflector) radiation patterns of the uniformly illuminated 20-cm diameter planar array feed alone are shown in Figure 10.6. The simulated half-power beamwidth is equal to 5.5° in the far-field, which agrees with the computed value. The simulated peak directivity of the feed array is 30.0 dBi, which compares closely with the calculated value of 30.5 dBi. The far-field radiation pattern of the array feed has nulls formed at $\pm 6.5^\circ$ and the first sidelobes are at the -17.2 dB level. In contrast, the feed array near-field amplitude at range distance 0.98m that illuminates the edge of

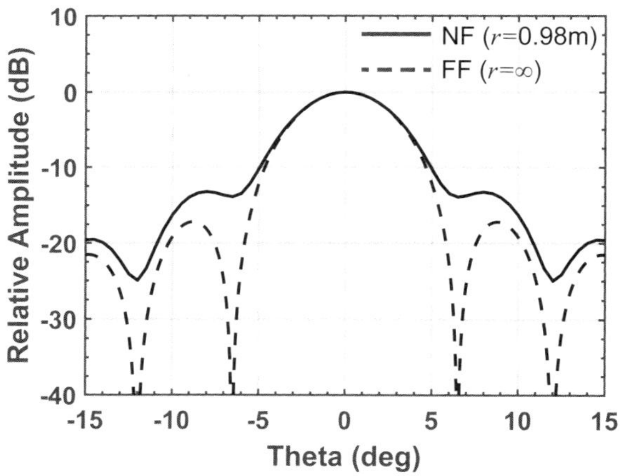

Figure 10.6 Comparison of the simulated near field (range is 0.98m) and far field (infinite range) radiation patterns of the planar array feed. The angle formed from the center of the feed array to the edge of the subreflector is $\pm 8^\circ$.

the subreflector is at the -13.2 dB level (edge illumination) and the far-field null at $\pm 6.5^\circ$ is filled in at the -13.8 dB level.

The simulated two-dimensional near-field distribution for the electric-field y component in the xz plane at $y = 0$ of the Gregorian reflector system is shown in Figure 10.7. In Figure 10.7, with the array feed operating in

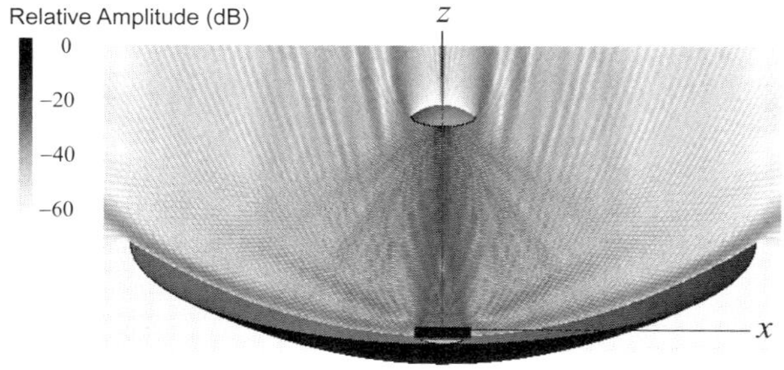

Figure 10.7 Simulated two-dimensional near-field distribution for the Gregorian reflector system.

transmit mode the field is concentrated toward the subreflector, and blockage by the subreflector is observed in the near field amplitude distribution behind the subreflector. The simulated surface currents on the subreflector and primary reflector are depicted in Figure 10.8. The surface current magnitude is

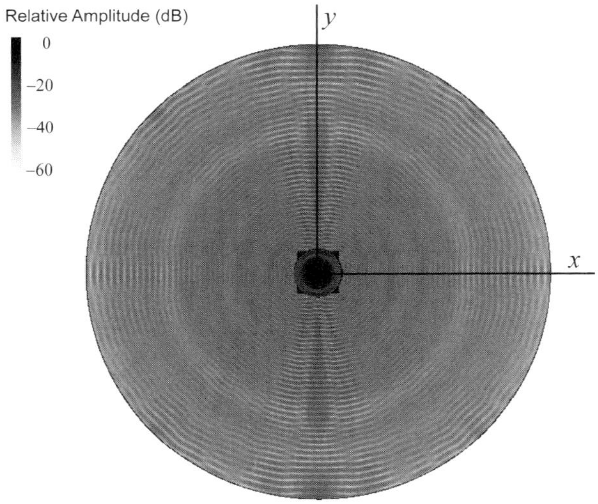

Figure 10.8 Simulated surface current density on the subreflector and primary reflector of the Gregorian reflector system.

observed to be centered and concentrated on the subreflector. The simulated E-plane and H-plane directivity patterns for the Gregorian reflector system over a $\pm 15°$ field of view at 16 GHz are shown in Figure 10.9. The simulated radiation pattern characteristics are as follows: peak directivity is 49.4 dBi, half-power beamwidth is $0.52°$, and first sidelobe level is -28.5 dB. For a 2.4 meter diameter aperture with 100% efficiency, the peak directivity calculated by Equation (1.670) at 16 GHz is 52 dBi, so the simulated aperture efficiency of this antenna design is 55%.

10.4 Summary

An axisymmetric array-fed confocal parabolic Gregorian reflector antenna system for potential inflatable deployment in space from a small satellite has been analyzed. The antenna utilizes a planar array located near the vertex of the primary parabolic reflector. Electromagnetic simulations based on the multilevel fast multipole method were used to analyze and optimize the antenna parameters for fixed on-axis peak directivity performance. Simulations of the radiation pattern performance of this dual-reflector system with a 2.4m diameter primary reflector operating at Ku band indicate that high gain and low sidelobes can be achieved with this design approach. All of the simulated results presented in the chapter assumed ideal aperture illumination and ideal parabolic surfaces for the primary and secondary reflectors. Future work in this area could consider the effects of reflector surface distortion

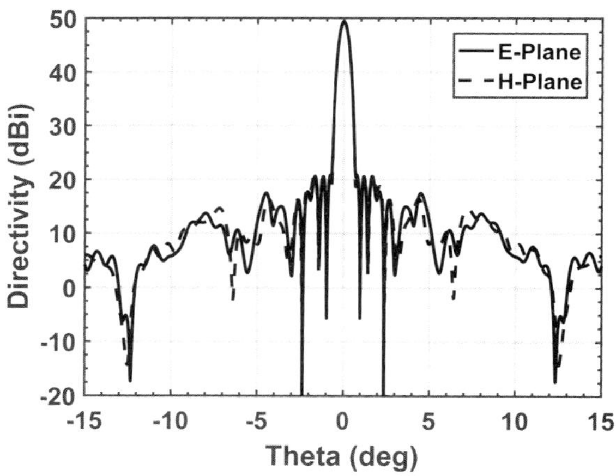

Figure 10.9 Comparison of the simulated E-plane and H-plane directivity patterns for the planar array fed axisymmetric Gregorian reflector system at 16 GHz.

along with reflector alignment errors, followed by calibration and distortion compensation using phased array feeding techniques.

References

[1] E.J. Ruggiero and D.J. Inman, "Gossamer Spacecraft: Recent Trends in Design, Analysis, Experimentation, and Control," *Journal of Spacecraft and Rockets,* vol. 43, no. 1, Jan.–Feb. 2006, pp. 10–24.

[2] E. Im, M. Thomson, H. Fang, J.C. Pearson, J. Moore, J. Lin, "Prospects of Large Deployable Reflector Antennas for a New Generation of Geostationary Doppler Weather Radar Satellites," *AIAA Space Conf.*, Long Beach, CA, 18-20 Sept. 2007, pp. 1–11.

[3] Y. Liu, H. Du, L. Liu, J. Leng, "Shape Memory Polymers and Their Composites in Aerospace Applications: A Review," *Smart Materials and Structures,* vol. 23, 2014, pp. 1–22.

[4] R.E. Freeland, G.D. Bilyeu, and G.R. Veal, "Development of Flight Hardware for a Large Inflatable-Deployable Antenna Experiment," *Acta Astronautica*, vol. 38, Nos. 4–8, 1996, pp. 251–260.

[5] Y. Xu and F. Guan, "Structure Design and Mechanical Measurement of Inflatable Antenna, *Acta Astronautica*," vol. 76, 2012, pp. 13–25.

[6] A. Babuscia, M. Van de Loo, Q. J. Wei, S. Pan, S. Mohan, and S. Seager, "Inflatable Antenna for CubeSat: Fabrication, Deployment and Results of Experimental Tests," *2014 IEEE Aerospace Conference,* 2014, pp. 1–12.

[7] A. Babuscia, T. Choi, C. Lee, and K-M. Cheung, "Inflatable Antennas and Arrays for Interplanetary Communication Using CubeSats and SmallSats," 2015 IEEE Aerospace Conference, 2015, pp. 1–9.

[8] C.J. Wilson, "Electronically Steerable Field Reflector Techniques," Technical Report No. RADC-TR-64-521, Feb. 1965.

[9] W.D. Fitzgerald, "Limited Electronic Scanning with an Offset-Feed Near-Field Gregorian System," MIT Lincoln Laboratory Technical Report 486, 24 Sept. 1971.

[10] J.A. Martinez-Lorenzo, A. Garcia-Pino, B. Gonzalez-Valdes, and C.M. Rappaport, "Zooming and Scanning Gregorian Confocal Dual Reflector Antennas," *IEEE Trans. on Antennas and Propagat.*, vol. 56, no. 9, 2008, pp. 2910–2919.

[11] A.J. Fenn and R.J. Richardson, "Analysis of an Adaptive Two-Reflector Phased-Array Fed System," *1980 IEEE Antennas and Propagation Society International Symposium,* 2–6 June 1980, pp. 134–137.

[12] A.J. Fenn and J.W. Jordan, "Design and Analysis of an Axisymmetric Phased Array Fed Gregorian Reflector System for Limited Scanning," *2016 IEEE Int. Symp. on Phased Array Systems and Technology,* 18 – 21 October 2016, Waltham, MA, pp. 1–3.

[13] W.C. Gibson, *The Method of Moments in Electromagnetics.* Boca Raton, Florida: Chapman & Hall, 2008.

[14] J. van Tonder; U. Jakobus, "Introduction of Curvilinear Higher-Order Basis Functions for MoM and MLFMM in FEKO," *Computational Electromagnetics Workshop (CEM),* 2013, pp. 13 – 14.

[15] C.A. Balanis, *Antenna Theory and Design, 3rd ed.* New York: Wiley, 2005.

11

Ultrawideband Dipole Array Antenna

11.1 Introduction

The effects of mutual coupling in dipole phased arrays, primarily with thin diameter dipoles, have been studied as described by Allen [1] and by Hansen [2]. Tightly-spaced array elements of various designs with strong mutual coupling are known to provide ultrawideband antenna characteristics [3–17]. A number of studies have investigated ultrawideband Vivaldi notch array elements [3, 4], but this type of element tends to require significant volume. The goal in the dipole arrays described in this chapter is to achieve good RF performance within a limited volume. An ultrawideband array with transverse electromagnetic (TEM) horn elements has been explored by McGrath and Baum [5]. Stepped-notch ultrawideband arrays with stripline feeds have been investigated by Brigham [6]. An ultrawideband low-profile UHF slot array has been explored by Lee [7]. Ultrawideband arrays have used flat dipole antenna elements with interlaced tips as explored by Munk [8] and Jones [9]. An ultrawideband interwoven spiral array over a ground plane has been explored by Tzanidis [10]. A planar ultrawideband modular antenna array has been investigated by Holland and Vouvakis [11]. The bandwidth limits for planar arrays over a ground plane have been explored by Doane [12]. Ultrawideband arrays of thin dipoles have been explored by Hansen [13]. The case of ultrawideband arrays of closely-spaced thick tubular dipole elements has been investigated [14–17].

Dipoles with swept-back arms were discussed in Chapters 2 and 3. This chapter concentrates on a straight-arm dipole design for ultrawideband application. Ultrawideband (UWB) dipole arrays are of interest for coverage in the VHF/UHF frequency range for communications or sensing applications. In this chapter, large-diameter open metallic tubular dipole elements are considered. This type of antenna-element design has good RF performance as a single element and in an array environment, and is applicable to airborne and ground-based analog and digital beamforming systems. In Section 11.2, the ultrawideband dipole array concept and design is discussed. The characteristics of ultrawideband dipole elements are analyzed using the method of moments for isolated elements in Section 11.3 and in Section 11.4 for an embedded element in a linear array. Prototype ultrawideband dipole array measurements are described in Section 11.5 and a summary is provided in Section 11.6.

11.2 Ultrawideband Dipole Array Design

The ultrawideband dipole array concept that is investigated in this chapter is depicted in Figure 11.1(a). The dipole antennas have a diameter D and overall length L. The ultrawideband array of tubular dipoles has tip-to-tip spacing s and ground-plane spacing (metal to metal) h. Close spacing between the tips of the dipoles provides enhanced mutual coupling in the array and is used

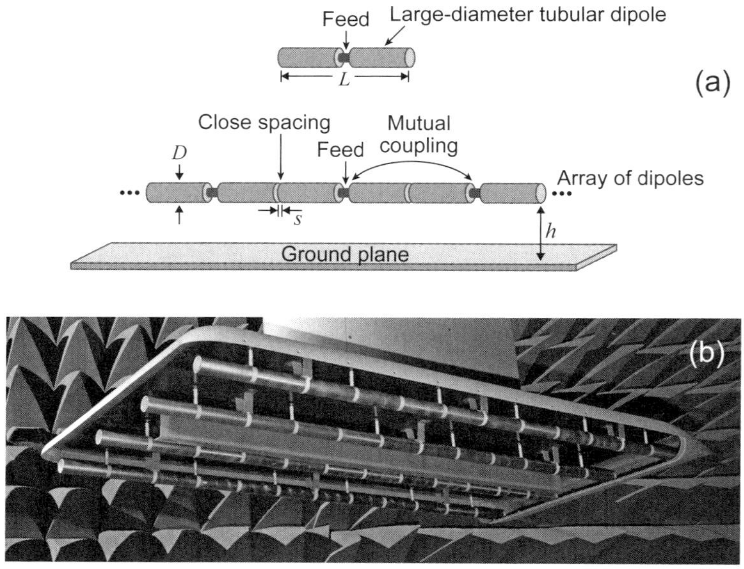

Figure 11.1 (a) Conceptual diagram of an ultrawideband dipole array. (b) prototype array composed of 24 ultrawideband dipoles.

to tune and extend the bandwidth of a single dipole element. Figure 11.1(b) shows a photograph of a prototype ultrawideband dipole array [17, p. 170].

The desired ultrawideband array performance can be achieved by providing dipole arms made with a large-diameter conducting tube and a central feed arrangement as shown conceptually for one dipole element in Figure 11.2. An antenna array can be designed and fabricated conveniently by supporting each of the dipole arms in the array with a dielectric (nonconducting) tube. Each dipole half can be formed by attaching a copper conductor (such as copper tape or thin-wall copper tubing) to the outside of the nonconducting tube. A coaxial feedline at right angles to the dipole axis can penetrate the dipole tube and connect to an ultrawideband transformer balun to provide a balanced feed over multiple octaves to the dipole terminals. An electrically conducting tube or honeycomb ground plane mounted parallel to the dipole array can be used as a reflector to provide mechanical support for the array and also provides some amount of front-to-back ratio. The conducting tube or ground plane (reflector) can contain multiple coaxial cable feedlines from other collinear array elements, and, in this manner, scattering from feedlines is reduced or avoided completely. A series of nonconducting tubes (perpendicular to the dipoles) can be used to provide mechanical support and maintain a desired electrical spacing between the dipole arms and the conducting ground plane tube. In the case of an airborne antenna array, the ends of the nonconducting tubes can be terminated with aerodynamic nose cones made of nonconducting material. Figure 11.3 depicts an ultrawideband dipole antenna with the E- and H-planes indicated. The UWB dipole is located with its axis in the x direction and the center feed port is located on the perimeter of the tubular dipole. With the feed port located off the axis of the dipole, depending on the antenna diameter the UWB dipole will only maintain approximate omnidirectional H-plane coverage. A single electrically thin dipole antenna in free space would tend to be narrowband and radiate in an omnidirectional manner in the H-plane. Electrically thin here refers to a dipole

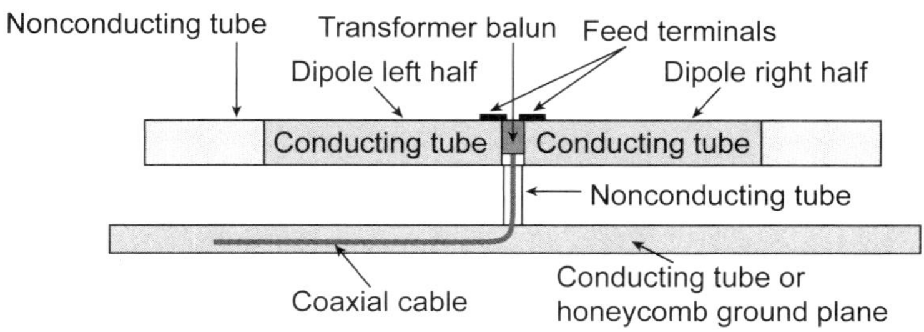

Figure 11.2 Conceptual ultrawideband dipole array element details.

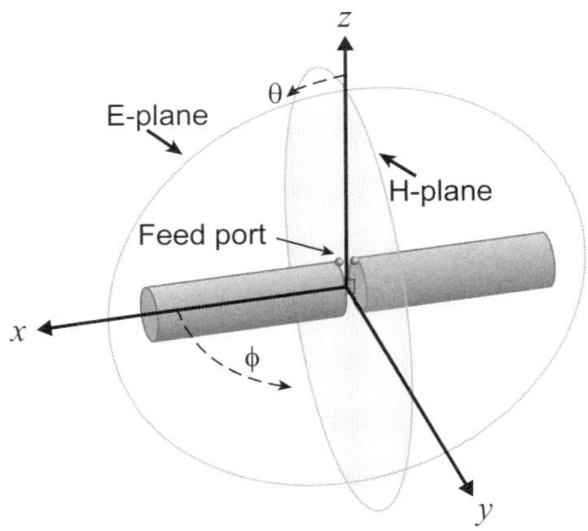

Figure 11.3 Ultrawideband dipole antenna element geometry.

with diameter less than about 0.01λ and electrically thick refers to a dipole diameter greater than about 0.01λ. Adding dipole elements to form a closely spaced collinear array can improve the tuning of the dipoles and provide increased antenna gain. Multiple rows of collinear arrays can be added to increase the antenna aperture size and gain as desired. A single collinear array of ultrawideband dipoles could be used as a focal line feed for a parabolic cylinder reflector to produce a large aperture size. In practice, the closely spaced dipoles can be mounted on a thin-wall dielectric tube, which spans the length of the collinear array. As will be shown, the dielelectric tube provides a convenient mechanical support for the array elements. Comparisons between open tubular-shaped and solid antennas have been made by Einarson [18]. A solid thick dipole is not required since the electric currents on the dipole are restricted to the outer surface layer of the dipole as governed by the skin depth in Equation (1.372). That is, a tubular dipole element is all that is required, which allows a low mass for practical implementations. Researchers have also investigated ultrawideband planar fat dipole antennas [19]. Thick dipole and monopole elements have been investigated in several studies [20–25].

From basic antenna theory, for an electrically short thin dipole antenna, the maximum effective aperture of the single element is equal to

$$A_{\text{em}}^{\text{short dipole}} = 3\lambda^2/(8\pi) = 0.119\lambda^2 \tag{11.1}$$

as discussed by Kraus [26] and is derived earlier in Section 1.10.2. Similarly, a one-half wavelength thin dipole has a maximum effective aperture equal to $0.13\lambda^2$. Thus, the maximum effective apertures of a short dipole (rounded off to $0.12\lambda^2$) and a one-half wavelength dipole ($0.13\lambda^2$) are within 10% of each other, and this variation can be neglected for a first-order analysis. Since the maximum effective aperture of a dipole one-half wavelength or less in length does not vary significantly, the maximum effective aperture of the dipole can be approximated as

$$A_{\text{em}}^{\text{dipole}} = 0.125\lambda^2. \tag{11.2}$$

The power received by an antenna is proportional to the effective aperture of the antenna. Referring to Equation (1.693), the realized gain G of an antenna is equal to the product of the directivity D, antenna radiation loss efficiency η_L, and transmission mismatch loss (ML) efficiency η_{ML}. In other words, the effective antenna gain is reduced by any mismatch loss for the feedline to antenna terminals connection, as well as by any resistive loss in the antenna, for example, caused by the finite electrical conductivity of the material used in fabricating the antenna. For a perfectly matched antenna, the antenna effective aperture is equal to the antenna radiation loss efficiency η_L times the maximum effective aperture A_{em}. Thus, the power received by an antenna is proportional to the antenna efficiency times the maximum effective aperture. The antenna efficiency takes account of the losses in the antenna. The antenna radiation loss efficiency can be calculated from the ratio of the antenna radiation resistance to the total input resistance, as discussed by Weeks [27]. The total input resistance is the sum of the radiation resistance plus all the resistance associated with dissipative losses. The directivity $D(\theta, \phi)$ of an antenna can be expressed as the ratio of the radiation intensity $U(\theta, \phi)$ to the average radiation intensity U_{ave} (refer to Equation (1.660))

$$D(\theta, \phi) = \frac{U(\theta, \phi)}{U_{\text{ave}}}. \tag{11.3}$$

Using the above relations and Equation (1.693), the gain of an antenna can be expressed in terms of the maximum effective area, radiation loss efficiency, transmission mismatch loss (ML) efficiency, and wavelength as

$$G = 4\pi A_{\text{em}}\, \eta_L\, \eta_{\text{ML}}/\lambda^2. \tag{11.4}$$

Thus, the gain relation for a dipole antenna is given approximately as

$$G_d = \frac{\pi}{2}\eta_L\, \eta_{\text{ML}} \tag{11.5}$$

which simplifies to

$$G_d = 1.57\, \eta_L\, \eta_{\text{ML}} \tag{11.6}$$

where the radiation loss efficiency η_L and mismatch loss depend on the dipole design. Without an impedance matching network, an electrically short dipole antenna in free space has poor radiating performance as quantified by a large mismatch loss (and possibly resistive loss), which yields a low effective (realized) gain. In all of the electromagnetic simulations that follow in this chapter, the dipole antennas and ground plane are assumed to be perfect conductors, and the radiation loss efficiency is 100% such that $\eta_L = 1$ in Equation (11.6). In this case with 100% efficiency, the realized gain of the dipole is then given by

$$G'_d = 1.57\eta_{\text{ML}}. \tag{11.7}$$

From Equation (11.7), the realized dipole gain in dBi should be about $10\log_{10} 1.57$ or 2 dB higher than the mismatch loss, and this behavior can be observed later in comparing Figures 11.7 and 11.8.

The realized gain is expected to decrease monotonically as the frequency decreases from the one-half wavelength resonance frequency of the dipole antenna. The gain of an electrically short linear antenna can be improved by surrounding the antenna with an array of antennas, which provides a tuning effect that significantly reduces mismatch loss. An example of the tuning effect resulting from array mutual coupling, for a thin dipole array, has been investigated by Hansen [13]. In [14–17], closely spaced large-diameter tubular dipole elements were investigated and ultrawideband performance was observed.

The radiation loss efficiency factor η_L of an isolated dipole antenna can be calculated as follows [28]. Referring to Section 1.14, the radiation loss efficiency of an antenna is expressed as

$$\eta_L = R_r/R_A \tag{11.8}$$

where R_r is the radiation resistance and

$$R_A = R_r + R_{\text{ohmic}} \tag{11.9}$$

is the total input resistance of the antenna. The ohmic resistance for an electrically short dipole of length L and diameter d is given by

$$R_{\text{ohmic}} = \frac{LR_s}{3\pi d} \tag{11.10}$$

where

$$R_s = \sqrt{\pi f \mu/\sigma} \tag{11.11}$$

is the surface resistance of the dipole, which depends on the frequency f, permeability μ, and electrical conductivity σ of the dipole material. Thus, since the ohmic resistance is inversely proportional to the dipole diameter, the efficiency can be increased by using a larger-diameter dipole.

For the examples investigated in this chapter, the dipoles analyzed are assumed to be perfectly conducting, and the ohmic resistance is zero; thus, the efficiency loss factor η_L is unity. For practical ultrawideband dipole antennas fabricated using copper conductor material ($\sigma = 5.7 \times 10^7$ S/m), the ohmic resistance is negligible at UHF frequencies, and the antenna efficiency is 100%. For example, assuming typical VHF/UHF ultrawideband dipole antenna parameters with L =12 inches (30.48 cm), $d = 1.625$ inches (4.1275 cm), and $\mu = \mu_o = 4\pi \times 10^{-7}$ H/m, the dipole antenna ohmic resistance is computed from Equations (11.10) and (11.11) as 0.0000625 ohms at 100 MHz, where the dipole is electrically short ($L = 0.1\lambda$). Thus, at 100 MHz, the total input resistance R_A given by Equation (11.9) is approximately equal to the radiation resistance, that is, $R_A \approx R_r$, such that the antenna efficiency η_L is approximately 100% using Equation (11.8).

At the lower end of the operating band for an ultrawideband antenna, the reflection coefficient and the corresponding mismatch loss tend to degrade, leading to a reduced antenna realized gain. At the upper end of the antenna operating frequency band, the maximum center-to-center electrical spacing d_{max} between the dipoles in an array is chosen to avoid grating lobes according to the following condition (see Equation (2.25)):

$$\frac{d_{max}}{\lambda_h} < \frac{1}{1 + |\sin\theta_s|} \tag{11.12}$$

where λ_h is the wavelength at the highest frequency and θ_s is the scan angle from broadside. For the array discussed in this chapter, for broadside scan $\theta_s = 0°$, from Equation (11.12) $d_{max} < \lambda_h$. The highest operating frequency is computed from the corresponding wavelength, λ_h, as

$$f_h = c/\lambda_h \tag{11.13}$$

where the speed of light is taken as $c = 3 \times 10^8$ meters per second.

11.3 Ultrawideband Dipole Compared to Thin Dipole: Free Space

In this section, the method of moments is used to compare the radiation characteristics of a large diameter tubular ultrawideband dipole and a thin-wire narrowband dipole [17]. Consider now the ultrawideband tubular dipole element depicted in Figure 11.3. The ultrawideband dipole is assumed to be composed of a perfect electrical conductor ($\sigma = \infty$) with infinitesimal wall thickness and tubular diameter 4.1275 cm (1.625 inches), overall dipole length 29.21 cm (11.5 inches), and feed gap spacing 1.27 cm (0.5 inch).

Over a decade of bandwidth, say from 0.1 GHz to 1.0 GHz the electrical diameter of this example UWB dipole varies from 0.014λ to 0.14λ. For a thin-wire linear dipole with the same length (29.21 cm), the resonant frequency at which the dipole is one-half wavelength long is approximately 500 MHz. As will be shown, when the dipole diameter takes on a significant dimension in terms of wavelength compared to a thin-wire dipole, the natural resonant frequency shifts to a significantly lower frequency and the bandwidth increases significantly.

The ultrawideband dipole was modeled here in a fine surface mesh using 340 triangular facets (triangle side length was 0.06λ at 1 GHz) for the method-of-moments solver in the commercial software package FEKO. The simulated input impedance of the ultrawideband tubular dipole in free space is shown in Figure 11.4, and the input impedance for the same length (29.21 cm) thin-wire dipole (wire diameter 0.002 cm) is shown in Figure 11.5. The imaginary part of the input impedance for the UWB dipole tends to be closer to a resonant condition over a large frequency range (from about 300 MHz to 900 MHz) compared to the thin-wire dipole. The simulated reflection coefficient (referenced to 50 ohms) of the ultrawideband dipole is compared to that of the thin-wire dipole in Figure 11.6. The UWB dipole reflection coefficient exhibits an ultrawideband response with two reflection coefficient minima (in dB) at 370 MHz and 815 MHz compared to the thin-wire dipole that has a single narrowband response at 500 MHz. The corresponding mismatch loss for the ultrawideband dipole and thin-wire dipole is compared in Figure 11.7. In the case of the thin dipole, the mismatch loss quickly degrades away from the resonant frequency at 500 MHz. In

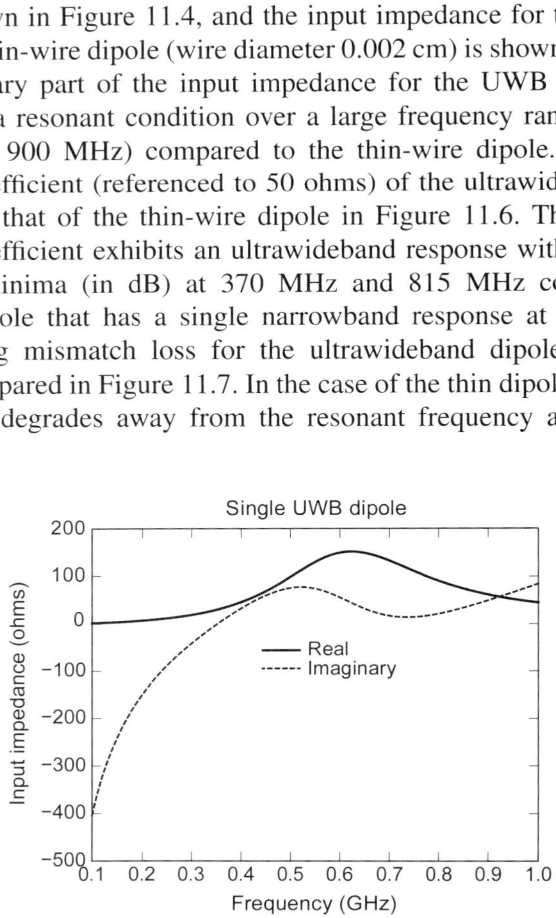

Figure 11.4 Simulated input impedance of an ultrawideband dipole in free space.

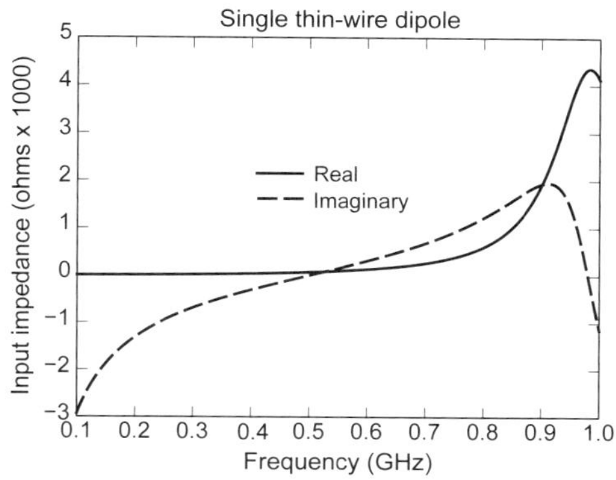

Figure 11.5 Simulated input impedance of a thin dipole in free space.

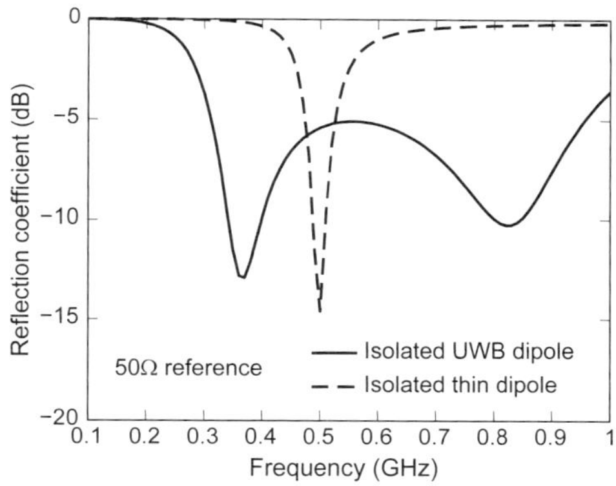

Figure 11.6 Comparison of the simulated reflection coefficient of a thin dipole and an ultrawideband dipole in free space.

contrast, for the UWB dipole the mismatch loss remains relatively low and constant from about 300 MHz to 900 MHz. The simulated UWB dipole boresight ($\theta = 0°$) realized gain versus frequency is compared against the thin dipole in Figure 11.8. The realized gain response of the UWB dipole is relatively constant from about 300 MHz to 900 MHz while the realized gain degrades rapidly away from resonance for the thin-wire dipole. The simulated E-plane and H-plane realized gain patterns for the UWB dipole antenna in free space are depicted in the polar diagrams in Figures 11.9 and 11.10, respectively. As the frequency increases and the dipole becomes electrically longer, the E-plane radiation pattern becomes more narrow. At 900 MHz, asymmetry in the E-plane radiation pattern is observed, which

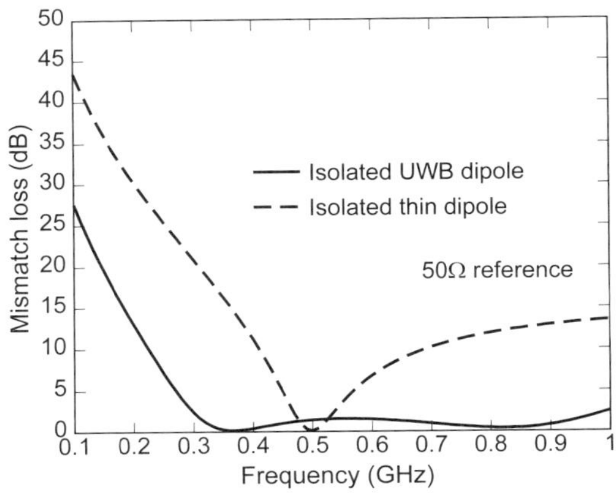

Figure 11.7 Comparison of the simulated mismatch loss of a thin dipole and an ultrawideband dipole in free space.

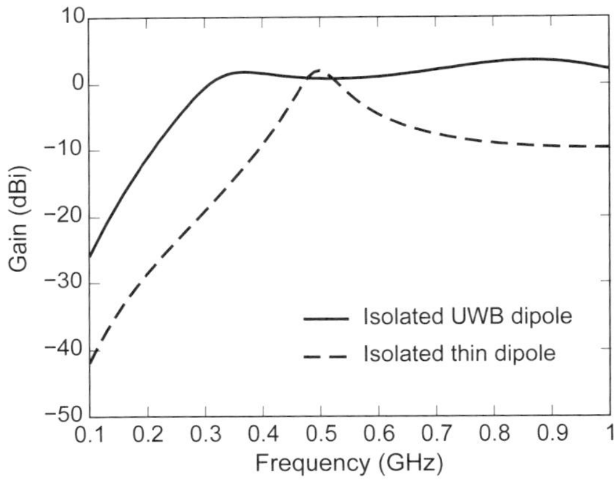

Figure 11.8 Comparison of the simulated boresight realized gain versus frequency of a thin dipole and an ultrawideband dipole in free space.

is due in this case to the feed port location on the perimeter of the dipole. The simulated H-plane realized gain patterns for the UWB dipole antenna in free space are depicted in the polar diagram in Figure 11.10, and a nearly omnidirectional characteristic is maintained at all frequencies shown. The next section considers the same UWB dipole in a linear array environment.

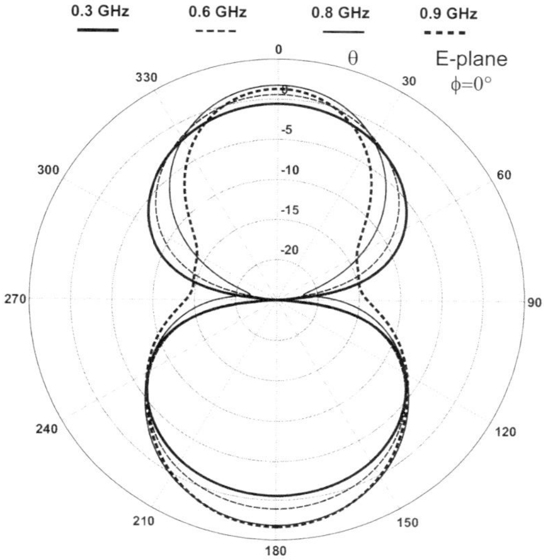

Figure 11.9 Comparison of the simulated E-plane realized gain patterns at multiple frequencies for an ultrawideband dipole in free space.

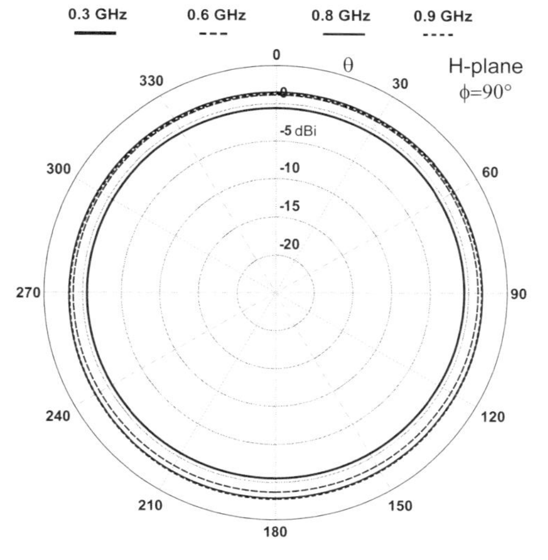

Figure 11.10 Comparison of the simulated H-plane realized gain patterns at multiple frequencies for an ultrawideband dipole in free space.

11.4 Ultrawideband Linear Array Simulation

In this section, the performance of a seven-element linear array of UWB dipoles is considered [17]. Figure 11.11 shows a meshed simulation model of

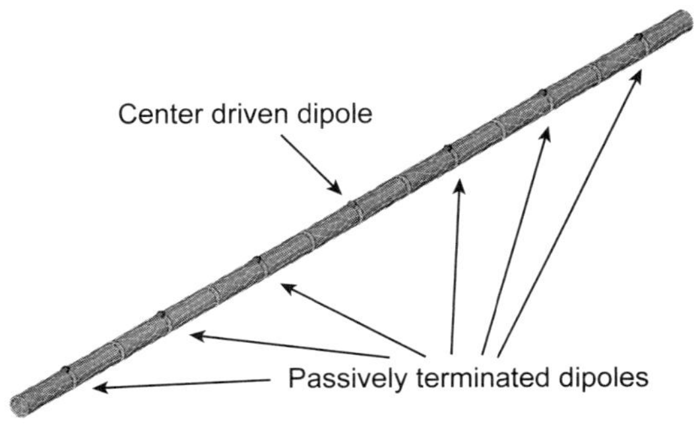

Figure 11.11 Meshed simulation model for a 7-element linear array of ultrawideband dipole elements.

the UWB dipole array. Each ultrawideband dipole is assumed to be composed of a perfect electrical conductor ($\sigma = \infty$) with infinitesimal wall thickness and tubular diameter 4.1275 cm (1.625 inches), overall dipole length 29.21 cm (11.5 inches), and feed gap spacing 1.27 cm (0.5 inch). The tip-to-tip spacing between the dipoles is 1.27 cm (0.5 inch). Method of moments simulated results are presented in Figures 11.12 to 11.17. The center element input impedance (real and imaginary) versus frequency with the center UWB element driven and the surrounding elements terminated in 50-ohm resistive loads is shown Figure 11.12. The input resistance is approximately 5 ohms at 100 MHz and 50 ohms at 1 GHz. The input resistance varies between about 40 ohms and 140 ohms from 300 MHz to 1 GHz. The input reactance is capacitive below the resonance that occurs at 380 MHz. A 50-ohm reference impedance was chosen to provide good performance at the lower frequencies.

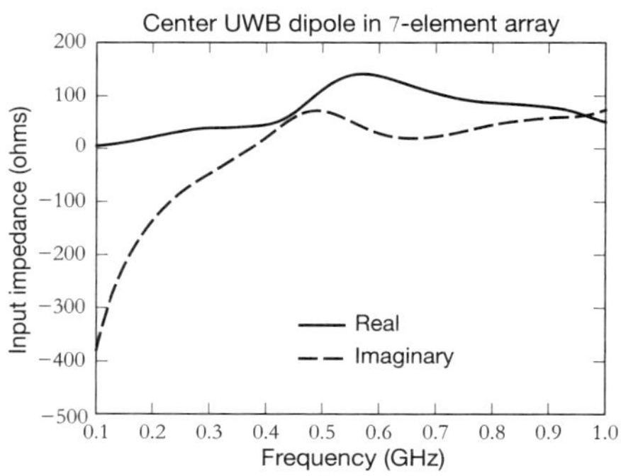

Figure 11.12 Simulated input impedance for the center-element driven of a 7-element linear array of ultrawideband dipole elements.

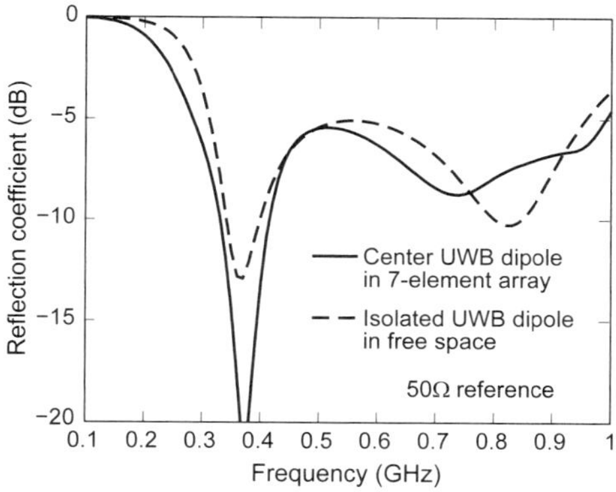

Figure 11.13 Comparison of the simulated reflection coefficient for the center-element driven of a 7-element linear array of passively loaded UWB dipole elements and a single isolated UWB dipole in free space.

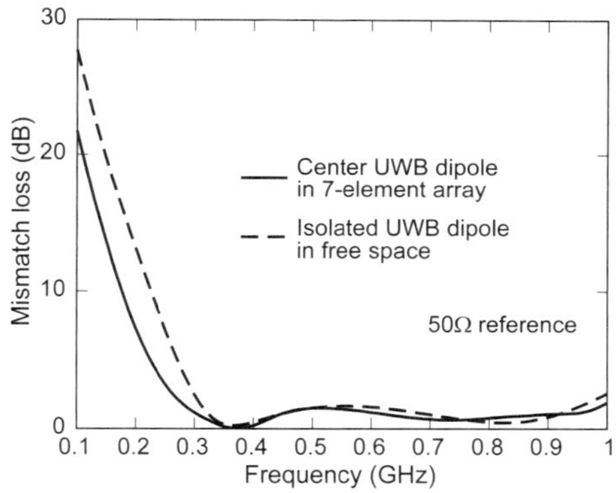

Figure 11.14 Comparison of the simulated mismatch loss for the center-element driven of a 7-element linear array of passively loaded UWB dipole elements and a single isolated UWB dipole in free space.

Figures 11.13 and 11.14 show the corresponding reflection coefficient and mismatch loss versus frequency relative to 50 ohms for both the center UWB element driven and an isolated UWB dipole. Below about 250 MHz, the mismatch loss is about 5 dB lower for the center element driven case compared to the isolated UWB dipole. In Figure 11.15, the embedded UWB dipole broadside gain versus frequency is compared with the isolated UWB dipole in free space. Below 250 MHz, the embedded UWB dipole broadside gain increases significantly up to about 5 dB compared to the isolated dipole, which is directly attributable to the reduction in mismatch loss in Figure 11.14.

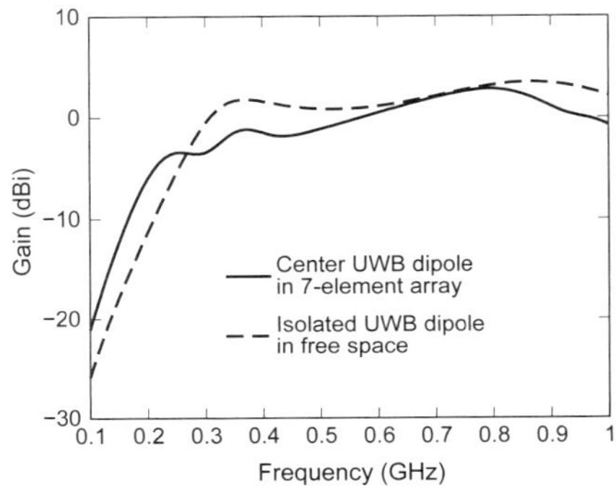

Figure 11.15 Comparison of the simulated boresight realized gain versus frequency for the center-element driven of a 7-element linear array of ultrawideband dipole elements and a single ultrawideband dipole in free space.

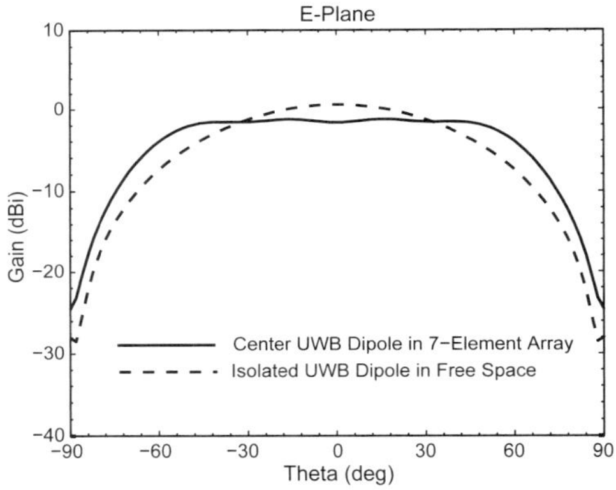

Figure 11.16 Simulated E-plane gain radiation pattern at 500 MHz for the center element of the seven-element passively terminated UWB dipole array compared to an isolated UWB dipole in free space.

In Figure 11.16, the embedded UWB dipole E-plane gain radiation pattern at 500 MHz is compared with the isolated UWB dipole in free space. The embedded UWB dipole E-plane gain radiation pattern is broadened compared to the isolated UWB dipole antenna. The simulated 3D radiation pattern (total gain normalized) at 500 MHz for the seven-element UWB dipole array in free space with uniform illumination and scan angle 30° is shown in Figure 11.17(a). In a phased array, the value of the reference impedance is a design parameter that can be chosen to help provide the desired scanning performance over a given frequency range. For example assuming a reference

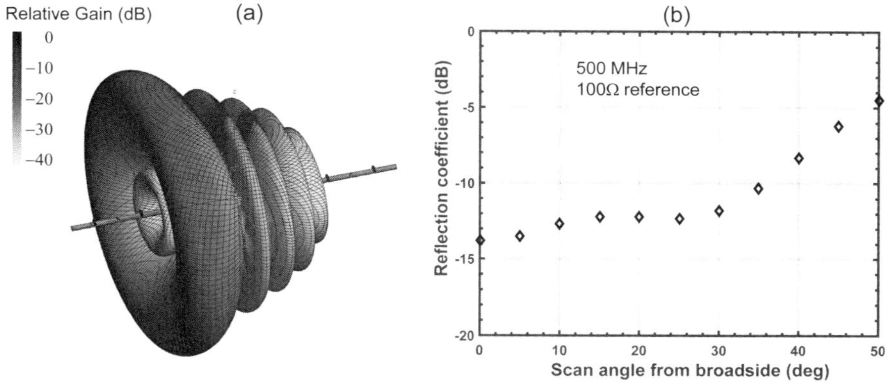

Figure 11.17 Simulated seven-element UWB dipole array in free space with uniform illumination at 500 MHz. (a) 3D radiation pattern for 30° scan, and (b) center element reflection coefficient versus scan angle.

impedance of 100 ohms, Figure 11.17(b) shows the simulated center element reflection coefficient versus scan angle from broadside at 500 MHz and good performance is observed.

11.5 Experimental Prototype UWB Array

A prototype ultrawideband dipole array design for operating over the 400 MHz to 900 MHz band (2.25:1 or 77% bandwidth) was evaluated in flight-testing [17] with the array mounted on the electrically conducting belly of an aircraft as shown in Figure 11.18. The array consisted of 72 UWB

Figure 11.18 Seventy-two element ultrawideband dipole array antenna installed on the belly of an aircraft.

dipoles (4.1275 cm [1.625 inch] diameter) arranged in a square lattice (25.4 cm [10 inch] center-to-center spacing) with six columns of 12 dipoles per column forming a 3.05 m by 1.32 m (120 inch by 52 inch) aperture for radiating or receiving with a downward, boresight beam peak. The dipoles had a half length of 10.795 cm (4.25 inches) and a feed gap of 1.27 cm (0.5 inch), with an overall dipole length of 9.0 inches (22.86 cm). A gap of 2.54 cm (1.0 inch) was used between the tips of adjacent dipoles. To keep a low profile, the dipoles were spaced 5.08 cm (2 inches) metal-to-metal from the ground plane. To maintain a balanced feed, the dipoles were fed with ultrawideband 50-ohm transformer baluns.

The overall 72-element UWB dipole array was composed of multiple subarray panels with the elements combined with analog power combiners. The subarrays were combined to form two independent phase centers in the aircraft cross-track direction, with each phase center composed of 36 dipole elements forming a 3×12 dipole array. Each of the two cross-track phase centers was fed with uniform illumination to form a broadside radiation pattern. The two phase centers could be combined to form sum and difference radiation patterns. Figure 11.19 shows the central subarray panel (one-third of the array aperture) under test in the MIT Lincoln Laboratory compact-range anechoic chamber, which has a 3.7m (12 ft) quiet zone [29]. This subarray panel has six linear arrays with four UWB dipoles per linear array. The reflection coefficient of a central UWB dipole element was measured with the surrounding subarray dipole elements terminated in 50-ohm resistive loads, and the data are shown in Figure 11.20. The peak reflection coefficient magnitude is -7 dB, which occurs at 400 MHz and the corresponding mismatch loss (gain loss) is 0.97 dB. Measured E-plane and H-plane radiation patterns of the uniformly illuminated subarray panel over the 400 MHz to 900 MHz band are shown in Figures 11.21 and 11.22, respectively. Since

Figure 11.19 Central 24-element subarray panel of the 72-element ultrawideband dipole array antenna under test.

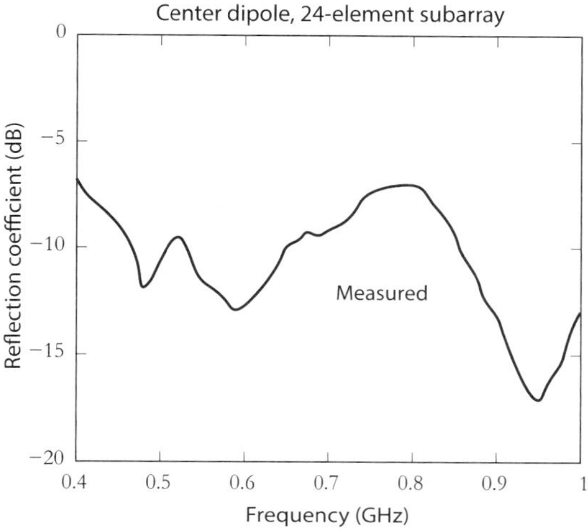

Figure 11.20 Measured reflection coefficient for a central dipole element of the center 24-element subarray panel of the 72-element UWB dipole array antenna.

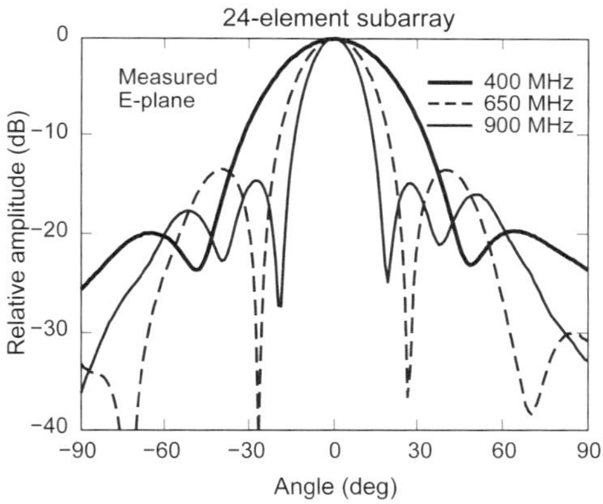

Figure 11.21 Measured E-plane radiation patterns for the uniformly illuminated 24-element subarray panel of the 72-element UWB dipole array antenna.

the electrical length of the aperture increases with increasing frequency, as expected the half-power beamwidth narrows as the frequency increases. In-flight radiation patterns of the ultrawideband dipole array were measured with a log-periodic antenna on a tripod on the ground with its peak radiation pattern aimed at zenith as the aircraft flew directly overhead [17]. Global Positioning System (GPS) coordinates were recorded on board the aircraft and

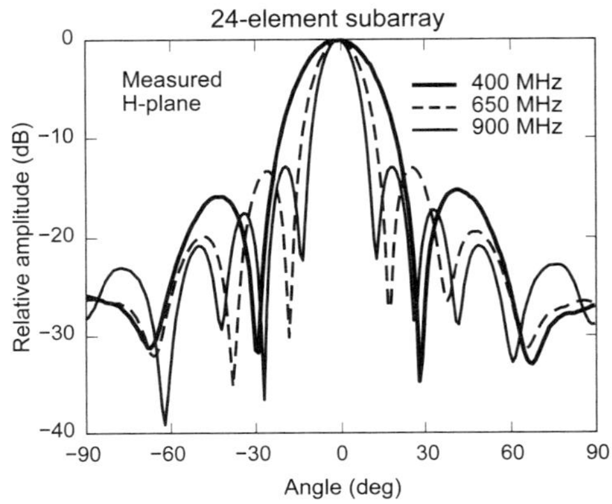

Figure 11.22 Measured H-plane radiation patterns for the uniformly illuminated 24-element subarray panel of the 72-element UWB dipole array antenna.

at the ground-based log-periodic antenna location to determine the relative observation angles during RF measurements. A CW signal stepped across the antenna bandwidth and the radiation patterns were measured. Measured and simulated array E-plane radiation patterns at 700 MHz of one of the 36-element phase centers is depicted in Figure 11.23, and good agreement is observed.

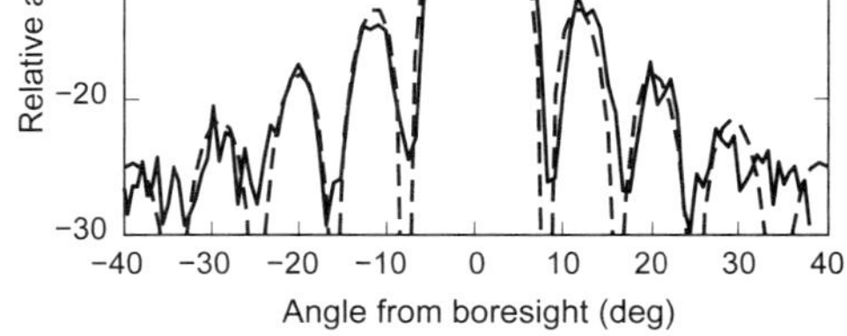

Figure 11.23 Simulated and in-flight measured E-plane radiation pattern at 700 MHz for one phase center (3 x 12 elements) of the 72-element ultrawideband dipole array antenna installed on the belly of an aircraft (see Figure 11.18).

11.6 Summary

Linearly polarized ultrawideband (UWB) dipole arrays have been discussed in this chapter. These arrays use large-diameter open metallic tubular dipole elements with or without a conducting ground plane. This type of antenna-element design has good RF performance as a single element and in an array environment, and is applicable to airborne and ground-based analog and digital beamforming. The method of moments was used to analyze single and multiple array elements in a linear array configuration. An example of a prototype two-dimensional array of ultrawideband dipoles was described.

References

[1] J.L. Allen, "Gain and Impedance Variation in Scanned Dipole Arrays," *IRE Trans. Antennas and Propagation*, vol. 10, no. 5, 1962, pp. 566–572.

[2] R.C. Hansen, *Microwave Scanning Antennas, vol. II: Array Theory and Practice.* Los Altos, Calif.: Peninsula Publishing, 1985.

[3] D.H. Schaubert, "Wide-Band Phased Arrays of Vivaldi Notch Antennas," *IEE 10th Int. Conf. on Antennas and Propagation*, 14–17 April 1997, pp. 1.6–1.12.

[4] M. Kragalott, W.R. Pickles, and M.S. Kluskens, "Design of a 5:1 Bandwidth Stripline Notch Array from FDTD Analysis," *IEEE Trans. Antennas and Propagation,* vol. 48, no. 11, 2000, pp. 1733–1741.

[5] D.T. McGrath and C.E. Baum, "Scanning and Impedance Properties of TEM Horn Arrays for Transient Radiation," *IEEE Trans. Antennas and Propagation*, vol. 47, no. 3, 1999, pp. 469–473.

[6] G.A. Brigham, M. Davidovitz, S.M. Duffy, and J.S. Herd, "Ultra-Wideband Step Notch Array Using Stripline Feed," *2007 IEEE Int. Symposium on Antennas and Propagation*, 2007, pp. 3660–3663.

[7] J.J. Lee, R. Koenig, D. Nagata, and L.L. Lai, "Compact Light Weight UHF Arrays Using Long Slot Apertures," *IEEE Trans. Antennas and Propagation,* vol. 54, no. 7, 2006, pp. 2009–2015.

[8] B.A. Munk, *Finite Antenna Arrays and FSS.* New Jersey: Wiley, 2003.

[9] M. Jones and J. Rawnick, "A New Approach to Broadband Array Design Using Tightly Coupled Elements," *IEEE Military Communications Conference, 2007*, pp. 1–7.

[10] I. Tzanidis, S. Kubilay, and J.L. Volakis, "Interwoven Spiral Array (ISPA) with a 10:1 Bandwidth on a Ground Plane," *IEEE Antennas and Wireless Propagation Letters,* vol. 10, 2011, pp. 115–118.

[11] S.S. Holland and M.N. Vouvakis, "The Planar Ultrawideband Modular Antenna (PUMA) Array," *IEEE Trans. Antennas and Propagation,* vol. 60, no. 1, 2012, pp. 130–140.

[12] J.P. Doane, K. Sertel, and J.L. Volakis, "Bandwidth Limits for Lossless Planar Arrays over Ground Plane," *Electronics Letters,* vol. 48, no. 10, 2012, pp. 540–542.

[13] R.C. Hansen, "Dipole Array Scan Performance Over a Wide-Band," *IEEE Trans. Antennas and Propagation,* vol. 47, no. 5, 1999, pp. 956–957.

[14] A.J. Fenn, "Wideband Dipole Array Antenna Element," United States provisional patent application Ser. No. 60/396,427, 17 July 2002.

[15] A.J. Fenn, *Adaptive Antennas and Phased Arrays for Radar and Communications.* Boston, Mass.: Artech House, 2008, pp. 351–358.

[16] A.J. Fenn, P.T. Hurst, J.D. Krieger, J.S. Sandora, and L.I. Parad, "Ultrawideband VHF/UHF Dipole Array Antenna," *2010 IEEE Int. Symposium on Phased Array Systems and Technology,* Waltham-Boston, Mass., 2010, pp. 79–82.

[17] A.J. Fenn and P.T. Hurst, *Ultrawideband Phased Array Antenna Technology for Sensing and Communications Systems,* Cambridge, Mass.: The MIT Press, 2015, pp. 135–179.

[18] O. Einarson, "A Comparison Between Tube-Shaped and Solid Antennas," *IEEE Trans. Antennas and Propagation,* vol. AP-14, 1966, pp. 31–37.

[19] Y.J. Park and J.H. Song, "Development of Ultra Wideband Planar Stepped-Fat Dipole Antenna," *Microwave and Optical Technology Letters,* vol. 48, no. 9, 2006, pp. 1698–1701.

[20] R.W.P. King and T.T. Wu, "The Thick Tubular Transmitting Antenna," *Radio Science*, vol. 2, Sept. 1967, pp. 1061–1065.

[21] D.C. Chang, "On the Electrically Thick Cylindrical Antenna," *Radio Science*, vol. 2, 1967, pp. 1043–1060.

[22] D.C. Chang, "On the Electrically Thick Monopole: Part I - Theoretical Solution," *IEEE Trans. Antennas and Propagation*, vol. 16, no. 1, 1968, pp. 58–64.

[23] D.C. Chang, "On the Electrically Thick Monopole: Part II - Experimental Study," *IEEE Trans. Antennas and Propagation,* vol. 16, no. 1, 1968, pp. 64–71.

[24] R.W.P. King, "Measured Admittances of Electrically Thick Monopoles," *IEEE Trans. Antennas and Propagation*, vol. 20, no. 6, 1972, pp. 763–766.

[25] P.G. Rogers and M. Gunn, "An Entire-Domain Galerkin Analysis of the Moderately Thick Dipole," *IEEE Trans. Antennas and Propagation*, pp. 117–121, 1980.

[26] J.D. Kraus, *Antennas, 2nd ed.* New York: McGraw-Hill, 1988, p. 44.

[27] W.L. Weeks, *Antenna Engineering.* New York: McGraw-Hill, 1968, p. 31.

[28] W.L. Stutzman and G.A. Thiele, *Antenna Theory and Design, 2nd ed.* New York: Wiley, 1998, pp. 43–48.

[29] M.W. Shields and A.J. Fenn, "A New Compact Range Facility for Antenna and Radar Target Measurements," *Lincoln Laboratory Journal*, vol. 16, no. 2, 2007, pp. 381–391.

12

Antenna and Electromagnetic Field Measurements

12.1 Introduction

Antenna, electromagnetic field, and radar cross section (RCS) measurements are often conducted indoors in shielded anechoic chambers [1-5]. Indoor measurements are generally desirable to avoid issues with weather, radiofrequency (RF) interference, and security. Outdoor measurements are also commonly conducted when the test antennas or radar test targets are too large or desired range distances too long for indoor measurements. A shielded anechoic chamber will typically consist of a room constructed of metal with the interior walls covered with microwave absorbing material. This type of metal shielded room is also referred to as a Faraday cage. The microwave absorber is generally made of foam that is loaded with RF lossy carbon material and it is chemically treated to be fire retardant. The metal walls with the added microwave absorber prevent radiofrequency signals from penetrating into or out of the anechoic chamber. With this metal shielding and RF absorbing material, external radiating sources will not interfere with RF measurements inside the anechoic chamber, and RF signals generated inside the anechoic chamber will not leak out and interfere with external RF systems. The magnitude of RF shielding can be quantified by radiating a CW signal from an antenna outside the anechoic chamber and receiving the signal level through the chamber walls with a second antenna inside the chamber that is connected to an RF spectrum analyzer. The relative RF shielding is quantified by comparing the received signal level through the chamber wall to a reference measurement with the same two antennas outside the chamber at the same antenna-to-antenna separation. A control room is usually attached to the anechoic chamber with a shielded panel containing bulkhead connectors to route RF signal cables, data cables, and control cables to the antenna under test. These cables must be shielded and filtered as needed to avoid external noise from leaking into or out of the anechoic chamber. The anechoic chamber will typically have an environmental monitoring and control system to achieve desired temperature and humidity levels, and it will have a fire-suppression system.

There are numerous designs of anechoic chambers for measurements of antennas, electromagnetic fields, and radar cross section. Multiple anechoic chambers of different designs are typically required to cover a full range of common antenna measurement frequencies spanning multiple decades from approximately 100 MHz to 100 GHz. Below about 100 MHz, outdoor antenna measurements are usually required.

The design of an anechoic chamber will depend on the operating frequency range and a number of other factors including the design of the antennas or test articles that need to be measured. Antenna and test article handling and positioning, RF instrumentation, and temperature and humidity control are important considerations in the overall chamber design. In this chapter, several types of anechoic chambers are described as well as some pertinent theory with an emphasis on phased array antenna measurements. Section 12.2 describes antenna test regions and typical indoor chamber concepts. An overview of the theory of planar near-field measurements is presented in Section 12.3. Focused near-field measurements of phased array antennas are discussed in Section 12.4. Section 12.5 describes an indoor RF Systems Test Facility at Lincoln Laboratory [6, 7] that has been updated and includes several anechoic chamber designs. This facility has replaced a prior Lincoln Laboratory antenna test facility [8].

12.2 Antenna Test Regions

The hemispherical volume in front of an antenna of maximum dimension D can be categorized by three regions: near, intermediate or Fresnel, and far zones. The required range distance R, for an antenna with diameter D, to the start of the far-field where the spherical wavefront will have a 22.5° phase taper is computed as

$$R_{\text{Far field}} = \frac{2D^2}{\lambda} \tag{12.1}$$

Due to spherical wavefront phase taper, the far-field distance given by Equation (12.1) introduces a reduction in antenna gain by 0.1 dB (compared to infinite range distance) for a uniformly illuminated aperture [1]. Normalizing the far-field distance by the aperture diameter yields

$$\frac{R_{\text{Far field}}}{D} = \frac{2D}{\lambda} \tag{12.2}$$

The required distance to the beginning of the Fresnel zone, or end of the near field region, is given by [9]

$$R_{\text{Fresnel}} = 0.62\sqrt{\frac{D^3}{\lambda}} \tag{12.3}$$

Normalizing this distance by the aperture diameter yields the near-zone range inequality

$$\frac{R_{\text{Near field}}}{D} \leq 0.62\sqrt{\frac{D}{\lambda}} \tag{12.4}$$

A diagram depicting the far-field, Fresnel, and near-field regions in terms of normalized distance is shown in Figure 12.1. Near-field testing techniques include compact range reflector [10], focused near-field phased array [11], and planar, cylindrical, and spherical near-field scanning [12], with typical normalized test distances as shown in Figure 12.1. The compact range reflector technique, described by example in Section 12.4.5, usually involves

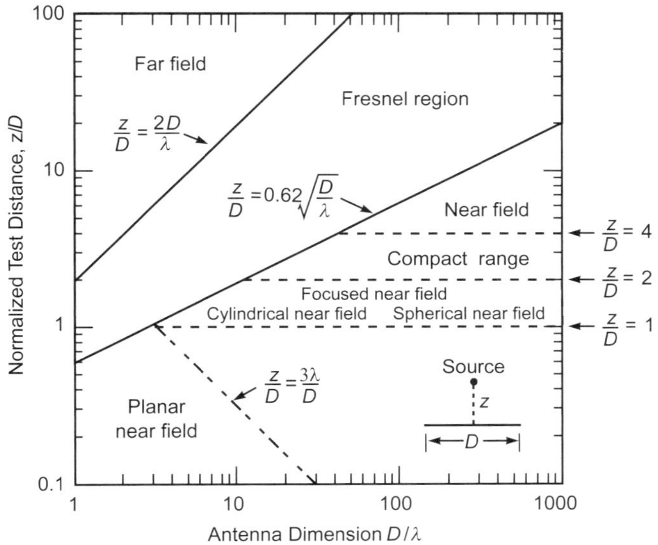

Figure 12.1 Diagram depicting the far-field, Fresnel, and near-field regions of an antenna.

a test range distance of between about two to four aperture diameters from the antenna under test. The focused near-field phased array technique and cylindrical near-field and spherical near-field scanning techniques typically involve a test distance between about one to two aperture diameters, but larger test distances can be used as well. The planar near-field measurement technique uses a test distance about three wavelengths or less from the antenna aperture. Note that in Figure 12.1, the dashed line for planar near-field measurements will extend below the horizontal axis of the plot, for example if the antenna dimension is $D = 300\lambda$ then the normalized test distance is $z/D = 0.01$. In the case where the physical aperture is a reflector antenna, the planar near-field scanning plane is positioned with respect to a projected aperture beyond the feed or subreflector.

Anechoic chambers can be configured for antenna or RCS measurements in a number of ways. Three common types of anechoic chambers are depicted in Figure 12.2. Typical far-field (spherical wave) anechoic chambers have a

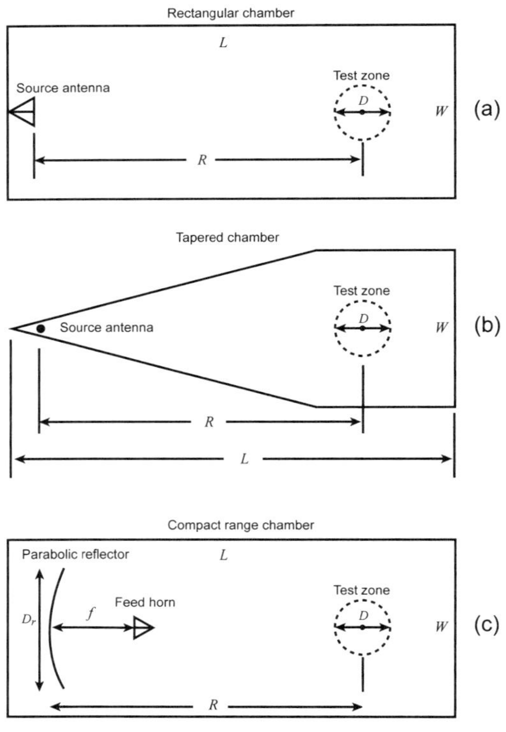

Figure 12.2 Three types of anechoic chambers for antenna or RCS measurements. (a) Rectangular chamber, (b) tapered chamber, and (c) compact range chamber.

rectangular shape as depicted in Figure 12.2(a), but in some cases a tapered chamber design is used as shown in Figure 12.2(b). Figure 12.2(c) depicts a rectangular chamber where instead of a spherical wavefront, a plane-wave is generated by a focused compact range parabolic reflector.

One of the important antenna parameters is the gain relative to isotropic. Standard gain antenna measurements are summarized in Figure 12.3. In Figure 12.3(a), a reference (standard) gain antenna will provide calibrated gain in dBi (decibels relative to isotropic) versus frequency. In a typical measurement setup (Figure 12.3(b)) the antenna under test will be compared with the standard gain antenna with the source antenna radiating the same power level. In Figure 12.3(c), the relative difference Δ in dB between the antenna under test (AUT) and the standard gain antenna with reference gain G_{ref} in dBi provides the measured AUT gain level in dBi, that is,

$$G_{\text{AUT}} = G_{\text{ref}} + \Delta \tag{12.5}$$

The three types of anechoic chambers with sources shown in Figure 12.2 allow real-time measurements and signal processing with either a spherical wavefront (rectangular chamber in Figure 12.2(a) and tapered chamber in Figure 12.2(b)) or with a planar wavefront in the compact range (Figure 12.2(c)). By using multiple sources, the rectangular chamber geometry can allow multiple spherical wavefronts, widely separated in angle, for real-time signal processing. In the situation where real-time signal processing is not required,

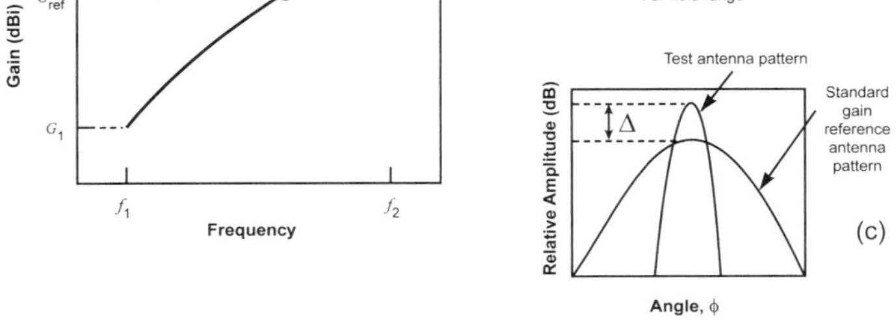

Figure 12.3 Standard gain measurements. (a) Gain calibration curve, (b) measurements setup with standard gain antenna and antenna under test, and (c) comparison of antenna under test and standard gain amplitude.

an alternate measurement method is to scan, with a probe antenna, the near-field amplitude and phase of the test antenna and compute the far-field gain patterns using a mathematical transform. Near-field scanning methods are depicted in Figure 12.4. Planar near-field scanning in rectangular (Cartesian) coordinates [12–24] is the most common method of the four measurement types shown in Figure 12.4, and theory for this method of scanning is addressed in Section 12.3. In the planar-scanning method, usually an open-end rectangular waveguide probe is moved in two Cartesian dimensions while the AUT remains fixed, which is most applicable to large heavy planar phased array antennas. In some planar scanning cases, the AUT can be moved in one dimension, while the waveguide probe is moved in the orthogonal dimension. Planar-polar near-field scanning can be accomplished by rotating the antenna and moving a near-field probe in a linear (radial) fashion [25]. With planar scanning, the measurement plane is truncated and thus the computed far-field radiation pattern will have a limited valid field of view. In cylindrical scanning [26, 27], the AUT is rotated in azimuth, and the near field probe is moved in a linear fashion forming a cylindrical surface enclosing the antenna. For cylindrical scanning, the transformed far-field data calculated will be valid over a full range of azimuth (ϕ) angles in the front and back region of the

<table>
<tr><th>Scanner Type</th><th>Scan Parameters</th><th>Geometry</th></tr>
<tr><td>Planar-rectangular</td><td>(x, y)</td><td>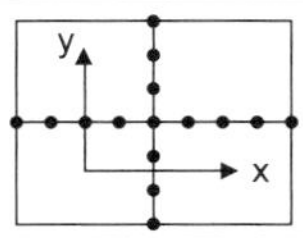
</td></tr>
<tr><td>Planar-polar</td><td>(R, ϕ)</td><td>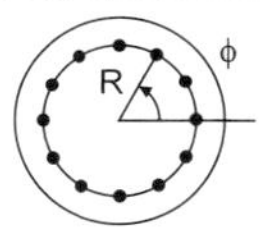
</td></tr>
<tr><td>Cylindrical</td><td>(z, ϕ)</td><td>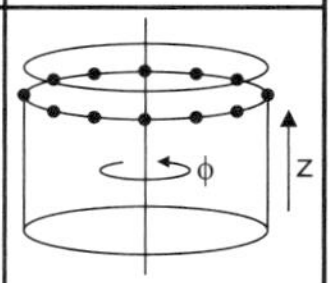
</td></tr>
<tr><td>Spherical</td><td>(θ, ϕ)</td><td>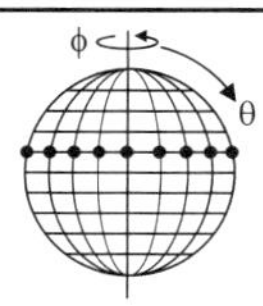
</td></tr>
</table>

Figure 12.4 Near-field scanning methods.

antenna, provided that the antenna support does not introduce significant field blockage. Furthermore, in cylindrical scanning, the extent of valid elevation angle data will be limited by truncation effects for the finite height (z distance) of the scan. With spherical near-field scanning [27–29], the measurements are conducted to fully encompass the AUT, thus the transformed far-field pattern calculated data will be valid over a full range of azimuth (ϕ) and elevation (θ) angles, except for any field blockage by the support mechanisms. In spherical near-field scanning, the AUT can be rotated with a two-axis positioner or the AUT can be rotated in azimuth while the near-field probe is moved along a circular arc.

It is worth noting that the planar near-field scanning technique, when implemented with the ability to position the near-field probe at an arbitrary (x, y, z) point, is particularly useful for calibrating the transmit/receive modules in a planar phased array antenna. For planar array elements arranged in either a rectangular or triangular lattice, the probe is moved in front of each radiating or receiving antenna element and the corresponding array element transmit/receive module is cycled through its amplitude and phase states to form a calibration look-up table from S21 mutual coupling measurements. These planar near-field probe S21 calibration measurements can be supplemented and replaced with array mutual coupling S21 measurements, or with S21 measurements between auxiliary near-field (or far-field) sources and the array elements for calibration once the phased array is deployed [30–38].

12.3 Planar Near-Field Measurements Theory

12.3.1 Introduction

As discussed above, the planar near-field scanning technique is commonly used in determining the far-field radiation patterns of antennas [12–24] and it is well suited for calibrating and evaluating the performance of planar phased array antennas. In a phased array antenna, low sidelobes are achieved using appropriate aperture amplitude taper together with precision element weighting and array calibration to achieve the desired accuracy in the array illumination function. The near field of the antenna is measured typically using a low-gain antenna probe that is moved to a regular grid of points on a planar surface as depicted in Figure 12.5. The near-field probe can be used either to transmit or receive the desired RF signal (usually a CW tone). After collecting the near-field data, a computer is used to mathematically transform the data, by means of the antenna plane wave spectrum, to a far-field pattern. The plane wave spectrum consists of radiating and evanescent (attenuated) plane waves as described below.

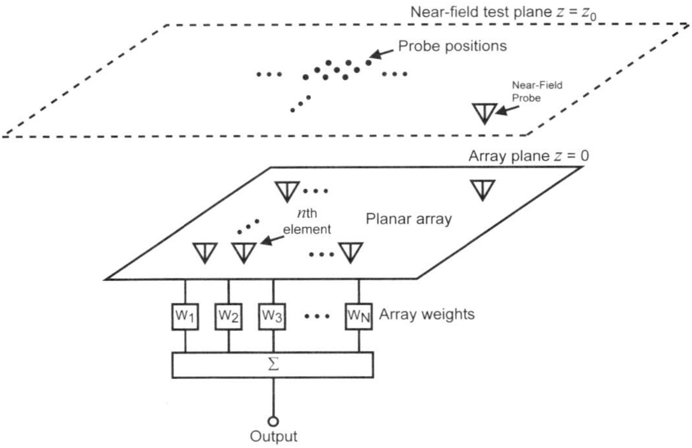

Figure 12.5 Phased array antenna and near-field test (scan) plane.

Planar near-field measurements are commonly performed in the radiating near-field region [12]. In this region (typically 2 to 3 wavelengths or more from the test antenna aperture), the evanescent contributions of the near field are sufficiently attenuated so that conventional one-half wavelength data sample spacing can be used. As the near-field test distance decreases, the near-field data sample spacing needs to decrease in order to take into account the increase that occurs in the evanescent component [19]. Data collection at a distance of approximately one wavelength or less is implemented in what has been referred to as the reactive or evanescent near-field region [12, 24]. The effect of the evanescent near field on the required sample spacing is a function of the aperture illumination and aperture diameter, as has been addressed theoretically by Wang [20]. For a low-sidelobe phased array antenna, effects of the near-field probe on the near-field data, plane wave spectrum, and transformed far-field pattern have been quantified [24].

12.3.2 Planar Near-Field Scanning Formulation

The purpose of this section is to briefly review the theory for determining antenna far-field radiation patterns from measured or simulated near-field probe data [15]. Let the time-harmonic vector electric field radiated from an antenna be denoted by $\boldsymbol{E}(x, y, z)$ and let the antenna aperture be located on the xy plane as shown in Figure 12.6(a). At a distance z in front of the antenna aperture the electric field can be represented as a superposition of plane waves [12–24]. Figure 12.6(b) depicts the plane wave spectrum, denoted as $\boldsymbol{A}$, in terms of the wavenumbers k_x, k_y, and k_z as described below.

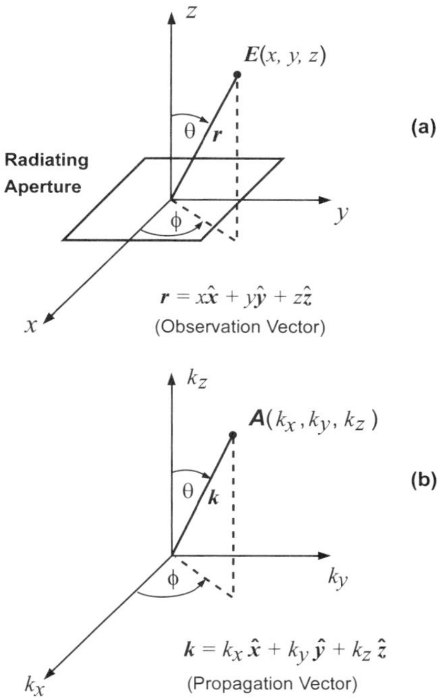

Figure 12.6 Coordinate systems for describing the radiation from an aperture antenna. (a) Electric field, and (b) plane wave spectrum.

From Maxwell's equations in free space and a vector identity, as derived in Chapter 1 (see Equation (1.349)) the homogenous vector wave equation is expressed as

$$\nabla^2 \boldsymbol{E} + k^2 \boldsymbol{E} = 0 \tag{12.6}$$

where $k = \omega\sqrt{\mu_o \epsilon_o} = 2\pi/\lambda$ is the wavenumber, with λ the wavelength, and ∇^2 is the Laplacian operator. A solution to the wave equation, for outgoing waves, with the $e^{j\omega t}$ time variation suppressed is of the form

$$\boldsymbol{E}(x, y, z) = \boldsymbol{A}(\boldsymbol{k})e^{-j\boldsymbol{k}\cdot\boldsymbol{r}} \tag{12.7}$$

where

$$\boldsymbol{r} = x\hat{\boldsymbol{x}} + y\hat{\boldsymbol{y}} + z\hat{\boldsymbol{z}} \tag{12.8}$$

is the observation position vector,

$$\boldsymbol{k} = k_x\hat{\boldsymbol{x}} + k_y\hat{\boldsymbol{y}} + k_z\hat{\boldsymbol{z}} \tag{12.9}$$

is the propagation vector, and as is depicted in Figure 12.2(b)

$$\boldsymbol{A}(\boldsymbol{k}) = A_x\hat{\boldsymbol{x}} + A_y\hat{\boldsymbol{y}} + A_z\hat{\boldsymbol{z}} \tag{12.10}$$

is commonly referred to as the plane-wave spectrum. The wavenumber components in terms of spherical coordinate angles θ and ϕ are expressed as

$$k_x = k\sin\theta\cos\phi \tag{12.11}$$

$$k_y = k\sin\theta\sin\phi \tag{12.12}$$

$$k_z = k\cos\theta \tag{12.13}$$

From Equations (12.11) to (12.13) it is observed that

$$k^2 = k_x^2 + k_y^2 + k_z^2 \tag{12.14}$$

In Equation (12.14), k_x and k_y are chosen to be the independent variables with k_z being the dependent variable, such that

$$k_z = \begin{cases} \sqrt{k^2 - k_x^2 - k_y^2} & k_x^2 + k_y^2 \le k^2 \\ -j\sqrt{k_x^2 + k_y^2 - k^2} & k_x^2 + k_y^2 > k^2 \end{cases} \tag{12.15}$$

Due to the exponential factor in Equation (12.7), the plane waves are considered propagating when $k_x^2 + k_y^2 \le k^2$ and they are evanescent (attenuated) when $k_x^2 + k_y^2 > k^2$. Only two components of the plane wave spectrum $\boldsymbol{A}$ are independent, A_x and A_y, with A_z chosen to be the dependent variable. Now since $\nabla \cdot \boldsymbol{E} = 0$, from Equations (12.7) and (12.10) it follows that

$$A_z(k_x, k_y) = -\frac{1}{k_z}(k_x A_x + k_y A_y) \tag{12.16}$$

The general solution for the electric field $\boldsymbol{E}$ can be expressed as a linear combination of the plane wave spectrum $\boldsymbol{A}$ over all values of k_x and k_y as the Fourier transform relationship

$$\boldsymbol{E}(x, y, z) = \int_{-\infty}^{\infty}\int_{-\infty}^{\infty} \boldsymbol{A}(k_x, k_y)e^{-j\boldsymbol{k}\cdot\boldsymbol{r}}dk_x dk_y \tag{12.17}$$

If the tangential electric field $\boldsymbol{E}_t$ is known over a planar surface at $z = z_o$, the plane wave spectrum can be determined from the inverse Fourier transform relationship

$$\boldsymbol{A}(k_x, k_y) = \frac{1}{4\pi^2}\int_{-\infty}^{\infty}\int_{-\infty}^{\infty} \boldsymbol{E}_t(x, y, z_o)e^{j\boldsymbol{k}\cdot\boldsymbol{r}}dx dy \tag{12.18}$$

For example, in electromagnetic simulations the tangential electric field of an antenna can be computed. Specifically, for $\boldsymbol{E}_t = E_x\hat{\boldsymbol{x}} + E_y\hat{\boldsymbol{y}}$ the independent components of $\boldsymbol{A}$ are

$$A_x(k_x, k_y) = e^{jk_z z_o} A_x(k_x, k_y, z_o) \tag{12.19}$$

$$A_y(k_x, k_y) = e^{jk_z z_o} A_y(k_x, k_y, z_o) \tag{12.20}$$

where

$$A_x(k_x, k_y, z_o) = \frac{1}{4\pi^2} \int_{-\infty}^{\infty} \int_{-\infty}^{\infty} E_x(x, y, z_o) e^{j(k_x x + k_y y)} dx dy \tag{12.21}$$

$$A_y(k_x, k_y, z_o) = \frac{1}{4\pi^2} \int_{-\infty}^{\infty} \int_{-\infty}^{\infty} E_y(x, y, z_o) e^{j(k_x x + k_y y)} dx dy \tag{12.22}$$

are the plane wave spectrum components computed from the electric field on the near-field scan plane. As suggested in Figure 12.7, the infinite integrals in Equation (12.21) and (12.22) are truncated according to the near-field scan length L, and the maximum angle for which the transformed data is accurate is given by [13, 17]

$$\theta_{\text{max}} = \tan^{-1}\left(\frac{L - l}{2H}\right) \tag{12.23}$$

For uniformly spaced near-field data, the resulting finite integrals are evaluated efficiently using the inverse fast Fourier transform (IFFT) algorithm. From Rhodes [23], in the far field of the antenna the electric field can be written in terms of the plane wave spectrum as

$$\boldsymbol{E}(r, \theta, \phi) = j2\pi k \cos\theta \frac{e^{-jkr}}{r} \boldsymbol{A}(k_x, k_y) \tag{12.24}$$

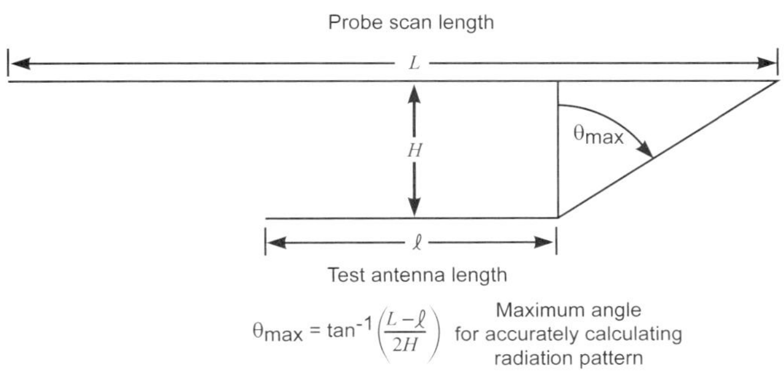

Figure 12.7 Diagram depicting the near-field scanning geometry and truncation for a test antenna.

To obtain E_θ and E_ϕ components from Equation (12.24) a standard conversion from rectangular components to spherical components is made.

In practice, near-field measured data are collected by utilizing a moveable antenna probe. For example, with the test antenna radiating, the voltage at the output of the probe is measured, rather than the electric field at a point in space. The probe radiation pattern shape affects the voltage measured at the output of the probe. Thus, probe compensation is required to accurately determine the test antenna radiation characteristics as summarized in the block diagram shown in Figure 12.8 At each near-field probe measurement position, two orthogonal probe orientations are required — one where the probe is $\hat{x}$ polarized and the other where the probe is $\hat{y}$ polarized. These two measurements can be accomplished with either a linearly polarized probe that is mechanically rotated, or with a dual-polarized probe. Let the far-zone electric field vector pattern of the near-field probe with x,y orientations in free space be expressed as

$$\boldsymbol{e}^x(\theta,\phi) = \hat{\theta} e_\theta^x(\theta,\phi) + \hat{\phi} e_\phi^x(\theta,\phi) \tag{12.25}$$

$$\boldsymbol{e}^y(\theta,\phi) = \hat{\theta} e_\theta^y(\theta,\phi) + \hat{\phi} e_\phi^y(\theta,\phi) \tag{12.26}$$

It is assumed that the probe radiation pattern is not affected by the presence of the test antenna and that the probe far-field pattern is obtained by independent measurements prior to near-field scanning.

Next, let the test antenna far-field pattern be given by

$$\boldsymbol{E}(\theta,\phi) = \hat{\theta} E_\theta(\theta,\phi) + \hat{\phi} E_\phi(\theta,\phi) \tag{12.27}$$

In Equation (12.27), E_θ and E_ϕ represent the two desired far-field components of the antenna under test. These components are determined by solving two simultaneous equations as described below.

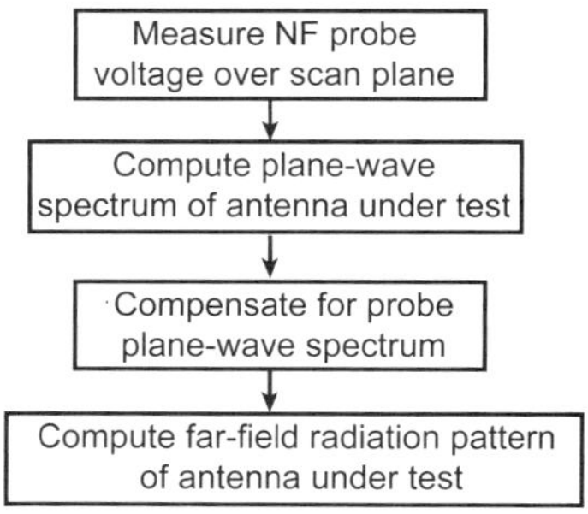

Figure 12.8 Block diagram depicting the measured near-field data to far-field transformation including probe compensation.

Let $\boldsymbol{a}^x$, $\boldsymbol{a}^y$ denote the probe plane wave spectrum for $\hat{x}$- and $\hat{y}$-polarized probe orientations, respectively. The received voltage at the array output, for a near-field transmitting probe at position (x, y, z_o) is given as [15]

$$v^x(x, y, z_o) = \frac{8\pi^2}{\omega\mu} \int_{-\infty}^{\infty} \int_{-\infty}^{\infty} k_z \boldsymbol{A}(k_x, k_y) \cdot \boldsymbol{a}^x(k_x, k_y) e^{-j\boldsymbol{k}\cdot\boldsymbol{r}_o} dk_x dk_y \tag{12.28}$$

$$v^y(x, y, z_o) = \frac{8\pi^2}{\omega\mu} \int_{-\infty}^{\infty} \int_{-\infty}^{\infty} k_z \boldsymbol{A}(k_x, k_y) \cdot \boldsymbol{a}^y(k_x, k_y) e^{-j\boldsymbol{k}\cdot\boldsymbol{r}_o} dk_x dk_y \tag{12.29}$$

In Equations (12.28) and (12.29), the received voltages $v^x(x, y, z_o)$ and $v^y(x, y, z_o)$, respectively, are equal to the superposition over all k_x, k_y of the dot product of the test antenna plane wave spectrum and probe plane wave spectrum. Taking the inverse Fourier transform of Equations (12.28) and (12.29) and defining a probe-distorted plane wave spectrum for x and y components as

$$A'^x(k_x, k_y, z_0) = \frac{1}{4\pi^2} \int_{-\infty}^{\infty} \int_{-\infty}^{\infty} v^x(x, y, z_o) e^{j(k_x x + k_y y)} dx dy \tag{12.30}$$

$$A'^y(k_x, k_y, z_0) = \frac{1}{4\pi^2} \int_{-\infty}^{\infty} \int_{-\infty}^{\infty} v^y(x, y, z_o) e^{j(k_x x + k_y y)} dx dy \tag{12.31}$$

yields the simplified expressions

$$k_z \boldsymbol{A}(k_x, k_y) \cdot \boldsymbol{a}^x = \frac{\omega\mu}{8\pi^2} e^{jk_z z_o} A'^x(k_x, k_y, z_0) \tag{12.32}$$

$$k_z \boldsymbol{A}(k_x, k_y) \cdot \boldsymbol{a}^y = \frac{\omega\mu}{8\pi^2} e^{jk_z z_o} A'^y(k_x, k_y, z_0) \tag{12.33}$$

As in Equations (12.21) and (12.22), the infinite integrals in (12.30) and (12.31) are replaced by finite integrals according to the actual near-field scan lengths. Next, analogous to Equation (12.24) the probe far-zone electric field is written in terms of the probe plane wave spectrum as

$$\boldsymbol{e}^x(r, \theta, \phi) = j2\pi k \cos\theta \frac{e^{-jkr}}{r} \boldsymbol{a}^x(k_x, k_y) \tag{12.34}$$

$$\boldsymbol{e}^y(r, \theta, \phi) = j2\pi k \cos\theta \frac{e^{-jkr}}{r} \boldsymbol{a}^y(k_x, k_y) \tag{12.35}$$

where $\boldsymbol{e}^x$,$\boldsymbol{e}^y$ denote the far-field vector patterns for the x-, y-directed probe, respectively. Substituting Equations (12.24), (12.34), and (12.35) into (12.32) and (12.33) and dropping the e^{-jkr}/r dependence in the far field yields

$$\boldsymbol{E}(\theta, \phi) \cdot \boldsymbol{e}^x(\theta, \phi) = C \cos\theta e^{jk_z z_o} A'^x \tag{12.36}$$

$$\boldsymbol{E}(\theta,\phi)\cdot\boldsymbol{e}^y(\theta,\phi) = C\cos\theta e^{jk_z z_o}A'^y \tag{12.37}$$

where the constant C is given by

$$C = -\frac{k\omega\mu_0}{2} \tag{12.38}$$

Equations (12.36) and (12.37) are solved simultaneously for the components E_θ and E_ϕ by utilizing Equations (12.25), (12.26), and (12.27) with the result

$$E_\theta = C\cos\theta e^{jk_z z_o}\left[\frac{e_\phi^y A'^x - e_\phi^x A'^y}{\Delta}\right] \tag{12.39}$$

$$E_\phi = C\cos\theta e^{jk_z z_o}\left[\frac{-e_\theta^y A'^x + e_\theta^x A'^y}{\Delta}\right] \tag{12.40}$$

where

$$\Delta = e_\theta^x e_\phi^y - e_\phi^x e_\theta^y \tag{12.41}$$

In evaluating the plane wave spectrum functions A'^x and A'^y, given by Equations (12.30) and (12.31), respectively, an N-point (including zero filling) IFFT is typically used.

12.3.3 Coordinate System Conversions

In the previous section, the near-field to far-field transformation was performed in spherical coordinates with the antenna under test in the xy plane. However, azimuth and elevation coordinates (a, e) are commonly used in practice as depicted in Figure 12.9. Thus, it is desirable to convert from spherical coordinates to azimuth and elevation coordinates in terms of angles and polarizations.

To convert from spherical coordinate angles (θ,ϕ) to azimuth and elevation angles (a, e) the required relations are determined as follows. The wavenumbers are expressed in terms of (θ,ϕ) and (a, e) angles as

$$k_x = k\sin\theta\cos\phi = k\cos e\sin a \tag{12.42}$$

$$k_y = k\sin\theta\sin\phi = k\sin e \tag{12.43}$$

$$k_z = k\cos\theta = k\cos e\cos a \tag{12.44}$$

It follows from Equations (12.43) and (12.42) that

$$e = \sin^{-1}(\sin\theta\sin\phi) \tag{12.45}$$

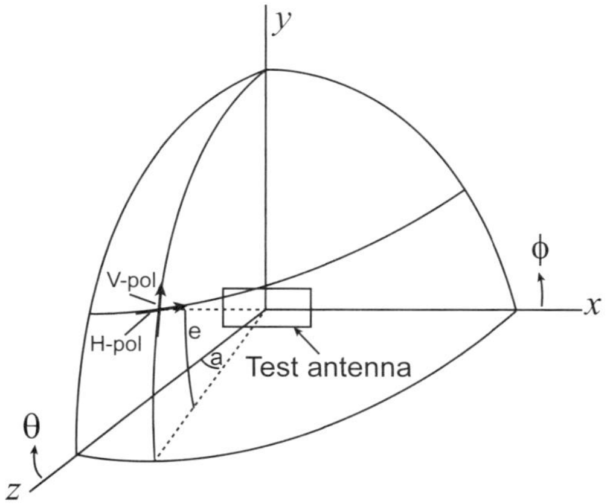

Figure 12.9 Antenna under test in the spherical coordinate system (θ, ϕ) and in the azimuth–elevation coordinate system (a, e).

$$a = \sin^{-1}\left(\frac{\sin\theta\cos\phi}{\cos e}\right) \tag{12.46}$$

Next, consider the conversion of polarization from one coordinate system to another for a general vector $\boldsymbol{A}$, which in this case represents a field vector conversion. For rectangular to spherical components (A_θ, A_ϕ):

$$A_\theta = A_x \cos\theta\cos\phi + A_y\cos\theta\sin\phi - A_z\sin\theta \tag{12.47}$$

$$A_\phi = -A_x\sin\phi + A_y\cos\phi \tag{12.48}$$

For spherical to rectangular components (A_θ, A_ϕ):

$$A_x = A_\theta\cos\theta\cos\phi - A_\phi\sin\phi \tag{12.49}$$

$$A_y = A_\theta\cos\theta\sin\phi + A_\phi\cos\phi \tag{12.50}$$

$$A_z = -A_\theta\sin\theta \tag{12.51}$$

For rectangular to elevation and azimuth components (A_e, A_a):

$$A_e = A_x\sin e\sin a + A_y\cos e - A_z\sin e\cos a \tag{12.52}$$

$$A_a = A_x\cos a - A_z\sin a \tag{12.53}$$

From spherical to elevation and azimuth components:

$$A_e = \frac{\cos e}{\cos^2\theta\sin^2\phi + \cos^2\phi}(A_\phi\cos\phi + A_\theta\cos\theta\sin\phi) \tag{12.54}$$

$$A_a = \frac{\cos e}{\cos^2\theta\sin^2\phi + \cos^2\phi}(A_\theta\cos\phi - A_\phi\cos\theta\sin\phi) \qquad (12.55)$$

For an antenna as shown in Figure 12.9, the A_e component corresponds to vertical polarization (V-pol) and the A_a component corresponds to horizontal polarization (H-pol).

12.4 Focused Near-Field Adaptive Nulling Concept

In the special case of a test antenna array, where the phase of each array element can be controlled either by analog or digital means, the array can be phase focused at arbitrary range distances. In focused near-field phased array testing, it is assumed that the near-field radiation pattern of the array should have the same or approximate characteristics as the far-field radiation pattern of the array. This means typically that a main beam and sidelobes should be formed in the near field, with amplitude levels approximately the same as in the far field. As depicted in Figure 12.10, to produce an array near-field pattern that is approximately equal to that of the far-field pattern, phase focusing can be used as demonstrated experimentally with

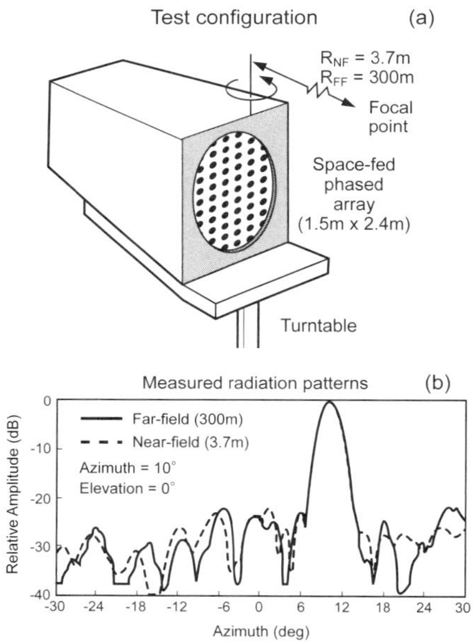

Figure 12.10 Focused near-field antenna pattern measurements. (a) Space-fed lens antenna, and (b) measured far-field and near-field radiation patterns versus azimuth angle.

a space-fed lens phased array by Scharfman and August [11, Figure 8, p. 348]. Figure 12.10(a) depicts the space-fed lens and test geometry where the focal point of the array was adjusted to the near field distance 3.7m (12 ft) and to the far-field distance 300m (1,000 ft). The C-band space-fed lens consisted of 1,721 radiating antenna elements in a 1.5m by 2.4m (5 ft by 8 ft) aperture. To focus in either the far field or near field, the relative phase shift of each antenna element was controlled by a 3-bit phase shifter. With this technique a CW calibration source antenna is located at a desired focal point of the array. The array can maximize the signal received from the calibration source by adjusting its phase shifters so that the spherical wavefront phase variation is removed. A main beam will be pointed at the array focal point. Sidelobes will exist at angles away from the main beam. The measured data in Figure 12.10(b) indicates good agreement between the far-field and focused near-field main beam shapes and sidelobe levels. It was also shown that the focused near-field main beam gain was in good agreement with standard focused far-field data [11]. The approximate equivalence of focused near-field radiation patterns and focused far-field radiation patterns occurs when the near-field focusing distance is at least one aperture diameter from the AUT. Besides antenna radiation pattern measurements, this focused near-field testing technique allows real-time measurements of adaptive phased array performance [39–45]. As depicted conceptually in an indoor anechoic chamber in Figure 12.11, the phased array antenna under test is focused using a CW calibration source. In the case of a radar system, digitally synthesized RF radiating sources such as target and clutter signals can be injected into the

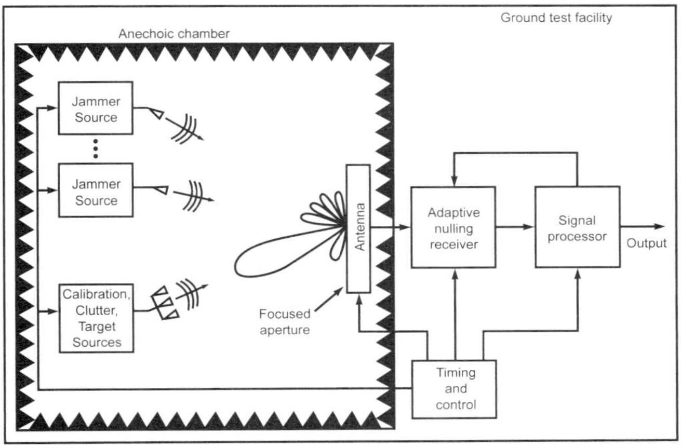

Figure 12.11 Focused near-field adaptive array antenna testing concept.

main beam region and interferers or jammers are placed on focused near-field sidelobes. The equivalent dynamic range of the main beam and sidelobes, allows standard adaptive array processing algorithms to be used [39–50]. Focused near-field adaptive nulling measurements were shown in Figure 2.56.

12.5 An RF Systems Test Facility

12.5.1 Introduction

Lincoln Laboratory operates an RF Systems Test Facility (RFSTF) [6, 7] that performs antenna, RCS, and EMI/EMC measurements, and has several shielded anechoic chambers as depicted in Figure 12.12. The RFSTF consists of six chambers as listed below:

- Tapered chamber,
- Millimeter wave chamber,
- System test chamber,
- Compact range,
- Small near-field scanner,
- Large near-field scanner.

The RF shielding level of these chambers is typically 90 dB or more based on measurements. An early-warning fire and detection system, and

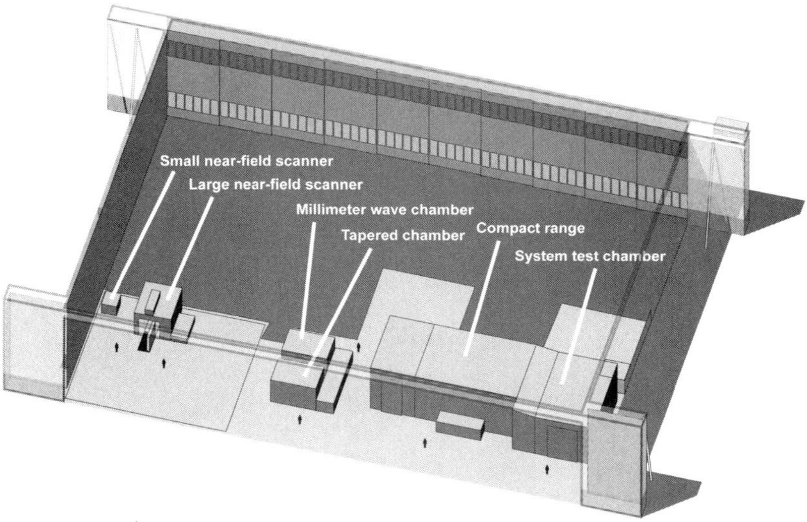

Figure 12.12 Diagram of the RF Systems Test Facility at MIT Lincoln Laboratory.

telescopic sprinkler head system are used as part of the fire protection system in these test chambers. The RFSTF has a rapid prototyping high-bay machine shop and general purpose RF instrumentation to support antenna system development and test. Network analyzers are used in performing S-parameter measurements to quantify test antenna reflection coefficient (S11) and mutual coupling (S21) between antennas. Spectrum analyzers are used in performing electromagnetic interference (EMI) measurements and shielding measurements. Each of the RFSTF anechoic chambers will be briefly described below.

12.5.2 Tapered Chamber with Ultrawideband Dual-Polarized Feed

The $L = 12.2$m (40 ft) long low-frequency tapered chamber (refer to the general diagram in Figure 12.2(b)) is EMI shielded and has a range length of $R = 8.5$m (28 ft) to the center of the test zone, with a rectangular section 4.3m by 4.3 m (14 ft by 14 ft) in cross section by 5.1m (16.8 ft) in length. Diagrams and photographs of key components of the tapered chamber are shown in Figures 12.13 to 12.18. The tapered chamber utilizes an ultrawideband dual-polarized conical tip feed [51] that is designed to cover approximately the 225 MHz to 3 GHz band, and higher frequencies, for both antenna and radar cross section measurements. The walls of the tapered chamber are covered with wedge absorber varying in height of 20 cm (8 inches) near the conical tip to 46 cm (18 inches) approaching the test zone. The test zone region uses 0.6m (24 inches) pyramidal absorber. The back wall is covered with

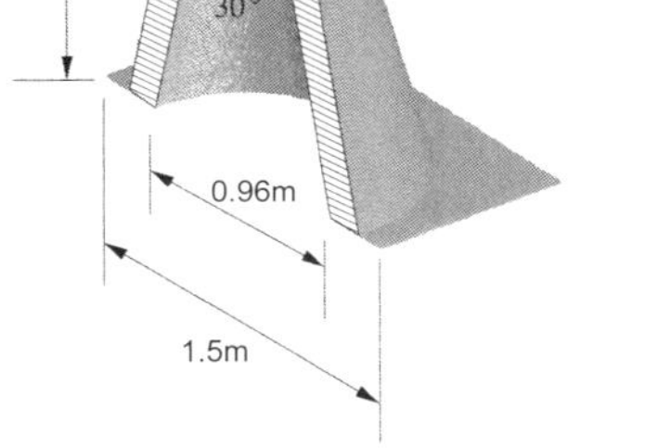

Figure 12.13 Sketch of the tapered chamber conical tip region with absorber.

Figure 12.14 Photograph of the conical tip region that provides the feed for the tapered anechoic chamber.

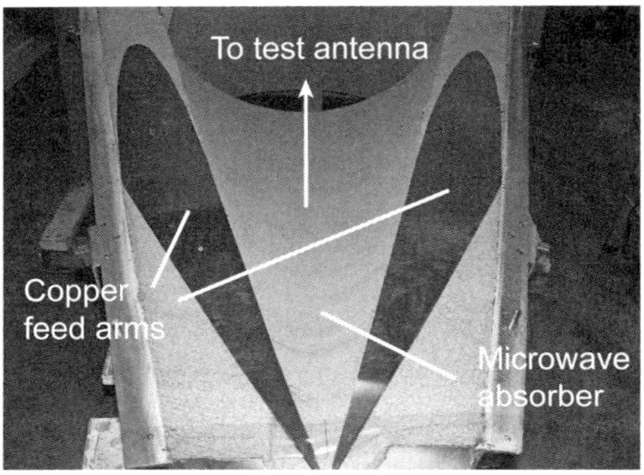

Figure 12.15 Photograph of one of the tapered chamber feeds installed in the conical tip region with absorber.

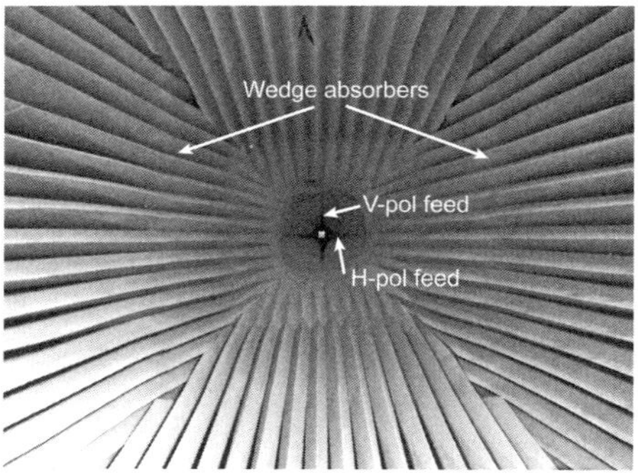

Figure 12.16 Photograph of the tapered chamber dual polarization feeds installed in the conical tip region with absorber.

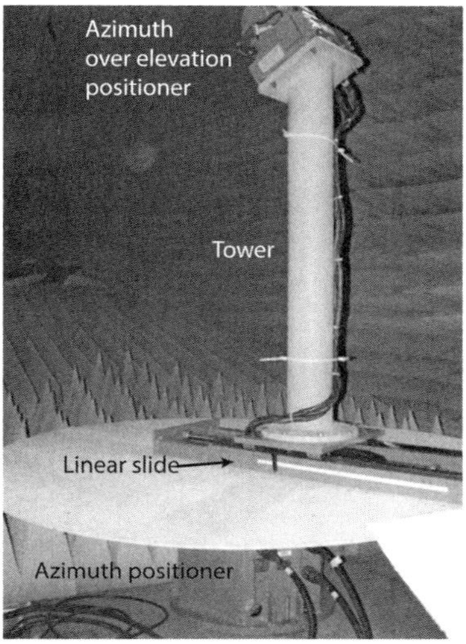

Figure 12.17 Photograph of the antenna positioner equipment in the tapered chamber. Microwave absorber at the base of the lower azimuth positioner was removed for clarity.

Figure 12.18 Photograph of a UHF Yagi array antenna mounted in the tapered chamber.

1.8m (6 ft) pyramidal absorber. As depicted in Figure 12.13, the shielded conical tip region is 1.8m in length with a 30° full cone angle. The inner surface is covered with microwave absorber. A photograph of the exterior view of the conical tip feed is shown in Figure 12.14. The square box attached to the conical tip houses two wideband 180° hybrids that provide a balanced feed to a pair of ultrawideband flared launchers. Figure 12.15 shows an interior view of one of the ultrawideband flared copper sheet launchers with microwave absorber before the conical tip feed was fully assembled. Figure 12.16 shows a view of the installed dual polarization ultrawideband feed in the tapered chamber. Wedge absorbers can be seen to provide a smooth transition toward the dual-polarization feed region. The antenna or under test positioner (Figure 12.17) consists, in part, of a 0.61m (24 inch) manual linear slide along the boresight axis over a lower azimuth positioner. A mast is mounted above the linear slide, and an azimuth (polarization) over elevation positioner is mounted at the top of this mast. In some measurements requiring only azimuth rotation, the AUT is supported on a foam column mounted above the linear slide. In Figure 12.18, a vertically polarized UHF Yagi 6-element array antenna [52] is attached directly to the linear slide, and is rotated using the lower azimuth positioner. In this figure, the layout of pyramidal absorber in the test zone region is observed.

12.5.3 Millimeter Wave Rectangular Chamber

The millimeter wave rectangular anechoic chamber (Figure 12.19) is EMI shielded and is used for testing small antennas (typically feed horns for reflector antennas) and small arrays from approximately 4 GHz to 100 GHz. The chamber has interior dimensions 9.1m (30 ft) length by 6.1m (20 ft) width by 5.5m (18 ft) height and a range length of $R = 6.7$m (22 ft) between the source antenna and antenna under test. To achieve good RF performance up to 100 GHz, the central portions of the walls, floor and ceiling are covered with unpainted 30.48 cm (12 inch) pyramidal absorber. Outside the central

Figure 12.19 Photograph of the millimeter wave chamber, looking toward the antenna under test positioner.

region, a blue-painted 20.32 cm (8 inch) pyramidal absorber is used. The AUT positioner is a polarization-over-elevation positioner mounted on a tower attached to a 0.61m (24 inch) manual linear slide along the boresight axis over an azimuth positioner, which is the same configuration used in the tapered chamber (refer to Figure 12.17). The AUT positioner is optically aligned to facilitate accurate measurements of millimeter wave antennas. A tracking laser interferometer was used in the optical alignment process to align the lower azimuth, slide, elevation, and polarization positioners.

12.5.4 System Test Rectangular Chamber

The rectangular System Test Chamber, shown in Figures 12.20 to 12.24, has dimensions 18.3m long by 12.2m wide by 10.7m height, (60 ft by 40 ft by 35 ft) and is RF shielded and covered uniformly with 1.22m (48 inch) pyramidal absorber for RF testing from approximately 100 MHz to 20 GHz.

This chamber is used, in part, for standard far-field antenna testing and spherical near-field testing. The antenna under test positioner and source tower are shown in Figures 12.20 and 12.21, respectively. The antenna under test positioner (Figure 12.20) consists of an azimuth over elevation positioner on a tower mounted on a linear slide above a lower azimuth positioner. The antenna under test shown in the photograph in Figure 12.21 is an X-band two-dimensional active electronically scanned array (AESA), with subarrays of Vivaldi flared notch radiators, that was used in radar system testing in the anechoic chamber including T/R module calibration, far-field radiation pattern measurements, and adaptive signal processing tests. The source tower is a linear positioner with a lifting capacity of 850 lbs that

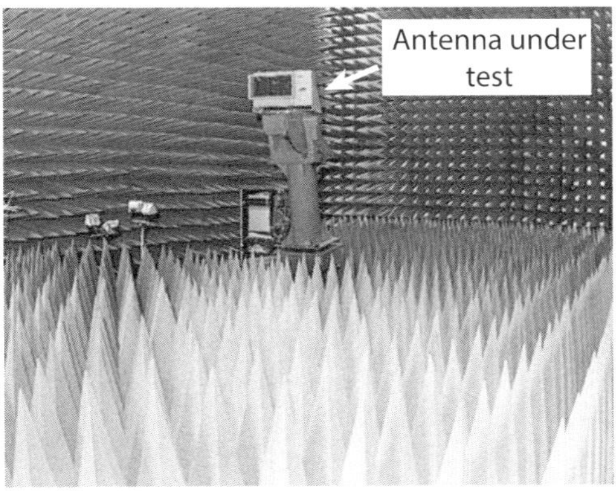

Figure 12.20 Photograph of the antenna under test positioner in the System Test Chamber.

Figure 12.21 Photograph of the source tower and source antenna in the System Test Chamber.

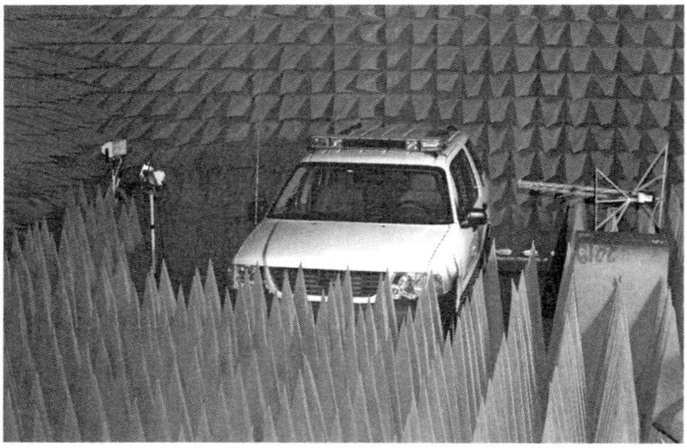

Figure 12.22 Photograph of a ground vehicle under test in the System Test Chamber.

allows the source antenna to reach a maximum height of 15 ft. The source tower has a polarization positioner with a DC to 26.5 GHz rotary joint. The example antenna mounted on the source tower in Figure 12.20 is a dual-polarized ultrawideband open-boundary quad-ridge horn that covers the frequency range 300 MHz to 6 GHz. To achieve accurate spherical near-field measurements, the coordinate systems of the source positioner and antenna under test positioner are laser aligned. The RF measurements system includes a 16-channel 0.1 GHz to 20 GHz SP16T switch (multiplexer) that is used to collect measured antenna under test data from multiple antenna elements or from subarrays. The facility is designed so that ground vehicles can be driven into the chamber for RF testing. Additionally, for RF testing, air vehicles or other test articles can be suspended from an overhead crane. Pipe penetrations

Figure 12.23 Photograph of an unmanned air vehicle under test in the System Test Chamber.

Figure 12.24 Photograph of a radome panel insertion loss test setup in the System Test Chamber.

through one of the chamber walls are provided to allow vehicle exhaust gases to be removed from the anechoic chamber. The pipe penetrations can be closed to maintain chamber RF shielding when not in use.

The anechoic chamber has large double doors and temporary ramps that allow vehicles to be driven into the chamber for antenna and electromagnetic interference (EMI) measurements. An example EMI test setup is shown in Figure 12.22, where a wideband log-periodic antenna is directed at a vehicle. Emissions from the RF systems in the vehicle would be collected and recorded in a spectrum analyzer. In the EMI measurement, the log-periodic antenna has a calibrated antenna factor that converts the measured signal voltage into an absolute electric-field amplitude in volts per meter according to Equation 1.589 as given in Chapter 1.

An overhead crane rail system is used in suspending test articles with non-conducting straps or ropes, for example an unmanned air vehicle as shown in Figure 12.23. This suspension system allows the air vehicle to be located toward the center of the anechoic chamber away from the walls that would otherwise affect the measurements of S-parameters including reflection coefficient and mutual coupling between antennas. The insertion loss of large radome panels can also be tested using the overhead crane as shown in Figure 12.24. For example, an initial network analyzer S21 mutual coupling measurement between two horns is conducted over the frequency band of interest with the radome panel suspended away from the horns. Then the radome panel is lowered between the two horns and the S21 measurement is repeated. The relative change in S21 (with and without the panel) provides the insertion loss in dB of the radome panel.

12.5.5 Blended Rolled-Edge Compact Range Chamber

The RFSTF compact range facility is described in Figures 12.25 to 12.32. An overall diagram of the facility is depicted in Figure 12.25 and it has an EMI shielded anechoic chamber with interior dimensions 20.1m length

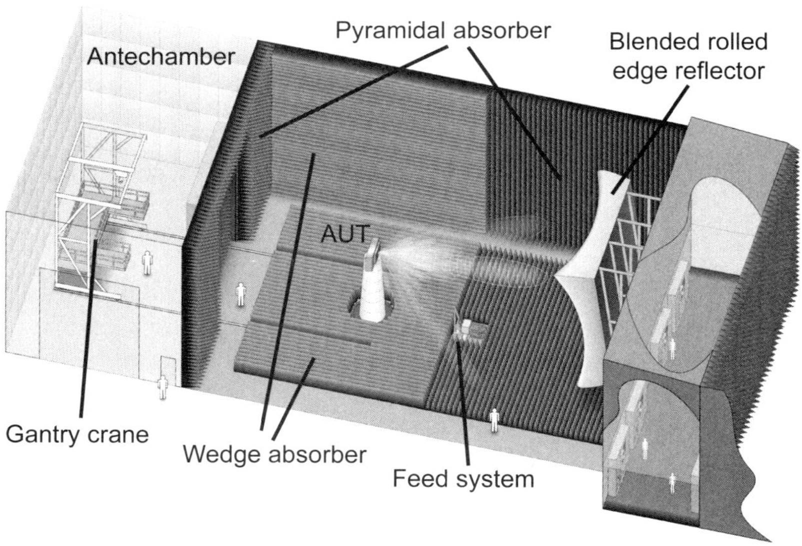

Figure 12.25 Diagram of the RFSTF compact range facility.

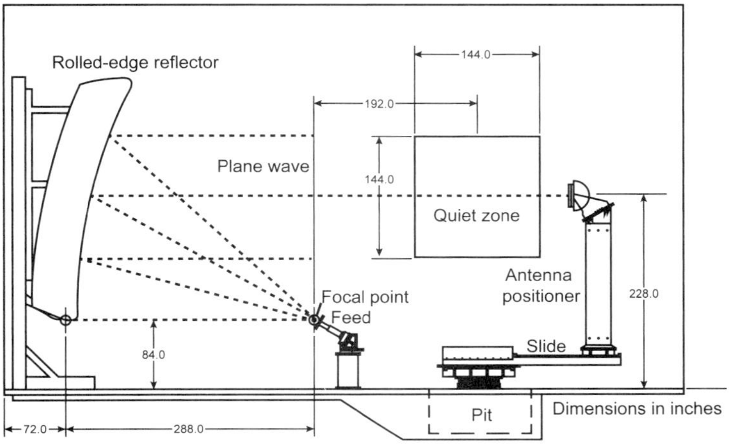

Figure 12.26 Diagram of the rolled edge compact range reflector and AUT positioner.

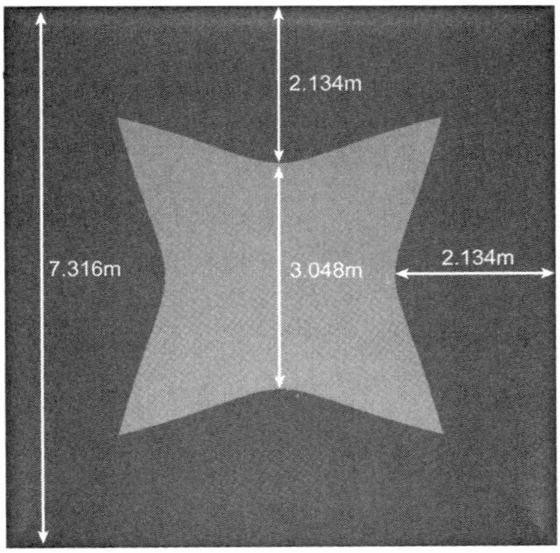

Figure 12.27 Diagram of the rolled edge compact range reflector.

by 13.4m width by 11.6m height (66 ft by 44 ft by 38 ft). The adjacent antenna/target handling room (antechamber in Figure 12.25) has dimensions 9.1m length by 13.4m width by 11.6m height (30 ft by 44 ft by 38 ft). Large double doors provide a test article access opening 9.1m (30 ft) high by 4.7m (15.3 ft) wide. The anechoic chamber temperature is maintained in the range of approximately 20°C to 22°C (68°F to 72°F) to ensure a

Figure 12.28 Photograph of the gantry crane in the antechamber connected to the compact range chamber.

Figure 12.29 Photograph of the multiple feed system for the compact range reflector.

stable environment for maintaining the reflector surface shape and for RF stability of the chamber for RCS background subtraction. The compact range reflector antenna (Figure 12.26) uses a blended rolled-edge design [53–57] to reduce diffraction from the edges of the reflector and improve the target zone amplitude and phase characteristics. The reflector is offset fed to avoid feed blockage effects. The rolled-edge parabolic reflector, feed, and target support structures are located on a common isolated concrete slab to mitigate potential vibration effects. The walls, floor, and ceiling of the chamber are covered with 0.91m (36 inch) pyramidal absorber between the reflector and the feed location. From the feed to the back wall (behind the test article positioner)

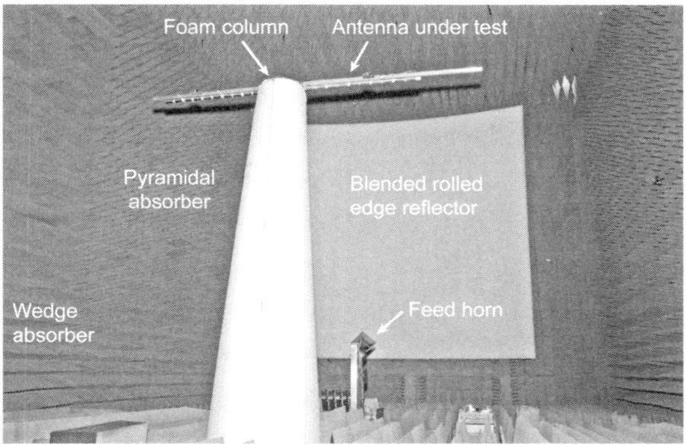

Figure 12.30 Photograph of a linear array under test on a foam column in the compact range chamber.

wedge absorber varying in height from 0.91m to 1.1m (36 inch to 44 inch) in a Chebyshev pattern [57, 58] is used. The back wall is covered with pyramidal absorber in a Chebyshev pattern with absorber height varying from 0.91m to 1.1m (36 inch to 44 inch).

The offset-fed blended rolled-edge reflector [6, 7, 55, 56] is a composite structure with a projected aperture of 7.3m by 7.3m (24 ft by 24 ft). The reflector depicted in the diagram shown in Figure 12.27 has a central parabolic section 3.048m by 3.048m (10 ft by 10 ft) with a blended section to generate a 3.7m by 3.7m by 3.7 m (12 ft by 12 ft by 12 ft) quiet zone. The focal length of the offset parabolic reflector is 24 ft. The electrically conductive surface of the reflector is a silver-based paint. The measured surface resistivity of the reflector is 0.11 ohms/square maximum over the surface of the reflector, which provides a highly electrically conductive surface. Reflector surface accuracy measurements were performed using a tracking laser interferometer [56]. The measured RMS surface accuracy of the reflector conductive surface was 0.0012 inches over the 10 ft by 10 ft parabolic section, 0.002 inches over the first transition section (0 to 28 inches outboard of the parabolic section), 0.0021 inches over the second transition section (28 to 56 inches outboard of the parabolic section), and 0.0031 inches over the third transition section (56 to 84 inches outboard of the parabolic section to the shadow boundary). A gantry with a crane and scissors lift (Figure 12.28) is used to move personnel, equipment, and test objects to and from the test positioner. The crane has a lifting capacity of 1134 kg (2500 lbs). The scissors lift has a lifting capacity of 454 kg (1000 lbs) for personnel and equipment. The gantry is electrically powered and rides on two rails recessed in the floor spaced approximately

Figure 12.31 Photograph of a two-dimensional array panel under test on an azimuth over elevation positioner in the compact range chamber.

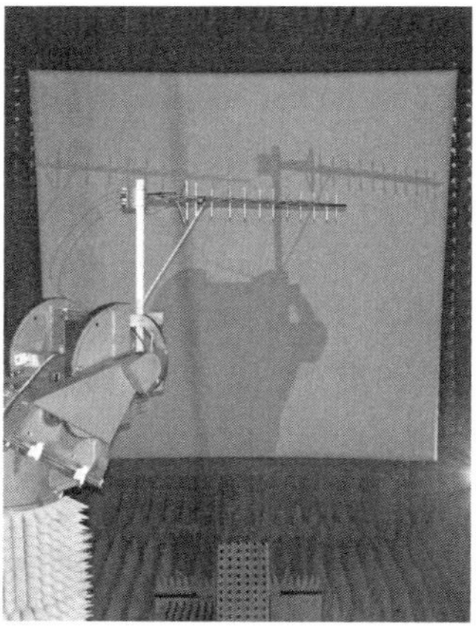

Figure 12.32 Photograph of a UHF high-gain Yagi antenna under test on an azimuth over elevation positioner in the compact range chamber.

4.1m (13.5 ft) apart. Stairs (not shown) allow personnel to safely access the gantry.

Seven dual linearly polarized feed antennas (Figure 12.29) are used to illuminate the reflector at the focal point in the following frequency bands as listed below:

- 0.4 GHz to 2 GHz (open-boundary quad-ridge horn)
- 2 GHz to 18 GHz (TEM horn lens)
- 18 GHz to 26 GHz (choke-ring circular waveguide)

- 26 GHz to 40 GHz (open-end circular waveguide)
- 40 GHz to 50 GHz (open-end circular waveguide)
- 55 GHz to 65 GHz (open-end circular waveguide)
- 90 GHz to 100 GHz (two types: open-end circular waveguide for antenna and RCS measurements with 3.7m target zone; and smooth wall conical horn for RCS measurements with a limited 1.5m (5 ft) target zone for increased sensitivity.)

The feed antennas are located on individual vertical slides on a platform that moves horizontally to allow the phase center of each horn to be placed at the focal point of the reflector. An absorber fence is placed behind the focal point feed location to reduce backlobe radiation from the feed into the target zone. A network analyzer based signal source and receiver system with time gating, along with a multichannel switch matrix, allows measurements of individual elements in an array or of multiple subarrays for digital beamforming.

The antenna positioner equipment normally used in this chamber is configured as follows (refer to the diagram in Figure 12.26). A lower azimuth over elevation positioner is located in a 1.37m (4.5 ft) deep pit with a diameter of 3.35m (11 ft). On top of this positioner is a work platform and spacers upon which an automated 1.8m (6 ft) linear slide is mounted above the floor absorber. The upper polarization over elevation positioner is mounted on a fiberglass tower that is attached to the linear slide. For radar cross section and some antenna measurements, the slide positioner, fiberglass tower, and upper positioner are removed using the gantry crane, and the lower azimuth over elevation positioner is used with various foam columns or a metal ogive to support and rotate test articles. Figure 12.30 shows an example of a UHF linear array of dipole antennas [52], with length 3m, mounted on a conical foam tower. Figure 12.31 shows an example of a two-dimensional UHF array of dipoles [52] mounted on the azimuth over elevation positioner on the fiberglass tower — measured data for this antenna were shown in Chapter 11 (Figures 11.20 to 11.22). A high-gain UHF Yagi dipole array, with length 2.4m, mounted on the azimuth over elevation positioner on the fiberglass tower is shown in Figure 12.32 — here, the lower slide was positioned so that the center of rotation is located at the midpoint of the boom supporting the Yagi elements. This positioning keeps the phase center of the Yagi antenna close to the center of rotation.

12.5.6 Small Near-Field Scanner

In the small near-field chamber, a 1.52m by 1.52m (5 ft by 5 ft) vertical planar near-field scanner is used primarily to calibrate and measure gain radiation patterns of small phased array aperture antennas. The scanner has four-axis automated control for x, y, z Cartesian coordinates and probe polarization for vertical and horizontal polarization measurements. The scanner is located in an anechoic chamber with interior dimensions 3.7m by 3.7m by 3.7m (12 ft by 12 ft by 12 ft) with the walls, floor and ceiling covered with 30.5 cm (12 inch) pyramidal absorber. Figure 12.33 shows a photograph of a rectangular waveguide probe positioned in front of an S-band phased array of dual-polarized square patch antennas. The probe antenna and vertical planar Cartesian coordinate scanner are surrounded by pyramidal absorber. The probe moves parallel to the array face, horizontally in the x direction and vertically in the y direction, and moves perpendicular to the array face in the z direction. The rectangular probe shown is linearly polarized and is rotated by means of a polarization positioner to measure the voltage response to horizontal and vertical electric-field components for the antenna under test. The radiation pattern of the near-field probe with surrounding absorber is pre-measured, and is used to compensate and generate the far-field radiation pattern of the antenna under test. The theory for the mathematical transformation of near-field voltage data to far-zone polarized field data was given in Section 12.3. During calibration of phased array transmit/receive (T/R) modules installed in the array, the near-field probe is positioned typically three wavelengths directly in front of each array antenna element. The T/R module associated with each array element is cycled through its amplitude and phase states while the surrounding T/R modules are switched to a passive load state. A lookup table is formed from the measured amplitude and phase data to set accurate amplitude and phase states for all the array elements to achieve desired main beam electronic steering. Once the phased

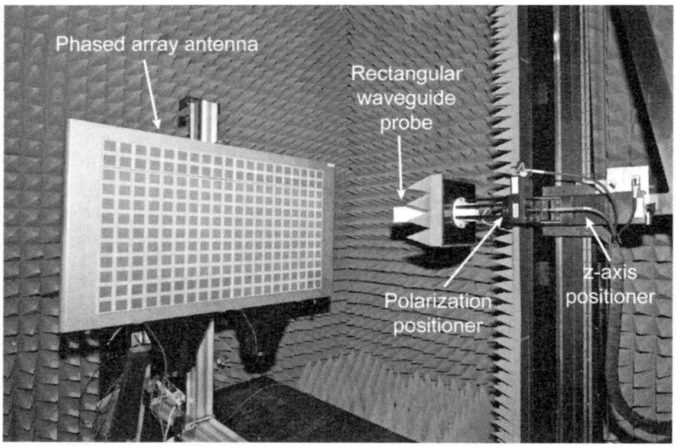

Figure 12.33 Photograph of a phased array antenna under test in the small near-field chamber.

array is calibrated, the full array is illuminated by the desired T/R module amplitude and phase settings and the planar scanner moves the probe and overscans the full aperture of the array to measure the near-field data and then, offline, a computer transforms the near-field data to far-field data. The antenna positioner in this chamber is additionally designed to rotate in azimuth to allow cylindrical near-field scanning. In this case the probe scanner moves the near field probe in the y direction and the antenna under test is rotated in the azimuth angle ϕ.

12.5.7 Large Near-Field Scanner

In the large near-field chamber (Figures 12.34 and 12.35), a 5.5m by 5.5m (18 ft by 18 ft) vertical planar near-field scanner is used to calibrate and measure gain radiation patterns of large phased array aperture antennas. Operation of this large scanner is similar to the small near-field scanner decribed in the previous section. The scanner has four-axis automated control for x, y, z cartesian coordinates and for probe polarization (vertical and horizontal polarization measurements). The anechoic chamber interior dimensions are approximately 9.1m (length) by 8.2m (width) by 7.6m (height) (30 ft by 27 ft by 25 ft). An overhead crane installed in the chamber has a lifting capacity of 3,629 kg (8,000 pounds). The walls, floor, and ceiling are covered with 46 cm (18 inch) pyramidal absorber with unpainted tips. The scanner is optically aligned and designed to perform accurate scans from about 700 MHz to 50 GHz. An antenna assembly area is located adjacent to the anechoic chamber. The test antenna is moved into and out of the chamber on a wheeled cart that is mounted on rails. An example of a large aperture antenna moving on the rails into the chamber is shown in Figure 12.35.

Figure 12.34 Photograph of the large near-field chamber.

Figure 12.35 Photograph of an antenna moving into the large near-field chamber.

12.6 Summary

This chapter has briefly described the subject of antenna measurements in indoor shielded anechoic chambers. Examples of different types of indoor anechoic chambers were given including rectangular, tapered, compact range, and near field scanners. Rectangular and tapered chambers typically will directly measure the far-field characteristics of antennas. In the case of a phased array, focused-near field testing is approximately equivalent to focused far-field testing. A compact range system with a large parabolic reflector measures the far-field performance of antennas and radar cross section of targets. Near-field scanning systems are used in calibrating phased array antennas, and they produce the far-field characteristics of antennas by means of a mathematical transformation and probe correction of measured near-field data. Antenna characteristics including gain, beamwidth, sidelobe levels, polarization, and other characteristics can be determined using an appropriate measurement technique.

References

[1] C.C. Cutler, A.P. King, and W.E. Kock, "Microwave Antenna Measurements," *Proceedings of the IRE*, vol. 35, no. 12, 1947, pp. 1462–1471.

[2] W.H. Kummer (chair), "IEEE Standard Test Procedures for Antennas," IEEE Std 149-1979 (Revised 2008), The Institute of Electrical and Electronics Engineers, Inc., New York: December 10, 2008.

[3] D.G. Svetanoff (chair), "IEEE Standard Method for Measuring the Effectiveness of Electromagnetic Shielding Enclosures," IEEE Std 299-2006 (Revision of IEEE Std 299-1997), The Institute of Electrical and Electronics Engineers, Inc. New York: February 28, 2007.

[4] J.S. Hollis, T.J. Lyon, and L. Clayton, *Microwave Antenna Measurements, 3rd edition* Scientific-Atlanta, Inc., Atlanta, Georgia, November 1985.

[5] J. Volakis, ed., *Antenna Engineering Handbook, 4th ed.* New York: McGraw-Hill, 2007, Chapter 50.

[6] A.J. Fenn, M.W. Shields, and G.A. Somers, "Introduction to the New MIT Lincoln Laboratory Suite of Ranges," *26th Annual Antenna Measurement Techniques Association Meeting & Symposium, AMTA 2004,* Stone Mountain Park, Georgia, 17–22 October 2004.

[7] M.W. Shields and A.J. Fenn, "A New Compact Range Facility for Antenna and Radar Target Measurements," *Lincoln Laboratory Journal*, vol. 16, no. 2, 2007, pp. 381–391.

[8] A. Cohen and A.W. Maltese, "The Lincoln Laboratory Antenna Test Range," *Microwave Journal,* vol. 4, Apr 1961, pp. 57-65.

[9] C.H. Walter, *Traveling Wave Antennas*, New York: Dover, 1970, pp. 36-50.

[10] R.C. Johnson, H.A. Ecker, R.A. Moore, "Compact Range Techniques and Measurements," *IEEE Trans. Antennas Propagat.*, vol. 17, No. 5, 1969, pp. 568–576.

[11] W.E. Scharfman, and G. August, "Pattern Measurements of Phased-Arrayed Antennas by Focusing into the Near Zone," in *Phased Array Antennas (Proc. of the 1970 Phased Array Antenna Symposium)*, A.A. Oliner and G.H. Knittel, (eds.), Dedham, MA: Artech House, 1972, pp. 344–350.

[12] A.D. Yaghjian, "An Overview of Near-Field Antenna Measurements," *IEEE Trans. Antennas Propagat.*, vol. 34, no. 1, 1986, pp. 30–45.

[13] A.C. Newell, "Error Analysis Techniques for Planar Near-Field Measurements," *IEEE Trans. Antennas Propagat.*, vol. 36, no. 6, 1988, pp. 754–768.

[14] D. Slater, *Near-Field Antenna Measurements*, Norwood, MA: Artech House, 1991.

[15] D.T. Paris, W.M. Leach, and E.B. Joy, "Basic Theory of Probe-Compensated Near-Field Measurements," *IEEE Trans. Antennas Propagat.*, vol. 26, no. 3, 1978, pp. 373–379.

[16] E.B. Joy, W.M. Leach, G.P. Rodrigue, and D.T. Paris, "Applications of Probe-Compensated Near-Field Measurements," *IEEE Trans. Antennas Propagat.*, vol. 26, no. 3, 1978, pp. 379–389.

[17] M.H. Francis (chair), "IEEE Recommended Practice for Near-Field Antenna Measurements," IEEE Std 1720-2012, The Institute of Electrical and Electronics Engineers, Inc., New York: December 5, 2012.

[18] A.C. Newell and M.L. Crawford, "Planar Near-Field Measurements on High Performance Array Antennas," National Bureau of Standards Report No. NBSIR 74-380, July 1974.

[19] E.B. Joy, and D.T. Paris, "Spatial Sampling and Filtering in Near-Field Measurements,"

IEEE Trans. Antennas Propagat., vol. 20, no. 3, 1972, pp. 253–261.

[20] J.H. Wang, "An Examination of the Theory and Practices of Planar Near-Field Measurement," *IEEE Trans. Antennas Propagat.*, vol. 36, no. 6, 1988, pp. 746–753.

[21] J.D. Hanfling, G.V. Borgiotti, and L. Kaplan, "The Backward Transform of the Near Field for Reconstruction of Aperture Fields," *1979 IEEE Antennas Propagat. Society Symposium Digest,* vol. 2, IEEE, New York, pp. 764–767.

[22] G. Borgiotti, "Fourier Transforms Method in Aperture Antennas Problems," *Alta Frequenza*, vol. 32, no. 11, 1963, pp. 808–816.

[23] D.R. Rhodes, *Synthesis of Planar Antenna Sources*, Oxford: Oxford University Press, 1974.

[24] A.J. Fenn, H.M. Aumann, and F.G. Willwerth, "Linear Array Characteristics with One–Dimensional Reactive–Region Near–Field Scanning: Simulations and Measurements," *IEEE Trans. Antennas and Propagat.*, vol. 39, no. 9, September 1991, pp. 1305–1311.

[25] Y. Rahmat-Samii, V. Galindo-Israel, R. Mittra, "A Plane-Polar Approach for Far-Field Construction from Near-Field Measurements," *IEEE Trans. on Antennas and Propagation*, vol. 28, no. 2, 1980, pp. 216–230.

[26] W.M. Leach Jr. and D.T. Paris, "Probe Compensated Near-Field Measurements on a Cylinder," *IEEE Transactions on Antennas and Propagation,* vol. AP-21, no. 4, July, 1973, pp. 435–445.

[27] M.A. Qureshi, C.H. Schmidt, and T.F. Eibert, "Adaptive Sampling in Spherical and Cylindrical Near-Field Antenna Measurements," *IEEE Antennas and Propagation Magazine*, vol. 55, no. 1, February 2013, pp. 243–249.

[28] D.W. Hess, "Spherical Near-Field Antenna Measurements: A Review of Correction Techniques," *The Second European Conf. on Antennas and Propagation, EuCAP 2007*, 2007 pp. 1–4.

[29] J.E. Hansen, *Spherical Near-Field Antenna Measurements,* London: Peter Peregrines Press, 1988.

[30] H.M. Aumann, A.J. Fenn, and F.G. Willwerth, "Phased Array Antenna Calibration and Pattern Prediction Using Mutual Coupling Measurements," *IEEE Trans. Antennas and Propagation*, vol. 37, no. 7, 1989, pp. 844–850.

[31] C. Shipley and D. Woods, "Mutual Coupling-Based Calibration of Phased Array Antennas," *1996 IEEE Int. Symp. on Phased Array Systems and Technology,* 1996, pp. 529–532.

[32] A. Agrawal and A. Jablon, "A Calibration Technique for Active Phased Array Antennas," *IEEE International Symposium on Phased Array Systems and Technology, 2003, pp. 223–228.*

[33] T. Gao, Y. Guo, J. Wang, X. Chen, "Large Active Phased Array Antenna Calibration Using MCM," *IEEE Antennas and Propagation Society Int. Symp. 2001*, vol.2, 2001, pp. 606–609.

[34] C. Fulton, W. Chappell, "Calibration Techniques for Digital Phased Arrays," *IEEE*

Int. Conf. on Microwaves Communications Antennas and Electronics Systems 2009, COMCAS 2009, pp. 1–10.

[35] F.S. Gerino, P.M. Belcaguy, "Mutual Coupling Internal Calibration Method Applied to a Polarimetric Phased Array Antenna Model," *Global Electromagnetic Compatibility Conf. (GEMCCON) 2016*, pp. 1–5.

[36] W. Chen, J.P. Lie, B.P. Ng, T. Wang, and M.H. Er, "Joint Gain/Phase and Mutual Coupling Array Calibration Technique with Single Calibrating Source," *Int. Journal of Antennas and Propagation,* vol. 2012.

[37] Y. Neidman, R. Shavit, A. Bronshtein, "Diagnostic of Phased Arrays With Fault Elements Using the Mutual Coupling Method," *IEEE International Conference on Microwaves Communications Antennas and Electronic Systems 2008, COMCAS 2008*, pp. 1–6.

[38] Y. Neidman, R. Shavit, A. Bronshtein, "Diagnostic of Phased Arrays With Faulty Elements Using the Mutual Coupling Method," IET Microwaves Antennas & Propagation, vol. 3, no. 2, 2009, pp. 235–241.

[39] A.J. Fenn, H.M. Aumann, F.G. Willwerth, and J.R. Johnson, "Focused Near-Field Adaptive Nulling: Experimental Investigation," *1990 IEEE Antennas and Propagation Society Int. Symp. Digest,* vol. 1, May 7-11, 1990, pp. 186–189.

[40] A.J. Fenn, *Adaptive Antennas and Phased Arrays for Radar and Communications.* Norwood, Mass.: Artech House, 2008.

[41] A.J. Fenn, "Theory and Analysis of Near Field Adaptive Nulling," *1986 IEEE AP-S Symposium Digest, vol. 2*, IEEE, New York, 1986, pp. 579–582.

[42] A.J. Fenn, "Theory and Analysis of Near Field Adaptive Nulling," *1986 Asilomar Conf. on Signals, Systems and Computers*, Computer Society Press of the IEEE, Washington, D.C., 1986, pp. 105–109.

[43] A.J. Fenn, "Evaluation of Adaptive Phased Array Far-Field Nulling Performance in the Near-Field Region," *IEEE Trans. Antennas Propagat.*, vol. 38, no. 2, 1990, pp. 173–185.

[44] A.J. Fenn, "Theoretical Near Field Clutter and Interference Cancellation for an Adaptive Phased Array Antenna," *1987 IEEE AP-S Symposium Digest, vol. 1*, IEEE, New York, 1987, pp. 46–49.

[45] A.J. Fenn, "Analysis of Phase-Focused Near-Field Testing for Multiphase-Center Adaptive Radar Systems," *IEEE Trans. Antennas Propagat.*, Vol. 40, No. 8, 1992, pp. 878–887.

[46] J.E. Hudson, *Adaptive Array Principles*, New York: Peter Peregrinus LTD, 1981, pp. 24–26.

[47] R.A. Monzingo and T.W. Miller, *Introduction to Adaptive Arrays*, New York: Wiley, 1980.

[48] Reed, I.S., J.D. Mallet and L.E. Brennnan, "Rapid Convergence in Adaptive Arrays," *IEEE Trans. Aerospace and Electronic Systems*, Vol. 10, 1974, pp. 853–863.

[49] Horowitz, L.L., et al., "Controlling Adaptive Antenna Arrays with the Sample Matrix

Inversion Algorithm," *IEEE Trans. Aerospace and Electronic Systems*, vol. 15, 1979, pp. 840–848.

[50] Manolakis, D.G., V.K. Ingle, and S.M. Kogon, *Statistical and Adaptive Signal Processing: Spectral Estimation, Signal Modeling, Adaptive Filtering, and Array Processing,* Norwood, MA: Artech House, 2005, pp. 659–669.

[51] K.H. Lee, C.C. Chen, and R. Lee, "Update on UWB Dual-Polarized Tapered Chamber Feed Development," *Antenna Measurements Techniques Association 2004 Meeting,* Stone Mountain, Georgia: 17–22 October 2004.

[52] A.J. Fenn and P.T. Hurst, *Ultrawideband Phased Array Antenna Technology for Sensing and Communications Systems*, Cambridge, MA: The MIT Press, 2015.

[53] I.J. Gupta, K.P. Ericksen, and W.D. Burnside, "A Method to Design Blended Rolled Edges for Compact Range Reflectors," IEEE Trans. Antennas and Propagation, vol. 39, no. 6, June 1990, pp. 853–861.

[54] T.H. Lee, and W.D. Burnside, "Performance Trade-off Between Serrated Edge and Blended Rolled Edge Compact Range Reflectors," *IEEE Trans. Antennas and Propagation,* vol. 44, no. 1, January 1996, pp. 87–96.

[55] T.H. Lee, Burnside, W.D., and I.J. Gupta, "Blended Rolled Edge Reflector Design for the New Compact Range at MIT Lincoln Laboratory," Antenna Measurements Techniques Association 2004 Meeting, Stone Mountain, Georgia: 17–22 October 2004.

[56] J.R. Proctor, J.R. Smith, D.R., Martin, P.F, Somers, G.A., Shields, M.W. and A.J. Fenn, "Compact Range Rolled Edge Reflector Design, Fabrication, Installation and Mechanical Qualification," *Antenna Measurements Techniques Association 2004 Meeting,* Stone Mountain, Georgia: 17–22 October 2004.

[57] R. Silz, "Design of the GE Aircraft Engine Compact Range Facility," *IEEE Int. Antennas and Propagation Society 2001 Digest,* vol. 4, 8–13 July 2001, pp. 432–435.

[58] J.-R.J. Gau, W.D. Burnside, and M. Gilreath, "Chebyshev Multilevel Absorber Design Concept," *IEEE Trans. Antennas and Propagation,* vol. 45, no. 8, August 1997, pp. 1286–1293.

About the Author

Alan J. Fenn is a senior staff member in the RF Technology Group in the Advanced Technology Division at Lincoln Laboratory, Massachusetts Institute of Technology. He has conducted over 40 years of research in the area of electromagnetics and antennas. He joined Lincoln Laboratory in 1981 and was a member of the Space Radar Technology Group from 1982 to 1991, where his primary research was in adaptive phased-array antenna design and testing. From 1992 to 1999 he was an assistant group leader in the Radio Frequency Technology Group, where he managed programs involving measurements of atmospheric effects on satellite communications. From 1978 to 1981, he was a senior engineer in the Antenna Systems Design/Analysis Group in the RF Systems Department at Martin Marietta Aerospace, Denver, Colorado. He received a B.S. from the University of Illinois at Chicago in 1974, and an M.S. in 1976 and a Ph.D. in 1978 from The Ohio State University, Columbus, all in electrical engineering.

Dr. Fenn was elected a Fellow of the IEEE in 2000 for his contributions to the theory and practice of adaptive phased-array antennas. He is currently serving as technical program chair of the 2018 IEEE International Symposium on Antennas and Propagation. He has served as an associate editor in the area of adaptive antennas for the IEEE Transactions on Antennas and Propagation. He was a corecipient of the 2016 IEEE Aerospace Conference, M. Charles Fogg Best Paper Award. He received the IEEE/URSI-sponsored 1994 International Symposium on Antennas (JINA 94) award. In 1990 he was a corecipient of the IEEE Antennas and Propagation Society's H.A. Wheeler Applications Prize Paper Award. In addition to this book, Dr. Fenn is an author of four other books as well as the author of numerous journal articles, patents, and short-course lectures and conference presentations on electromagnetics and antennas.

Index

Recent Titles in the Artech House Antennas and Electromagnetics Analysis Library

Christos Christodoulou, Series Editor

Adaptive Array Measurements in Communications, M. A. Halim

Advanced Computational Electromagnetic Methods and Applications, Wenhua Yu, Wenxing Li, Atef Elsherbeni, Yahya Rahmat-Samii, Editors

Advances in Computational Electrodynamics: The Finite-Difference Time-Domain Method, Allen Taflove, editor

Advances in FDTD Computational Electrodynamics: Photonics and Nanotechnology, Allen Taflove, editor; Ardavan Oskooi and Steven G. Johnson, coeditors

Analysis Methods for Electromagnetic Wave Problems, Volume 2, Eikichi Yamashita, editor

Antenna Design for Cognitive Radio, Youssef Tawk, Joseph Costantine, and Christos Christodoulou

Antenna Design with Fiber Optics, A. Kumar

Antenna Engineering Using Physical Optics: Practical CAD Techniques and Software, Leo Diaz and Thomas Milligan

Antennas and Propagation for Body-Centric Wireless Communications, Second Edition, Peter S. Hall and Yang Hao, editors

Antennas and Site Engineering for Mobile Radio Networks, Bruno Delorme

Analysis of Radome-Enclosed Antennas, Second Edition, Dennis J. Kozakoff

Applications of Neural Networks in Electromagnetics, Christos Christodoulou and Michael Georgiopoulos

The Art and Science of Ultrawideband Antennas, Second Edition, Hans G. Schantz

AWAS for Windows Version 2.0: Analysis of Wire Antennas and Scatterers, Antonije R. Djordjević, et al.

Broadband Microstrip Antennas, Girsh Kumar and K. P. Ray

Broadband Patch Antennas, Jean-François Zürcher and Fred E. Gardiol

CAD of Microstrip Antennas for Wireless Applications, Robert A. Sainati

The CG-FFT Method: Application of Signal Processing Techniques to Electromagnetics, Manuel F. Cátedra, et al.

Computational Electrodynamics: The Finite-Difference Time-Domain Method, Third Edition, Allen Taflove and Susan C. Hagness

Electromagnetics and Antenna Technology, Alan J. Fenn

Electromagnetic Modeling of Composite Metallic and Dielectric Structures, Branko M. Kolundzija and Antonije R. Djordjević

Electromagnetic Waves in Chiral and Bi-Isotropic Media, I. V. Lindell, et al.

Electromagnetics, Microwave Circuit and Antenna Design for Communications Engineering, Peter Russer

Engineering Applications of the Modulated Scatterer Technique, Jean-Charles Bolomey and Fred E. Gardiol

Fast and Efficient Algorithms in Computational Electromagnetics, Weng Cho Chew, et al., editors

Frequency-Agile Antennas for Wireless Communications, Aldo Petosa

Fresnel Zones in Wireless Links, Zone Plate Lenses and Antennas, Hristo D. Hristov

Handbook of Antennas for EMC, Thereza MacNamara

Handbook of Reflector Antennas and Feed Systems, Volume I: Theory and Design of Reflectors, Satish Sharma, Sudhakar Rao, and Lotfollah Shafai, editors

Handbook of Reflector Antennas and Feed Systems, Volume II: Feed Systems, Lotfollah Shafai, Satish Sharma, and Sudhakar Rao, editors

Handbook of Reflector Antennas and Feed Systems, Volume III: Applications of Reflectors, Sudhakar Rao, Lotfollah Shafai, and Satish Sharma, editors

Introduction to Antenna Analysis Using EM Simulators, Hiroaki Kogure, Yoshie Kogure, and James C. Rautio

Iterative and Self-Adaptive Finite-Elements in Electromagnetic Modeling, Magdalena Salazar-Palma, et al.

LONRS: Low-Noise Receiving Systems Performance and Analysis Toolkit, Charles T. Stelzried, Macgregor S. Reid, and Arthur J. Freiley

Measurement of Mobile Antenna Systems, Second Edition, Hiroyuki Arai

Microstrip Antenna Design Handbook, Ramesh Garg, et al.

Microwave and Millimeter-Wave Remote Sensing for Security Applications, Jeffrey A. Nanzer

Mobile Antenna Systems Handbook, Third Edition, Kyohei Fujimoto, editor

Multiband Integrated Antennas for 4G Terminals, David A. Sánchez-Hernández, editor

New Foundations for Applied Electromagnetics: The Spatial Structure of Fields, Said Mikki and Yahia Antar

Noise Temperature Theory and Applications for Deep Space Communications Antenna Systems, Tom Y. Otoshi

Phased Array Antenna Handbook, Third Edition, Robert J. Mailloux

For further information on these and other Artech House titles, including previously considered out-of-print books now available through our In-Print-Forever® (IPF®) program, contact:

Artech House
685 Canton Street
Norwood, MA 02062
Phone: 781-769-9750
Fax: 781-769-6334
e-mail: artech@artechhouse.com

Artech House
16 Sussex Street
London SW1V HRW UK
Phone: +44 (0)20 7596-8750
Fax: +44 (0)20 7630 0166
e-mail: artech-uk@artechhouse.com

Find us on the World Wide Web at: www.artechhouse.com